NUMERICAL MATHEMATICS AND SCIENTIFIC COMPUTATION

Series Editors

G. H. GOLUB   Ch. SCHWAB

W. A. LIGHT   E. SÜLI

NUMERICAL MATHEMATICS AND SCIENTIFIC COMPUTATION

Books in the series

Monographs marked with an asterisk (*) appeared in the series 'Monographs in Numerical Analysis' which is continued by the current series.

For a full list of titles please visit
http://www.oup.co.uk/academic/science/maths/series/nmsc

* J. H. Wilkinson: *The Algebraic Eigenvalue Problem*
* I. Duff, A.Erisman, and J. Reid: *Direct Methods for Sparse Matrices*
* M. J. Baines: *Moving Finite Elements*
* J. D. Pryce: *Numerical Solution of Sturm–Liouville Problems*

C. Schwab: *p- and hp- Finite Element Methods: Theory and Applications in Solid and Fluid Mechanics*
J. W. Jerome: *Modelling and Computation for Applications in Mathematics, Science, and Engineering*
A. Quarteroni and A. Valli: *Domain Decomposition Methods for Partial Differential Equations*
G. Em Karniadakis and S. J. Sherwin: *Spectral/hp Element Methods for Computational Fluid Dynamics*
I. Babuška and T. Strouboulis: *The Finite Element Method and its Reliability*
B. Mohammadi and O. Pironneau: *Applied Shape Optimization for Fluids*
S. Succi: *The Lattice Boltzmann Equation for Fluid Dynamics and Beyond*
P. Monk: *Finite Element Methods for Maxwell's Equations*
A. Bellen and M. Zennaro: *Numerical Methods for Delay Differential Equations*
J. Modersitzki: *Numerical Methods for Image Registration*
M. Feistauer, J. Felcman, and I. Straškraba: *Mathematical and Computational Methods for Compressible Flow*
W. Gautschi: *Orthogonal Polynomials: Computation and Approximation*
M. K. Ng: *Iterative Methods for Toeplitz Systems*
M. Metcalf, J. Reid, and M. Cohen: *Fortran 95/2003 Explained*
G. Em Karniadakis and S. Sherwin: *Spectral/hp Element Methods for Computational Fluid Dynamics, Second Edition*
D. A. Bini, G. Latouche, and B. Meini: *Numerical Methods for Structured Markov Chains*
H. Elman, D. Silvester, and A. Wathen: *Finite Elements and Fast Iterative Solvers: with Applications in Incompressible Fluid Dynamics*
M. Chu and G. Golub: *Inverse Eigenvalue Problems: Theory, Algorithms, and Applications*
J.-F. Gerbeau, C. Le Bris, and T. Lelièvre: *Mathematical Methods for the Magnetohydrodynamics of Liquid Metals*
G. Allaire and A. Craig: *Numerical Analysis and Optimization: An Introduction to Mathematical Modelling and Numerical Simulation*
K. Urban: *Wavelet Methods for Elliptic Partial Differential Equations*
B. Mohammadi and O. Pironneau: *Applied Shape Optimization for Fluids, Second Edition*
K. Boehmer: *Numerical Methods for Nonlinear Elliptic Differential Equations: A Synopsis*
M. Metcalf, J. Reid, and M. Cohen: *Modern Fortran Explained*
J. Liesen and Z. Strakoš: *Krylov Subspace Methods: Principles and Analysis*

Numerical Methods for Delay Differential Equations

Alfredo Bellen and Marino Zennaro
Dipartimento di Scienze Matematiche
Università di Trieste, Italy

OXFORD
UNIVERSITY PRESS

OXFORD
UNIVERSITY PRESS

Great Clarendon Street, Oxford OX2 6DP
United Kingdom

Oxford University Press is a department of the University of Oxford. It furthers the University's objective of excellence in research, scholarship, and education by publishing worldwide. Oxford is a registered trade mark of Oxford University Press in the UK and in certain other countries

First published 2003
First published in paperback 2013

British Library Cataloguing in Publication Data
Data available

Library of Congress Cataloguing in Publication Data
Data available

ISBN 978-0-19-850654-6
ISBN 978-0-19-967137-3 (Pbk)

To Nevy and Doris

PREFACE

In this book we cover initial value problems for Volterra functional differential equations characterized by the presence of discretely distributed delays, possibly of neutral type, usually called "delay differential equations".

The interest of applied mathematicians for the numerical solution of such a class of problems has increased considerably in the last decades and the number of papers dealing with different aspects of their numerical integration now amounts to several hundreds. Nevertheless, there is no book describing, analyzing, unifying and, where necessary, extending and improving the various approaches and techniques appearing in the literature.

Indeed, this is the main aim of this book, which is intended for a wide variety of readers, including mathematicians, physicists, engineers, economists and other scientists ranging from those who are most interested in the theoretical aspects of numerical methods for ordinary and delay differential equations to those who are just looking for a suitable technique in order to simulate their own model by numerically solving some specific equations. The second aim is to bridge the gap between the basic knowledge on discrete and continuous numerical methods for ordinary differential equations in view of possible applications to dense output, error estimation, discontinuous equations, problems with driving equations and, of course, delay differential equations and more general Volterra differential and integro-differential equations.

Although we have reported many concepts and results on accuracy and stability of numerical methods for ordinary differential equations (mainly in Chapters 3, 5 and 8), for a fruitful reading of the book, at least a basic level knowledge of these topics is recommended.

In the introductory Chapter 1 we formalize the classes of problems treated and focus on some of the most significant qualitative differences between delay equations and ordinary equations, as well as on how such differences reflect in their numerical treatment.

In Chapter 2 we discuss the regularity of the solutions by analyzing how the discontinuities, caused by different types of delays, propagate along the solutions and the impact this lack of smoothness has on the design of efficient numerical methods. Far from being exhaustive, the chapter also reports some existence and uniqueness results for the most known classes of delay equations.

Chapter 3 deals with a general formulation and convergence results for discrete and continuous methods for ordinary equations, followed by an introductory review of the most used numerical methods for delay equations, including a fast "historical" excursus.

Chapter 4 is devoted to the error (convergence) analysis of the most usual technique, based on the use of a discrete method for ordinary equations endowed with some interpolant, which we call "the standard approach". It is also shown that, despite the fact that any discrete method is, in principle, suitable for the standard approach, one-step methods (essentially Runge-Kutta methods) are preferable to multistep methods. At the end, a list of available codes is given.

Chaper 5 is a self-contained presentation of the "continuous Runge-Kutta methods". Besides being the bricks for the construction of the standard approach for delay equations, as developed in the rest of the book, this chapter provides, on a small scale, one of the first systematic collections of results on continuous Runge-Kutta methods useful for treating discontinuous differential equations, systems with driving equations, dense output and more general Volterra functional differential equations.

Chapter 6 specializes the theory presented in Chapter 4 for the class of continuous Runge-Kutta methods and provides several additional results on superconvergence for "constrained mesh" methods.

Chapter 7 treats the stepsize control mechanism. As with ordinary differential equation solvers, this issue plays a central role in the production of efficient numerical algorithms for delay equations. In particular, we discuss the importance of the choice of the continuous extension in the estimation of the local error and its influence on the response of the global error to the user supplied tolerance. Finally, we illustrate the implementation of RADAR5, the most recent code designed for a very general class of stiff delay equations, for two critical examples.

The last chapters, Chapters 8, 9 and 10, are entirely devoted to the important issue of stability. They are largely theoretical, discussing quite technical aspects; nevertheless, in order to produce efficient numerical algorithms for particular classes of "stiff" delay problems, knowledge of most of their contents is crucial.

Chapter 8 is relevant to ordinary differential equations and provides an essential review of stability concepts and results for Runge-Kutta methods, including the first systematic presentation of the so-called "stability with respect to forcing terms". Chapters 9 and 10 address the stability analysis of some classes of test delay equations and of Runge-Kutta methods for their stable integration, respectively. In particular, due to the many-sided stability requirement and to the large variety of significantly different test equations, several definitions of stability for numerical methods are introduced in Chapter 10. In the corresponding stability analysis of continuous Runge-Kutta methods, some order barriers are proved that make the standard approach unsuitable for the most general delay-dependent asymptotic stability property of linear delay systems. A completely different approach, based on restating the delay equation as an "abstract Cauchy problem" and, equivalently, as a partial differential equation with suitable initial/boundary conditions, is then considered. These alternative approaches, still under investigation, overcome the mentioned order barriers and show good potential for the stable integration of an even larger class of Volterra functional differential equations. Chapter 10 ends with some specific additional stability issues which have been developed in the literature or are still in progress.

Throughout the book we provide many examples and discuss the results by means of illustrative numerical experiments. We also supply many algorithms (often in the form of pseudo-codes), so that the interested reader can write his own computer programs without too much effort.

We wish to thank all who have helped us either by useful suggestions and discussions or by providing material for examples, illustrations, bibliographic references, etc. and, in general, by encouraging us to pursue the realization of this book (in particular, C.T.H. Baker, H. Brunner, N. Guglielmi, E. Hairer, V. Kolmanovskii and S. Maset). We want to give particular thanks to K. Burrage, who gave a detailed reading of an advanced version of the book.

Finally, we want to thank our very understanding wives for their support and great deal of patience while drafting the book.

Trieste A. B.
July 2002 M. Z.

CONTENTS

1

INTRODUCTION

Many real-life phenomena in physics, engineering, biology, medicine, economics, etc. can be modeled by an *initial value problem* (IVP), or *Cauchy problem*, for *ordinary differential equations* (ODEs) of the type

$$\begin{cases} y'(t) = g\big(t, y(t)\big), & t \geq t_0, \\ y(t_0) = y_0, \end{cases} \tag{1.0.1}$$

where the function $y(t)$, called the state variable, represents some physical quantity that evolves over time.

However, in order to make the model more consistent with the real phenomenon, it is sometimes necessary to modify the right-hand side of (1.0.1) to include also the dependence of the derivative y' on past values of the state variable y. The most general form of such models is given by the *retarded functional differential equation*

$$y'(t) = f(t, y_t), \quad t \geq t_0,$$

where $y_t = y(t+\theta)$, $\theta \in [-r, 0]$, is a function belonging to the Banach space $C = C^0([-r,0], \mathbb{R}^d)$ of continuous functions mapping the interval $[-r, 0]$ into $\mathbb{R}^d$, and $f : \Omega \longrightarrow \mathbb{R}^d$ is a given function of the set $\Omega \subset \mathbb{R} \times C$ into $\mathbb{R}^d$.

In this context $y'(t)$ stands for the right-hand derivative $y'(t)^+$, and the initial value problem is

$$\begin{cases} y'(t) = f(t, y_t), & t \geq t_0, \\ y_{t_0} = y(t_0 + \theta) = \Phi(\theta), \end{cases} \tag{1.0.2}$$

where $\Phi(\theta) \in C$ represents the *initial point* or the *initial data.*

Equation (1.0.2), also called the *Volterra functional differential equation*, includes both *distributed delay differential equations*, where f depends on y computed on a continuum, possibly unbounded ($r = +\infty$), set of past values, and *discrete delay differential equations*, where only a finite number of past values of the state variable y are involved. Despite the latter being special cases of the former, they are suitable to describe a wide class of phenomena in many branches of applied mathematics and we shall confine our interest to them. Throughout the book they will be referred to as *delay differential equations* (DDEs) or *difference differential equations*.

The general theory of DDEs is widely developed and we refer the reader to the classical books by Bellman and Cooke [40], Hale [121], Driver [78], El'sgol'ts and Norkin [81] and to the more recent books by Hale and Verduyn Lunel [123], Kolmanovskii and Myshkis [172], Kolmanovskii and Nosov [173], Diekmann, van Gils, Verduyn-Lunel and Walther [75] and Kuang [183], which also include many real-life examples of DDEs and more general retarded functional differential equations.

1.1 DDEs versus ODEs: some examples

Throughout the book, the initial value problem (1.0.2) will be expressed in a more friendly manner by

$$\begin{cases} y'(t) = f\big(t, y(t-\tau_1), \ldots, y(t-\tau_n)\big), & t \geq t_0, \\ y(t) = \phi(t), \quad t \leq t_0. \end{cases} \tag{1.1.1}$$

Here, according to the complexity of the phenomenon, the *delays* (or *lags*) τ_i, which always are non-negative, may be just constants (the *constant delay* case), or functions of t, $\tau_i = \tau_i(t)$ (the *variable* or *time dependent delay* case), or even functions of t and y itself, $\tau_i = \tau_i(t, y(t))$ (the *state dependent delay* case). In order to simplify the notation, the function $\phi(t)$ is understood to be defined in $[\rho, t_0]$, where

$$\rho = \min_{1 \leq i \leq n} \Big\{ \min_{t \geq t_0} (t - \tau_i) \Big\}.$$

In particular, for state dependent delays, the bound ρ cannot be determined a priori.

An interesting and quite common case is given by $n = 2$ and $\tau_1 \equiv 0$ for which (1.1.1) takes the standard form

$$\begin{cases} y'(t) = f\big(t, y(t), y(t-\tau)\big), & t \geq t_0, \\ y(t) = \phi(t), \quad t \leq t_0. \end{cases} \tag{1.1.2}$$

Since for some $t \geq t_0$ it can be that $t - \tau < t_0$, a first difference between equations (1.0.1) and (1.1.2) is that the solution of the latter is usually determined by an *initial function* $\phi(t)$ rather than by a simple initial value y_0, as happens for the former. In general, the right-hand derivative $y'(t_0)^+$, that is $f(t_0, \phi(t_0), \phi(t_0 - \tau))$, does not equal the left-hand derivative $\phi'(t_0)^-$ and hence the solution y is not smoothly linked to the initial function $\phi(t)$ at the point t_0, where only C^0-continuity can be assured. Moreover, such a derivative jump discontinuity propagates (see Chapter 2) from the initial point t_0 along the integration interval and gives rise to subsequent discontinuity points where the solution is smoothed out more and more. As a consequence, even if the functions $f(t, y, x)$, $\tau(t, y)$ and $\phi(t)$ in (1.1.2) are C^∞-continuous, in general the solution $y(t)$ is simply C^1-continuous in $[t_0, t_f]$.

Example 1.1.1 Consider the equation

$$\begin{cases} y'(t) = -y(t-1), & t \geq 0, \\ y(t) = 1, \quad t \leq 0, \end{cases} \tag{1.1.3}$$

whose solution is depicted in Figure 1.1. Since $y'(0)^- = 0$ and $y'(0)^+ = -y(-1) = -1$, the derivative function $y'(t)$ has a jump at $t = 0$. The second derivative $y''(t)$ is given by

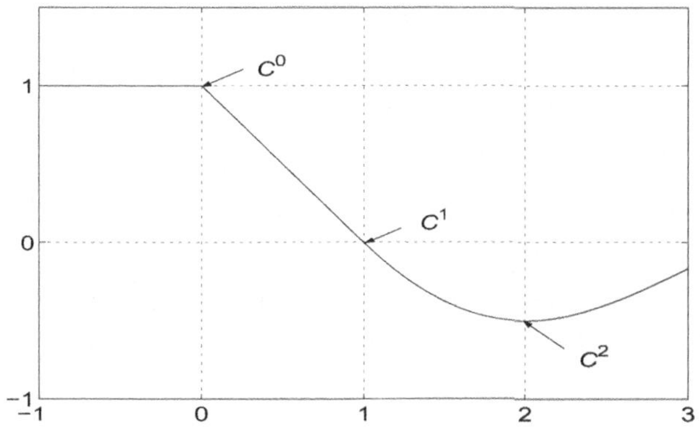

FIG. 1.1. Solution of (1.1.3).

$$y''(t) = -y'(t-1),$$

and therefore it has a jump at $t = 1$. The third derivative $y'''(t)$ is given by

$$y'''(t) = -y''(t-1) = y'(t-2),$$

and hence it has a jump at $t = 2$, and so forth at multiples of the delay $t = 3, 4, \ldots$ ◇

The presence of an initial function in the problem (1.1.2) has various other unexpected consequences on the solutions. Some of them are illustrated by the following examples.

Example 1.1.2 Unlike the ordinary equations, there is no longer injectivity between the set of initial data and the set of solutions $y(t)$, $t \geq t_0$. In fact, the equation

$$y'(t) = y(t-1)\big(y(t) - 1\big), \qquad t \geq 0,$$

has the constant solution $y(t) = 1$ in $[0, +\infty)$ for any initial function $\phi(t)$ defined in $[-1, 0]$ such that $\phi(0) = 1$. ◇

The next two examples show that, in the state dependent delay case, the lack of regularity of the initial function $\phi(t)$ may cause a loss of uniqueness for the solution of (1.1.2) or its termination after some bounded interval.

Example 1.1.3 As an example of non-uniqueness, consider the equation

$$\begin{cases} y'(t) = y\big(t - |y(t)| - 1\big) + \frac{1}{2}, & t \geq 0, \\ y(t) = \phi(t), & t \leq 0, \end{cases} \tag{1.1.4}$$

where

$$\phi(t) = \begin{cases} 1, & t < -1, \\ 0, & -1 \leq t \leq 0. \end{cases} \tag{1.1.5}$$

It is easy to see that in $[0, 2]$ both functions

$$y(t) = \tfrac{3}{2}t$$

and

$$y(t) = \tfrac{1}{2}t$$

are solutions of (1.1.4). ◇

Example 1.1.4 (see El'sgol'ts and Norkin [81]) As an example of termination of the solution, consider the equation

$$\begin{cases} y'(t) = -y\big(t - 2 - y(t)^2\big) + 5, & t \geq 0, \\ y(t) = \phi(t), \quad t \leq 0, \end{cases} \tag{1.1.6}$$

where

$$\phi(t) = \begin{cases} \frac{9}{2}, & t < -1, \\ \\ -\frac{1}{2}, & -1 \leq t \leq 0. \end{cases} \tag{1.1.7}$$

The solution in $[0, \frac{125}{121}]$ is given by

$$y(t) = \begin{cases} \frac{1}{2}(t-1), & 0 \leq t \leq 1, \\ \\ \frac{11}{2}(t-1), & 1 \leq t \leq \frac{125}{121}. \end{cases} \tag{1.1.8}$$

It is not difficult to see that the solution cannot be continued beyond the point $t = \frac{125}{121}$. In fact, at $t = \frac{125}{121}$ the deviated argument $t - 2 - y(t)^2$ is equal to -1 and therefore, in a right neighborhood of such a point, $y(t - 2 - y(t)^2)$ is given by one of the two values of $\phi(t)$. Thus the solutions of (1.1.6) should take the form

$$y(t) = c\Big(t - \tfrac{125}{121}\Big) + \tfrac{2}{11},$$

with

$$c = \tfrac{1}{2} \quad \text{if } t - 2 - y(t)^2 < -1$$

and

$$c = \tfrac{11}{2} \quad \text{if } t - 2 - y(t)^2 \geq -1.$$

Now, each choice of c leads to a solution $y(t)$ that contradicts the assumption made on $t - 2 - y(t)^2$ and hence the solution does not exist for $t > \frac{125}{121}$. It is worth remarking that, from a numerical point of view, termination of the solution is a very delicate issue. In fact, it may result in surprising and misleading behavior in the implementation of the numerical method. For example, in a right neighborhood of the termination point $t_N = \frac{125}{121}$, where $y_N \approx \frac{2}{11}$, the *forward Euler* method reads

$$y_{n+1} = y_n + h_{n+1}(-\tfrac{9}{2} + 5) \quad \text{if } t_n - 2 - y_n^2 < -1$$

and

$$y_{n+1} = y_n + h_{n+1}(\tfrac{1}{2} + 5) \quad \text{if } t_n - 2 - y_n^2 \geq -1,$$

and for no reason it stops integrating at any $n \geq N$. The resulting approximation is plotted in Figure 1.2 where, for $t \geq \frac{125}{121}$, a *ghost solution* appears that

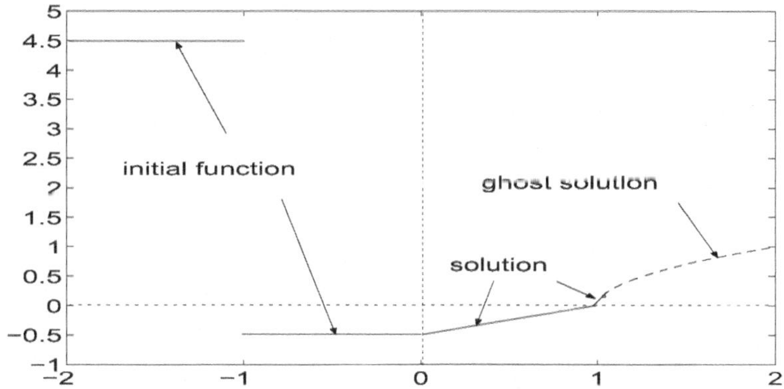

FIG. 1.2. Solutions (solid) and ghost solution (dashed) of (1.1.6).

approximates the function $\sqrt{t-1}$. Such a function is not a solution of (1.1.6) but fulfills the equation $t - 2 - y(t)^2 = -1$. In fact, the numerical solution is forced to attain values y_n such that the delayed arguments $t_n - 2 - y_n^2$ oscillate around the equilibrium value -1.

On the other hand, the *backward Euler* method reads

$$y_{n+1} = y_n + h_{n+1}(-\tfrac{9}{2} + 5) \quad \text{if } t_{n+1} - 2 - y_{n+1}^2 < -1$$

and

$$y_{n+1} = y_n + h_{n+1}(\tfrac{1}{2} + 5) \quad \text{if } t_{n+1} - 2 - y_{n+1}^2 \geq -1.$$

It is not difficult to see that, for small values of h_{n+1}, no solution y_{n+1} exists. This agrees perfectly with the theory, but may result in a very large number of rejected steps in the root-finding mechanism before the overall procedure stops. ◇

In more general models, the derivative $y'(t)$ may depend on y and y' itself at some past value $t - \tau$. In this case (1.0.1) changes into the form

$$\begin{cases} y'(t) = f\big(t, y(t), y(t-\tau), y'(t-\tau)\big), & t \geq t_0, \\ y(t) = \phi(t), \quad t \leq t_0, \end{cases} \tag{1.1.9}$$

where the function $\phi(t)$ is assumed to be at least C^1-continuous. Equation (1.1.9) is called a *delay differential equation* of *neutral* type (NDDE). As before, the initial function $\phi(t)$ does not link smoothly with the solution $y(t)$ at t_0, where the sole continuity is guaranteed. This discontinuity point is spread forward on a set of discontinuity points where the solution, unlike the non-neutral case, remains solely of class C^0. Therefore, unless the *splicing condition*

$$\phi'(t_0)^- = f\big(t_0, \phi(t_0), \phi(t_0 - \tau), \phi'(t_0 - \tau)\big)$$

holds, the solution of (1.1.9) must be understood in the "almost everywhere" generalized sense.

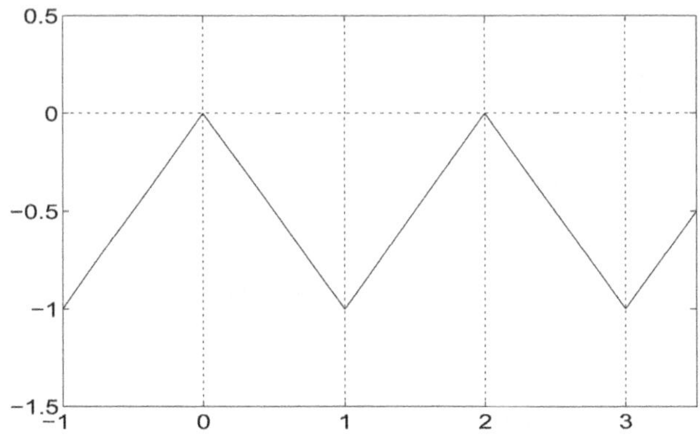

FIG. 1.3. Solutions of (1.1.10).

Example 1.1.5 Consider the equation

$$\begin{cases} y'(t) = -y'(t-1), & t \geq 0, \\ y(t) = t, & t \leq 0, \end{cases} \tag{1.1.10}$$

whose solution is depicted in Figure 1.3.

Since $y'(0)^- = 1$ and $y'(0)^+ = -y'(-1) = -1$, the derivative $y'(t)$ has a jump discontinuity at $t = 0$. Moreover, since $y'(t) = -y'(t-1)$ for every $t \geq 0$, the derivative $y'(t)$ is discontinuous at $t = 1$ and at all its multiples $t = 2, 3, \ldots$ as well. ◇

For equations (1.1.2) and (1.1.9), where the delay does not depend on $y(t)$, the theory of existence and uniqueness of solutions does not present substantial additional difficulties with respect to the ordinary case (1.0.1) as long as the deviated argument $t - \tau(t)$ is uniformly strictly positive as, for example, in the case of constant delay τ. Nevertheless, even in this case, DDEs possess a dynamical structure that is richer than ODEs. In fact, whereas bounded solutions of autonomous ODEs may oscillate only if the system has at least two components and may behave chaotically only if there are at least three components (Poincaré–Bendixon theorem), DDEs may already exhibit oscillatory and even chaotic behavior in the scalar case.

Example 1.1.6 Consider the following *delay logistic equation*:

$$y'(t) = ay(t)\big(1 - y(t-1)\big), \tag{1.1.11}$$

which models the dynamics of populations (see Kuang [183]). It improves the Verhulst–Pearl model $y'(t) = ay(t)\big(1 - y(t)\big)$ in that the *growing factor* $1 - y(t)$ does not act instantaneously but after some time lag.

Whereas the solutions of the Verhulst–Pearl equation are monotonic, the positive solutions of (1.1.11) are monotonic for $a \in (0, 1/e)$, oscillate for $a \in [1/e, \pi/2)$ and approach periodic orbits for $a > \pi/2$ (Figures 1.4 and 1.5). ◇

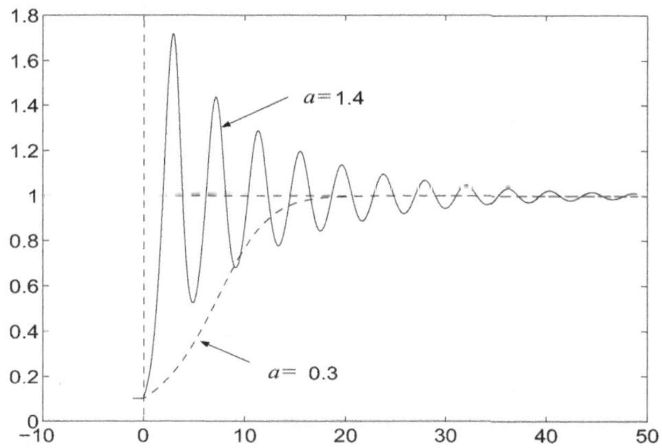

FIG. 1.4. Solutions of (1.1.11) with $y(t) = 0.1$, $t \leq 0$, for $a = 1.4$ and 0.3.

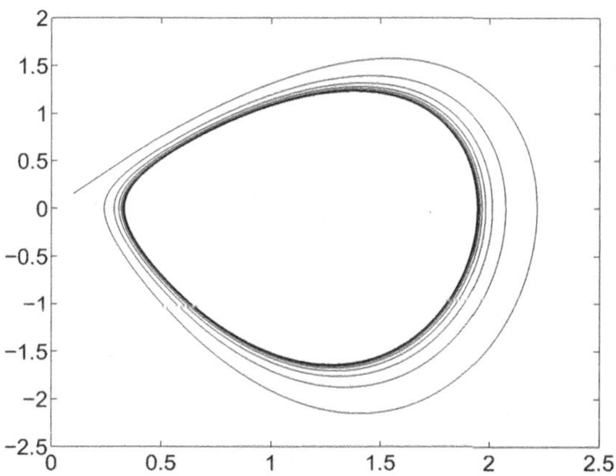

FIG. 1.5. Solution of (1.1.11) for $a = 1.7$ in the phase plane.

Example 1.1.7 Consider the following equation, presented by Mackey and Glass [193] in connection with the release of mature cells into the blood stream:

$$y'(t) = \frac{by(t-\tau)}{1+[y(t-\tau)]^n} - ay(t). \tag{1.1.12}$$

For certain values of the parameters and of the delay, the solution is oscillatory, and sometimes it even oscillates chaotically. This is the case in patients with leukemia. In particular, for $a = 0.1$, $b = 0.2$, $n = 10$ and $\tau = 20$, the model behaves chaotically, as shown in Figure 1.6. ◇

Finally, observe that the presence of a delayed term may drastically change the qualitative behavior of the solution by acting as a stabilizer or a destabilizer of models governed by ODEs.

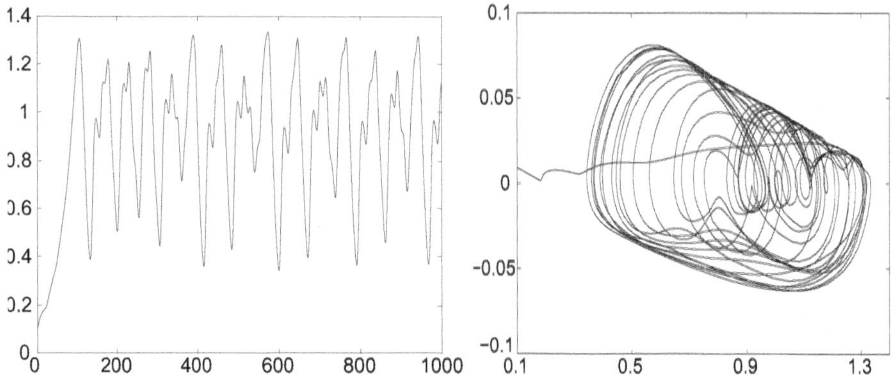

FIG. 1.6. Solutions of equation (1.1.12) for $t \leq 1000$ in the (t, y)-plane (left) and in the phase plane (right).

Example 1.1.8 Consider the linear scalar equation

$$\begin{cases} y'(t) = \lambda y(t) + \mu y(t-1), & t \geq 0, \\ y(t) = -t + 1, & t \leq 0, \end{cases} \tag{1.1.13}$$

with real constant coefficients λ and μ. It is known that, for $\mu = 0$, the solution of equation (1.1.13), which reads

$$\begin{cases} y'(t) = \lambda y(t), & t \geq 0, \\ y(0) = 1, \end{cases} \tag{1.1.14}$$

vanishes asymptotically for any negative λ, whereas it blows up for any positive λ. Moreover, in the former case it remains bounded by the initial value 1. On the other hand, for $\mu \neq 0$ the delayed term $\mu y(t-1)$ in (1.1.13) acts as a *forcing term* and the above-mentioned properties of the solution might not hold. In particular, for any $\mu > 0$ there exists $\lambda < 0$ for which the solution does not vanish asymptotically, and another $\lambda < 0$ for which the solution does, but is not bounded by the initial value $y(0) = 1$. Figure 1.7 illustrates such situations for $\mu = 4$ and for $\lambda = -3.5$ and -5. Also, for $\lambda = 0.5$ and $\mu = -1$ the delayed term $-y(t-1)$ acts as a *stabilizer* of the model whose solution behaves stably despite the positivity of λ (see Figure 1.8). ◇

Example 1.1.9 An example showing that small delays can produce a large effect is given by the NDDE

$$y'(t) = -1.5y'(t-\tau) + \lambda y(t), \quad \lambda < 0,$$

the solution of which is asymptotically stable for $\tau = 0$ and unstable for any $\tau > 0$. In this case the delayed term acts as a *destabilizer*. ◇

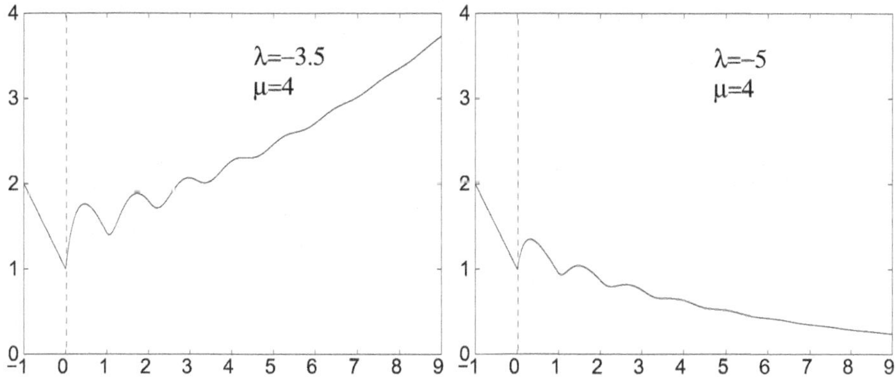

FIG. 1.7. Stable and unstable solutions of (1.1.13) for $\lambda < 0$.

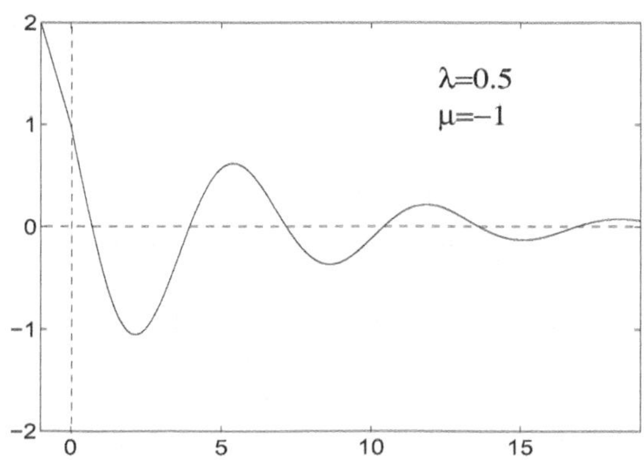

FIG. 1.8. Stable solution of (1.1.13) for $\lambda > 0$.

1.2 Numerical solution of DDEs: is ODE theory enough?

In order to illustrate some basic features of numerical methods for DDEs and the differences they exhibit with respect to ODE methods, consider the constant delay differential equation

$$\begin{cases} y'(t) = f\big(t, y(t), y(t-1)\big), & t \geq 0, \\ y(t) = \phi(t), \quad t \leq 0. \end{cases} \tag{1.2.1}$$

The most natural, but not unique, approach for solving (1.2.1) numerically is to assign integration steps less than or equal to the delay $\tau = 1$ and to integrate step by step the ODEs obtained from (1.2.1) by substituting the delayed term $y(t-1)$ by a function $\eta(t-1)$ given, according to the value of t, either by the initial function $\phi(t-1)$ or by a continuous extension of the approximate solution previously computed by the method itself. Thus, at the $(n+1)$st step the equation to be solved is

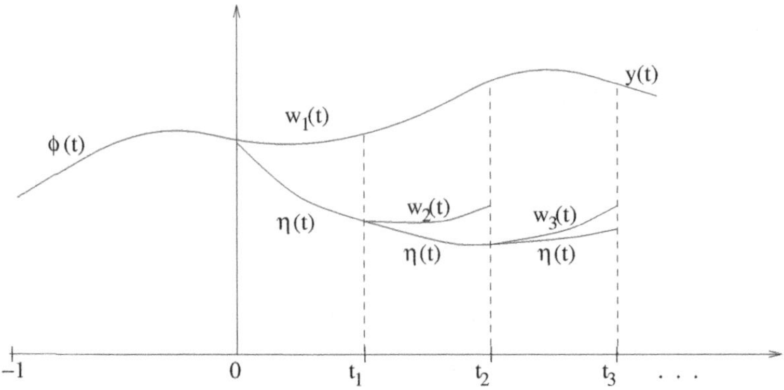

FIG. 1.9. Approximate solution of (1.2.1) by the continuous ODE method.

$$\begin{cases} w'_{n+1}(t) = f\big(t, w_{n+1}(t), x(t-1)\big), & t_n \le t \le t_{n+1}, \\ w_{n+1}(t_n) = y_n, \end{cases} \tag{1.2.2}$$

where

$$x(s) = \begin{cases} \phi(s) & \text{for } s \le 0, \\ \eta(s) & \text{for } 0 \le s \le t_n. \end{cases}$$

The integration formula provides the value y_{n+1} and the approximate solution η of (1.2.1) is then continued in $[t_n, t_{n+1}]$ in such a way that $\eta(t_{n+1}) = y_{n+1}$.

A peculiarity of this approach is that, whereas the numerical ODE method furnishes approximate values of the solution at nodal points only, implementation of the numerical method for the solution of (1.2.2) may require knowledge of the approximate solution $\eta(t)$ at some points $t-1$ possibly other than the nodal points. Therefore, in general, the DDE methods will be based on *continuous extensions* of numerical ODE schemes. This can be done either by a posteriori *interpolation* of the values y_n given by the underlying *discrete* ODE method or, preferably, by *continuous* ODE methods, that is methods that provide step by step a continuous approximation of the solution (see Figure 1.9). As we shall see, the success of the resulting DDE method, in terms of accuracy and stability, depends on the particular choice of the discrete method and of the continuous extension as well.

We assume the reader is familiar with the theory of ODEs and with the (discrete) methods for their numerical solution (see, for example, Lambert [184], Butcher [62], Hairer, Nørsett and Wanner [117], Hairer and Wanner [118], Dekker and Verwer [74]). Over the past fifteen years several papers on the theory of continuous extensions of numerical ODE methods and their applications to DDEs have appeared. Some of those results, along with further recent results on the stability of numerical ODE methods, are reported in this book and will play a central role in the accuracy and stability analysis of numerical DDE methods.

We have already pointed out that the presence of a delayed term can drastically modify some boundedness or stability properties and, in general, the dynamics of the simpler ODE model. Now we want to illustrate, by means of some examples, that also in the implementation of numerical schemes some desirable accuracy and stability properties of the underlying ODE method can be destroyed when the method is applied to a DDE.

1.2.1 *Order failure*

In order to illustrate the possible loss of accuracy, consider the class of constant coefficient linear equations

$$\begin{cases} y'(t) = ay(t) - \frac{\pi}{2}e^{a}y(t-1), & t \geq 0, \\ y(t) = \phi(t) = e^{at}\sin(\frac{\pi}{2}t), & t \leq 0, \end{cases} \tag{1.2.3}$$

whose solutions, $y(t) = e^{at}\sin(\frac{\pi}{2}t)$, are of class C^{∞} in $[-1, +\infty)$.

According to (1.2.2), for $n = 0, 1, \ldots$, we shall solve the ODE

$$\begin{cases} w'_{n+1}(t) = aw_{n+1}(t) - \frac{\pi}{2}e^{a}x(t-1), & t_n \leq t \leq t_{n+1}, \\ w_{n+1}(t_n) = y_n, \end{cases} \tag{1.2.4}$$

where

$$x(s) = \begin{cases} \phi(s) = e^{as}\sin(\frac{\pi}{2}s) & \text{for } s \leq 0, \\ \eta(s) & \text{for } 0 \leq s \leq t_n. \end{cases}$$

A good class of methods for integrating (1.2.4) is given by *collocation at ν Gaussian points*, which can be viewed as ν-stage Runge–Kutta methods of order 2ν. Since they are projection methods based on piecewise polynomial approximations of degree ν, they also furnish a continuous extension $\eta(t)$ of uniform accuracy order $\nu+1$. Therefore, Gaussian collocation appears to be an attractive class of continuous methods for integrating DDEs such as (1.2.3).

For $\nu = 1$ the method is known as the *midpoint rule*, and for the general equation (1.2.2) it takes the form

$$y_{n+1} = y_n + hf\left(t_n + \frac{h}{2}, \frac{y_n + y_{n+1}}{2}, x\left(t_n + \frac{h}{2} - 1\right)\right). \tag{1.2.5}$$

Application of (1.2.5) to (1.2.4) with constant integration stepsize $h = 1/(m-\delta)$, $m \geq 2$, m integer, and $0 \leq \delta < 1$, yields

$$y_{n+1} = \begin{cases} y_n + h\Big(a\frac{y_n+y_{n+1}}{2} - e^{a}\frac{\pi}{2}\phi(t_n + \frac{h}{2} - 1)\Big), & t_n + \frac{h}{2} - 1 \leq 0, \\ y_n + h\Big(a\frac{y_n+y_{n+1}}{2} - e^{a}\frac{\pi}{2}\eta(t_n + \frac{h}{2} - 1)\Big), & t_n + \frac{h}{2} - 1 > 0, \end{cases}$$

where $\eta(t_n + \frac{h}{2} - 1)$ is given by linear interpolation, that is

$$\eta\Big(t_n + \frac{h}{2} - 1\Big) = \eta\Big(t_n + \frac{h}{2} - (m-\delta)h\Big)$$

$$= \begin{cases} (\frac{1}{2}-\delta)y_{n-m} + (\delta+\frac{1}{2})y_{n-m+1}, & 0 \le \delta \le \frac{1}{2}, \\ (\frac{3}{2}-\delta)y_{n-m+1} + (\delta-\frac{1}{2})y_{n-m+2}, & \frac{1}{2} < \delta < 1. \end{cases}$$

Summing up, the midpoint rule for (1.2.3) takes the form

$$y_{n+1} = \frac{(1+\frac{1}{2}ha)y_n - \frac{\pi}{2}he^{a(n+\frac{1}{2})h}\sin\left(\pi/2((n+\frac{1}{2})h-1)\right)}{1-\frac{1}{2}ha} \tag{1.2.6}$$

for $n \le m-\delta-\frac{1}{2}$ and

$$y_{n+1} = \begin{cases} \dfrac{(1+\frac{1}{2}ha)y_n - \frac{\pi}{2}e^a h\big((\frac{1}{2}-\delta)y_{n-m} + (\delta+\frac{1}{2})y_{n-m+1}\big)}{1-\frac{1}{2}ha}, & 0 \le \delta \le \frac{1}{2}, \\ \dfrac{(1+\frac{1}{2}ha)y_n - \frac{\pi}{2}e^a h\big((\frac{3}{2}-\delta)y_{n-m+1} + (\delta-\frac{1}{2})y_{n-m+2}\big)}{1-\frac{1}{2}ha}, & \frac{1}{2} < \delta < 1, \end{cases} \tag{1.2.7}$$

for $n > m-\delta-\frac{1}{2}$.

For $\nu > 1$ the method takes a more complicated form involving the solution of a linear system in $\mathbb{R}^\nu$ and the use of a polynomial of degree ν at each integration step. Let us restrict ourselves to the investigation of $\nu = 1$ and $\nu = 2$ and recall that, whereas the uniform and discrete accuracy orders of the midpoint rule ($\nu = 1$) for ODEs are both equal to 2, the two-stage Gaussian collocation ($\nu = 2$) has nodal order equal to 4 and uniform order equal to 3 (see Section 5.2.1).

In order to check the nodal accuracy order p of the resulting DDE methods, consider the solution of (1.2.3) in the interval $[0,10]$ and remember that the solution is of class C^∞. Therefore, no discontinuities are present which can spoil the accuracy order of the method. Let us denote by e_h the maximum absolute error at the nodal points for the integration stepsize $h = 1/(m-\delta)$. By halving the stepsize, the asymptotic value of the ratio $r_h = e_h/e_{\frac{h}{2}}$ is expected to be a power of 2, where the exponent is the order p.

The behavior of r_h is shown in Figure 1.10 for integer ($\delta = 0$) and non-integer ($\delta > 0$) values of $m-\delta$, corresponding to integration stepsizes h which are and which are not submultiples of the delay $\tau = 1$. In both cases the midpoint rule turns out to preserve the expected nodal order 2. On the contrary, the two-stage Gaussian collocation method exhibits nodal order 4 or 3 according to whether or not $m-\delta$ is integer, while the uniform approximation is accurate to order 3. More generally, we shall see that for arbitrary stepsizes the DDE method based on ν-stage Gaussian collocation exhibits nodal and uniform accuracy orders equal to the uniform order $\nu+1$ of the continuous ODE method whereas, for a constant stepsize $h = 1/m$, with m integer, it preserves the nodal accuracy order 2ν. All these facts will be completely explained and cleared up in Chapter 6.

1.2.2 *Stability failure*

As for the stability analysis of DDE methods, consider the class of constant coefficient linear test equations

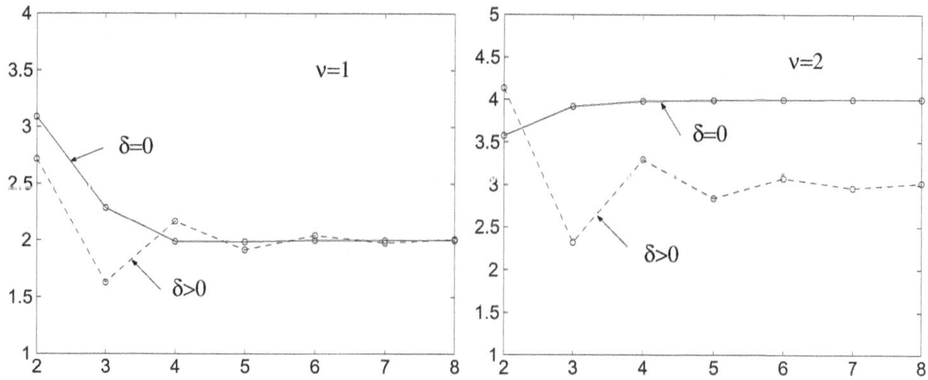

FIG. 1.10. Logarithmic plots of ratios r_h, $h = 1/(m-\delta)$, $m - \delta = 2^i * k$, as a function of i, for the numerical solutions of equation (1.2.3) with $a = 1$ obtained by collocation at $\nu = 1$ and $\nu = 2$ Gaussian points. The values of $m - \delta$ are determined by $k = 1$ (solid) and $k = 5/3$ (dotted).

$$\begin{cases} y'(t) = \lambda y(t) - \frac{4}{5}\lambda y(t-1), & t \geq 0, \\ y(t) = 1, & t \leq 0, \end{cases} \tag{1.2.8}$$

whose solutions, depicted in Figure 1.11 for some values of λ, are asymptotically stable for any $\lambda < 0$.

The midpoint rule (1.2.5), extended by linear interpolation, for equation (1.2.8) takes the form

$$y_{n+1} = \frac{(1 + \frac{1}{2}h\lambda)y_n - \frac{4}{5}h\lambda}{1 - \frac{1}{2}h\lambda}$$

for $n \leq m - \delta - \frac{1}{2}$ and

$$y_{n+1} = \begin{cases} \dfrac{(1+\frac{1}{2}h\lambda)y_n - \frac{4}{5}h\lambda\big((\frac{1}{2}-\delta)y_{n-m} + (\delta+\frac{1}{2})y_{n-m+1}\big)}{1-\frac{1}{2}h\lambda}, & 0 \leq \delta \leq \frac{1}{2}, \\ \dfrac{(1+\frac{1}{2}h\lambda)y_n - \frac{4}{5}h\lambda\big((\frac{3}{2}-\delta)y_{n-m+1} + (\delta-\frac{1}{2})y_{n-m+2}\big)}{1-\frac{1}{2}h\lambda}, & \frac{1}{2} < \delta < 1, \end{cases} \tag{1.2.9}$$

for $n > m - \delta - \frac{1}{2}$.

It is well known that the midpoint rule is *A-stable*, that is for all equations $y' = \lambda y$, $\Re(\lambda) < 0$, it provides numerical solutions that vanish asymptotically for any integration stepsize $h > 0$. Since the solutions of (1.2.8) are asymptotically stable for any $\lambda < 0$ (see Chapter 9), one expects the same behavior for the numerical solution given by the midpoint rule independently of the stepsize h. On the contrary, whereas application of (1.2.9) provides stable numerical solutions of (1.2.8) whenever $m - \delta$ is integer ($\delta = 0$), somewhat surprisingly, non-integer values of $m - \delta$ may produce numerical instability. Figure 1.12 shows the numerical solutions of (1.2.8) with $\lambda = -50$, given by (1.2.9) for $m - \delta = 10$ and

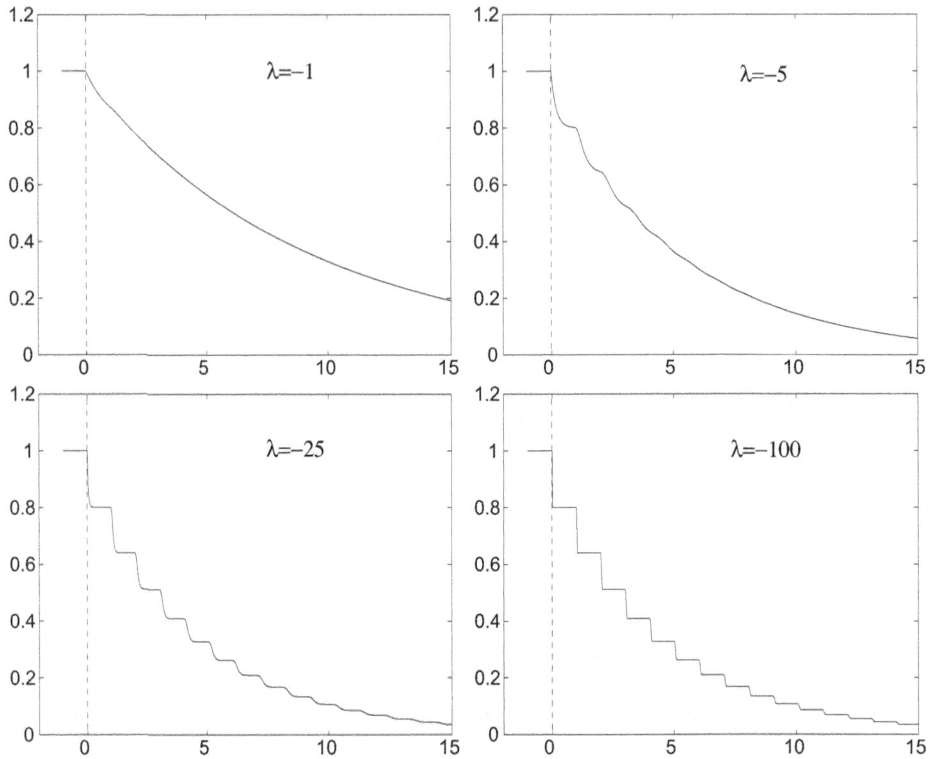

FIG. 1.11. Solutions of (1.2.8) for some $\lambda < 0$.

$m - \delta = 12.5$. Despite the latter being obtained by a smaller integration stepsize, it behaves unstably.

Now consider another very popular A-stable method, namely the *trapezoidal rule*, which, for the general equation (1.2.2), is

$$y_{n+1} = y_n + \frac{h}{2}\big(f(t_n, y_n, x(t_n - 1)) + f(t_{n+1}, y_{n+1}, x(t_{n+1} - 1))\big). \quad (1.2.10)$$

Application of (1.2.10) to (1.2.8) with constant stepsize $h = 1/(m - \delta)$, m integer, $m \geq 2$, and $0 \leq \delta < 1$, and linear interpolation between nodal points yields

$$y_{n+1} = \frac{(1 + \frac{1}{2}h\lambda)y_n - \frac{4}{5}h\lambda}{1 - \frac{1}{2}h\lambda}$$

for $n \leq m - 2$,

$$y_{n+1} = \frac{(1 + \frac{1}{2}h\lambda)y_n - \frac{1}{2}h\frac{4}{5}\lambda(2 - \delta + \delta y_1)}{1 - \frac{1}{2}h\lambda}$$

for $n = m - 1$ and

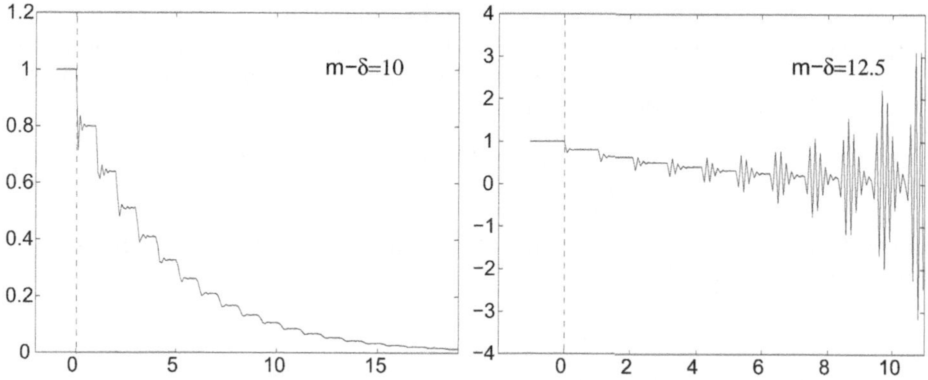

FIG. 1.12. Numerical solutions of equation (1.2.8) with $\lambda = -50$ by the midpoint rule with $h = 1/(m - \delta)$ for integer and non-integer $m - \delta$.

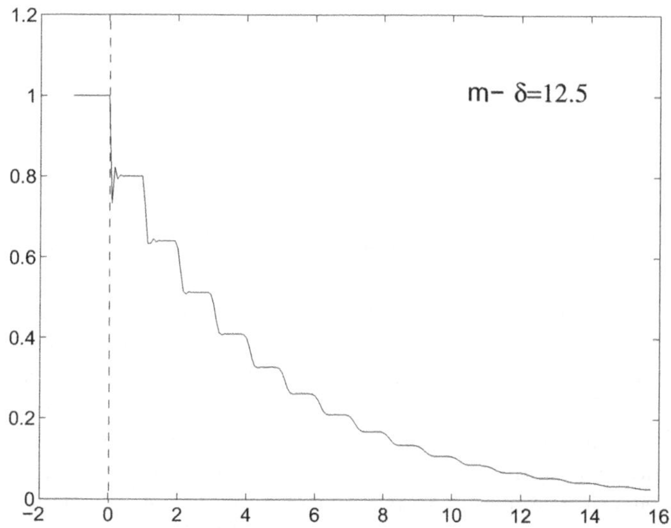

FIG. 1.13. Numerical solution of equation (1.2.8) with $\lambda = -50$ by the trapezoidal rule with $m - \delta = 12.5$.

$$y_{n+1} = \frac{(1 + \frac{1}{2}h\lambda)y_n - \frac{1}{2}h\frac{4}{5}\lambda\big((1-\delta)y_{n-m} + y_{n-m-1} + \delta y_{n-m+2}\big)}{1 - \frac{1}{2}h\lambda} \tag{1.2.11}$$

for $n \geq m$.

Unlike the midpoint rule, the trapezoidal rule also provides stable solutions for integration stepsizes that are not submultiples of the delay. Figure 1.13 illustrates the solution behavior of (1.2.11) for $\lambda = -50$ and $m - \delta = 12.5$, where the midpoint rule failed.

Although the trapezoidal rule appears to be more robust than the midpoint rule for constant coefficient linear DDEs, they may both be inadequate for the larger class of variable coefficient linear DDEs, even for integer values of $m - \delta$. In order to illustrate this further discrepancy between the methods as applied to ODEs and DDEs, consider the equation

$$\begin{cases} y'(t) = \lambda(t)y(t) - \frac{4}{5}\lambda(t)y(t-1), & t \geq 0, \\ y(t) = t + 1, & t \leq 0, \end{cases} \tag{1.2.12}$$

where $\lambda(t) = -50\sin^2(\frac{2\pi}{3}(t - \frac{1}{4}))$, whose solution, plotted in Figure 1.14, is asymptotically stable.

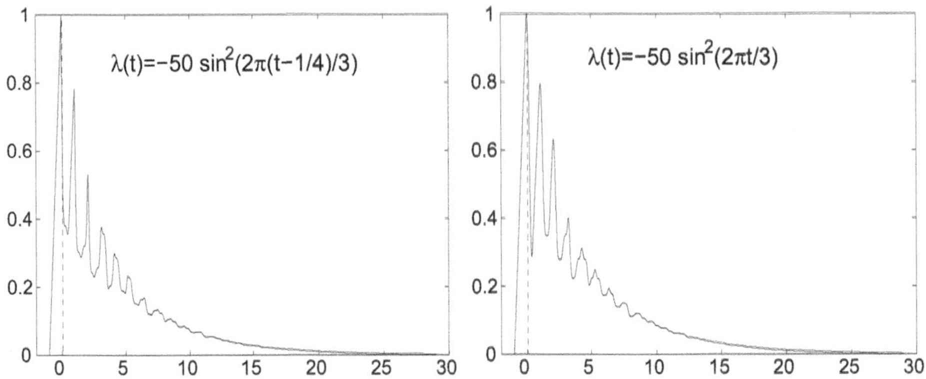

FIG. 1.14. Solutions of equation (1.2.12) with some $\lambda(t) \leq 0$.

Now, for the variable coefficient linear equation $y'(t) = \lambda(t)y(t)$, $y(0) = y_0$, it is known that the solutions go to zero for any real function $\lambda(t) \leq 0$ such that $\int_0^t \lambda(s)\mathrm{d}s \to -\infty$ as $t \to +\infty$. It is also known that the midpoint rule is *algebraically stable* and, consequently, the numerical solution is bounded by the initial value $|y_0|$ for any constant stepsize $h > 0$. Unlike the case of constant coefficients, now the midpoint rule gives rise to unstable solutions even for some constant integration stepsizes that are submultiples of the delay. Results for $h = 0.5$, corresponding to $m - \delta = 2$, are plotted in Figure 1.15 (left).

As the final negative result in this section, we show that the trapezoidal rule may also behave unstably for equations such as (1.2.12). To this end, consider equation (1.2.12) with the slightly different coefficient $\lambda(t) = -50\sin^2(\frac{2\pi}{3}t)$, whose solution is plotted in Figure 1.14 (right). The result, for the same integration stepsize $h = 0.5$, is plotted in Figure 1.15 (right).

It is worth pointing out that the last two examples, designed to spoil the algebraic stability properties of the ODE methods as applied to DDEs, are pathological for the methods we considered. A slight modification of either the coefficient $\lambda(t)$ in (1.2.12) or the stepsize may lead to models for which the midpoint rule and the trapezoidal rule become stable. The basic idea underlying the search for such counterexamples is due to Torelli [247]. Numerical results given by the midpoint rule and the trapezoidal rule for the last two examples with stepsize

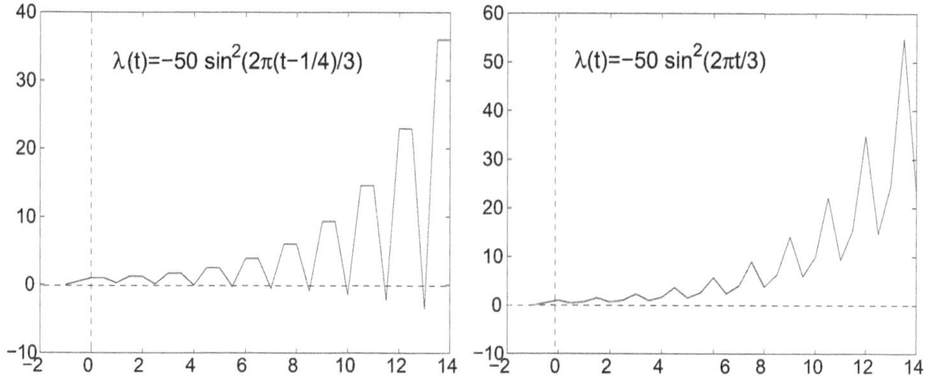

FIG. 1.15. Numerical solutions of equation (1.2.12) with some $\lambda(t) \leq 0$ obtained by the midpoint rule (left) and the trapezoidal rule (right) with integer $m - \delta = 2$.

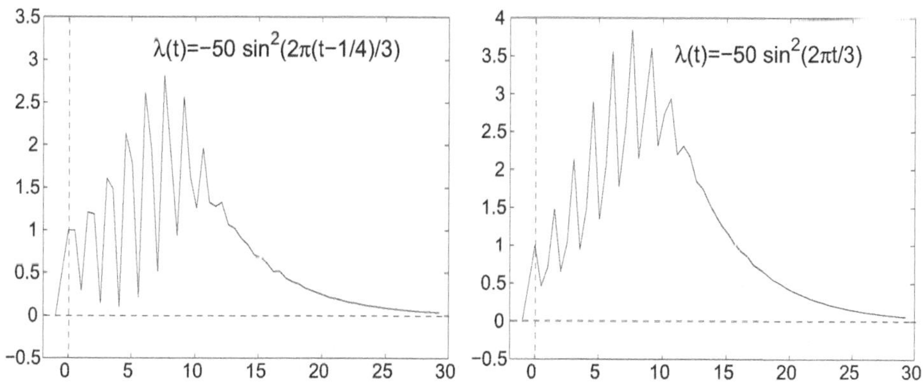

FIG. 1.16. Numerical solutions of equation (1.2.12) with some $\lambda(t) \leq 0$ obtained by the midpoint rule (left) and the trapezoidal rule (right) with $m - \delta = 1.98$.

$h = 0.505$, corresponding to $m - \delta = 1.98$, are depicted in Figure 1.16 where, after some spurious oscillations, they appear asymptotically stable.

1.2.3 *A good method for DDEs*

In order to get a second-order approximation for a class of stable problems including (1.2.12) which is stable for any stepsize, we may choose the two-stage *Lobatto IIIC* method with linear interpolation between nodal points.

For equation (1.2.12) with constant integration stepsize $h = 1/(m - \delta)$, m integer, $m \geq 2$, and $0 \leq \delta < 1$, the method reads

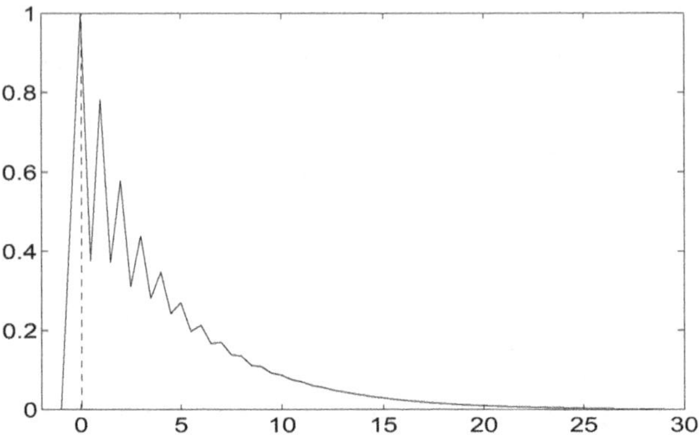

FIG. 1.17. Numerical solution obtained by the two-stage Lobatto IIIC method for equation (1.2.12) with $\lambda(t) = -50sin^2(\frac{2\pi}{3}(t-\frac{1}{4}))$ and with $m - \delta = 2$.

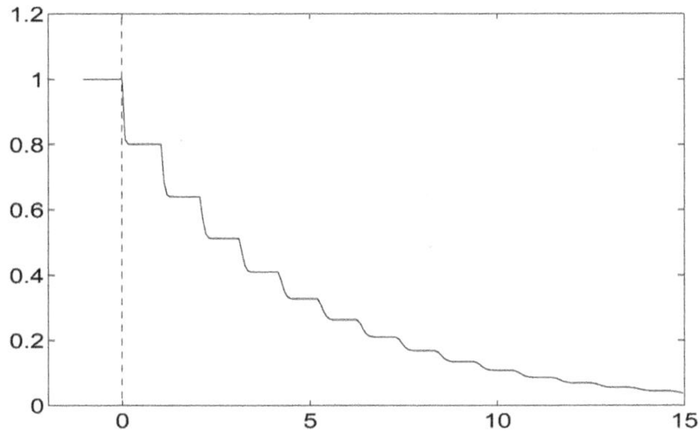

FIG. 1.18. Numerical solution of equation (1.2.8) with $\lambda = -50$ by the two-stage Lobatto IIIC method with $m - \delta = 12.5$.

$$
\begin{aligned}
Y_1^n &= y_n + \tfrac{1}{2}h\big(\lambda(t_n)Y_1^n - \lambda(t_{n+1})Y_2^n - \tfrac{4}{5}\lambda(t_n)\eta(t_n-1) + \tfrac{4}{5}\lambda(t_{n+1})\eta(t_{n+1}-1)\big),\\
Y_2^n &= y_n + \tfrac{1}{2}h\big(\lambda(t_n)Y_1^n + \lambda(t_{n+1})Y_2^n - \tfrac{4}{5}\lambda(t_n)\eta(t_n-1) - \tfrac{4}{5}\lambda(t_{n+1})\eta(t_{n+1}-1)\big),\\
y_{n+1} &= Y_2^n,
\end{aligned}
$$

where

$$
\left.\begin{aligned}
\eta(t_n-1) &= t_n\\
\eta(t_{n+1}-1) &= t_{n+1}
\end{aligned}\right\} \quad \text{for } n \le m-2,
$$

$$
\left.\begin{aligned}
\eta(t_n-1) &= t_n\\
\eta(t_{n+1}-1) &= 1-\delta+\delta y_1
\end{aligned}\right\} \quad \text{for } n = m-1,
$$

$$
\left.\begin{aligned}
\eta(t_n-1) &= (1-\delta)y_{n-m} + \delta y_{n-m+1}\\
\eta(t_{n+1}-1) &= (1-\delta)y_{n-m+1} + \delta y_{n-m+2}
\end{aligned}\right\} \quad \text{for } n \ge m.
$$

For $\lambda(t) = -50\sin^2(\frac{2\pi}{3}(t-\frac{1}{4}))$, it yields stable solutions for any stepsize. In particular, for $h = 0.5$, where the midpoint rule failed, the solution remains stable (see Figure 1.17).

The superiority of the Lobatto IIIC method also with respect to the trapezoidal rule is illustrated in Figure 1.18, where the solution of equation (1.2.8) with $\lambda = -50$ is plotted with the same stepsize as in Figure 1.13. The spurious oscillations exibited by the trapezoidal rule near the corners have disappeared.

On the basis of the previous examples showing the failure of accurate and stable methods as applied to the simplest class of scalar linear equations with constant delay, we conclude this section by stressing that *integration of DDEs cannot be based on the mere adaptation of some standard ODE code to the presence of delayed terms. Integration of DDEs actually requires the use of specifically designed methods, according to the nature of the equation and the behavior of the solution.*

2

EXISTENCE AND REGULARITY OF SOLUTIONS OF DDES

In this chapter we analyze the degree of smoothness of the solution for systems of DDEs

$$\begin{cases} y'(t) = f\Big(t, y(t), y\big(t - \tau(t, y(t))\big)\Big), & t_0 \le t \le t_f, \\ y(t) = \phi(t), \quad t \le t_0, \end{cases} \tag{2.0.1}$$

where $f : [t_0, t_f] \times \mathbb{R}^d \times \mathbb{R}^d \longrightarrow \mathbb{R}^d$, and for systems of NDDEs

$$\begin{cases} y'(t) = f\Big(t, y(t), y\big(t - \tau(t, y(t))\big), y'\big(t - \sigma(t, y(t))\big)\Big), & t_0 \le t \le t_f, \\ y(t) = \phi(t), \quad t \le t_0, \end{cases} \tag{2.0.2}$$

where $f : [t_0, t_f] \times \mathbb{R}^d \times \mathbb{R}^d \times \mathbb{R}^d \longrightarrow \mathbb{R}^d$. In particular, we shall fix our attention on the nature and location of derivative jump discontinuities, if any, and their propagation along the integration interval under different hypotheses on the delays τ and σ.

Moreover, for the sake of completeness, we shall report some results on existence and uniqueness of the solution for the initial value problems above. Whereas for constant and time dependent delays these problems have been widely investigated, a complete and satisfactory analysis for the most general case of state dependent delay and, in particular, for NDDEs is not available yet. For NDDEs with time dependent delays, results often refer to the following equation:

$$\begin{cases} \frac{\mathrm{d}}{\mathrm{d}t}\Big(y(t) - G\big(t, y(t - \tau(t))\big)\Big) = H\Big(t, y(t), y\big(t - \tau(t)\big)\Big), & t_0 \le t \le t_f, \\ y(t) = \phi(t), \quad t \le t_0, \end{cases} \tag{2.0.3}$$

where ϕ is simply continuous, which is obtainable (see Hale [121]) from (2.0.2) whenever $\sigma(t) = \tau(t)$, $\tau'(t) \neq 1$, and the right-hand-side function f has the form

$$f(t, u, v, z) = g(t, v)z + h(t, u, v).$$

In this case

$$G(t, v) = \frac{1}{1 - \tau'(t)} \int_0^v g(t, s)\mathrm{d}s$$

and

$$H(t, u, v) = h(t, u, v) - \frac{\partial G}{\partial t}(t, v).$$

A solution of (2.0.3) is a function $y(t)$ that is continuous in $[\min_{t \ge t_0}(t - \tau(t)), t_f]$, coincides with ϕ on $[\min_{t \ge t_0}(t - \tau(t)), t_0]$ and is such that $y(t) - G\big(t, y(t - \tau(t))\big)$, not $y(t)$, is continuously differentiable on $[t_0, t_f]$ and that satisfies equation (2.0.3) on $[t_0, t_f]$. Equation (2.0.3) is simpler than (2.0.2) and yet sufficiently general to include many applications.

2.1 Location of discontinuities and smoothing of the solution

We have already pointed out in Chapter 1 that the presence of deviated arguments in y, or y', can cause the appearance of jump discontinuities in y' or in higher derivatives of y at subsequent points. On the other hand, it is known that every step by step numerical method for the initial value problems achieves its own accuracy order provided that the solution is sufficiently smooth at each step interval $[t_n, t_{n+1}]$. More precisely, for a method to be of order p, we usually ask the solution to be at least C^{p+1}-continuous on $[t_n, t_{n+1}]$. Therefore, for the success of the method, it is crucial to include among the mesh points all the discontinuities of $y^{(s)}$ at least for $s = 0, 1, ..., p+1$. Aspects of the location of discontinuity points and the piecewise smoothness of solutions of DDEs were investigated, from a theoretical point of view, by Neves and Feldstein [206] for the general case of state dependent delay. Subsequently, the problem was addressed from a more practical point of view by Feldstein and Neves [97], Paul [216], Willé and Baker [270, 273] and Corwin, Sarafyan and Thompson [69]. This section is devoted to analyzing how the discontinuity points propagate through the integration interval $[t_0, t_f]$ and the increase in smoothness that occurs at any discontinuity point with respect to its *ancestor*, the discontinuity point it originates from.

The number and location of discontinuity points essentially depends on the behavior of the so-called *deviated arguments*

$$\alpha(t) = t - \tau\big(t, y(t)\big)$$

and

$$\beta(t) = t - \sigma\big(t, y(t)\big),$$

viewed as functions of t either for constant or variable or even state dependent delays τ and σ. Moreover, we shall assume $\alpha(t) \le t$ and $\beta(t) \le t$ because the delays are always non-negative. In particular, if $\alpha(t) \ge t_0$ and $\beta(t) \ge t_0$ for all $t \ge t_0$, then no values of y are needed in (2.0.1) and (2.0.2) behind t_0 and, therefore, no discontinuities propagate from t_0. Then the solution is regular according to the regularity of f, α and β. This is the case, for example, in the following equation, sometimes called *generalized pantograph equation* or *equation with proportional delay*:

$$\begin{cases} y'(t) = f\big(t, y(t), y(qt), y'(pt)\big), & t \ge 0, \\ y(0) = y_0, \end{cases} \tag{2.1.1}$$

where $0 < q < 1$ and $0 < p < 1$.

2.1.1 *Primary and secondary discontinuities*

Consider the scalar instance of equation (2.0.1) (i.e. $d = 1$) and assume that for the deviated argument $\alpha(t) = t - \tau$ it is $\alpha(t) < t_0$ for some points $t \in [t_0, t_f]$. Moreover, assume that the solution $y(t)$ does not link smoothly to the initial function $\phi(t)$ at t_0, i.e. $\phi'(t_0)^- \ne y'(t_0)^+ = f\big(t_0, \phi(t_0), \phi(\alpha(t_0))\big)$. If the functions

f, ϕ and α are continuous, then it is obvious that $y'(t)$ is also continuous for any $t > t_0$. On the other hand, if f, ϕ and α are differentiable, then $y''(t)$ exists for any t except for the points $\xi_{1,i}(> t_0)$ such that

$$\alpha(\xi_{1,i}) = t_0$$

and

$$\alpha'(\xi_{1,i}) \neq 0,$$

i.e. for the simple roots, if any, of the equation

$$\alpha(t) = t_0.$$

In fact, for any smooth function $f(t, y, x)$ we can formally write

$$\begin{aligned} y''(t)^{\pm} &= \frac{\partial f}{\partial t}\Big(t, y(t), y\big(\alpha(t)\big)\Big) + \frac{\partial f}{\partial y}\Big(t, y(t), y\big(\alpha(t)\big)\Big)y'(t) \\ &\quad + \frac{\partial f}{\partial x}\Big(t, y(t), y\big(\alpha(t)\big)\Big)y'\big(\alpha(t)\big)^{\pm}\alpha'(t), \end{aligned} \tag{2.1.2}$$

and hence

$$\begin{aligned} y''(\xi_{1,i})^{+} &= \frac{\partial f}{\partial t}(\xi_{1,i}, y(\xi_{1,i}), y(t_0)) + \frac{\partial f}{\partial y}(\xi_{1,i}, y(\xi_{1,i}), y(t_0))y'(\xi_{1,i}) \\ &\quad + \frac{\partial f}{\partial x}(\xi_{1,i}, y(\xi_{1,i}), y(t_0))y'(t_0)^{+}\alpha'(\xi_{1,i}) \end{aligned} \tag{2.1.3}$$

and

$$\begin{aligned} y''(\xi_{1,i})^{-} &= \frac{\partial f}{\partial t}(\xi_{1,i}, y(\xi_{1,i}), y(t_0)) + \frac{\partial f}{\partial y}(\xi_{1,i}, y(\xi_{1,i}), y(t_0))y'(\xi_{1,i}) \\ &\quad + \frac{\partial f}{\partial x}(\xi_{1,i}, y(\xi_{1,i}), y(t_0))\phi'(t_0)^{-}\alpha'(\xi_{1,i}). \end{aligned} \tag{2.1.4}$$

Since $\alpha'(\xi_{1,i}) \neq 0$ and $\phi'(t_0)^{-}$ is assumed to be different from $y'(t_0)^{+}$, y'' does not exist at $\xi_{1,i}$ and its prolongation by $y''(\xi_{1,i}) = y''(\xi_{1,i})^{+}$ has a jump discontinuity.

These jump discontinuities in y'' are called 1-*level primary discontinuities*. By differentiating (2.1.2), one easily checks that each 1-level primary discontinuity point $\xi_{1,i}$ gives rise in turn to 2-*level primary discontinuities* in y''' at any point $\xi_{2,j}(> \xi_{1,i})$ which is a simple root of

$$\alpha(t) = \xi_{1,i} \qquad \text{for some } i.$$

In general, any k-*level primary discontinuity* point $\xi_{k,i}$ gives rise to $(k+1)$-*level primary discontinuities* in $y^{(k+2)}$ at subsequent points $\xi_{k+1,j}$, where the solution of (2.0.1) becomes smoother and smoother as the primary discontinuity level increases. This increase in the regularity of $y(t)$ will be referred to as *smoothing of the solution*. On the contrary, the same argument applied to (2.0.2) reveals that for NDDEs smoothing does not occur and, in general, the solution remains C^0-continuous at any primary discontinuity point.

In the particular case that a discontinuity point is a root of

$$\alpha(t) = \xi_{j,i} \tag{2.1.5}$$

for some j, i with *odd* multiplicity $z \geq 3$, smoothing of the solution is faster than for $z = 1$ and possibly applies to neutral equations too. In fact, by (2.1.3) and (2.1.4) it is evident that for $\alpha'(\xi_{1,i}) = 0$ the solution y is at least of class C^2 at $\xi_{1,i}$. This phenomenon, called *generalized smoothing*, was investigated by Neves and Feldstein [206] for DDEs and is specified in the following theorem.

Theorem 2.1.1 (For DDEs) *If $\xi_{j,i}$ is a primary discontinuity point where the function $y(t)$ has continuous derivatives up to the order $\omega - 1$, then $y(t)$ is continuously differentiable at the propagated point $\xi_{j+1,k}$ at least to the order $z \cdot \omega$, provided $\xi_{j+1,k}$ is a root of (2.1.5) with odd multiplicity z.*

We have already observed that, in general, for NDDEs smoothing of the solution does not occur. More precisely, the solution $y(t)$ preserves, at any propagated point, the same regularity as its ancestor and hence as the initial point t_0. Nevertheless, Neves and Thompson [207] observed that, for $\tau = \sigma$, generalized smoothing may actually occur, according to the following theorem, provided that the *splicing condition*

$$\phi'(t_0)^- = y'(t_0)^+ = f\big(t_0, \phi(t_0), \phi(\alpha(t_0)), \phi'(\alpha(t_0))\big)$$

holds.

Theorem 2.1.2 (For NDDEs) *If $\xi_{j,i}$ is a primary discontinuity point where the function $y(t)$ has continuous derivatives up to the order $\omega - 1$, then $y(t)$ is continuously differentiable at the propagated point $\xi_{j+1,k}$ at least to the order $z \cdot (\omega - 1)$, provided that $\xi_{j+1,k}$ is a root of (2.1.5) with odd multiplicity z.*

Other discontinuities can appear if the functions f, τ and ϕ in (2.0.1) and (2.0.2) have some discontinuities with respect to t in some of their derivatives. Then such discontinuities are also propagated by the deviated arguments $\alpha(t)$ and $\beta(t)$ according to the primary discontinuity propagation rule and are called *secondary discontinuities*. As with the primary discontinuities, to preserve the accuracy order of a numerical method they must also be included in the mesh.

Henceforth, primary and secondary discontinuities will often be referred to as *discontinuities*. However, for the sake of simplicity, we assume that all the functions in (2.0.1) and (2.0.2) are C^∞-continuous. Therefore, in the interior of each interval between consecutive primary discontinuity points the solution $y(t)$ is C^∞-continuous as well, and no secondary discontinuities are present.

Definition 2.1.3 *A discontinuity point ξ is said to be of* order k *if $y^{(s)}(\xi)$ exists for $s = 0, \ldots, k$ and $y^{(k)}$ is Lipschitz continuous at ξ.*

Of course, for DDEs, any p-level primary discontinuity point has order $k \geq p$.

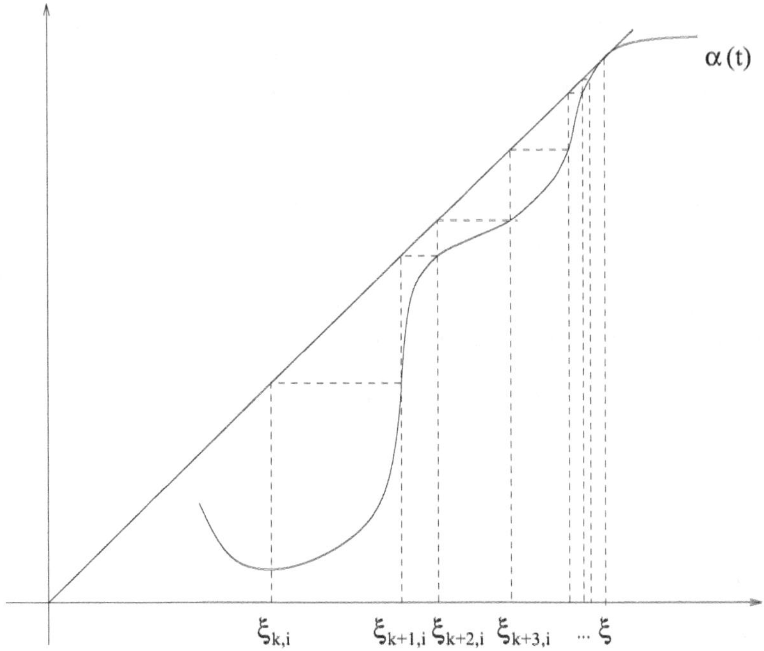

FIG. 2.1. Accumulation of discontinuity points in a left neighborhood of a vanishing delay point ξ.

Apart from the trivial case of constant delay, where the discontinuity points are given by $t_0+k\tau$, $k = 1, 2, \ldots$, the behavior of the deviated arguments $\alpha(t)$ and $\beta(t)$ may be more complicated and even unpredictable for equations with state dependent delay. By speculating on the properties of the deviated argument, one might build up equations with discontinuities that propagate in an arbitrarily wild manner on $[t_0, t_f]$, where possibly $t_f = +\infty$. Since it is crucial to include at least the discontinuity points of low order into the mesh, it is worth analyzing how they may propagate along the integration interval by pointing out some particular situations.

2.1.2 *Vanishing and non-vanishing time dependent delays*

First let us investigate, for (2.0.1) and for (2.0.2) with $\tau = \sigma$, how the primary discontinuities are located near the points where the delay $\tau(t)$ vanishes. In this case, called the *vanishing delay* case, a point $\xi > t_0$ is assumed to exist such that $\alpha(\xi) = \xi$. Owing to the continuity of $\alpha(t)$, it is evident that, for any k-level discontinuity point $\xi_{k,i} < \xi$ such that $\alpha(\xi_{k,i}) < \xi_{k,i}$, a $(k+1)$-level discontinuity point, say $\xi_{k+1,j}$, exists such that $\alpha(\xi_{k+1,j}) < \xi_{k+1,j}$ and $\xi_{k,i} < \xi_{k+1,j} < \xi$. In other words, there are infinitely many discontinuity points in any left neighborhood of ξ (see Figure 2.1).

On the other hand, for DDEs smoothing of the solution takes place, and then a left neighborhood of ξ exists which includes only discontinuity points of arbitrarily large order, that is where the solution is as smooth as needed. This is not the case for NDDEs, where smoothing of the solution does not take place.

In order to avoid the clustering of discontinuities due to vanishing delays, the following hypothesis will often be assumed to be satisfied:

(H_1) There exists a constant $\tau_0 > 0$ such that $\tau = t - \alpha(t) \geq \tau_0$ for all $t \in [t_0, t_f]$.

It is obvious that, under hypothesis (H_1), the distance between a discontinuity point and its ancestor is at least τ_0. Therefore, in any bounded interval $[t_0, t_f]$ the number of discontinuity points is finite.

2.1.3 *Bounded and unbounded time dependent delays*

Here we investigate how the primary discontinuities propagate according to the general rule

$$\alpha(\xi_{k,j}) = \xi_{k-1,i} \qquad \text{for some } i, \tag{2.1.6}$$

where, for $k > 0$ and any j, $\xi_{k,j}$ is a k-level primary discontinuity point and $\xi_{0,1} = t_0$ is the only 0-level primary discontinuity point. In particular, when the integration interval is unbounded, i.e. $t_f = +\infty$, it is worth distinguishing between models with a *bounded* or an *unbounded after-effect*, that is with a bounded or an unbounded delay function τ. We shall consider the following hypotheses:

(H_2) $\lim_{t \to +\infty} \alpha(t) = +\infty$.

(H_3) There exists a constant $\tau_1 > 0$ such that $\tau = t - \alpha(t) \leq \tau_1$ for all $t \in [t_0, t_f]$.

With respect to the propagation of discontinuities, (H_2) means that the solution is smoothed out indefinitely and, as k increases, the discontinuity points $\xi_{k,i}$ must either coalesce to a vanishing delay point (see Figure 2.1) or diverge to $+\infty$ (see Figure 2.2). In both cases, any k-level primary discontinuity point is met after a sufficiently large t.

As for the boundedness hypothesis (H_3), it evidently implies (H_2) but not vice versa. For instance, in the pantograph equation $y'(t) = f(t, y(t), y(qt))$, $t \geq 0$, we have

$$\alpha(t) = qt, \quad 0 < q < 1,$$

and

$$\tau(t) = (1 - q)t,$$

which are both unbounded.

In particular, when (H_3) holds, the model is said to have *fading memory*. This means that for any t, after a sufficiently long but uniformly bounded elapsed time interval, the solution value $y(t)$ will not influence the right-hand side of (2.0.1) or (2.0.2). In other words, to integrate the DDE it is sufficient to store a finite segment of the last past history.

On the contrary, for unbounded delays $\tau(t)$ the deviated argument $\alpha(t) = t - \tau(t)$ may be bounded or not. When both $\tau(t)$ and $\alpha(t)$ are unbounded, as with

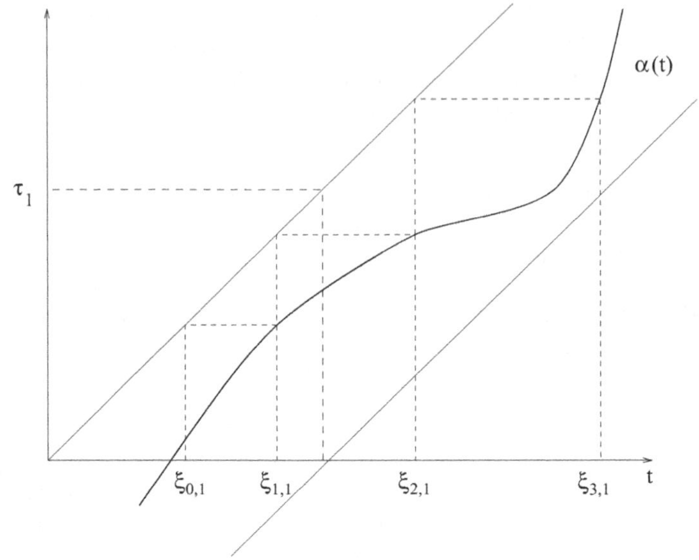

FIG. 2.2. Divergent sequence of discontinuity points.

the pantograph equation, the solution $y(t)$ eventually depends on an arbitrarily large segment of the past history which must be stored for numerical integration.

If the deviated argument is instead bounded, say $\alpha(t) \leq M$, then infinitely many primary discontinuities might lie to the right of M, but none of them, if any, can increase its level in $[M, +\infty)$ (see Figure 2.3). This prevents smoothing of the solution beyond a certain class of regularity. Moreover, the solution $y(t)$ definitively depends on the history up to M.

In many applications the following hypothesis is satisfied:

(H_4) The deviated argument $\alpha(t)$ is a strictly increasing function for all $t \in [t_0, t_f]$.

This is the case, for instance, for all models with constant delay τ. In general, if (H_4) holds and $\alpha(t_0) < t_0$, then the primary discontinuity points form an increasing sequence $\xi_1 < \xi_2 < \cdots < \xi_j < \cdots$, where, for any j, $\xi_j = \xi_{j,1}$ is the sole j-level discontinuity point. That is, as time passes the solution becomes smoother and smoother. On the contrary, oscillating deviated arguments can cause the interlacing of different level discontinuities, as shown in Figure 2.4.

However, in many cases it is sufficient to locate the *principal discontinuity points* defined as follows.

Definition 2.1.4 *The one-index subset of primary discontinuity points $\bar{\xi}_i$ defined inductively by $\bar{\xi}_0 = t_0$ and, for $i \geq 0$, by the minimum root $\bar{\xi}_{i+1}$ of*

$$\alpha(t) = \bar{\xi}_i$$

with odd multiplicity, is called the set of principal discontinuity points.

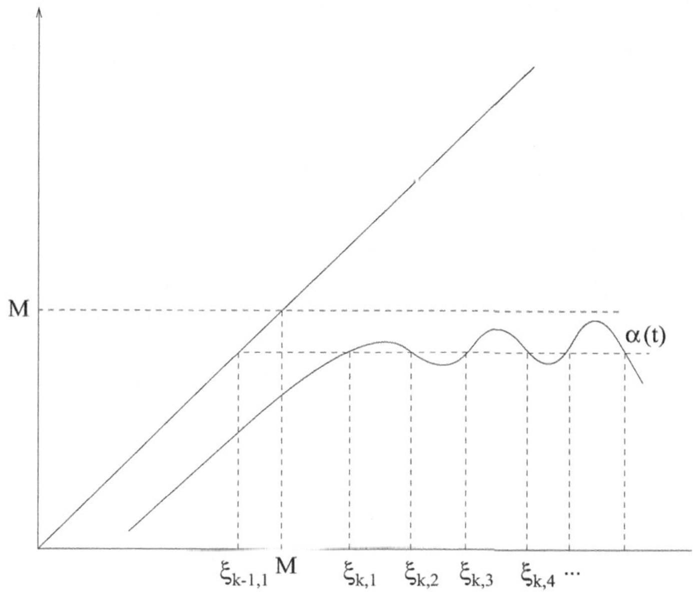

FIG. 2.3. Discontinuity points of the same level generated by a bounded deviated argument.

The principal discontinuity points are of interest because, for any i,

$$\alpha(t) \leq \bar{\xi}_i \quad \forall t \in [\bar{\xi}_i, \bar{\xi}_{i+1}].$$

Note that $\bar{\xi}_i$ is nothing but $\min_j \xi_{i,j}$. In particular, if (H_4) holds, then all the primary discontinuity points are principal.

2.1.4 *Multiple delays*

Sometimes in applications we find DDEs or NDDEs where the right-hand side depends on more than one retarded argument, that is equations such as

$$\begin{cases} y'(t) = f\Big(t, y(t), y\big(\alpha_1(t)\big), \ldots, y\big(\alpha_r(t)\big)\Big), & t \geq t_0, \\ y(t) = \phi(t), \quad t \leq t_0, \end{cases}$$

or

$$\begin{cases} y'(t) = f\Big(t, y(t), y\big(\alpha_1(t)\big), \ldots, y\big(\alpha_r(t)\big), y'\big(\beta_1(t)\big), \ldots, y'\big(\beta_r(t)\big)\Big), & t \geq t_0, \\ y(t) = \phi(t), \quad t \leq t_0, \end{cases}$$

respectively.

In principle, the presence of many delays does not introduce any particular additional difficulties over those in the one-delay case. The main complication lies in the possibly more chaotic proliferation of the discontinuity points. For example, even if all the deviated arguments are strictly increasing, each discontinuity

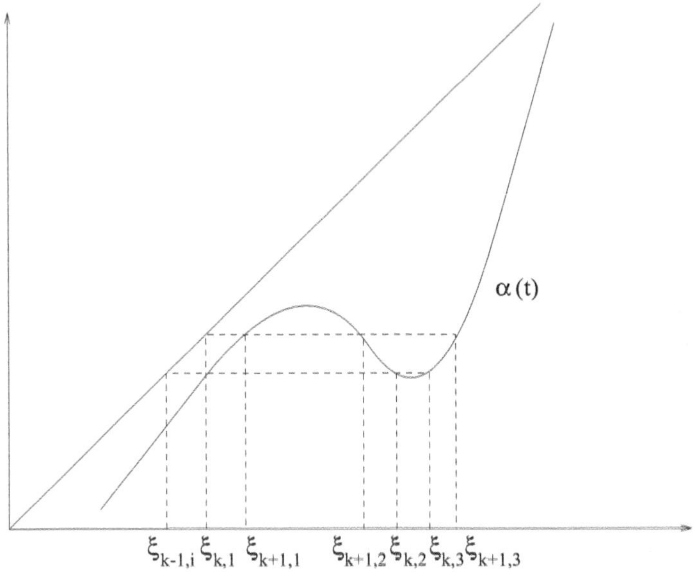

FIG. 2.4. Interlacing of different level discontinuity points.

point generates, in turn, another r such points. Thus, if $\xi_{k-1,j}$ is a discontinuity point of level $k-1$, then we find the corresponding discontinuity points of level k by solving the r equations

$$\alpha_i(\xi_{k,(j-1)r+i}) = \xi_{k-1,j}, \quad i = 1, \ldots, r.$$

It may happen that two or more discontinuity points, possibly of different levels, coincide. Similar chaotic proliferation occurs for the principal discontinuities. However, in most cases the difference with respect to the one-delay case is technical rather than conceptual. Nevertheless, in the analysis and implementation of numerical methods, it is often convenient to restate the equations in the following *simplified form* with no loss of generality:

$$\begin{cases} y'(t) = f\Big(t, y(t), y\big(\alpha(t)\big)\Big), & t \geq t_0, \\ y(t) = \phi(t), \quad t \leq t_0, \end{cases} \tag{2.1.7}$$

where $y = [y_1, \ldots, y_d]^T \in \mathbb{R}^d$, $\alpha = [\alpha_1, \ldots, \alpha_d]^T \in \mathbb{R}^d$ and $y(\alpha(t))$ stands for the vector $[y_1(\alpha_1(t)), \ldots, y_d(\alpha_d(t))]^T$. In other words, every component is allowed to have at most one delay. This is an undesirable restrictive condition which is violated in many real-life models. However, all DDEs may be transformed into the simplified form by the following transformation, introduced by Neves [204] and termed the *method of augmented arguments*.

Assume, for simplicity, that the sole violation to the simplified form condition is that the variable $y_i(t)$ depends on two different deviated arguments $\alpha_i(t)$ and $\rho_i(t)$ in the jth component of the system as follows:

$$y_j'(t) = f_j\Big(t, y_1(t), \ldots, y_d(t), y_1\big(\alpha_1(t)\big), \ldots, y_i\big(\alpha_i(t)\big), \ldots, y_d\big(\alpha_d(t)\big), y_i\big(\rho_i(t)\big)\Big). \tag{2.1.8}$$

Now consider the new fictitious variable $y_{d+1}(t)$, rename the deviated argument $\rho_i(t)$ as $\alpha_{d+1}(t)$ and then substitute into equation (2.1.8) to obtain

$$y_j'(t) = f_j\Big(t, y_1(t), \ldots, y_d(t), y_1\big(\alpha_1(t)\big), \ldots, y_d\big(\alpha_d(t)\big), y_{d+1}\big(\alpha_{d+1}(t)\big)\Big).$$

Finally, augment the system by the additional equation

$$y_{d+1}'(t) = f_i\Big(t, y_1(t), \ldots, y_d(t), y_1\big(\alpha_1(t)\big), \ldots, y_d\big(\alpha_d(t)\big)\Big), \quad \text{if } j \neq i,$$

or

$$y_{d+1}'(t) = f_i\Big(t, y_1(t), \ldots, y_d(t), y_1\big(\alpha_1(t)\big), \ldots, y_{d+1}\big(\alpha_{d+1}(t)\big)\Big), \quad \text{if } j = i,$$

endowed by the initial condition

$$y_{d+1}(t) = \phi_i(t), \qquad t \leq t_0,$$

which makes $y_{d+1}(t) = y_i(t)$ for all $t \geq t_0$.

In general, the enlarged system takes the simplified form (2.1.7) in some dimension $d' > d$ with d' equal to the number of different delays present in the original system.

Apart from the evident programming simplification in the DDE solver, the advantage in using the simplified form is that it makes it much easier to extract from the set of primary discontinuities $\xi_{k,s}$ a totally ordered subset of principal discontinuities $\bar{\xi}_j$ such that, for any j,

$$\alpha_i(t) \leq \bar{\xi}_j \quad \forall t \in [\bar{\xi}_j, \bar{\xi}_{j+1}] \;\text{ and }\; \forall i \leq d$$

as with the one-delay case. Such a set is obtained by simply collecting and ordering the set of the principal discontinuities relevant to each individual deviated argument $\alpha_i(t)$ in the simplified form DDE (2.1.7).

Definition 2.1.5 *The* principal discontinuity points *for the solutions of DDEs in the simplified form (2.1.7) are given by the set of points $\bar{\xi}_{i,j}$, $1 \leq i \leq d$, such that $\bar{\xi}_{i,0} = t_0$ and, for $j \geq 0$, $\bar{\xi}_{i,j+1}$ is the minimum root of*

$$\alpha_i(t) = \bar{\xi}_{i,j}$$

with odd multiplicity.

In order to avoid confusion, note that the first index in the principal discontinuity point $\bar{\xi}_{i,j}$ is no longer related to its level, as a generic primary discontinuity point, but to the component of the deviated argument $\alpha(t)$ it depends on.

2.1.5 *State dependent delays*

Finally, we briefly consider the difficulties related to the case when the delay has the form $\tau(t, y(t))$. First of all, note that hypothesis (H_1), which aims to avoid vanishing delays, is now modified to

(H_1^*) There exists a constant $\tau_0 > 0$ such that $\tau(t, z) \geq \tau_0$ for all $t \in [t_0, t_f]$ and $z \in \mathbb{R}^d$.

Of course, the monotonicity hypothesis (H_4) cannot be considered.

In order to locate the discontinuities, in principle one should apply the general propagation rule (2.1.6) which, for the case of state dependent delay, is

$$\xi_{k,j} - \tau\big(\xi_{k,j}, y(\xi_{k,j})\big) = \xi_{k-1,i} \quad \text{for some } i, \tag{2.1.9}$$

and solve it for $\xi_{k,j}$. Because the delay is dependent of $y(t)$, this cannot be done a priori without knowledge of the solution. Moreover, it is evident that even assuming some approximation of $y(t)$ is available, we must be satisfied with an approximation of the discontinuity point $\xi_{k,j}$. However, as Willé and Baker remarked in [273], finding the multiple roots is an inherently ill-conditioned problem. In conclusion, the impossibility of locating the discontinuity points a priori makes the implementation and convergence analysis of numerical methods for (2.0.1) and (2.0.2) a rather complicated task. The problem will be faced in Section 4.3.

2.1.6 *Propagation of discontinuities in systems*

The work of Neves and Feldstein [206] on the propagation of discontinuities was extended by Willé and Baker [269, 270] to systems of DDEs and Volterra integro-differential equations. We shall follow their line and notations.

For simplicity, consider equations (2.0.1) and (2.0.2) with $d > 1$ and a unique delay $\tau = \sigma$. If we assume that each component y_i' depends on all components of the delayed solution $y(t - \tau)$, the propagation of discontinuities, as well as the smoothing and generalized smoothing of the solution, takes place according to the results of the foregoing subsections. On the contrary, if some components are not coupled to all delayed components, the way the discontinuities propagate is different and some higher order smoothing of solutions might actually occur.

Assume that in (2.0.1) each component y_i' depends on certain components $y_{k_j}(t)$, $j = 1, \ldots, s_i$, and $y_{l_j}(t - \tau)$, $j = 1, \ldots, r_i$. In this case we say that the component y_i of the solution is *strongly* coupled to the components y_{k_j}, $j = 1, \ldots, s_i$, and *weakly* coupled to the components y_{l_j}, $j = 1, \ldots, r_i$. Willé and Baker [270] illustrated the coupling by means of an oriented graph, called a *dependency network*, interconnecting the nodes, labeled by $i = 1, \ldots, d$, by solid and dotted lines for strong and weak coupling of corresponding components y_i, respectively. For example, for the system

$$\begin{cases} y_1'(t) = f_1\big(y_2(t), y_3(t-1)\big), \\ y_2'(t) = f_2\big(y_3(t)\big), \\ y_3'(t) = f_3\big(y_1(t), y_2(t-1)\big), \end{cases} \quad t \geq 0, \tag{2.1.10}$$

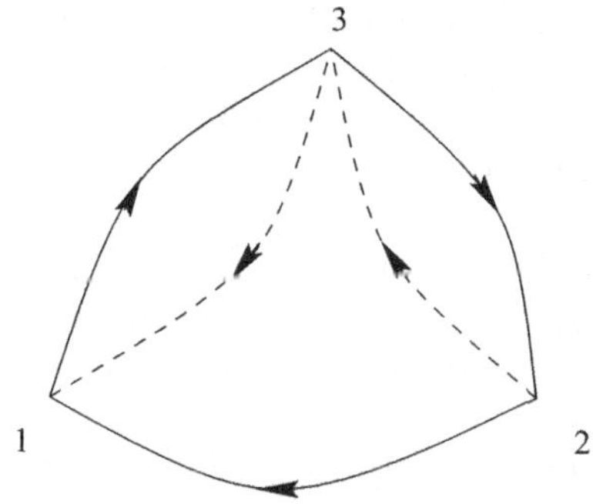

FIG. 2.5. Dependency network of the system (2.1.10).

the dependency network behaves as in Figure 2.5. Assume that the components y_3 and y_2 do not satisfy the splicing condition so that discontinuities occur in y_3' and y_1' at $t = 0$. In order to analyze how such discontinuities spread onto subsequent points $t = 1, 2, \ldots$, consider the dynamical version of the dependency network as in Figure 2.6, where the strong couplings take place at the same discontinuity level, that is at the same time, whereas weak couplings take place between consecutive levels, that is from $t-1$ to t. At every level, the order of the

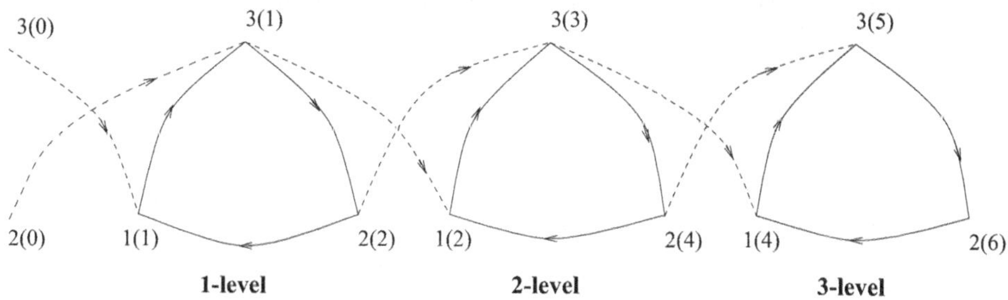

FIG. 2.6. Dynamic dependency network of the system (2.1.10).

propagated discontinuity is bracketed next to each node number and defines the order of discontinuity of the node, that is the maximum order of the continuous derivative of the corresponding component. At every level, discontinuities are "imported" through the nodes of the previous level which are weakly coupled to nodes of the current level. Between two consecutive levels, every weak coupling increases the order of a propagated discontinuity by 1. For variable delays, according to Theorem 2.1.1 on generalized smoothing, oriented arcs corresponding to weak couplings are weighted by the multiplicity z of the discontinuity point as roots of the propagation equation (2.1.6). In particular, if $z > 1$, then the coupling produces an increase in order equal to $(z-1)\omega + 1$, $\omega - 1$ being the order of the ancestor. Strong couplings, which are relevant to the same level, always have weight $z = 1$, and hence increase the order by 1. Therefore, the order of discontinuity attained by a node at each level is then given by the minimum order provided by all the couplings (oriented arcs) entering the node. In conclusion, the increase in the overall order of discontinuity of the vector function y is at least equal to 1 for each level, and possibly higher. In particular, for the system (2.1.10) there is a jump in smoothing at level 3, where the overall discountinuity order passes from 2 to 4.

Similarly for NDDEs, the lack of some components of the delayed term $y'(t-\tau)$ may cause smoothing of the solution even when no splicing conditions hold. In this case there is a third kind of coupling among the components in the system. We shall say that the component y_i is *weakly** coupled to the component y_j if, in equation (2.0.2), $y_i'(t)$ depends on $y_j'(t-\tau)$. This occurrence will be expressed in the dependency network by a weighted oriented arc, drawn by a thicker dotted line, from node j to node i and will undergo the rule on generalized smoothing stated in Theorem 2.1.2. It is evident that through this arc the discontinuity is propagated with no smoothing of the solution. Consider, for example, the system

$$\begin{cases} y_1'(t) = f_1\big(t, y_1(t), y_2(t)\big), \\ y_2'(t) = f_2\big(t, y_1(t), y_2(t), y_1(t-1), y_1'(t-1)\big), \end{cases} \quad t \geq t_0, \tag{2.1.11}$$

for which the dependency network is given in Figure 2.7. Observe that the overall order of discontinuity increases by one unit for each level despite the equation being neutral and the splicing condition not being assumed.

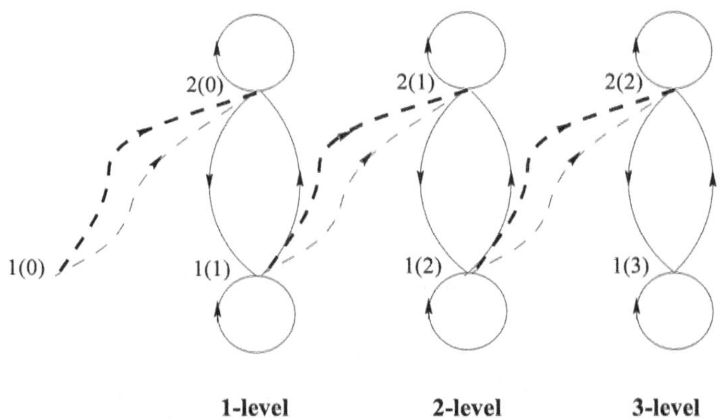

FIG. 2.7. Dynamic dependency network of the system (2.1.11).

2.2 Existence and uniqueness of solutions

As with ordinary differential equations, existence and uniqueness theorems for the IVPs (2.0.1) and (2.0.2) are essentially based on the continuity of the functions $f(t, u, v)$ and $f(t, u, v, w)$ with respect to t and Lipschitz continuity with respect to u, v and w.

In particular, for the equation

$$\begin{cases} y'(t) = f\Big(t, y(t), y\big(t - \tau(t, y(t))\big), y'\big(t - \sigma(t, y(t))\big)\Big), & t \geq t_0, \\ y(t) = \phi(t), \quad t \leq t_0, \end{cases}$$

the local existence and uniqueness analysis is almost trivial whenever (H_1^*) holds for both delays, that is

$$\inf_{[t_0,t_f]\times\mathbb{R}^d} \tau(t, x) = \tau_0 > 0$$

and

$$\inf_{[t_0,t_f]\times\mathbb{R}^d} \sigma(t, x) = \sigma_0 > 0.$$

In fact, in the interval $[t_0, t_0+H]$, $H = \min\{\tau_0, \sigma_0\}$, the equation to solve reduces to the ODE

$$\begin{cases} y'(t) = f\Big(t, y(t), \phi\big(t - \tau(t, y(t))\big), \phi'\big(t - \sigma(t, y(t))\big)\Big), & t \geq t_0, \\ y(t_0) = \phi(t_0), \end{cases}$$

for which well-known standard results may be used to prove existence and uniqueness of the solution. In particular, the continuity of $f(t, u, v, w)$ with respect to t and Lipschitz continuity with respect to u, v and w, along with the Lipschitz continuity of ϕ, ϕ', τ and σ, guarantee the local existence and uniqueness of the solution in $[t_0, t_0 + \delta]$ for some $\delta > 0$.

For the existence of the solution in a finite interval $[t_0, t_f]$ we may proceed by successive integrations on the intervals $[t_0 + iH, t_0 + (i+1)H]$, $i = 0, 1, \ldots$, where the equations to be solved are still ordinary, the solution being known up to $[t_0 + iH]$. This method is known in the literature as the *method of steps* and is one of the basic methods for the theoretical analysis as well as numerical integration of DDEs and NDDEs under the hypothesis (H_1^*).

When the delays τ or σ vanish at some point t^*, the method of steps does not apply in a neighborhood of t^* and results on existence and uniqueness become more difficult to prove.

Now we give some known results for equations (2.0.1) under more general conditions on the delays. They are obtained as particular instances of theorems proved in Hale [122] and El'sgol'ts and Norkin [81].

Theorem 2.2.1 (Local existence) *Consider the equation*

$$\begin{cases} y'(t) = f\Big(t, y(t), y\big(t - \tau(t)\big)\Big), & t_0 \leq t < t_f, \\ y(t_0) = y_0, \end{cases} \tag{2.2.1}$$

and assume that the function $f(t, u, v)$ is continuous on $A \subseteq [t_0, t_f)\times\mathbb{R}^d\times\mathbb{R}^d$ and locally Lipschitz continuous with respect to u and v. Moreover, assume that the delay function $\tau(t) \geq 0$ is continuous in $[t_0, t_f)$, $\tau(t_0) = 0$ and, for some $\xi > 0$, $t - \tau(t) > t_0$ in the interval $(t_0, t_0 + \xi]$. Then the problem (2.2.1) has a unique solution in $[t_0, t_0 + \delta)$ for some $\delta > 0$ and this solution depends continuously on the initial data.

It can be shown that, under the same hypotheses, the solution can be continued until a *maximal solution* defined in the interval $[t_0, b)$, with $t_0 < b \leq t_f$. This allows us to prove the following global existence theorem.

Theorem 2.2.2 (Global existence) *If, under the hypotheses of Theorem 2.2.1, the unique maximal solution of (2.2.1) is bounded, then it exists on the entire interval* $[t_0, t_f)$.

In order to apply the global existence theorem, we need an a priori bound for the solution. This is given in the following corollary.

Corollary 2.2.3 *Besides the hypotheses of Theorem 2.2.1, assume that the function* $f(t, u, v)$ *satisfies the condition*

$$\|f(t,u,v)\| \leq M(t) + N(t)(\|u\| + \|v\|)$$

in $[t_0, t_f) \times \mathbb{R}^d \times \mathbb{R}^d$, *where* $M(t)$ *and* $N(t)$ *are continuous positive functions on* $[t_0, t_f)$. *Then the solution of (2.2.1) exists and is unique on the entire interval* $[t_0, t_f)$.

The previous results hold for multiple delays and furnish, in particular, the global existence, uniqueness and continuous dependence on the initial data for the solution of the linear DDE

$$\begin{cases} y'(t) = \sum_{i=0}^{r} A_i(t) y(t - \tau_i(t)) + g(t), & t \geq t_0, \\ y(t) = \phi(t), & t \leq t_0, \end{cases}$$

for any continuous functions $A_i(t)$, $i = 0, \ldots, r$, $\phi(t)$ and $g(t)$, and for any set of continuous delays $\tau_i(t) \geq 0$.

The following result, proved by Driver [77], extends the existence result of Theorem 2.2.1 to the more general case of state dependent delays.

Theorem 2.2.4 (Local existence) *Consider the equation*

$$\begin{cases} y'(t) = f\Big(t, y(t), y\big(t - \tau(t, y(t))\big)\Big), & t \geq t_0, \\ y(t) = \phi(t), & t \leq t_0. \end{cases} \tag{2.2.2}$$

Let $U \subseteq \mathbb{R}^d$ *and* $V \subseteq \mathbb{R}^d$ *be neighborhoods of* $\phi(t_0)$ *and* $\phi(t_0 - \tau(t_0, \phi(t_0)))$, *respectively, and assume that the function* $f(t, u, v)$ *is continuous with respect to* t *and Lipschitz continuous with respect to* u *and* v *in* $[t_0, t_0 + h] \times U \times V$ *for some* $h > 0$. *Moreover, assume that the initial function* $\phi(t)$ *is Lipschitz continuous for* $t \leq t_0$ *and that the delay function* $\tau(t, y) \geq 0$ *is continuous with respect to* t *and Lipschitz continuous with respect to* y *in* $[t_0, t_0 + h] \times U$. *Then the problem (2.2.2) has a unique solution in* $[t_0, t_0 + \delta)$ *for some* $\delta > 0$ *and this solution depends continuously on the initial data.*

Driver proved an analogous result for equations with multiple delays.

As for NDDEs of the type

$$\begin{cases} y'(t) = f\Big(t, y(t), y\big(t - \tau(t)\big), y'\big(t - \sigma(t)\big)\Big), & t \geq t_0, \\ y(t) = \phi(t), & t \leq t_0, \end{cases} \tag{2.2.3}$$

things are slightly different and additional conditions must be imposed on the function $f(t, u, v, w)$. For example, observe that, if $\tau(t_0) = \sigma(t_0) = 0$, then equation (2.2.3) yields $y'(t_0) = f\big(t_0, y(t_0), y(t_0), y'(t_0)\big)$. Therefore, if the initial value $y(t_0)$ is such that the equation

$$z = f\big(t_0, y(t_0), y(t_0), z\big) \tag{2.2.4}$$

has no solutions for z, then the IVP (2.2.3) has no solutions through the point $(t_0, y(t_0))$. The following theorem, due to Kamenskii [164], holds for the case of vanishing delay.

Theorem 2.2.5 (Local existence) *Consider the NDDE*

$$\begin{cases} y'(t) = f\Big(t, y(t), y\big(t-\tau(t)\big), y'\big(t-\tau(t)\big)\Big), & t \geq t_0, \\ y(t_0) = y_0, \\ y'(t_0) = y_0', \end{cases} \tag{2.2.5}$$

where $\tau(t_0) = 0$ and, for some $\xi > 0$, $t - \tau(t) > t_0$ in the interval $(t_0, t_0 + \xi]$. Assume $y'(t_0)$ to be a real solution of (2.2.4) and let $\mu_\xi = \inf_{[t_0, t_0+\xi]} \tau'(t) > 0$. Moreover, assume the function $f(t, u, v, w)$ to be continuous with respect to t and Lipschitz continuous with respect to u, v and w in a neighborhood of the point $(t_0, y(t_0), y(t_0), y'(t_0))$. If L, the Lipschitz constant with respect to w, satisfies

$$L(1 - \mu_\xi) < 1,$$

then a unique continuously differentiable solution of (2.2.5) exists in $[t_0, t_0 + \delta)$ for some $\delta > 0$.

More general theorems dealing with NDDEs of the form (2.0.2) and (2.0.3) with state dependent delays are also available. We refer the interested reader to the already mentioned books of Bellman and Cooke [40], Hale [121], Driver [78], El'sgol'ts and Norkin [81], Hale and Verduyn Lunel [123], Kolmanovskii and Myshkis [172], Nosov [173], Diekmann et al. [75], Kuang [183] and the references therein.

Results on existence and uniqueness of solutions for a larger class of retarded functional differential equations are presented in a more recent survey paper by Bellen et al. [25], where the peculiar phenomenon of bifurcation and possible termination of the solution is also investigated from both the theoretical and the numerical point of view.

3

A REVIEW OF DDE METHODS

We saw in Chapter 1 that the IVP for DDEs can be correctly formulated as

$$\begin{cases} y'(t) = f(t, y_t), & t \geq t_0, \\ y_{t_0} = y(t_0 + \theta) = \Phi(\theta), \end{cases} \tag{3.0.1}$$

in the Banach space $C = C^0([-r, 0], \mathbb{R}^d)$ of continuous functions $y_t = y(t + \theta)$, $\theta \in [-r, 0]$, mapping the interval $[-r, 0]$ into $\mathbb{R}^d$. The characterization of such equations is that, for every t, the derivative $y'(t)$ depends on the *state function* y_t, rather than on the state value $y(t)$ as happens for ODEs. This makes the difference between the two classes of equations and leads to various complications in their solution from both the theoretical and numerical points of view. In particular, for the numerical integration of (3.0.1), even being satisfied with a discretized solution $y(t_n), n = 1, 2, \ldots$, its computation depends on the state functions $y(t_n + \theta)$ for all or some values of θ, which are sometimes unknown and often unpredictable. This is the main difficulty with the numerical integration of functional differential equations.

Several approaches, quite different from one another, have been proposed for the numerical solution of (3.0.1). Some of them are based on an a priori transformation of the original problem into another one which is equivalent but simpler in some sense, such as, for example, a Volterra integral equation or a system of ODEs for which robust methods are available, or partial differential equations (PDEs) with particular initial/boundary conditions. Other approaches start directly from the problem in the formulation

$$\begin{cases} y'(t) = f\Big(t, y(t), y\big(t - \tau(t, y(t))\big)\Big), & t_0 \leq t \leq t_f, \\ y(t) = \phi(t), \quad t \leq t_0, \end{cases}$$

or

$$\begin{cases} y'(t) = f\Big(t, y(t), y\big(t - \tau(t, y(t))\big), y'\big(t - \sigma(t, y(t))\big)\Big), & t_0 \leq t \leq t_f, \\ y(t) = \phi(t), \quad t \leq t_0, \end{cases}$$

by combining discrete ODE methods with interpolation. For DDEs of neutral type the problem is often transformed into a system of ODEs coupled to an algebraic recursion with delay (see Section 3.3.2).

In this chapter we briefly outline some quite different approaches that have been proposed in the literature. Some of them will be described fully and investigated in the forthcoming chapters.

An exhaustive collection of methods and related references, up to the beginning of 1970s, is given in Cryer [71] and, subsequently, in Bellen [22] and Meinardus and Nürnberger [200] for papers up to the 1980s, and, more recently, in Zennaro [283], Baker, Paul and Willé [14, 15] and Baker [6, 7]. Many papers address the more general problem (3.0.1) of Volterra functional differential equations and results for DDEs can be stated as a particular case. For a large set of problems and bibliographic references, see Brunner [56].

3.1 The first approaches

The first approaches to the numerical solution of DDEs in the simpler non state dependent delay case go back to the 1950s and were characterized by the direct application of well-known formulae for ODEs, essentially *linear multistep* (LM) methods, to the equation

$$\begin{cases} y'(t) = f\Big(t, y(t), y\big(t-\tau(t)\big)\Big), & t_0 \le t \le t_f, \\ y(t) = \phi(t), & t \le t_0. \end{cases} \tag{3.1.1}$$

In particular, a set $\Delta = \{t_0, t_1, \ldots, t_n, \ldots, t_N = t_f\}$ of mesh points was assumed to exist such that, for all $t_n \in \Delta$, either $t_n - \tau(t_n) < t_0$ or $t_n - \tau(t_n) \in \Delta$. Once such a mesh is available, any discrete method that makes use of nodal points only can be implemented directly for delay and neutral delay equations. For example, the *forward Euler* method for equation (3.1.1) is

$$y_{n+1} = y_n + h_{n+1} f(t_n, y_n, y_q),$$

whereas for the equation

$$\begin{cases} y'(t) = f\Big(t, y(t), y\big(t-\tau(t)\big), y'\big(t-\tau(t)\big)\Big), & t_0 \le t \le t_f, \\ y(t) = \phi(t), & t \le t_0, \end{cases}$$

it is

$$\begin{aligned} y_{n+1} &= y_n + h_{n+1} f(t_n, y_n, y_q, y'_q), \\ y'_n &= f(t_n, y_n, y_q, y'_q), \end{aligned}$$

for some integer $q < n$.

This approach, introduced by El'sgol'ts [80], imposes a severe constraint on the mesh which, in some cases, makes the method impracticable. In fact, for the problem

$$\begin{cases} y'(t) = y(t/2), & 0 \le t \le 1, \\ y(0) = 1, \end{cases} \tag{3.1.2}$$

for any $t \in \Delta$, $t/2$ should belong to the mesh Δ and, thus, an initial stepsize does not exist. Moreover, since the grid points must satisfy $t_{n+1} = 2t_n$ for $n \ge 1$, we have $h_{n+1} = t_n$. Therefore, the last stepsize is always equal to $\frac{1}{2}$ and convergence cannot even be considered. At any rate, apart from the vanishing delay case, a mesh Δ can always be constructed on any bounded interval $[t_0, t_f]$

with arbitrarily small stepsize. The best strategy for doing so is to start from the first point t_0, to locate all discontinuities that lie in the integration interval $[t_0, t_f]$ and to put them into Δ. Afterwards, fill in Δ by starting from the last point t_f and proceed backward with the desired maximum stepsize. Any new mesh point t_n gives rise to a preceding point $t_n - \tau(t_n)$, possibly $> t_0$, also to be included in the mesh Δ. So, for any maximum stepsize h, we get a mesh Δ with finitely many elements which is suitable for convergence.

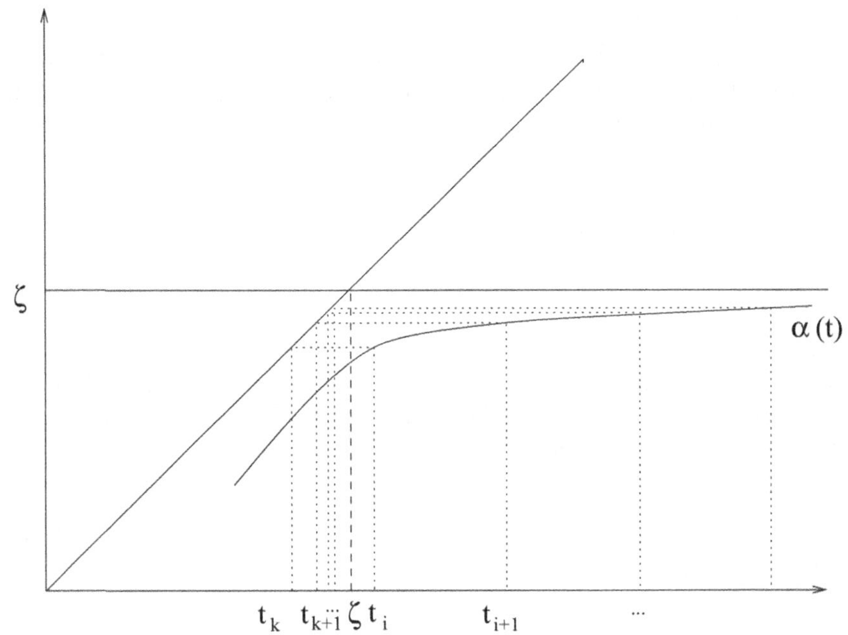

FIG. 3.1. Undesirable clustering of mesh points near ζ.

This approach is not always satisfactory because, for some deviated arguments $\alpha(t) = t - \tau(t)$, the mesh point distribution inside $[t_0, t_f]$ may be quite irregular and locally redundant. Consider the case of a bounded deviated argument with a horizontal asymptote ζ, such as in Figure 3.1, for which the larger t_f the denser Δ to the left of ζ.

The previous formulae easily extend to *Adams*-like formulae and, more generally, to LM formulae, but also to some *Runge–Kutta* (RK) formulae with abscissae 0 and 1 such as the *Heun* method. However, extension of the previous approach to higher order multistep formulas is not possible when the solution is not smooth enough. Zwerkina [286, 287, 288] proposed a class of methods, based on a modified version of the Taylor expansion, of the form

$$\sum_{j=0}^{k} \alpha_j y_{n+j} = h \sum_{j=0}^{k} \beta_j f(t_{n+j}, y_{n+j}, y_{n+j-m}) + T(\delta_n), \quad n = 0, 1, \ldots,$$

where the corrections $T(\delta_n)$ depend on the jumps δ_n in the derivatives of $y(t)$ at the mesh points t_n and are such that the order of the method is preserved. The method, subsequently extended by Hutchinson [143], is called the *modified method* and is applied only at the beginning of the integration procedure until the solution is smoothed out. For neutral equations, where the smoothing effect does not take place, the modified method must be used throughout the integration interval.

Another approach, aimed to avoid complications due to the discontinuity of the solution, was proposed by Gorbunov and Popov [104]. They introduced a new variable $x(t) = y(t) - D(t)$, where $D(t)$ is a piecewise polynomial function such that $x(t)$ is as smooth as necessary. Then the new problem is solved for $x(t)$ by the unmodified LM method.

To our knowledge, it was Feldstein [95] who first proposed freeing the mesh selection from the delay and to use extranodal points for the approximation of the delayed term $y(t - \tau(t))$. In particular, he investigated the convergence of some modified Euler methods for the equation

$$\begin{cases} y'(t) = f\Big(t, y(t), y\big(\alpha(t)\big)\Big), & t_0 \le t \le t_f, \\ y(t_0) = y_0, \end{cases} \tag{3.1.3}$$

where

$$t_0 \le \alpha(t) \le t.$$

Assume $h = \frac{t_f - t_0}{m}$, m a positive integer. For any point $t_n = t_0 + nh$, define the integer values $q(n) = \lfloor \frac{\alpha(t_n) - t_0}{h} \rfloor$ and $r(n) = \frac{\alpha(t_n) - t_0}{h} - q(n)$. Note that

- $\alpha(t_n) = t_{q(n)} + hr(n)$;
- $0 \le r(n) < 1$;
- $0 \le q(n) \le \frac{\alpha(t_n) - t_0}{h} \le n$, and $\alpha(t_n) = t_n$ if and only if $q(n) = n$, in which case $r(n) = 0$.

The first formula proposed by Feldstein is

$$\begin{aligned} y_{n+1} &= y_n + hf(t_n, y_n, z_n), \\ y_0 &= y(t_0), \end{aligned}$$

where

$$z_n = y_{q(n)} + hr(n)f(t_{q(n)}, y_{q(n)}, z_{q(n)}) = (1 - r(n))y_{q(n)} + r(n)y_{q(n)+1},$$

which corresponds to the approximation of $y(\alpha(t))$ by piecewise linear interpolation. Alternatively, Feldstein proposed

$$z_n = y_{q(n)}, \tag{3.1.4}$$

which corresponds to the piecewise constant approximation of $y(\alpha(t))$. Both methods converge for equation (3.1.3), which includes (3.1.2). The previous methods trivially extend to the more general equation (3.1.1) with delay which satisfies $\alpha(t) = t - \tau(t) \leq t$ by simply setting $z_n = \phi(\alpha(t_n))$ whenever $\alpha(t_n) \leq t_0$.

A general theory for higher order methods has been successively provided by Cryer and Tavernini [73] and by Tavernini [242, 243], who considered one-step and multistep methods for the more general class of Volterra functional differential equations (3.0.1). Their approach, as well as most of the methods that appeared subsequently, can be framed in a more general class of methods based on continuous ODE methods (see Bellen et al. [25]).

3.2 Preliminary results on continuous ODE methods

Let $\|\cdot\|$ be a given norm on $\mathbb{R}^d$. Let us consider the standard IVP

$$\begin{cases} y'(t) = g(t, y(t)), & t_0 \leq t \leq t_f, \\ y(t_0) = y_0, \end{cases} \tag{3.2.1}$$

where $g(t, y) \in C^0([t_0, t_f] \times \mathbb{R}^d, \mathbb{R}^d)$ is continuous with respect to t and globally Lipschitz continuous with respect to y in the given norm $\|\cdot\|$ of $\mathbb{R}^d$, that is

$$\|g(t, y_1) - g(t, y_2)\| \leq L\|y_1 - y_2\| \quad \forall t \in [t_0, t_f] \quad \text{and} \quad \forall y_1, y_2 \in \mathbb{R}^d, \tag{3.2.2}$$

for some Lipschitz constant $L > 0$. Moreover, let $\Delta = \{t_0, \ldots, t_N = t_f\}$ be a given mesh and $h_{n+1} = t_{n+1} - t_n$, $n = 0, \ldots, N-1$, be the corresponding stepsizes. Consider on Δ the very general k-step method

$$y_{n+1} = \alpha_{n,1} y_n + \cdots + \alpha_{n,k} y_{n-k+1} + h_{n+1} \Phi(y_n, \ldots, y_{n-k+1}; g, \Delta_n), \tag{3.2.3}$$

$n \geq k - 1$, where $\Delta_n = \{t_{n-k+1}, \ldots, t_n, t_{n+1}\}$ and where the *increment function* Φ satisfies a global Lipschitz condition with respect to the y arguments with the Lipschitz constant which depends only on the right-hand-side function $g(t, y)$ and, more precisely, on the constant L in (3.2.2). The initial values $y_0, \ldots, y_{k-1}$ are assumed to be given and the coefficients $\alpha_{n,i}$ are assumed to be uniformly bounded with respect to n. We shall also assume that the increment function Φ is continuous with respect to the functional argument g in the following sense: there exist a stepsize $h_g > 0$ and a constant $\gamma_g > 0$, continuously dependent on L, but independent of the nodal values $y_n, \ldots, y_{n-k+1}$ and of the mesh points Δ_n, such that, for all stepsizes $h_{n-k+2}, \ldots, h_{n+1} \leq h_g$,

$$\begin{aligned} &\|\Phi(y_n, \ldots, y_{n-k+1}; \tilde{g}, \Delta_n) - \Phi(y_n, \ldots, y_{n-k+1}; g, \Delta_n)\| \\ &\quad \leq \gamma_g \sup_{t_{n-k+1} \leq t \leq t_{n+1},\ y \in \mathbb{R}^d} \|\tilde{g}(t, y) - g(t, y)\| \end{aligned} \tag{3.2.4}$$

holds for all $\tilde{g} \in C^0([t_0, t_f] \times \mathbb{R}^d, \mathbb{R}^d)$. Note that condition (3.2.4) is stronger than continuity with respect to the functional argument g in that, for fixed g, it implies a sublinear growth of Φ with respect to the increment $\tilde{g} - g$.

Formula (3.2.3) includes one-step and multistep RK methods as well as LM methods. Whereas for LM methods the properties required for the increment function Φ may be easily proved, for RK methods this is proved in Chapter 5.

We also remark that, for multistep methods, i.e. when $k \geq 2$, the uniform boundedness of the various constants and coefficients mentioned above are assured only if we make the following assumption on the meshes Δ we use.

Assumption 3.2.1 *There exists a positive constant Ω, independent of the mesh Δ, such that*

$$\Omega^{-1} h_{n+1+i} \leq h_{n+1} \leq \Omega h_{n+1+i}, \quad i = -k+1, \dots, -1,$$

for all n.

Observe that, for one-step methods, i.e. when $k = 1$, no restriction is imposed on the meshes.

Next, we consider some continuous extension $\eta(t)$ of the method (3.2.3) on $[t_n, t_{n+1}]$ which can also be chosen by following two alternative strategies, namely *multistep interpolation* and *one-step interpolation*, defined as follows.

Definition 3.2.2 *We call* continuous extension, *or* interpolant, *of the ODE method (3.2.3) the piecewise polynomial function $\eta(t)$ given by the restrictions on each interval $[t_n, t_{n+1}]$ of any interpolant based on values computed in a possibly larger interval $[t_{n-i_n}, t_{n+j_n+1}]$, $i_n, j_n \geq 0$, of the form*

$$\begin{aligned}\eta(t_n + \theta h_{n+1}) = {} & \beta_{n,1}(\theta) y_{n+j_n} + \cdots + \beta_{n,j_n+i_n+1}(\theta) y_{n-i_n} \\ & + h_{n+1} \Psi(y_{n+j_n}, \dots, y_{n-i_n}; \theta, g, \Delta'_n), \quad 0 \leq \theta \leq 1, \quad (3.2.5)\end{aligned}$$

where $\Delta'_n = \{t_{n-i_n}, \dots, t_{n+j_n}, t_{n+j_n+1}\}$, satisfying the continuity conditions

$$\eta(t_n) = y_n \quad \textit{and} \quad \eta(t_{n+1}) = y_{n+1}. \tag{3.2.6}$$

If $i_n = j_n = 0$, then the interpolation procedure is based on values belonging to the sole interval $[t_n, t_{n+1}]$ and is referred to as one-step interpolation. *Otherwise, the interpolation procedure is called* multistep interpolation.

In short, any ODE method (3.2.3) endowed with its continuous extension (3.2.5) will be called a continuous ODE method.

Observe that, for $j_n > 0$, the interpolation (3.2.5) cannot be computed simultaneously using the discrete formula (3.2.3), but only after discrete integration has proceeded up to the forthcoming mesh point t_{n+j_n+1}. Simultaneous computation is possible if and only if $j_n = 0$.

Remark 3.2.3 *Since the continuity condition (3.2.6) must hold for every equation, it results in a deeper relationship between the ODE method (3.2.3) and its continuous extension (3.2.5). That is, the continuous extension can be expressed by formula (3.2.5) in such a way that*

$$\beta_{n,j}(0) = \begin{cases} 1 & \textit{for } j = 1 + j_n, \\ 0 & \textit{otherwise}, \end{cases}$$

$$\Psi(y_{n+j_n}, \dots, y_{n-i_n}; 0, g, \Delta'_n) = 0$$

and

$$\beta_{n,j}(1) = \begin{cases} \alpha_{n,j-j_n} & \textit{for } 1 + j_n \leq j \leq k + j_n, \\ 0 & \textit{otherwise}, \end{cases}$$

$$\Psi(y_{n+j_n},\ldots,y_{n-i_n};1,g,\Delta'_n)=\Phi(y_n,\ldots,y_{n-k+1};g,\Delta_n).$$

In particular, for $\theta = 1$ formula (3.2.5) collapses to formula (3.2.3).

The *increment function* Ψ is assumed to satisfy a global Lipschitz condition with respect to the y arguments with the Lipschitz constant which depends only on the constant L in (3.2.2). The polynomial coefficients $\beta_{n,i}(\theta)$ are uniformly bounded with respect to θ and Δ'_n. Moreover, as with the discrete method, we also assume that the increment function Ψ is continuous with respect to the functional argument g as follows: there exist a stepsize $h_g > 0$ and a constant $\delta_g > 0$, continuously dependent on L, but independent of the nodal values $y_{n-i_n},\ldots,y_{n+j_n}$, of the mesh points Δ'_n and of $\theta \in [0,1]$, such that, for all stepsizes $h_{n-i_n+1},\ldots,h_{n+j_n+1} \leq h_g$,

$$\begin{aligned}\|\Psi(y_{n+j_n},\ldots,y_{n-i_n};\theta,\tilde{g},\Delta'_n)-\Psi(y_{n+j_n},\ldots,y_{n-i_n};\theta,g,\Delta'_n)\|\\ \leq \delta_g \sup_{t_{n-i_n}\leq t\leq t_{n+j_n+1},\ y\in\mathbb{R}^d}\|\tilde{g}(t,y)-g(t,y)\|\end{aligned} \tag{3.2.7}$$

holds for all $\tilde{g} \in C^0([t_0,t_f]\times\mathbb{R}^d,\mathbb{R}^d)$.

As in the case of the discrete method (3.2.3), the uniform boundedness of the various constants and coefficients mentioned above is assured only if we make the following assumption on the meshes Δ we use.

Assumption 3.2.4 *There exists a positive constant Ω, independent of the mesh Δ, such that*

$$\Omega^{-1}h_{n+1+i}\leq h_{n+1}\leq \Omega h_{n+1+i},\qquad i=-i_n+1,\ldots,j_n,$$

for all n.

Observe that, for one-step interpolants, i.e. when $i_n = j_n = 0$, no restriction is imposed on the meshes.

In some cases the interpolation is given quite naturally by the method itself. This is the case for the *Adams* methods in the class LM, and for the *collocation* methods in the class RK, which are conceived as polynomial approximation methods rather than simply discrete methods. Anyway, in principle, interpolation is independent of the discrete method and the one-step and multistep approaches can be combined.

The following definition of *consistent* method is not standard but is equivalent to the usual definition which makes use of the exact solution $y(t)$. We have adopted it mainly because it is more appropriate for analyzing the convergence properties of the forthcoming methods for DDEs, but also because it is the one that is normally considered in the estimation of the local error for the stepsize control.

Definition 3.2.5 *We say that the ODE method (3.2.3) is* consistent of order *(or, equivalently, has* order*) p if $p \geq 1$ is the largest integer such that, for all*

C^p-continuous right-hand-side functions g in (3.2.1) and for all mesh points, we have that

$$\|z_{n+1}(t_{n+1}) - \tilde{y}_{n+1}\| = O(h_{n+1}^{p+1}),$$

uniformly with respect to y_n^* in any bounded subset of $\mathbb{R}^d$ and to $n = 0, \ldots, N-1$, where $z_{n+1}(t)$ is the local solution to the local problem

$$\begin{cases} z'_{n+1}(t) = g(t, z_{n+1}(t)), & t_n \le t \le t_{n+1}, \\ z_{n+1}(t_n) = y_n^*, \end{cases} \tag{3.2.8}$$

and

$$\begin{aligned} \tilde{y}_{n+1} = {} & \alpha_{n,1} z_{n+1}(t_n) + \cdots + \alpha_{n,k} z_{n+1}(t_{n-k+1}) \\ & + h_{n+1} \Phi(z_{n+1}(t_n), \ldots, z_{n+1}(t_{n-k+1}); g, \Delta_n). \end{aligned} \tag{3.2.9}$$

We say that the interpolant (3.2.5) of the ODE method (3.2.3) is consistent of uniform order *(or, equivalently, has* uniform order*) q if $q \ge 1$ is the largest integer such that, for all C^q-continuous right-hand-side functions g and for all mesh points, we have that*

$$\max_{t_n \le t \le t_{n+1}} \|z_{n+1}(t) - \tilde{\eta}(t)\| = O(h_{n+1}^{q+1}), \tag{3.2.10}$$

where

$$\begin{aligned} \tilde{\eta}(t_n + \theta h_{n+1}) = {} & \beta_{n,1}(\theta) z_{n+1}(t_{n+j_n}) + \cdots + \beta_{n,j_n+i_n+1}(\theta) z_{n+1}(t_{n-i_n}) \\ & + h_{n+1} \Psi(z_{n+1}(t_{n+j_n}), \ldots, z_{n+1}(t_{n-i_n}); \theta, g, \Delta'_n). \end{aligned} \tag{3.2.11}$$

Remark 3.2.6 *Because of the continuity condition (3.2.6), it is clear that the uniform order q of the interpolant (3.2.5) and the order p of the method (3.2.3) satisfy the inequality*

$$1 \le q \le p.$$

Remark 3.2.7 *In order to simplify the presentation and avoid possible misunderstandings, when talking of a continuous ODE method (3.2.3), (3.2.5), we shall often refer to the order p of consistency of the ODE method (3.2.3) as its discrete or nodal order (of consistency) and to the uniform order q of consistency of the interpolant (3.2.5) as its uniform order (of consistency).*

Moreover, we shall sometimes stress the discrete nature of the ODE method (3.2.3) (in contrast with the continuous nature of its interpolant (3.2.5)) by referring to it as a discrete ODE method.

The proofs of the results in this section may seem rather complicated from a notational point of view. This is due to the fact that we want to keep them within the general framework of the multistep method (3.2.3) and multistep interpolation (3.2.5). Things would be significantly simpler if we restricted ourself to treating one-step methods and one-step interpolants. At any rate, in order to simplify and unify the notation, for given vectors $a_i \in \mathbb{R}^d$, $i = 1, \ldots, k$, we shall denote by $[a_1, \ldots, a_k]^T$ the vector of $\mathbb{R}^{kd}$ formed by stacking the column vectors a_i into one long vector, whose formally correct notation ought to be $[a_1^T, \ldots, a_k^T]^T$.

Theorem 3.2.8 *Let the ODE method (3.2.3) be consistent of order $p \geq 1$ and, for each n, let*

$$C_n = \begin{bmatrix} \alpha_{n,1} & \alpha_{n,2} & \cdots & \alpha_{n,k-1} & \alpha_{n,k} \\ 1 & 0 & \cdots & 0 & 0 \\ 0 & 1 & \cdots & 0 & 0 \\ \vdots & \vdots & & \vdots & \vdots \\ 0 & 0 & \cdots & 1 & 0 \end{bmatrix} \tag{3.2.12}$$

be the companion matrix of the polynomial

$$p_n(\lambda) = \lambda^k - \sum_{i=1}^{k} \alpha_{n,i}\lambda^{k-i}. \tag{3.2.13}$$

If

- *there exists a norm $\|\cdot\|_*$ on $\mathbb{R}^k$, independent of both n and Δ, such that, for the corresponding induced matrix norm, the 0-stability condition*

$$\|C_n\|_* \leq 1 \tag{3.2.14}$$

  holds;
- *the right-hand-side function $g(t, y)$ in (3.2.1) is C^p-continuous;*
- *the set of starting values $y_0, \ldots, y_{k-1}$ approximate the exact solution to the order p;*

then the ODE method (3.2.3) is convergent of order (or, equivalently, has global order) p on any bounded interval $[t_0, t_f]$, that is

$$\max_{1 \leq n \leq N} \|y(t_n) - y_n\| = O(h^p), \tag{3.2.15}$$

where $h = \max_{1 \leq n \leq N} h_n$.

Moreover, if the interpolant (3.2.5) is consistent of uniform order q, then the continuous ODE method (3.2.3), (3.2.5) is uniformly convergent of order (or, equivalently, has uniform global order) $q' = \min\{p, q+1\}$, that is

$$\max_{t_0 \leq t \leq t_f} \|y(t) - \eta(t)\| = O(h^{q'}). \tag{3.2.16}$$

Proof. The convergence result (3.2.15) for the discrete method is obtained by the following standard arguments based on the conventional definition of consistency corresponding to the choice $y_n^*=y(t_n)$ in (3.2.8), that is $z_{n+1}(t) = y(t)$. Thus (3.2.9) and the order p of the method yield

$$\begin{aligned} y(t_{n+1}) &= \alpha_{n,1}y(t_n) + \cdots + \alpha_{n,k}y(t_{n-k+1}) \\ &\quad + h_{n+1}\Phi(y(t_n), \ldots, y(t_{n-k+1}); g, \Delta_n) + \epsilon_{n+1}, \end{aligned} \tag{3.2.17}$$

with

$$\|\epsilon_{n+1}\| \leq c h_{n+1}^{p+1} \tag{3.2.18}$$

for some constant $c > 0$ for all $n = k-1, \ldots, N-1$. Therefore, with the blockwise notation $\mathbf{y}_n = [y_n, y_{n-1}, \ldots, y_{n-k+1}]^T$ and $\mathbf{y}(t_n) = [y(t_n), y(t_{n-1}), \ldots, y(t_{n-k+1})]^T$, by subtracting (3.2.3) from (3.2.17) we get

$$\mathbf{y}(t_{n+1}) - \mathbf{y}_{n+1} = \mathcal{C}_n(\mathbf{y}(t_n) - \mathbf{y}_n) + h_{n+1}\Gamma_n + E_{n+1}, \quad n = k-1, \ldots, N-1, \tag{3.2.19}$$

where

$$\mathcal{C}_n = C_n \otimes I_d,$$

C_n being the companion matrix (3.2.12), I_d the $d \times d$ identity matrix and $\otimes$ the tensor product, and

$$\Gamma_n = \left(\Phi(\mathbf{y}(t_n); g, \Delta_n) - \Phi(\mathbf{y}_n; g, \Delta_n), 0, \ldots, 0\right)^T,$$

$$E_{n+1} = (\epsilon_{n+1}, 0, \ldots, 0)^T,$$

0 being the null vector in $\mathbb{R}^d$.

It is easy to verify that the 0-stability condition (3.2.14) is inherited by the family of matrices $\mathcal{C}_n$ with respect to the following norm on $\mathbb{R}^{kd}$, which is determined by the norm $\|\cdot\|_*$ appearing in (3.2.14):

$$|||\mathbf{x}||| = \max_{1 \leq i \leq d} \|x^{(i)}\|_*, \tag{3.2.20}$$

where the k-vectors $x^{(i)}$ are defined by

$$x^{(i)} = (x_i, x_{i+d}, \ldots, x_{i+(k-1)d})^T, \quad i = 1, \ldots, d. \tag{3.2.21}$$

Therefore, by (3.2.19) we get

$$|||\mathbf{y}(t_{n+1}) - \mathbf{y}_{n+1}||| \leq |||\mathbf{y}(t_n) - \mathbf{y}_n||| + h_{n+1}|||\Gamma_n||| + |||E_{n+1}|||,$$

$n = k-1, \ldots, N-1$, and then, by the equivalence of the norms in finite dimensional spaces, by the global Lipschitz continuity of the increment function Φ with respect to the y arguments (in the given norm $\|\cdot\|$ on $\mathbb{R}^d$) and by (3.2.18), there exists a constant $Q > 0$ such that

$$|||\mathbf{y}(t_{n+1}) - \mathbf{y}_{n+1}||| \leq (1 + h_{n+1}Q)|||\mathbf{y}(t_n) - \mathbf{y}_n||| + c' h_{n+1} h^p,$$

$n = k-1, \ldots, N-1$, for some constant $c' > 0$. Standard arguments on the last recurrence relation yield

$$|||\mathbf{y}(t_{n+1}) - \mathbf{y}_{n+1}||| \leq e^{Q(t_f - t_0)}|||\mathbf{y}(t_{k-1}) - \mathbf{y}_{k-1}||| + \frac{e^{Q(t_f - t_0)} - 1}{Q} c' h^p,$$

$n = k-1, \ldots, N-1$. Thus, again by the equivalence of the norms, since the starting values approximate the solution $y(t)$ up to order p, (3.2.15) is proved.

As for the global uniform order result (3.2.16), assume again $y_n^* = y(t_n)$ in (3.2.8), so that (3.2.11) yields

$$y(t_n + \theta h_{n+1}) = \beta_{n,1}(\theta)y(t_{n+j_n}) + \ldots + \beta_{n,j_n+i_n+1}(\theta)y(t_{n-i_n}) \\ +h_{n+1}\Psi(y(t_{n+j_n}), \ldots, y(t_{n-i_n}); \theta, g, \Delta_n') + \epsilon_{n+1}(\theta), \quad (3.2.22)$$

where, according to (3.2.10),

$$\max_{0\le n\le N-1}\max_{0\le\theta\le 1}\|\epsilon_{n+1}(\theta)\| = O(h^{q+1}). \quad (3.2.23)$$

By subtracting (3.2.5) from (3.2.22), the estimates (3.2.15) and (3.2.23), the assumed uniform boundedness of $\beta_{n,i}(\theta)$ and the Lipschitz continuity of Ψ with respect to the y arguments yield

$$\max_{0\le n\le N-1}\max_{0\le\theta\le 1}\|y(t_n + \theta h_{n+1}) - \eta(t_n + \theta h_{n+1})\| \le O(h^p) + O(h^{q+1}).$$

Thus (3.2.16) is proved as well. ■

Remark 3.2.9 *As with the order of consistency, in order to simplify the presentation and avoid possible misunderstandings, we shall often refer to the global order p of the ODE method (3.2.3) as the discrete or nodal global order of the continuous ODE method (3.2.3), (3.2.5).*

Remark 3.2.10 *If the ODE method (3.2.3) is implemented with constant stepsize h, then the coefficients $\alpha_{n,i}$ are independent of the mesh Δ and, hence, all the polynomials $p_n(\lambda)$ in (3.2.13) are equal to the same polynomial $p(\lambda)$ and their companion matrices C_n are equal to the same matrix C. It is well known that, in this case, the 0-stability condition (3.2.14) is equivalent to the following root condition:*

$$p(\lambda) = 0 \quad \Longrightarrow \quad \{|\lambda| \le 1 \text{ with } \lambda \text{ simple if } |\lambda| = 1\}.$$

Remark 3.2.11 *Theorem 3.2.8 can also be proved under the slightly weaker condition on the right-hand-side function $g(t, y)$ in (3.2.1) that it is just of class C^{p-1} with a Lipschitz continuous $(p-1)$st derivative. This fact is crucial in the treatment of certain DDEs, namely DDEs with state dependent delay and NDDEs in the form (2.0.3), where the right-hand side is only Lipschitz continuous but not differentiable (see Sections 4.3 and 4.5.1). For the preservation of the order 1 of Euler's method, see Gear [100].*

3.3 The standard approach via continuous ODE methods

The most popular approach for solving

$$\begin{cases} y'(t) = f\Big(t, y(t), y(t - \tau(t, y(t)))\Big), & t_0 \le t \le t_f, \\ y(t) = \phi(t), \quad t \le t_0, \end{cases} \quad (3.3.1)$$

where $f : [t_0, t_f] \times \mathbb{R}^d \times \mathbb{R}^d \longrightarrow \mathbb{R}^d$, is based on the use of continuous ODE methods and consists in proceeding step by step across a mesh $\Delta = \{t_0, t_1, \ldots, t_n, \ldots,$

$t_N = t_f\}$ as follows. Once the approximation y_n is obtained at t_n, the subsequent $(n+1)$st step consists in solving, by method (3.2.3), the equation

$$\begin{cases} w'_{n+1}(t) = f\Big(t, w_{n+1}(t), x\big(t - \tau(t, w_{n+1}(t))\big)\Big), & t_n \le t \le t_{n+1}, \\ w_{n+1}(t_n) = y_n, \end{cases} \tag{3.3.2}$$

where

$$x(s) = \begin{cases} \phi(s) & \text{for } s \le t_0, \\ \eta(s) & \text{for } t_0 \le s \le t_n, \\ w_{n+1}(s) & \text{for } t_n \le s \le t_{n+1}, \end{cases}$$

and $\eta(t)$ is the interpolant given by (3.2.5).

Observe that if $s = t - \tau(t, w_{n+1}(t)) \le t_n \ \forall t \in [t_n, t_{n+1}]$, then $x(s)$ equals the interpolant $\eta(s)$ given by (3.2.5). Such an interpolant can be considered to be known if the interpolation is based on nodes $\le t_n$. In this case (3.3.2) reduces to an ODE. The local solution $w_{n+1}(t)$ is then approximated by the discrete method (3.2.3) and the approximation of $w_{n+1}(t_{n+1})$ is set equal to y_{n+1}.

On the contrary, when $s = t - \tau(t, w_{n+1}(t)) > t_n$ for some $t \in [t_n, t_{n+1}]$ determined by the discrete method (3.2.3), the value $x(s)$ equals $w_{n+1}(s)$ and is unknown. Therefore, (3.3.2) can no longer be viewed as an ODE. However, $x(s)$ is still approximated by the interpolant $\eta(t)$ in the underlying interval $[t_n, t_{n+1}]$, implicitly defined by (3.2.5) with $j_n = 0$, that is by

$$\begin{aligned} \eta(t_n + \theta h_{n+1}) = {} & \beta_{n,1}(\theta) y_n + \cdots + \beta_{n,i_n+1}(\theta) y_{n-i_n} \\ & + h_{n+1} \Psi(y_n, \ldots, y_{n-i_n}; \theta, g_\eta, \Delta'_n), \quad 0 \le \theta \le 1, \end{aligned} \tag{3.3.3}$$

where

$$g_\eta(t, y) = f\Big(t, y, \eta\big(t - \tau(t, y)\big)\Big).$$

We have set $j_n = 0$ because, for the computation of (3.3.3) in the current interval $[t_n, t_{n+1}]$, we must avoid using values that are still to be computed in subsequent steps. It is worth remarking that, in this case, use of the continuous extension makes the overall method implicit even if the discrete method we are using is explicit. We call this occurrence *overlapping*. It is evident that, in this case, the implementation of the continuous ODE method requires a completely different algorithm.

From now on we shall call the above mechanism the *standard approach* and, in Chapter 4, we shall prove that, even when overlapping occurs, it is well defined, that is (3.3.3) is uniquely solvable for both time dependent and state dependent delays.

We remark that Cryer and Tavernini [73] and Tavernini [242, 243] developed a more general approach acting directly on the equation in the form (3.0.1) that, in the absence of overlapping, may be framed in the standard approach and that, in the presence of overlapping, treats the unknown functional argument y_t in a completely different way which, unlike the standard approach, allows the construction of entirely explicit methods.

It is clear that, in the standard approach, a large variety of methods is available for (3.3.2) according to the particular choices we can make for the discrete ODE method and for its continuous extension. Usually, the ODE methods are chosen in the well-known classes of RK and LM methods, but possibly also in the more general class of k-step methods (3.2.3).

From a practical point of view, finding which mesh interval $[t_{n-m}, t_{n-m+1}]$ the deviated argument $s = t - \tau$ belongs to for a given $t \in [t_n, t_{n+1}]$ is not difficult for constant or time dependent delays. On the contrary, this is not an easy task when the delay τ is state dependent and, therefore, the deviated argument s depends on $w_{n+1}(t)$ which is unknown. In particular, this prevents answering a priori the following two questions which are important for theoretical and practical reasons:

(Q_1) Whether or not, for equation (3.3.2), overlapping occurs and, in particular, whether or not the approximated delayed function $\eta(t - \tau)$ is known at a given point t of the current interval $[t_n, t_{n+1}]$.

(Q_2) Whether or not, for the stepsize h_{n+1}, the current interval $[t_n, t_{n+1}]$ includes some discontinuity point ξ and, more specifically, how to tune the stepsize h_{n+1} in order to have $t_{n+1} = \xi$, as we would like to have in the proximity of ξ .

As far as question (Q_1) is concerned, overlapping can be avoided for a sufficiently small stepsize by assuming the hypothesis (H_1) or, for state dependent delay, the more restrictive form (H_1^*) introduced in Sections 2.1.2 and 2.1.5:

(H_1) There exists a constant $\tau_0 > 0$ such that $\tau = t - \alpha(t) \geq \tau_0$ for all $t \in [t_0, t_f]$.

(H_1^*) There exists a constant $\tau_0 > 0$ such that $\tau(t, z) \geq \tau_0$ for all $t \in [t_0, t_f]$ and $z \in \mathbb{R}^d$.

Indeed, (H_1^*) prevents the state dependent delay from vanishing even if it is computed in a perturbation of the true local solution $w_{n+1}(t)$. Moreover, observe that (H_1^*) is equivalent to (H_1) whenever the delay is not state dependent.

Under the hypothesis (H_1^*) on the delay, for sufficiently small stepsize, namely $h_{n+1} = t_{n+1} - t_n \leq \tau_0$, overlapping does not occur and the function $\eta(s)$ is known for every $s = t - \tau(t, z)$ with $t \in [t_n, t_{n+1}]$ and for all $z \in \mathbb{R}^d$. In fact,

$$t - t_n \leq \tau_0 \leq \tau(t, z)$$

for all $t \in [t_n, t_{n+1}]$ and, hence,

$$t - \tau(t, z) \leq t_n.$$

Thus overlapping is avoided for any approximation of the local solution $w_{n+1}(t)$.

On the other hand, when $\tau_0 < h_0$, the minimum stepsize the method would like to use, it might well be that overlapping occurs. In fact, in such a case, for some n there might exist $t \in [t_n, t_{n+1}]$ such that

$$t - t_n > \tau_0 \quad \text{and} \quad \tau(t, w_{n+1}(t)) = \tau_0,$$

which yield

$$t - \tau(t, w_{n+1}(t)) > t_n.$$

When the hypothesis (H_1), or (H_1^*), does not hold, the delay τ necessarily vanishes at some point ξ (see Section 2.1.2) and thus overlapping inevitably occurs whenever the integration interval $[t_n, t_{n+1}]$ includes ξ.

As for question (Q_2), which is particularly hard for state dependent delays, the location of discontinuity points has been discussed by various authors from both the theoretical and implementational points of view. Two approaches have been pursued in the literature.

The first, which Baker et al. [14] refer to as the *tracking of discontinuities* (see also Section 4.3.1), is based on finding the discontinuities $\xi_{k,j}$ satisfying

$$\xi_{k,j} - \tau\big(\xi_{k,j}, w_{n+1}(\xi_{k,j})\big) = \xi_{k-1,i} \quad \text{for some } i$$

(see (2.1.9)) and to include them as mesh points. Along this line are the early papers by Tavernini [242, 244], Feldstein and Neves [97], Bock and Schlöder [46] and the more recent papers by Willé and Baker [270, 271, 273], Baker and Paul [12], Enright and Hu [92], Hayashi and Enright [126], Paul [216] and Karoui and Vaillancourt [166, 167].

An alternative approach that relies on stepsize control was proposed by Oppelstrup [212], Neves [204, 205], Thompson [207], Enright [87], Enright and Hayashi [90] and Guglielmi and Hairer [112]. These methods, in contrast to the previous ones, give up tracking the discontinuities, which are instead assumed to be automatically included in the mesh by suitable variable stepsize strategies based on the estimation of the local error or on the computation of the defect. In general, the codes are simpler but undergo a larger number of rejected steps and may lead to a sequence of very small stepsizes in the neighborhood of a low order discontinuity ξ, as one easily guesses if the method repeatedly fails in placing the end point of the current interval at the discontinuity ξ.

Figure 3.2 displays the continuous approximation and the nodal points ($\diamond$) of the solution of equation (1.1.6) obtained by the variable stepsize code RADAR5 developed by Guglielmi and Hairer [112]. For the tolerance $TOL = 10^{-6}$, after the nodal point $t = 0.7775$ the code computes 17 steps, 8 of which rejected, for approaching and passing the discontinuity point $t = 1$.

3.3.1 *DDEs of neutral type*

When the DDE is of neutral type, i.e.

$$\begin{cases} y'(t) = f\Big(t, y(t), y\big(t - \tau(t, y(t))\big), y'\big(t - \sigma(t, y(t))\big)\Big), & t_0 \le t \le t_f, \\ y(t) = \phi(t), \quad t \le t_0, \end{cases} \tag{3.3.4}$$

the approach, based on the use of a continuous ODE method, must take care of the additional term $y'\big(t - \sigma(t, y(t))\big)$. Therefore, besides the continuous extension $\eta(t)$ for approximating $y(t)$, the method must provide a continuous approximation, say $\lambda(t)$, for the derivative $y'(t)$ as well. So, the standard approach consists in proceeding step by step across a mesh $\Delta = \{t_0, t_1, \ldots, t_n, \ldots, t_N = t_f\}$ as

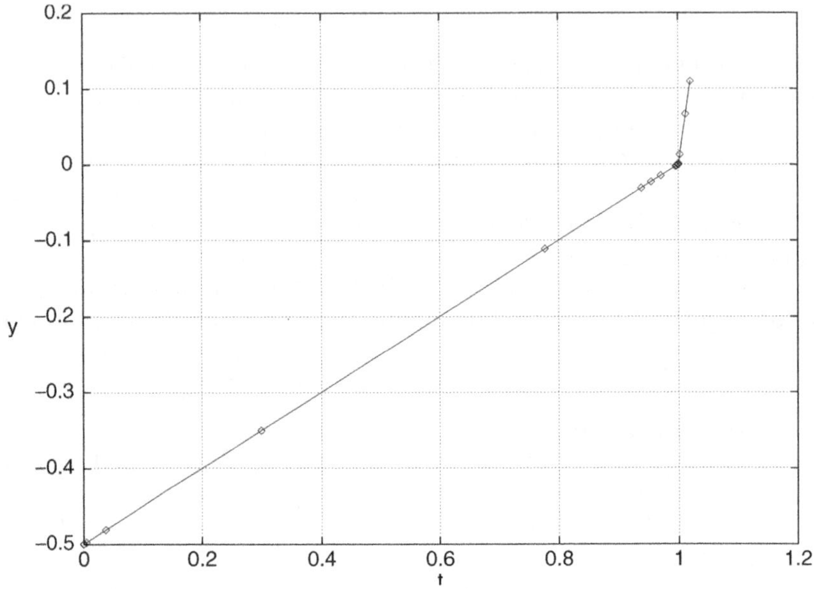

FIG. 3.2. Numerical solution of Example 1.1.4 in the interval $[0, 1.02]$ obtained by the code RADAR5 for $TOL = 10^{-6}$.

follows. Once the approximation y_n is obtained at t_n, the subsequent $(n+1)$st step consists in solving the local problem

$$\begin{cases} w'_{n+1}(t) = f\Big(t, w_{n+1}(t), x\big(t - \tau(t, w_{n+1}(t))\big), z\big(t - \sigma(t, w_{n+1}(t))\big)\Big), \\ \qquad\qquad\qquad\qquad\qquad\qquad\qquad\qquad t_n \le t \le t_{n+1}, \\ w_{n+1}(t_n) = y_n, \end{cases} \tag{3.3.5}$$

where

$$x(s) = \begin{cases} \phi(s) & \text{for } s \le t_0, \\ \eta(s) & \text{for } t_0 \le s \le t_n, \\ w_{n+1}(s) & \text{for } t_n \le s \le t_{n+1}, \end{cases}$$

and

$$z(s) = \begin{cases} \phi'(s) & \text{for } s \le t_0, \\ \lambda(s) & \text{for } t_0 \le s \le t_n, \\ w'_{n+1}(s) & \text{for } t_n \le s \le t_{n+1}, \end{cases}$$

where $\eta(t)$ is the interpolant given by (3.2.5) and $\lambda(t)$ is an approximation of $y'(t)$ for which we have essentially two options available: the first, quite obvious, is

$$\lambda(t) = \eta'(t), \tag{3.3.6}$$

and the second is

$$\lambda(t) = \mathcal{P}\Big(f\Big(\cdot, \eta(\cdot), \eta\big(\cdot - \tau(\cdot, \eta(\cdot))\big), \lambda\big(\cdot - \sigma(\cdot, \eta(\cdot))\big)\Big)\Big)(t), \tag{3.3.7}$$

where, in each mesh interval $[t_k, t_{k+1}]$, $\mathcal{P}$ is an interpolation operator in a suitable polynomial space of degree possibly other than $\deg(\eta')$ and nodes in $[t_{k-i_k}, t_{k+j_k+1}]$.

Observe that, if $s = t - \tau(t, w_{n+1}(t)) \leq t_n$ and $s' = t - \sigma(t, w_{n+1}(t)) \leq t_n$ $\forall t \in [t_n, t_{n+1}]$, then $x(s)$ and $z(s')$ equal $\eta(s)$ and $\lambda(s')$, respectively. As with the non-neutral case, the interpolants η and λ can be considered to be known if they are based on nodes $\leq t_n$. In this case, (3.3.5) reduces to an ODE. The local solution $w_{n+1}(t)$ is then approximated by the discrete method (3.2.3) and the approximation of $w_{n+1}(t_{n+1})$ is set equal to y_{n+1}.

On the contrary, when $s = t - \tau(t, w_{n+1}(t)) > t_n$ and/or $s' = t - \sigma(t, w_{n+1}(t)) > t_n$ for some $t \in [t_n, t_{n+1}]$ determined by the discrete method (3.2.3), the values $x(s)$ and/or $z(s')$ are unknown. In this case, they are still approximated by $\eta(t)$ and $\lambda(t)$ in the current interval $[t_n, t_{n+1}]$. Similarly to the non-neutral case, $\eta(t)$ is implicitly defined by

$$\begin{aligned}\eta(t_n + \theta h_{n+1}) = {} & \beta_{n,1}(\theta) y_n + \cdots + \beta_{n,j_n+i_n+1}(\theta) y_{n-i_n} \\ & + h_{n+1} \Psi(y_n, \ldots, y_{n-i_n}; \theta, g_{\eta,\lambda}, \Delta_n'), \quad 0 \leq \theta \leq 1, \end{aligned} \tag{3.3.8}$$

where

$$g_{\eta,\lambda}(t, y) = f\Big(t, y, \eta(t - \tau(t, y)), \lambda(t - \sigma(t, y))\Big),$$

and $\lambda(t)$ is given by (3.3.6) or (3.3.7) with $j_k = 0$.

Papers following this approach are due to Castleton and Grimm [67], Jackiewicz [153, 154, 155, 157, 158], Jackiewicz and Lo [160], Bellen, Jackiewicz and Zennaro [31], Hayashi [125], Guglielmi and Hairer [112].

3.3.2 *Other formulations of DDEs of neutral type*

Delay differential equations with neutral terms are often formulated in terms of a system of delay differential algebraic equations (DDAEs) without delayed derivatives. For instance, equation (3.3.4) may always be stated in the form

$$\begin{cases} y'(t) = z(t), \\ z(t) = f\Big(t, y(t), y(t - \tau(t, y(t))), z(t - \sigma(t, y(t)))\Big), & t_0 \leq t \leq t_f, \\ y(t) = \phi(t), \quad t \leq t_0, \\ z(t) = \phi'(t), \quad t \leq t_0. \end{cases} \tag{3.3.9}$$

From a numerical point of view, the system (3.3.9) appears more attractive than (3.3.4) in that no continuous approximation is needed for $y'(t)$. On the other hand, this simplification is paid for with the inclusion of the algebraic component $z(t)$ into the system that must be solved by some DAE method extended by continuous approximations for $y(t)$ and $z(t)$ as well. Note that, although similar and in some cases even equivalent, this approach provides a numerical solution which, in general, is different from the one obtained by means of the standard approach with either option (3.3.6) or (3.3.7). In this setting, Guglielmi and Hairer [112] considered the more general equation

$$\begin{cases} My'(t) = f\Big(t, y(t), y(t - \tau_1(t, y(t))), \dots, y(t - \tau_r(t, y(t)))\Big), & t_0 \le t \le t_f, \\ y(t) = \phi(t), \quad t \le t_0, \end{cases}$$

where the possibly singular matrix M leads to all kinds of DDAEs, finite element discretizations of PDEs with delays, singularly perturbed problems, etc.

The numerical solution of some special classes of DDAEs with constant delay has been investigated by Ascher and Petzold [5] for BDF and RK methods. The general theory of numerical methods for DDAEs is beyond the scope of this book and is not developed here. However, there are cases where the DDAE may still be treated by the standard approach based on continuous ODE methods, in that the differential and the algebraic components are decoupled.

In particular, when the delay is not state dependent and the neutral equation is given, or may be tranformed, into the form

$$\begin{cases} \frac{\mathrm{d}}{\mathrm{d}t}\Big(y(t) - G(t, y(t - \tau(t)))\Big) = H\Big(t, y(t), y(t - \tau(t))\Big), & t_0 \le t \le t_f, \\ y(t) = \phi(t), \quad t \le t_0 \end{cases}$$

(see (2.0.3)), it is natural to solve it for the function

$$\Phi(t) = y(t) - G\big(t, y(t - \tau(t))\big).$$

In this case the equation takes the form of the DDAE

$$\begin{cases} \Phi'(t) = K\Big(t, \Phi(t), y(t - \tau(t))\Big), & \\ y(t) = \Phi(t) + G\Big(t, y(t - \tau(t))\Big), & t_0 \le t \le t_f, \\ \Phi(t_0) = \phi(t_0) - G\Big(t_0, \phi(t_0 - \tau(t_0))\Big), & \\ y(t) = \phi(t), \quad t \le t_0, & \end{cases} \tag{3.3.10}$$

to be solved for $\Phi(t)$ and $y(t)$, where

$$K(t, u, v) = H\big(t, u + G(t, v), v\big).$$

For delays satisfying hypothesis (H_1), the system (3.3.10) is basically an ODE and can be solved step by step by a continuous method. In fact, for sufficiently small stepsizes, overlapping is avoided and the algebraic component $y(t)$ may be computed separately, by recursion, after the differential component $\Phi(t)$. On the contrary, the components $y(t)$ and $z(t)$ in (3.3.9) are strongly coupled for any delay and for any stepsize, and the system must be solved by some continuous DAE method. Once the approximation y_n is obtained, the $(n+1)$st step of the standard approach reduces to the numerical solution, for $\Lambda_{n+1}(t)$ and subsequently for $w_{n+1}(t)$, of the local problem

$$\begin{cases} \Lambda'_{n+1}(t) = K\Big(t, \Lambda_{n+1}(t), \eta(t - \tau(t))\Big), & t_n \le t \le t_{n+1}, \\ w_{n+1}(t) = \Lambda_{n+1}(t) + G\big(t, \eta(t - \tau(t))\big), & \\ \Lambda_{n+1}(t_n) = y_n - G\Big(t_n, \eta(t_n - \tau(t_n))\Big), & \end{cases}$$

where $\eta(t-\tau(t))$ is assumed to be known for $t \in [t_n, t_{n+1}]$. The ODE for $\Lambda_{n+1}(t)$ is solved by the discrete method (3.2.3) and, subsequently, extended by the interpolant $\lambda(t)$ given by (3.2.5). Finally, the approximate solution $\eta(t)$ is extended to the current interval $[t_n, t_{n+1}]$ by the continuous approximation of $w_{n+1}(t)$. It is obtained by evaluating the algebraic equation $\eta(t) = \lambda(t) + G\big(t, \eta(t-\tau(t))\big)$ or, preferably, by a suitable *local* approximation of the right-hand side. The need for a local approximation is due to the recursive nature of the algebraic equation, which, *per se*, would require tracing the approximate solution $\eta(t)$ back to the initial function $\phi(t)$ for each evaluation point $t \in [t_n, t_{n+1}]$.

Note that the solution $\Phi(t)$ in (3.3.10) is of class $C^1[t_0, t_f]$ with the Lipschitz continuous derivative $\Phi'(t)$ at every discontinuity point and that the discontinuities propagate only in the second derivative $\Phi''(t)$.

An alternative approach, still avoiding using uniform approximations of the derivative of the solution $y(t)$, has been introduced and developed by Bellen, Guglielmi and Ruehli [26, 27], Bellen, Guglielmi and Zennaro [29, 30] and Bellen, Maset and Torelli [33] for equations of the form

$$\begin{cases} y'(t) = F\Big(t, y(t), G\big(t, y(t-\tau(t)), y'(t-\tau(t))\big)\Big), & t_0 \le t \le t_f, \\ y(t) = \phi(t), \quad t \le t_0. \end{cases}$$

Here, it is natural to define

$$\Psi(t) = G\big(t, y(t-\tau(t)), y'(t-\tau(t))\big),$$

which leads to the DDAE

$$\begin{cases} y'(t) = F\big(t, y(t), \Psi(t)\big), \quad t_0 \le t \le t_f, \\ \Psi(t) = H\Big(t, y\big(t-\tau(t)\big), \Psi\big(t-\tau(t)\big)\Big), \quad t-\tau(t) > t_0, \\ y(t_0) = \phi(t_0), \\ \Psi(t) = G\Big(t, \phi\big(t-\tau(t)\big), \phi'\big(t-\tau(t)\big)\Big), \quad t-\tau(t) \le t_0, \end{cases} \tag{3.3.11}$$

to be solved for $\Psi(t)$ and $y(t)$, where

$$H(t, y, x) = G\Big(t, y, F\big(t-\tau(t), y, x\big)\Big).$$

For delays satisfying hypothesis (H_1), the system can be solved first for the algebraic component $\Psi(t)$ and then for the differential component $y(t)$. Here, the $(n+1)$st step of the standard approach reduces to the numerical solution, for $\Lambda_{n+1}(t)$ and subsequently for $w_{n+1}(t)$, of the local problem

$$\begin{cases} w_{n+1}'(t) = F\big(t, w_{n+1}(t), \Lambda_{n+1}(t)\big), \quad t_n \le t \le t_{n+1}, \\ \Lambda_{n+1}(t) = H\Big(t, \eta\big(t-\tau(t)\big), \lambda\big(t-\tau(t)\big)\Big), \quad t-\tau(t) > t_0, \\ w_{n+1}(t_n) = y_n, \\ \Lambda_{n+1}(t) = G\Big(t, \phi\big(t-\tau(t)\big), \phi'\big(t-\tau(t)\big)\Big), \quad t-\tau(t) \le t_0, \end{cases}$$

where $\eta(t-\tau(t))$ and $\lambda(t-\tau(t))$ are assumed to be known for $t \in [t_n, t_{n+1}]$. The approximation of $w_{n+1}(t_{n+1})$ is set equal to y_{n+1} and, subsequently, the continuous numerical solution $\eta(t)$ is extended to the current interval $[t_n, t_{n+1}]$ by the continuous numerical approximation of $w_{n+1}(t)$. The approximate algebraic component $\lambda(t)$ is set equal to $\Lambda_{n+1}(t)$ or, in order to avoid tracing it back to the initial function $G\Big(t, \phi(t-\tau(t)), \phi'(t-\tau(t))\Big)$, to a *local* uniform approximation of the right-hand side $H\Big(t, \eta(t-\tau(t)), \lambda(t-\tau(t))\Big)$.

3.4 Bellman's method of steps

The most elegant approach aimed at avoiding the need for interpolation in the numerical solution of

$$\begin{cases} y'(t) = f\Big(t, y(t), y(t-\tau(t))\Big), & t_0 \le t \le t_f, \\ y(t) = \phi(t), \quad t \le t_0, \end{cases} \tag{3.4.1}$$

is known as the *method of steps*. It was first developed by Bellman [39] for constant delay and, successively, by Bellman and Cooke [41] for variable delays subject to the hypotheses (H_1) and (H_4) of Section 2.1. In our opinion, Bellman's method is not the most convenient approach for solving DDEs numerically, but it is certainly attractive because it allows the use of variable stepsizes without any interpolation. Moreover, it furnishes a powerful approach for proving the convergence and superconvergence of numerical methods in some special cases that will be considered later.

In order to simplify the discussion, we begin with the case of constant delay τ. It is clear that the discontinuity points are $\xi_k = t_0 + k\tau$. In the first interval $[t_0, t_0+\tau]$ the DDE (3.4.1) has the form

$$\begin{cases} y'(t) = f\big(t, y(t), \phi(t-\tau)\big), \\ y(t_0) = \phi(t_0). \end{cases}$$

In the second interval $[t_0+\tau, t_0+2\tau]$ we can define $y_1(t) = y(t-\tau)$ and $y_2(t) = y(t)$, and thus we can write the DDE (3.4.1) as the $2d$-dimensional system of ODEs

$$\begin{cases} y_1'(t) = f\big(t-\tau, y_1(t), \phi(t-2\tau)\big), \\ y_2'(t) = f\big(t, y_2(t), y_1(t)\big), \\ y_1(t_0+\tau) = \phi(t_0), \\ y_2(t_0+\tau) = y(t_0+\tau). \end{cases}$$

In general, in the interval $[t_0+(k-1)\tau, t_0+k\tau]$ we can write the DDE (3.4.1) as the kd-dimensional system of ODEs

$$\begin{cases} y_i'(t) = f\big(t-(k-i)\tau, y_i(t), y_{i-1}(t)\big), & i = 1, \ldots, k, \\ y_i(t_0+(k-1)\tau) = y(t_0+(i-1)\tau), & i = 1, \ldots, k, \end{cases} \tag{3.4.2}$$

where we have set $y_0(t) = \phi(t-k\tau)$ and $y_i(t) = y(t-(k-i)\tau)$, $i = 1, \ldots, k$.

Passing from k to $k+1$ in (3.4.2) means shifting the integration interval from $[t_0+(k-1)\tau, t_0+k\tau]$ to $[t_0+k\tau, t_0+(k+1)\tau]$ and extending the solution from $[t_0, t_0 + k\tau]$ to $[t_0, t_0 + (k+1)\tau]$ by adding the component $y_{k+1}(t) = y(t)$ in the current interval. Therefore, we can choose a standard numerical method for ODEs and solve, for increasing k, the larger and larger systems (3.4.2). For each step k, the numerical solution of the kd-dimensional system (3.4.2) aims to provide an approximate value of $y_k(t_0 + k\tau) = y(t_0 + k\tau)$ to be used as the initial value of the new component $y_{k+1}(t) = y(t)$ in the next step. The process ends when, for some k, $t_0 + k\tau \geq t_f$.

In this way we are obliged to solve a system of ever-growing dimension and, in principle, to recompute many times the same pieces of solution related to the previous intervals. On the other hand, the reduction to a system of ODEs avoids the typical complications due to the presence of the delayed argument, that is storing and interpolating the computed solution throughout the interval $[t_0, t_f - \tau]$.

As for the convergence of the process, observe that, at each step, any numerical method of global order p provides a numerical approximation of order p for any perturbation of the initial values in (3.4.2) of order $\geq p$. Therefore, in the next step, we have to solve an IVP with initial values perturbed to the same order as that of the method, and so forth. The whole process performs to order p for any finite number of steps.

Now consider the more general case of a variable delay $\tau(t)$ and define, as usual, the function $\alpha(t) = t - \tau(t)$. Again, in the first interval $[\xi_0, \xi_1]$ the DDE (3.4.1) has the form

$$\begin{cases} y'(t) = f\Big(t, y(t), \phi\big(\alpha(t)\big)\Big), \\ y(t_0) = \phi(t_0). \end{cases}$$

In the second interval $[\xi_1, \xi_2]$ we can define $y_1(t) = y(\alpha(t))$ and $y_2(t) = y(t)$. Thus we have $y_1'(t) = \alpha'(t)y'(\alpha(t))$ and, therefore, we can write the DDE (3.4.1) as the $2d$-dimensional system of ODEs

$$\begin{cases} y_1'(t) = \alpha'(t) f\Big(\alpha(t), y_1(t), \phi\big(\alpha(\alpha(t))\big)\Big), \\ y_2'(t) = f\big(t, y_2(t), y_1(t)\big), \\ y_1(\xi_1) = \phi(\xi_0), \\ y_2(\xi_1) = y(\xi_1). \end{cases}$$

In general, by setting $\alpha_0(t) = t$ and $\alpha_i(t) = \alpha(\alpha_{i-1}(t)), i = 1, ..., k$, in the interval $[\xi_{k-1}, \xi_k]$ we can write the DDE (3.4.1) as the kd-dimensional system of ODEs

$$\begin{cases} y_i'(t) = \alpha_{k-i}'(t) f\big(\alpha_{k-i}(t), y_i(t), y_{i-1}(t)\big), & i = 1, \dots, k, \\ y_i(\xi_{k-1}) = y(\xi_{i-1}), \quad i = 1, \dots, k, \end{cases} \tag{3.4.3}$$

where we have set $y_0(t) = \phi(\alpha_k(t))$ and $y_i(t) = y(\alpha_{k-i}(t))$, $i = 1, \dots, k$.

As for the numerical solution, the procedure is the same as for the case of constant delay, except that we have to use the analytic expression for the derivative $\alpha'(t)$ of the functional argument.

Remark 3.4.1 *The previous formulae are based on a change of variable such that, at the kth step, we must solve the system of k equations (3.4.3) in the* *interval* $[\xi_{k-1}, \xi_k]$. *The original papers by Bellman [39] and Bellman and Cooke [41] deal with a different change of variable so that every system is solved in the same fixed interval* $[t_0, \xi_1]$. *In that case the derivative of the inverse of* $\alpha(t)$ *is needed.*

3.5 DDEs as abstract Cauchy problems

Consider the following constant coefficient linear system of DDEs

$$\begin{cases} y'(t) = Ly(t) + My(t-\tau), & t \geq 0, \\ y(t) = \phi(t), & -\tau \leq t \leq 0, \end{cases} \tag{3.5.1}$$

where L and M are $d \times d$ matrices, $\tau > 0$ and $\phi \in C^0([-\tau, 0], \mathbb{R}^d)$.

In order to set a DDE as an abstract Cauchy problem, we first recall some elementary facts about semigroup theory (see, for example, Pazy [220]).

Let X be a Banach space that we call the *state space*. A family $\{T(t)\}_{t \geq 0}$ of linear bounded operators on X is called a C^0-*semigroup* if the following three conditions are satisfied:

- $T(0) = I$;
- $T(t+s) = T(t)T(s) \quad \forall t, s \geq 0$;
- $\lim_{t \downarrow 0} T(t)x - x = 0 \quad \forall x \in X$.

The *infinitesimal generator* $\mathcal{A}$ of the semigroup is the operator defined by

$$D(\mathcal{A}) = \left\{ x \in X \;\middle|\; \exists \lim_{t \downarrow 0} \frac{T(t)x - x}{t} \right\},$$

$$\mathcal{A}x = \lim_{t \downarrow 0} \frac{T(t)x - x}{t}, \qquad x \in D(\mathcal{A}).$$

The domain $D(\mathcal{A})$ of the infinitesimal generator is a dense subspace of X.

For fixed $x \in X$, the function

$$u(t, x) = T(t)x, \quad t \geq 0,$$

is continuous. If $x \in D(\mathcal{A})$, then $u(\cdot, x)$ is the solution of the *abstract Cauchy problem*

$$\begin{cases} \frac{\mathrm{d}}{\mathrm{d}t} U(t) = \mathcal{A}U(t), & t \geq 0, \\ U(0) = x, \end{cases} \tag{3.5.2}$$

where $U : [0, \infty) \to D(\mathcal{A})$. If $x \notin D(\mathcal{A})$, then the abstract Cauchy problem (3.5.2) has no solutions.

On the other hand, for every $x \in X$ the function $u(\cdot, x)$ is the solution of the *abstract Volterra integral equation*

$$U(t) = x + \mathcal{A} \int_0^t U(s)\,\mathrm{d}s,$$

where $U : [0, \infty) \to X$ and the integral is a Riemann integral. So, when $x \notin D(\mathcal{A})$, the function $u(\cdot, x)$ is called a *generalized solution*, or a *mild solution*, of (3.5.2).

Now, let us consider equation (3.5.1) and, as usual, denote by $y_t(\theta)$ the function

$$y_t(\theta) = y(t+\theta), \quad \theta \in [-\tau, 0],$$

with $y(t)$ the solution of (3.5.1). Moreover, consider the state space

$$X = C^0\left([-\tau, 0], \mathbb{R}^d\right),$$

equipped with the maximum norm

$$|||x||| = \max_{-\tau \le \theta \le 0} \|x(\theta)\|, \quad x \in X,$$

where $\|\cdot\|$ is the norm chosen on $\mathbb{R}^d$. It is known (see, for example, Diekmann et al. [75]) that the family $\{T(t)\}_{t\ge 0}$ of linear bounded operators on the Banach space X defined by

$$T(t)\phi = y_t, \quad \phi \in X,$$

where $y(t)$ is the solution of (3.5.1) with initial function $\phi(t)$, is a C^0-semigroup. The infinitesimal generator $\mathcal{A}$ of the semigroup is given by

$$D(\mathcal{A}) = \{x \in X \mid x' \in X \text{ and } x'(0) = Lx(0) + Mx(-\tau)\},$$

$$\mathcal{A}x = x', \quad x \in D(\mathcal{A}).$$

In this setting, we can restate (3.5.1) as the abstract Cauchy problem

$$\begin{cases} \frac{\mathrm{d}}{\mathrm{d}t} U(t) = (U(t))', & t \ge 0, \\ U(0) = \phi, \end{cases} \tag{3.5.3}$$

where $U : [0, \infty) \to D(\mathcal{A})$ and, according to the previous notation, $(U(t))' = \mathcal{A}U(t)$. The function

$$U(t) = y_t$$

is the solution (mild solution if $\phi \notin D(\mathcal{A})$) of (3.5.3).

We can also interpret the abstract Cauchy problem (3.5.3) as a PDE. In fact, if $\phi \in D(\mathcal{A})$, then by (3.5.3) we have that the hyperbolic PDE

$$\frac{\partial u(t,\theta)}{\partial t} = \frac{\partial u(t,\theta)}{\partial \theta}, \quad t \ge 0, \quad -\tau \le \theta \le 0, \tag{3.5.4}$$

with the boundary condition

$$\frac{\partial u}{\partial \theta}(t, 0) = Lu(t, 0) + Mu(t, -\tau), \quad t \ge 0,$$

and the initial value

$$u(0, \theta) = \phi(\theta), \quad -\tau \le \theta \le 0,$$

has the solution

$$u(t, \theta) = y(t + \theta),$$

where $y(t)$ is the solution of (3.5.1).

The approximate solution of (3.5.1) is then obtained by discretizing the operator $\mathcal{A}U(t) = (U(t))'$ in (3.5.3), or equivalently in (3.5.4), into a system of ODEs and then by solving it numerically. It corresponds to the semi-discretization in space of the PDE (3.5.4), also termed the *method of lines* (MOL).

In order to discretize the operator $\mathcal{A}$, for a given integer $N \geq 1$, consider the mesh

$$\Omega_N = \left\{ nh \;\middle|\; n = -N, \ldots, -1, 0 \right\}, \quad h = \frac{\tau}{N},$$

on the interval $[-\tau, 0]$ and the relevant discrete state space

$$X_N = \left(\mathbb{R}^d\right)^{\Omega_N} \cong \mathbb{R}^{d(N+1)},$$

equipped with the maximum norm

$$|||x_N||| = \max_{\theta \in \Omega_N} \|x(\theta)\|, \quad x_N \in X_N.$$

As an example of the discretization of (3.5.3), consider the $d(N+1)$-dimensional linear system of ODEs

$$\begin{cases} \frac{\mathrm{d}}{\mathrm{d}t} U_N(t) = \mathcal{A}_N U_N(t), & t \geq 0, \\ U_N(0) = \phi_N, \end{cases} \tag{3.5.5}$$

where $\phi_N = [\phi(-\tau), \ldots, \phi(-h), \phi(0)]^T \in \mathbb{R}^{d(N+1)}$ and

$$\mathcal{A}_N = \begin{bmatrix} B_N \otimes I_d \\ M\ 0 \ldots 0\ L \end{bmatrix}.$$

Here,

$$B_N = \frac{1}{h} \begin{bmatrix} -1 & 1 & & & \\ & \ddots & \ddots & & \\ & & \ddots & \ddots & \\ & & & -1 & 1 \\ & & & & -1\ 1 \end{bmatrix}$$

is an $N \times (N+1)$ matrix, I_d is the $d \times d$ identity matrix and $\otimes$ denotes the tensor product. Note that (3.5.5) is obtained from (3.5.3) by replacing the exact value of the derivative by the first-order forward difference approximation

$$(U(t))'(\theta) \simeq \frac{U(t)(\theta + h) - U(t)(\theta)}{h}$$

at the points $\theta \in \Omega_N \setminus \{0\}$, whereas at the point $\theta = 0$ it is assumed that

$$(U(t))'(0) = LU(t)(0) + MU(t)(-\tau). \tag{3.5.6}$$

The method described above was analyzed from a convergence and asymptotic stability point of view by Bellen and Maset [32] and Maset [194].

For smooth solutions, higher order schemes leading to different matrices B_N have been considered by Maset [198] and Koto [177]. The higher order discretization of the derivative $(U(t))'(\theta)$ in (3.5.3) for points $\theta \neq 0$ of the enlarged mesh

$$\Omega_{N,\nu} = \{0\} \cup \left\{ \theta_n - c_i h \;\middle|\; n = -N+1, \ldots, 0, \quad i = 1, \ldots, \nu \right\}$$

can be provided, for example, by the stage values K_{n+1}^i of a ν-stage RK method (A, b) (see Chapter 5) as applied to the degenerate ODE $z'(\theta) = (U(t))'(\theta)$, $z(0) = (U(t))(0)$ (see Maset [198]). If the RK matrix A is invertible, $b = [a_{\nu 1}, \ldots, a_{\nu\nu}]^T$ and $0 < c_1 < \cdots < c_\nu = 1$, then the $N \times (N+1)$ matrix B_N modifies to the $\nu N \times (1 + \nu N)$ matrix

$$B_{N,\nu} = \frac{1}{h} \begin{bmatrix} w_1 & w_{11} & \cdots & w_{1\nu} & & & & & & \\ \vdots & \vdots & & \vdots & & & & & & \\ w_\nu & w_{\nu 1} & \cdots & w_{\nu\nu} & & & & & & \\ & & & w_1 & w_{11} & \cdots & w_{1\nu} & & & \\ & & & \vdots & \vdots & & \vdots & & & \\ & & & w_\nu & w_{1\nu} & \cdots & w_{\nu\nu} & & & \\ & & & & & & \ddots & & & \\ & & & & & & & w_1 & w_{11} \cdots & w_{1\nu} \\ & & & & & & & \vdots & \vdots & \vdots \\ & & & & & & & w_\nu & w_{\nu 1} \cdots & w_{\nu\nu} \end{bmatrix},$$

where $w = [w_1, \ldots, w_\nu]^T = A^{-1}e$, $e = [1, \ldots, 1]^T$ and $W = [w_{ij}]_{i,j=1}^{\nu} = -A^{-1}$. As before, the derivative at $\theta = 0$ is still given by (3.5.6). Now the ODE system is

$$\begin{cases} \frac{\mathrm{d}t}{\mathrm{d}t} U_{N,\nu}(t) = \mathcal{A}_{N,\nu} U_{N,\nu}(t), \quad t \geq 0, \\ U_{N,\nu}(0) = \phi_{N,\nu}, \end{cases}$$

where $\phi_{N,\nu} = [\phi(\theta)]_{\theta \in \Omega_{N,\nu}}^T$ and

$$\mathcal{A}_{N,\nu} = \begin{bmatrix} B_{N,\nu} \otimes I_d \\ M\ 0 \ldots 0\ L \end{bmatrix}. \tag{3.5.7}$$

As with Bellman's method, the overall approach avoids interpolation at the cost of some redundancy of computation. Moreover, the resulting ODE system (3.5.5) may be integrated with variable stepsizes, including stepsizes larger than the delay τ.

Solving linear DDEs and NDDEs by passing through the abstract Cauchy problem (3.5.3) and the discretization of $\mathcal{A}U(t)$ has already been done in the literature. In particular, Banks and Kappel [17] and Kappel and Kunish [165] considered the abstract Cauchy problem in a different state space and then discretized it by using projections into suitable spline function spaces.

A second possible approach to the approximation of (3.5.3) or (3.5.4) is based on semi-discretization in time, also termed the *transversal method of lines* (TMOL). Since it corresponds to the discretization of the abstract Cauchy problem (3.5.2), it can be seen as an abstract method. In particular, Maset [196] considered the implicit Euler method and called it the *abstract implicit Euler method.* When it is applied to (3.5.3) with constant stepsize Δt, it reads

$$U^{k+1} = U^k + \Delta t \mathcal{A} U^{k+1}, \quad k = 0, 1, 2, \ldots, \quad U^0 = \phi.$$

Since $U^{k+1} \in D(\mathcal{A})$ and $\mathcal{A}U^{k+1} = \left(U^{k+1}\right)'$, each step corresponds to the two-point boundary value problem

$$U^{k+1}(\theta) = U^k(\theta) + \Delta t \left(U^{k+1}\right)'(\theta), \quad -\tau \le \theta \le 0, \tag{3.5.8}$$

with the boundary condition

$$\left(U^{k+1}\right)'(0) = LU^{k+1}(0) + MU^{k+1}(-\tau), \tag{3.5.9}$$

which, by using (3.5.8) itself, can be equivalently rewritten as

$$U^{k+1}(0) = U^k(0) + \Delta t \left(LU^{k+1}(0) + MU^{k+1}(-\tau)\right). \tag{3.5.10}$$

In general, the abstract implicit Euler method applies to any non-linear autonomous retarded functional differential equations and the non-linearity is simply reflected in the non-linearity of the boundary condition (3.5.9) or (3.5.10).

We note that the proposed abstract implicit Euler method is known in linear semigroup theory. A convergence proof, without any consideration of the order of convergence, can be found in Pazy [220, Theorem 8.3].

3.6 Waveform relaxation methods

A further approach aimed at avoiding the theoretical and practical complications due to the presence of delay terms in the Cauchy problem is the *waveform relaxation* (WR) method. It was introduced by Lalarasmee, Ruehli and Sangiovanni-Vincentelli [185] for the time domain analysis of large-scale differential systems modelling electrical circuits, and later analyzed by Miekkala and Nevanlinna [201], Skeel [232] and others on a firm mathematical basis.

In the context of functional differential equations, the WR method may be used to reduce the functional equations into a sequence of simpler equations by assigning all or some of the functional arguments from the previous iteration. For example, given an initial guess function $y_0(t)$, the state dependent DDE

$$y'(t) = f\left(t, y(t), y(t - \tau(t, y(t)))\right)$$

may be solved by integrating the sequence of simpler time dependent DDEs

$$y'_{k+1}(t) = f\left(t, y_{k+1}(t), y_{k+1}\left(t - \tau(t, y_k(t))\right)\right),$$

or even ODEs of the form

$$y'_{k+1}(t) = f\big(t, y_{k+1}(t), y_k\,(t - \tau(t, y_{k+1}(t)))\big)$$

or

$$y'_{k+1}(t) = f\big(t, y_{k+1}(t), y_k\,(t - \tau(t, y_k(t)))\big).$$

Also, the neutral equation

$$y'(t) = f\,(t, y(t), y(t-\tau), y'(t-\tau))$$

can be integrated by solving the sequence of non-neutral equations

$$y'_{k+1}(t) = f\,(t, y_{k+1}(t), y_{k+1}(t-\tau), y'_k(t-\tau))\,.$$

WR methods may also avoid any kind of overlapping which can be intrinsic in the functional arguments or determined by integration stepsizes larger than the delay.

The convergence analysis for the numerical implementation of the waveform iterates (discretized WR) requires the preliminary convergence analysis of the exact waveform relaxation (continuous-time WR) sequence as well as of its perturbation. The convergence of exact iterates is usually superlinear and the overall procedure might be more efficient than the standard approach.

However, the subject is beyond the scope of this book and we refer the interested reader to the papers by Bjørhus [44], Jackiewicz, Kwapisz and Lo [159], Bartoszewski and Kwapisz [18], Zubik-Kowal and Vandewalle [285] and the bibliography therein.

Finally, it is worth pointing out that WR methods, as well as other functional iterative schemes, have been used for the approximation of DDEs with boundary conditions, such as periodicity conditions or initial function and final value conditions (see Bellen [20], Bellen and Zennaro [34, 35] and Zennaro [276] and the survey papers by Bellen [22] and Meinardus and Nürnberger [200]).

4

THE STANDARD APPROACH VIA CONTINUOUS ODE METHODS

In this chapter we go deeper into the standard approach introduced in Section 3.3. In particular, we prove the well-posedness of the method and analyze its order of convergence for various kinds of DDEs and NDDEs.

As for DDEs, the most general problem we consider is

$$\begin{cases} y'(t) = f\Big(t, y(t), y\big(t - \tau(t, y(t))\big)\Big), & t_0 \le t \le t_f, \\ y(t) = \phi(t), \quad t \le t_0, \end{cases}$$

where $f(t, u, v) \in C^0([t_0, t_f] \times \mathbb{R}^d \times \mathbb{R}^d, \mathbb{R}^d)$ is continuous with respect to t and globally Lipschitz continuous with respect to u and v in a given norm $\|\cdot\|$ of $\mathbb{R}^d$, that is

$$\|f(t, u_1, v_1) - f(t, u_2, v_2)\| \le L\|u_1 - u_2\| + M\|v_1 - v_2\|$$

for all $t \in [t_0, t_f]$, $u_1, u_2, v_1, v_2 \in \mathbb{R}^d$ and for some Lipschitz constants $L > 0$ and $M > 0$.

As for NDDEs, we consider instead

$$\begin{cases} y'(t) = f\Big(t, y(t), y\big(t - \tau(t, y(t))\big), y'\big(t - \sigma(t, y(t))\big)\Big), & t_0 \le t \le t_f, \\ y(t) = \phi(t), \quad t \le t_0, \end{cases}$$

where $f(t, u, v, w) \in C^0([t_0, t_f] \times \mathbb{R}^d \times \mathbb{R}^d \times \mathbb{R}^d, \mathbb{R}^d)$ is continuous with respect to t and globally Lipschitz continuous with respect to u, v and w, that is

$$\|f(t, u_1, v_1, w_1) - f(t, u_2, v_2, w_2)\| \le L\|u_1 - u_2\| + M\|v_1 - v_2\| + N\|w_1 - w_2\|$$

for all $t \in [t_0, t_f]$, $u_1, u_2, v_1, v_2, w_1, w_2 \in \mathbb{R}^d$ and for some Lipschitz constants $L > 0$, $M > 0$ and $N > 0$.

In Section 3.3 it was pointed out that the standard approach leads to substantially different procedures according to whether or not *overlapping* occurs (see question (Q_1)). We have seen that, in each mesh interval $[t_n, t_{n+1}]$, this occurrence may or may not be determined by the stepsize h_{n+1} according to the validity of the hypothesis (H_1) or (H_1^*). More precisely, if (H_1) holds, then overlapping is avoided for sufficiently small stepsizes. Nevertheless, it may occur for large stepsizes even if (H_1) holds. However, if (H_1) does not hold, then overlapping inevitably occurs in any neighborhood of a vanishing delay point. The well-posedness and convergence analysis of the standard approach will be examined separately for various cases characterized by the validity or not of (H_1) or (H_1^*).

We also pay particular attention to the problem of the location of the discontinuities (see question (Q_2)), especially for equations with state dependent delays, and of their inclusion into the mesh, mainly in connection with the possible effects on the order of convergence.

Finally, we briefly review methods proposed in the literature and related codes.

4.1 DDEs with constant or non-vanishing time dependent delay

We start the convergence analysis of the standard approach by analyzing the DDE

$$\begin{cases} y'(t) = f\Big(t, y(t), y\big(t - \tau(t)\big)\Big), & t_0 \leq t \leq t_f, \\ y(t) = \phi(t), \quad t \leq t_0, \end{cases} \tag{4.1.1}$$

where the delay $\tau = \tau(t)$ satisfies hypothesis (H_1).

In order to avoid overlapping, in this section we exploit hypothesis (H_1) by using sufficiently small stepsizes, i.e. $\leq \tau_0$, and assume being able to compute the discontinuity points to be included in the mesh Δ a priori. However, implementations with completely free meshes, possibly with stepsizes $\geq \tau_0$ which may cause overlapping, are also very common. They will be discussed in the next section, which is devoted to the more general case of DDEs with arbitrary time dependent delays.

Let the mesh $\Delta = \{t_0, t_1, \ldots, t_n, \ldots, t_N = t_f\}$ include the principal discontinuity points (see Definition 2.1.4) as well as the discontinuity points of order $\leq p$ in $[t_0, t_f]$ (see Definition 2.1.3). Let us denote them by $\xi_0 = t_0 < \xi_1 < \cdots < \xi_s$ so that, on each interval $[\xi_0, \xi_1], [\xi_1, \xi_2], \ldots, [\xi_s, t_f]$, the solution $y(t)$ is of class at least C^{p+1} and, with $\xi_{s+1} = t_f$, each $[\xi_i, \xi_{i+1}]$ is mapped to the left of ξ_i by the deviated argument $\alpha(t) = t - \tau(t)$. Under these hypotheses the sequence of ODEs

$$\begin{cases} z'(t) = f\Big(t, z(t), \eta\big(t - \tau(t)\big)\Big), & \xi_i \leq t \leq \xi_{i+1}, \\ z(\xi_i) = \eta(\xi_i), \end{cases}$$

is solved by the method (3.2.3) and, subsequently, interpolated by (3.2.5) for $i = 0, \ldots, s$, where $\eta(s)$ is given by either the initial function $\phi(s)$ for $s \leq t_0$ or by the previously computed numerical solution itself for $s > t_0$.

In analogy to Bellman's method, the foregoing procedure, based on the recursive integration of ODEs accross the *macro intervals* $[\xi_{i-1}, \xi_i]$, is still called the *method of steps* and is summarized in the following pseudo-code.

Algorithm 4.1.1

1 *Locate all the principal discontinuity points and the discontinuity points of order* $\leq p$

$$\xi_1, \ldots, \xi_s (< t_f),$$

and set $\xi_0 = t_0$, $\xi_{s+1} = t_f$.

2 *Solve the equation*

$$\begin{cases} z'(t) = f\Big(t, z(t), \phi\big(t - \tau(t)\big)\Big), & \xi_0 \le t \le \xi_1, \\ z(\xi_0) = \phi(\xi_0), \end{cases}$$

by any discrete ODE method.

3 *For $i = 1$ to s:*

- *compute and store the continuous extension $\eta(t)$ for $t \in [\xi_{i-1}, \xi_i]$ by interpolating data from the same macro interval $[\xi_{i-1}, \xi_i]$ only;*
- *solve the equation*

$$\begin{cases} z'(t) = f\Big(t, z(t), \eta\big(t - \tau(t)\big)\Big), & \xi_i \le t \le \xi_{i+1}, \\ z(\xi_i) = \eta(\xi_i), \end{cases}$$

by the discrete ODE method.

4 *Plot the approximate solution $\eta(t)$ in $[\xi_0, \xi_{s+1}]$.*

5 *End.*

The convergence of the method of steps is given by the following theorem.

Theorem 4.1.2 *Consider the delay differential equation (4.1.1) where $f(t, y, x)$ is C^p-continuous in $[t_0, t_f] \times \mathbb{R}^d \times \mathbb{R}^d$, the delay $\tau(t)$ is C^p-continuous in $[t_0, t_f]$ and fulfills hypothesis (H_1), and the initial function $\phi(t)$ is C^p-continuous. Moreover, assume that:*

(h_1) *The mesh $\Delta = \{t_0, t_1, \ldots, t_n, \ldots, t_N = t_f\}$ includes all the principal discontinuity points and the discontinuity points of order $\le p$ lying in $[t_0, t_f]$, denoted by $\xi_1, \ldots, \xi_s (< t_f)$.*

(h_2) *The ODE method (3.2.3) is consistent of order p, satisfies the stability condition (3.2.14) and, for $k > 1$, is restarted after each discontinuity point $\xi_i, i = 0, 1, \ldots, s$, by a method of order $\ge p - 1$.*

(h_3) *The interpolant (3.2.5) is consistent of uniform order q.*

(h_4) *For each n, the interval $[t_{n-i_n}, t_{n+j_n+1}]$ where the interpolation takes place is included in $[\xi_i, \xi_{i+1}]$ for some index $0 \le i \le s$.*

Then the resulting method of steps has discrete global order and uniform global order $q' = \min\{p, q+1\}$; that is

$$\max_{1 \le n \le N} \|y(t_n) - y_n\| = O(h^{q'})$$

and

$$\max_{t_0 \le t \le t_f} \|y(t) - \eta(t)\| = O(h^{q'}),$$

where $h = \max_{1 \le n \le N} h_n$.

The proof is based on the following lemmas stating the continuous dependence on the data for the numerical solutions of the IVP (3.2.1).

Lemma 4.1.3 *Let $\{y_n^\alpha\}_{n=0}^N$ and $\{y_n^\beta\}_{n=0}^N$ be the numerical solutions of*

$$y'(t) = g\big(t, y(t)\big), \quad a \le t \le b,$$

with initial conditions $y(a)$ equal to α and β, respectively, obtained by the method (3.2.3) with the $\alpha_{n,i}$'s satisfying the stability condition (3.2.14) and starting values $\mathbf{y}_{k-1}^\alpha = [y_{k-1}^\alpha, y_{k-2}^\alpha, \ldots, y_0^\alpha]^T$, $y_0^\alpha = \alpha$ and $\mathbf{y}_{k-1}^\beta = [y_{k-1}^\beta, y_{k-2}^\beta, \ldots, y_0^\beta]^T$, $y_0^\beta = \beta$. Then there exists a constant $M > 0$ independent of the mesh Δ such that

$$\max_{k \le n \le N} \|y_n^\alpha - y_n^\beta\| \le M \max_{0 \le s \le k-1} \|y_s^\alpha - y_s^\beta\|. \tag{4.1.2}$$

Analogously, for the continuous extensions $\eta^\alpha(t)$ and $\eta^\beta(t)$ given by the method (3.2.5), there exists a constant $S > 0$ such that

$$\max_{a \le t \le b} \|\eta^\alpha(t) - \eta^\beta(t)\| \le S \max_{0 \le s \le k-1} \|y_s^\alpha - y_s^\beta\|. \tag{4.1.3}$$

Proof. The discrete numerical method for the two IVPs reads

$$y_{n+1}^\alpha = \alpha_{n,1} y_n^\alpha + \cdots + \alpha_{n,k} y_{n-k+1}^\alpha + h_{n+1} \Phi(y_n^\alpha, \ldots, y_{n-k+1}^\alpha; g, \Delta_n)$$

and

$$y_{n+1}^\beta = \alpha_{n,1} y_n^\beta + \cdots + \alpha_{n,k} y_{n-k+1}^\beta + h_{n+1} \Phi(y_n^\beta, \ldots, y_{n-k+1}^\beta; g, \Delta_n)$$

for $n = k-1, \ldots, N-1$. By the same arguments and notations as in the proof of Theorem 3.2.8, we get

$$|||\mathbf{y}_{n+1}^\alpha - \mathbf{y}_{n+1}^\beta||| \le |||\mathbf{y}_n^\alpha - \mathbf{y}_n^\beta||| + h_{n+1} |||\Gamma_n|||, \quad n = k-1, \ldots, N-1,$$

where

$$\Gamma_n = [\Phi(\mathbf{y}_n^\alpha; g, \Delta_n) - \Phi(\mathbf{y}_n^\beta; g, \Delta_n), 0, \ldots, 0]^T.$$

Then, by the equivalence of the norms in finite dimensional spaces and by the global Lipschitz continuity of the increment function Φ with respect to the y arguments (in the given norm $\|\cdot\|$ on $\mathbb{R}^d$), there exists a constant $L > 0$ such that

$$|||\mathbf{y}_{n+1}^\alpha - \mathbf{y}_{n+1}^\beta||| \le (1 + h_{n+1} L) |||\mathbf{y}_n^\alpha - \mathbf{y}_n^\beta|||, \quad n = k-1, \ldots, N-1.$$

Standard arguments on the last recurrence relation yield

$$|||\mathbf{y}_{n+1}^\alpha - \mathbf{y}_{n+1}^\beta||| \le \mathrm{e}^{L(b-a)} |||\mathbf{y}_{k-1}^\alpha - \mathbf{y}_{k-1}^\beta|||, \quad n = k-1, \ldots, N-1,$$

and thus, again by the equivalence of the norms, (4.1.2) is proved.

Similarly, for the continuous extensions we have

$$\begin{aligned}\eta^\alpha(t_n + \theta h_{n+1}) &= \beta_{n,1}(\theta)y^\alpha_{n+j_n} + \cdots + \beta_{n,j_n+i_n+1}(\theta)y^\alpha_{n-i_n} \\ &\quad + h_{n+1}\Psi(y^\alpha_{n+j_n}, \ldots, y^\alpha_{n-i_n}; \theta, g, \Delta'_n), \quad 0 \le \theta \le 1,\end{aligned}$$

and

$$\begin{aligned}\eta^\beta(t_n + \theta h_{n+1}) &= \beta_{n,1}(\theta)y^\beta_{n+j_n} + \cdots + \beta_{n,j_n+i_n+1}(\theta)y^\beta_{n-i_n} \\ &\quad + h_{n+1}\Psi(y^\beta_{n+j_n}, \ldots, y^\beta_{n-i_n}; \theta, g, \Delta'_n), \quad 0 \le \theta \le 1,\end{aligned}$$

and hence, by the uniform boundedness of the coefficients $\beta_{n,i}(\theta)$ and of the Lipschitz constant of the increment function Ψ with respect to the y arguments, there exist constants $P > 0$ and $T > 0$ such that

$$\begin{aligned}&\|\eta^\alpha(t_n + \theta h_{n+1}) - \eta^\beta(t_n + \theta h_{n+1})\| \\ &\qquad \le P \max_{n-i_n \le s \le n+j_n} \|y^\alpha_s - y^\beta_s\| + h_{n+1}T \max_{n-i_n \le s \le n+j_n} \|y^\alpha_s - y^\beta_s\|\end{aligned}$$

for all $n \le N-1$ and $0 \le \theta \le 1$. Therefore, (4.1.3) follows from (4.1.2). ■

Lemma 4.1.4 *Let $\{y_n\}_{n=0}^N$ and $\{z_n\}_{n=0}^N$ be the numerical solutions of the IVPs*

$$\begin{cases} y'(t) = f\big(t, y(t), u(t)\big), & a \le t \le b, \\ y(a) = \alpha, \end{cases}$$

and

$$\begin{cases} z'(t) = f\big(t, z(t), v(t)\big), & a \le t \le b, \\ z(a) = \alpha, \end{cases}$$

respectively, obtained by the method (3.2.3) with the $\alpha_{n,i}$'s satisfying the stability condition (3.2.14) and starting values $\mathbf{y}_{k-1} = [y_{k-1}, y_{k-2}, \ldots, y_0]^T$, $y_0 = \alpha$, $\mathbf{z}_{k-1} = [z_{k-1}, z_{k-2}, \ldots, z_0]^T$ and $z_0 = \alpha$. Then there exist a stepsize $h_0 > 0$ and a constant $M > 0$ independent of the mesh Δ, such that, for all $h \le h_0$,

$$\max_{1 \le n \le N} \|y_n - z_n\| \le M \max_{1 \le s \le k-1} \|y_s - z_s\| + M \max_{a \le t \le b} \|u(t) - v(t)\|.$$

Analogously, for all $h \le h_0$ their continuous extensions $\eta(t)$ and $\zeta(t)$ satisfy the inequality

$$\max_{a \le t \le b} \|\eta(t) - \zeta(t)\| \le S \max_{1 \le s \le k-1} \|y_s - z_s\| + S \max_{a \le t \le b} \|u(t) - v(t)\|$$

for some constant $S > 0$.

Proof. We introduce the simplified notation $f_u(t,y) = f(t,y,u)$ and $f_v(t,y) = f(t,y,v)$. By the same arguments as in the proof of Theorem 3.2.8, we get

$$|||\mathbf{y}_{n+1} - \mathbf{z}_{n+1}||| \leq |||\mathbf{y}_n - \mathbf{z}_n||| + h_{n+1}|||\Gamma_n|||, \quad n = k-1, \ldots, N-1,$$

where

$$\Gamma_n = [\Phi(\mathbf{y}_n; f_u, \Delta_n) - \Phi(\mathbf{z}_n; f_v, \Delta_n), 0, \ldots, 0]^T.$$

Now, the equivalence of the norms in the finite dimensional spaces, the global Lipschitz continuity of Φ with respect to the y arguments, the continuity condition (3.2.4) on Φ and the global Lipschitz continuity of $f(t,y,x)$ with respect to x yield the existence of the constants $C > 0$, $P > 0$ and $Q > 0$ such that, for $h \leq h_0$,

$$\begin{aligned}
|||\Gamma_n||| &\leq C\|\Phi(\mathbf{y}_n; f_u, \Delta_n) - \Phi(\mathbf{z}_n; f_v, \Delta_n)\| \\
&\leq C(\|\Phi(\mathbf{y}_n; f_u, \Delta_n) - \Phi(\mathbf{z}_n; f_u, \Delta_n)\| + \|\Phi(\mathbf{z}_n; f_u, \Delta_n) - \Phi(\mathbf{z}_n; f_v, \Delta_n)\|) \\
&\leq Q|||\mathbf{y}_n - \mathbf{z}_n||| + C\gamma \sup_{a \leq t \leq b,\ y \in \mathbb{R}^d} \|f_u(t,y) - f_v(t,y)\| \\
&\leq Q|||\mathbf{y}_n - \mathbf{z}_n||| + P \max_{a \leq t \leq b} \|u(t) - v(t)\|.
\end{aligned}$$

Note that, according to the properties of the increment functions Φ and Ψ, the stepsize h_0 and the constants P and Q depend on the Lipschitz constant of the function $f_u(t,y)$ with respect to y, and hence are independent of u.

Summarizing, we have

$$|||\mathbf{y}_{n+1} - \mathbf{z}_{n+1}||| \leq (1 + h_{n+1}Q)|||\mathbf{y}_n - \mathbf{z}_n||| + h_{n+1}P \max_{a \leq t \leq b} \|u(t) - v(t)\|$$

for all $n = k-1, \ldots, N-1$. Usual arguments on the last recurrence relation yield

$$|||\mathbf{y}_{n+1} - \mathbf{z}_{n+1}||| \leq e^{Q(b-a)}|||\mathbf{y}_{k-1} - \mathbf{z}_{k-1}||| + \frac{e^{Q(b-a)} - 1}{Q} P \max_{a \leq t \leq b} \|u(t) - v(t)\|$$

for all $n = k-1, \ldots, N-1$, and the first part of the proof is complete.

As for the continuous extensions, we have

$$\begin{aligned}
\eta(t_n + \theta h_{n+1}) - \zeta(t_n + \theta h_{n+1}) = {} & \beta_{n,1}(\theta)(y_{n+j_n} - z_{n+j_n}) \\
& + \cdots + \beta_{n,j_n+i_n+1}(\theta)(y_{n-i_n} - y_{n-i_n}) \\
& + h_{n+1}(\Psi(y_{n+j_n}, \ldots, y_{n-i_n}; \theta, f_u, \Delta_n') \\
& - \Psi(z_{n+j_n}, \ldots, z_{n-i_n}; \theta, f_v, \Delta_n')), \quad 0 \leq \theta \leq 1.
\end{aligned}$$

Hence, by the uniform boundedness of the coefficients $\beta_{n,i}(\theta)$ and by the Lipschitz continuity properties of Ψ, we get, similarly to the previous lemma, completion of the proof. ■

Remark 4.1.5 *For one-step methods the bounds furnished by Lemma 4.1.4 reduce to*

$$\max_{1\le n\le N}\|y_n - z_n\| \le M \max_{a\le t\le b}\|u(t)-v(t)\|$$

and

$$\max_{a\le t\le b}\|\eta(t)-\zeta(t)\| \le S \max_{a\le t\le b}\|u(t)-v(t)\|.$$

Proof of Theorem 4.1.2. The proof will be carried out by induction on the intervals $[\xi_i, \xi_{i+1}]$, $i = 0, 1, \ldots, s$ where, by hypothesis (h_1), the solution $y(t)$ is of class at least C^{p+1} and, $\forall t \in [\xi_i, \xi_{i+1}]$, $t-\tau(t) \le \xi_i$. On $[\xi_0, \xi_1]$ the problem to solve is

$$\begin{cases} y'(t) = f\Big(t, y(t), \phi\big(t-\tau(t)\big)\Big), & t_0 \le t \le \xi_1, \\ y(t_0) = \phi(t_0), \end{cases}$$

and hence, by Theorem 3.2.8 with $g(t,y) = f\big(t, y, \phi(t-\tau(t))\big)$, the assertion is true, that is

$$\max_{\xi_0\le t\le\xi_1}\|y(t)-\eta(t)\| = O(h^{q'}).$$

Suppose the assertion is true until ξ_i. We will prove it is true on $[\xi_i, \xi_{i+1}]$, where the problem to solve is

$$\begin{cases} w'(t) = f\Big(t, w(t), \eta\big(t-\tau(t)\big)\Big), & \xi_i \le t \le \xi_{i+1}, \\ w(\xi_i) = \eta(\xi_i). \end{cases} \tag{4.1.4}$$

Since $\eta(t)$ is a piecewise polynomial function, $\eta(t-\tau(t))$ is not regular enough on $[\xi_i, \xi_{i+1}]$, and therefore Theorem 3.2.8 with $t_0 = \xi_i$, $t_f = \xi_{i+1}$ and $g(t,y) = f\big(t, y, \eta(t-\tau(t))\big)$ does not apply directly to equation (4.1.4). On the other hand, it applies to

$$\begin{cases} x'(t) = f\Big(t, x(t), y\big(t-\tau(t)\big)\Big), & \xi_i \le t \le \xi_{i+1}, \\ x(\xi_i) = y(\xi_i), \end{cases} \tag{4.1.5}$$

with $g(t,x) = f\big(t, x, y(t-\tau(t))\big)$, where the solution is just $y(t)$ and the delayed term $y(t-\tau(t))$ is smooth. Consequently, by virtue of conditions (h_2) and (h_4), its continuous numerical solution $\zeta(t)$ fulfills

$$\max_{\xi_i\le t\le\xi_{i+1}}\|y(t)-\zeta(t)\| = O(h^{q'}). \tag{4.1.6}$$

Now consider the auxiliary problem

$$\begin{cases} z'(t) = f\Big(t, z(t), \eta\big(t-\tau(t)\big)\Big), & \xi_i \le t \le \xi_{i+1}, \\ z(\xi_i) = y(\xi_i). \end{cases} \tag{4.1.7}$$

By Lemma 4.1.3 applied to equations (4.1.7) and (4.1.4) with $a = \xi_i$, $b = \xi_{i+1}$, $\alpha = y(\xi_i)$, $\beta = \eta(\xi_i)$, $g(t,y) = f\big(t, y, \eta(t-\tau(t))\big)$ and the starting values $\mathbf{y}_{k-1}^{\alpha}$

and $\mathbf{y}_{k-1}^{\beta}$ furnished by the initializing method of order $\geq p-1$, the relevant continuous numerical solutions $\rho(t)$ and $\eta(t)$ satisfy

$$\max_{\xi_i \leq t \leq \xi_{i+1}} \|\rho(t) - \eta(t)\| \leq S \max_{0 \leq s \leq k-1} \|y_s^{\alpha} - y_s^{\beta}\|.$$

By inductive hypothesis we have

$$\max_{\xi_0 \leq t \leq \xi_i} \|y(t) - \eta(t)\| = O(h^{q'}). \tag{4.1.8}$$

Therefore, Lemma 4.1.3 itself applied to the initializing methods yields

$$\max_{0 \leq s \leq k-1} \|y_s^{\alpha} - y_s^{\beta}\| \leq M \|y(\xi_i) - \eta(\xi_i)\| = O(h^{q'}),$$

and hence

$$\max_{\xi_i \leq t \leq \xi_{i+1}} \|\rho(t) - \eta(t)\| = O(h^{q'}). \tag{4.1.9}$$

Moreover, consider again the original problem (4.1.5) in the current interval $[\xi_i, \xi_{i+1}]$ and its numerical solution $\zeta(t)$. By Lemma 4.1.4 applied to equations (4.1.5) and (4.1.7) with $a = \xi_i$, $b = \xi_{i+1}$, $\alpha = y(\xi_i)$, $u(t) = y(t - \tau(t))$, $v(t) = \eta(t - \tau(t))$ and the starting values $\mathbf{y}_{k-1} = \mathbf{y}_{k-1}^{\alpha}$ and $\mathbf{z}_{k-1}$ obtained by the initializing methods, for h sufficiently small we get

$$\begin{aligned} \max_{\xi_i \leq t \leq \xi_{i+1}} \|\rho(t) - \zeta(t)\| &\leq S \max_{1 \leq s \leq k-1} \|y_s - z_s\| \\ &\quad + S \max_{\xi_i \leq t \leq \xi_{i+1}} \|y(t - \tau(t)) - \eta(t - \tau(t))\| \\ &\leq S \max_{1 \leq s \leq k-1} \|y_s - z_s\| \\ &\quad + S \max_{t \leq \xi_i} \|y(t) - \eta(t)\|. \end{aligned}$$

On the other hand, since the constant S depends only on the solution $y(t)$ of the original problem (4.1.5), by Lemma 4.1.4 itself applied to the initializing methods and by (4.1.8) we can conclude that

$$\max_{\xi_i \leq t \leq \xi_{i+1}} \|\rho(t) - \zeta(t)\| = O(h^{q'}).$$

Therefore, by (4.1.9) we obtain

$$\max_{\xi_i \leq t \leq \xi_{i+1}} \|\eta(t) - \zeta(t)\| = O(h^{q'})$$

and eventually, by (4.1.6), the proof is complete. ■

It is clear that including the discontinuity points of order $\leq p$ in the mesh Δ was motivated by the need for piecewise regularity of the solution $y(t)$. Including *all* the principal discontinuity points was instead required by the method of steps for treating the DDE as a sequence of ODEs. So, it is not difficult to guess that, in order to attain the accuracy order p, it is sufficient to include only the jump discontinuity points of order $\leq p$ in the mesh and to proceed with stepsizes that prevent overlapping. This could be desirable in that avoiding to locate a number of discontinuity points may considerably reduce the computational cost of the overall procedure, especially when the deviated argument to be passed into the root-finding procedure is very time consuming and the number of such points is very large.

The procedure may actually be carried out by simply requiring that the mesh $\Delta = \{t_0, t_1, \ldots, t_n, \ldots, t_N = t_f\}$ includes all the discontinuity points ξ_i of order up to p and avoids overlapping, that is for any $n < N-1$, $t - \tau(t) \leq t_n$ for all $t \in [t_n, t_{n+1}]$. The convergence of the procedure is guaranteed by the following theorem.

Theorem 4.1.6 *Theorem 4.1.2 still holds if conditions* (h_1) *and* (h_3) *are replaced by*

(h_1^*) *The mesh* $\Delta = \{t_0, t_1, \ldots, t_n, \ldots, t_N = t_f\}$ *includes all the discontinuity points* ξ_i *of order* $\leq p$ *and* $t - \tau(t) \leq t_n$ *for all* $t \in [t_n, t_{n+1}]$ *and for all* $n = 0, \ldots, N-1$.

(h_3^*) *The interpolant (3.2.5), with* $j_n = 0$, *is consistent of uniform order* q.

Since not all the principal discontinuity points are included in the mesh, the proof cannot follow the line of the method of steps. In this case the proof must be carried out step by step through the intervals $[t_n, t_{n+1}]$ by bounding the propagation of the error in $[t_0, t_f]$ via the stability and consistency of the method. However, the proof is omitted in that it is completely included in the proof relevant to the more general case of vanishing delay with possible overlapping developed in the next section.

4.2 DDEs with arbitrary time dependent delay

In this section we analyze the standard approach for more general time dependent delays and with no bounds on the integration stepsize. So, for example, we allow the deviated argument to be of *proportional* type, that is $\alpha(t) = qt$, $t \geq t_0 = 0$, $0 < q < 1$, as in the pantograph equation (2.1.1), where no discontinuities are present in $[0, +\infty)$ and overlapping occurs in any neighborhood of 0 for all initial stepsizes. Also, we allow the delay $\tau(t)$ to vanish at some point $\xi^* > t_0$, so that infinitely many discontinuity points are possibly located in any left neighborhood of ξ^* and, similarly to the pantograph equation, the DDE cannot be reduced to an ODE in any neighborhood of such a point. Finally, we also consider the case of non-vanishing delays with overlapping caused by an integration stepsize larger than the delay.

In all these cases the standard approach consists in solving, for $n = 0, 1, \ldots, N-1$, the local problem

$$\begin{cases} w_{n+1}'(t) = f\Big(t, w_{n+1}(t), x\big(t - \tau(t)\big)\Big), & t_n \leq t \leq t_{n+1}, \\ w_{n+1}(t_n) = y_n, \end{cases} \tag{4.2.1}$$

where

$$x(s) = \begin{cases} \phi(s) & \text{for } s \le t_0, \\ \eta(s) & \text{for } t_0 \le s \le t_n, \\ w_{n+1}(s) & \text{for } t_n \le s \le t_{n+1}, \end{cases}$$

so that the function $x(t - \tau(t))$ is possibly partially or totally unknown.

If overlapping occurs, then equation (4.2.1) is not an ODE and the formal application of the continuous method (3.2.3), (3.2.5) with $j_n = 0$ yields the following equation, to be solved for the extension of $\eta(s)$ on $[t_n, t_{n+1}]$:

$$\eta(t_n + \theta h_{n+1}) = \beta_{n,1}(\theta)y_n + \cdots + \beta_{n,1+i_n}(\theta)y_{n-i_n} \\ + h_{n+1}\Psi(y_n, \ldots, y_{n-i_n}; \theta, g_\eta, \Delta'_n), \quad 0 \le \theta \le 1, \tag{4.2.2}$$

where

$$g_\eta(t, y) = f(t, y, \eta(t - \tau(t))). \tag{4.2.3}$$

We have set $j_n = 0$ because, for the computations of (4.2.2) in the current interval $[t_n, t_{n+1}]$, we must avoid using values which are still to be computed in subsequent steps.

Observe that, on the basis of Remark 3.2.3, the nodal approximation

$$y_{n+1} = \alpha_{n,1}y_n + \cdots + \alpha_{n,k}y_{n-k+1} + h_{n+1}\Phi(y_n, \ldots, y_{n-k+1}; g_\eta, \Delta_n) \tag{4.2.4}$$

is given by formula (4.2.2) for $\theta = 1$ and that solving (4.2.2) and (4.2.3) for $\eta(t)$ means finding $\eta(t_n + \theta h_{n+1})$ for some values of θ determined by the particular choice of the method.

Note also that, for $t - \tau(t) \le t_n$, the term $\eta(t - \tau(t))$ is known, whereas for $t - \tau(t) > t_n$ formula (4.2.2) may be implicit even if the underlying ODE method (4.2.4) is explicit. In fact, if overlapping occurs, then the term $\eta(t - \tau(t))$ required in Φ is given just by (4.2.2) itself for

$$\theta = \frac{t - \tau(t) - t_n}{h_{n+1}}.$$

Example 4.2.1 Consider the equation

$$\begin{cases} y'(t) = f\Big(t, y(t), y(\tfrac{t}{2})\Big), & t \ge 0, \\ y(0) = y_0. \end{cases} \tag{4.2.5}$$

The Heun method with linear interpolation yields

$$\begin{cases} y_1 = y_0 + \frac{h}{2}\Big(f(0, y_0, y_0) + f\big(h, y_0 + hf(0, y_0, y_0), \eta(\tfrac{h}{2})\big)\Big), \\ \eta(\theta h) = (1 - \theta)y_0 + \theta y_1, \end{cases}$$

that is

$$y_1 = y_0 + \frac{h}{2}\Big(f(0, y_0, y_0) + f\big(h, y_0 + hf(0, y_0, y_0), X\big)\Big),$$

where $X = \eta(\frac{h}{2})$ is implicitly defined by

$$X = y_0 + \frac{h}{4}\Big(f(0, y_0, y_0) + f\big(h, y_0 + hf(0, y_0, y_0), X\big)\Big). \ \diamond$$

Example 4.2.2 Consider again equation (4.2.5). Here we want to integrate it by the backward Euler method with linear interpolation

$$\begin{cases} y_1 = y_0 + hf\big(h, y_1, \eta(\frac{h}{2})\big), \\ \eta(\theta h) = (1-\theta)y_0 + \theta y_1, \end{cases}$$

that is

$$y_1 = y_0 + hf(h, Y, X),$$

where $X = \eta(\frac{h}{2})$ and $Y = y_1$ are implicitly defined by

$$\begin{cases} Y = y_0 + hf(h, Y, X), \\ X = y_0 + \frac{h}{2}f(h, Y, X). \end{cases}$$

Observe that, whereas the stage Y is intrinsic to the backward Euler method, the additional stage X is due to the delayed term $y(\frac{h}{2})$. However, it is worth remarking that the system has only apparently increased. In fact, for the variable

$$K = f(h, Y, X),$$

the system reduces to

$$K = f\Big(h, y_0 + hK, y_0 + \frac{h}{2}K\Big)$$

and the continuous approximation is given by

$$\eta(\theta h) = y_0 + \theta hK.$$

This may be done in general and, in particular, it will be illustrated for RK methods in Chapter 6. $\diamond$

Now the problem is to prove, in general, the existence of a suitable polynomial $\eta(t)$ satisfying (4.2.2) and (4.2.3).

Theorem 4.2.3 (Well-posedness) *Equations (4.2.2) and (4.2.3) admit, for sufficiently small stepsize h_{n+1}, a unique solution $\eta(t)$ in a suitable polynomial function space.*

Proof. The proof is obtained by applying the contraction mapping principle to the operator $\mathcal{F}$ which transforms the polynomial function $\eta(s)$, $s \in [t_n, t_{n+1}]$, in (4.2.3) into the polynomial function $\mathcal{F}(\eta)(s)$, $s \in [t_n, t_{n+1}]$, defined by the right-hand side of (4.2.2) with

$$\theta = \frac{s - t_n}{h_{n+1}}.$$

To this end, consider a new function $\lambda(s)$ that extends $\eta(s)$ in $[t_n, t_{n+1}]$ and let us estimate

$$\begin{aligned} \|\mathcal{F}(\lambda)(s) - \mathcal{F}(\eta)(s)\| = h_{n+1}\|&\Psi(y_n, \ldots, y_{n-i_n}; \theta, g_\lambda, \Delta_n') \\ &-\Psi(y_n, \ldots, y_{n-i_n}; \theta, g_\eta, \Delta_n')\|. \end{aligned}$$

By (3.2.7), for $h_{n+1} \leq h$ sufficiently small,

$$\|\mathcal{F}(\lambda)(s) - \mathcal{F}(\eta)(s)\| \leq \delta h_{n+1} \sup_{t_{n-i_n} \leq t \leq t_{n+1},\ y \in \mathbb{R}^d} \|g_\lambda(t, y) - g_\eta(t, y)\|,$$

where both h and the continuity constant δ, which depend on the Lipschitz constant of the function $g_\eta(t, y)$ with respect to y, turn out to be independent of η. Therefore, for any pair η, λ, we have that

$$\|\mathcal{F}(\lambda)(s) - \mathcal{F}(\eta)(s)\| \leq \delta h_{n+1} M \sup_{t_{n-i_n} \leq t \leq t_{n+1},\ y \in \mathbb{R}^d} \|\lambda(t - \tau(t)) - \eta(t - \tau(t))\|,$$

where M is the Lipschitz constant of the function $f(t, y, x)$ with respect to x.

In conclusion,

$$\max_{t_n \leq s \leq t_{n+1}} \|\mathcal{F}(\lambda)(s) - \mathcal{F}(\eta)(s)\| \leq \delta h_{n+1} M \max_{t_n \leq s \leq t_{n+1}} \|\lambda(s) - \eta(s)\|,$$

where $\eta(s)$ and $\lambda(s)$ are equal to each other for $s \leq t_n$. Thus, for $h_{n+1} \leq h$ sufficiently small, $\mathcal{F}$ is a contraction and, consequently, a unique polynomial $\eta(t)$ exists in $[t_n, t_{n+1}]$ which solves (4.2.2) and (4.2.3). ■

Since for $h_{n+1} \leq h$ sufficiently small the operator $\mathcal{F}$ that transforms the polynomial $\eta(t)$ in (4.2.3) into the right-hand side of (4.2.2) is a contraction of the space of polynomials of suitable degree defined in the mesh interval $[t_n, t_{n+1}]$, the approximate solution $\eta(t)$ can be obtained by the standard Picard iteration

$$\eta^{k+1} = \mathcal{F}(\eta^k).$$

A convenient initial guess η^0 may be obtained, for instance, by extrapolation from the previous interval.

We recall that, of course, iterations are needed only if overlapping occurs. Otherwise, just one iteration gives the approximate solution $\eta(t)$ in $[t_n, t_{n+1}]$.

The proof of convergence for the DDE method needs the following lemma.

Lemma 4.2.4 *Let $\zeta(t)$ and $\eta(t)$ be the numerical solutions of the IVPs*

$$\begin{cases} z'_{n+1}(t) = f\Big(t, z_{n+1}(t), u\big(t-\tau(t)\big)\Big), & t_n \le t \le t_{n+1}, \\ z_{n+1}(t_n) = z_n, \end{cases}$$

and

$$\begin{cases} w'_{n+1}(t) = f\Big(t, w_{n+1}(t), v\big(t-\tau(t)\big)\Big), & t_n \le t \le t_{n+1}, \\ w_{n+1}(t_n) = w_n, \end{cases}$$

respectively, obtained by the method (3.2.3) with starting values $\mathbf{z}_n = [z_n, z_{n-1}, \ldots, z_{n-k+1}]^T$ and $\mathbf{w}_n = [w_n, w_{n-1}, \ldots, w_{n-k+1}]^T$, and by the interpolant (3.2.5) with nodal values $\tilde{\mathbf{z}}_n = [z_n, z_{n-1}, \ldots, z_{n-i_n}]^T$ and $\tilde{\mathbf{w}}_n = [w_n, w_{n-1}, \ldots, w_{n-i_n}]^T$.

Then there exist a stepsize $h_0 > 0$ and constants $P > 0$, $Q > 0$, $R > 0$, $S > 0$ and $T > 0$, independent of the mesh Δ, such that, for $h \le h_0$,

$$|||\mathbf{z}_{n+1} - \mathbf{w}_{n+1}||| \le (1 + h_{n+1}Q)|||\mathbf{z}_n - \mathbf{w}_n||| + h_{n+1}P \max_{t \le t_{n+1}} \|u(t) - v(t)\|, \quad (4.2.6)$$

where $\mathbf{z}_{n+1} = [\zeta(t_{n+1}), z_n, \ldots, z_{n-k+2}]^T$, $\mathbf{w}_{n+1} = [\eta(t_{n+1}), w_n, \ldots, w_{n-k+2}]^T$ and $|||\cdot|||$ is the norm on $\mathbb{R}^{kd}$ defined by (3.2.20) and (3.2.21), and

$$\max_{t_n \le t \le t_{n+1}} \|\zeta(t) - \eta(t)\| \le (T + h_{n+1}S) \max_{n-i_n \le s \le n} \|z_s - w_s\| \\ + h_{n+1}R \max_{t \le t_{n+1}} \|u(t) - v(t)\|. \quad (4.2.7)$$

Proof. By subtracting the two formulae (3.2.3) for $z_{n+1} = \zeta(t_{n+1})$ and $w_{n+1} = \eta(t_{n+1})$, and by using the blockwise notation as in the proof of Lemma 4.1.4, we get

$$|||\mathbf{z}_{n+1} - \mathbf{w}_{n+1}||| \le |||\mathbf{z}_n - \mathbf{w}_n||| + h_{n+1}|||\Gamma_n|||,$$

where

$$\Gamma_n = [\Phi(\mathbf{z}_n; f_u, \Delta_n) - \Phi(\mathbf{w}_n; f_v, \Delta_n), 0, \ldots, 0]^T,$$

with $f_u(t,y) = f(t, y, u(t - \tau(t)))$ and $f_v(t,y) = f(t, y, v(t-\tau(t)))$. Thus there exist constants $C > 0$, $P > 0$ and $Q > 0$ such that, for $h \le h_0$,

$$\begin{aligned} |||\Gamma_n||| &\le C\|\Phi(\mathbf{z}_n; f_u, \Delta_n) - \Phi(\mathbf{w}_n; f_v, \Delta_n)\| \\ &\le C(\|\Phi(\mathbf{z}_n; f_u, \Delta_n) - \Phi(\mathbf{w}_n; f_u, \Delta_n)\| + \|\Phi(\mathbf{w}_n; f_u, \Delta_n) - \Phi(\mathbf{w}_n; f_v, \Delta_n)\|) \\ &\le Q|||\mathbf{z}_n - \mathbf{w}_n||| + C\gamma \sup_{t_{n-k+1} \le t \le t_{n+1},\ y \in \mathbb{R}^d} \|f_u(t,y) - f_v(t,y)\| \\ &\le Q|||\mathbf{z}_n - \mathbf{w}_n||| + P \sup_{t_{n-k+1} \le t \le t_{n+1}} \|u(t - \tau(t)) - v(t - \tau(t))\|. \end{aligned}$$

Summarizing, since $\tau(t) \ge 0$, we get (4.2.6) and the first part of the proof is complete.

As for the continuous extensions, we have

$$\begin{aligned}\zeta(t_n+\theta h_{n+1})-\eta(t_n+\theta h_{n+1}) = {} & \beta_{n,1}(\theta)(z_n-w_n)\\ & +\cdots+\beta_{n,i_n+1}(\theta)(z_{n-i_n}-w_{n-i_n})\\ & +h_{n+1}\big(\Psi(\tilde{\mathbf{z}}_n;\theta,f_u,\Delta'_n)\\ & -\Psi(\tilde{\mathbf{w}}_n;\theta,f_v,\Delta'_n)\big),\quad 0\le\theta\le 1.\end{aligned}$$

Therefore, similar arguments as before lead to (4.2.7). ■

Theorem 4.2.5 *Consider the DDE*

$$\begin{cases} y'(t)=f\Big(t,y(t),y\big(t-\tau(t)\big)\Big), & t_0\le t\le t_f,\\ y(t)=\phi(t), \quad t\le t_0, \end{cases} \tag{4.2.8}$$

where $f(t,y,x)$ is C^p-continuous in $[t_0,t_f]\times\mathbb{R}^d\times\mathbb{R}^d$, the delay $\tau(t)$ is C^p-continuous in $[t_0,t_f]$ and $\phi(t)$ is C^p-continuous. Moreover, assume that:

(h'_1) *The mesh $\Delta=\{t_0,t_1,\ldots,t_n,\ldots,t_N=t_f\}$ includes the discontinuity points $\xi_1<\xi_2<\cdots<\xi_s$ of order $\le p$.*

(h'_2) *The ODE method (3.2.3) is consistent of order p, satisfies the stability condition (3.2.14) and, for $k>1$, is restarted after each discontinuity point $\xi_i, i=0,1,\ldots,s$, by a method of order $\ge p-1$.*

(h'_3) *The interpolant (3.2.5), with $j_n=0$, is consistent of uniform order q.*

(h'_4) *For each n, the interval $[t_{n-i_n},t_{n+1}]$ where the interpolation takes place is included in $[\xi_i,\xi_{i+1}]$ for some index $0\le i\le s$.*

Then the resulting DDE method has discrete global order and uniform global order $q'=\min\{p,q+1\}$; that is

$$\max_{1\le n\le N}\|y(t_n)-y_n\|=O(h^{q'})$$

and

$$\max_{t_0\le t\le t_f}\|y(t)-\eta(t)\|=O(h^{q'}),$$

where $h=\max_{1\le n\le N}h_n$.

Proof. In connection to the DDE (4.2.8), for each $n=k-1,\ldots,N-1$, consider the local ODE

$$\begin{cases} z'_{n+1}(t)=f\Big(t,z_{n+1}(t),y\big(t-\tau(t)\big)\Big), & t_n\le t\le t_{n+1},\\ z_{n+1}(t_n)=y(t_n), \\ y(t)=\phi(t), \quad t\le t_0, \end{cases} \tag{4.2.9}$$

whose solution evidently is $z_{n+1}(t)=y(t)$, i.e. the solution of (4.2.8). Moreover, consider the auxiliary local problem

$$\begin{cases} w'_{n+1}(t) = f\Big(t, w_{n+1}(t), \eta\big(t - \tau(t)\big)\Big), & t_n \le t \le t_{n+1}, \\ w_{n+1}(t_n) = \eta(t_n), \\ \eta(t) = \phi(t), \quad t \le t_0, \end{cases} \tag{4.2.10}$$

where, for $s \le t_{n+1}$, $\eta(s)$ is the continuous numerical solution given by the DDE method itself. Then define $\mathbf{y}_n = [y(t_n), y(t_{n-1}), \ldots, y(t_{n-k+1})]^T$ and $\boldsymbol{\eta}_n = [\eta(t_n), \eta(t_{n-1}), \ldots, \eta(t_{n-k+1})]^T$.

By Lemma 4.2.4 with $u(x) = y(x)$, $v(x) = \eta(x)$, $\mathbf{z}_n = \mathbf{y}_n$ and $\mathbf{w}_n = \boldsymbol{\eta}_n$, for $h \le h_0$, the numerical solutions $\zeta(t)$ and $\eta(t)$ of (4.2.9) and (4.2.10), respectively, satisfy the inequality

$$|||\mathbf{z}_{n+1} - \boldsymbol{\eta}_{n+1}||| \le (1 + h_{n+1}Q)|||\mathbf{y}_n - \boldsymbol{\eta}_n||| + h_{n+1}P \max_{t \le t_{n+1}} \|y(t) - \eta(t)\|, \tag{4.2.11}$$

where $\mathbf{z}_{n+1} = (\zeta(t_{n+1}), y(t_n), \ldots, y(t_{n-k+2}))^T$, and

$$\max_{t_n \le t \le t_{n+1}} \|\zeta(t) - \eta(t)\| \le (T + h_{n+1}S) \max_{n - i_n \le s \le n} \|y(t_s) - \eta(t_s)\| + h_{n+1}R \max_{t \le t_{n+1}} \|y(t) - \eta(t)\|. \tag{4.2.12}$$

Now consider the inequality

$$|||\mathbf{y}_{n+1} - \boldsymbol{\eta}_{n+1}||| \le |||\mathbf{y}_{n+1} - \mathbf{z}_{n+1}||| + |||\mathbf{z}_{n+1} - \boldsymbol{\eta}_{n+1}|||. \tag{4.2.13}$$

Owing to hypothesis (h'_1), the solution $z_{n+1}(t)$ of (4.2.9) is sufficiently smooth. Therefore, by hypothesis (h'_2) (see Definition 3.2.5), (4.2.11) yields

$$|||\mathbf{y}_{n+1} - \boldsymbol{\eta}_{n+1}||| \le M h_{n+1}^{p+1} + (1 + h_{n+1}Q)|||\mathbf{y}_n - \boldsymbol{\eta}_n||| + h_{n+1}P \max_{t \le t_{n+1}} \|y(t) - \eta(t)\|.$$

Therefore, with

$$e_n = \max_{k-1 \le i \le n} |||\mathbf{y}_i - \boldsymbol{\eta}_i|||$$

and

$$E_n = \max_{t \le t_n} \|y(t) - \eta(t)\|$$

for $n = k-1, \ldots, N$, we obtain

$$e_{n+1} \le M h_{n+1}^{p+1} + (1 + h_{n+1}Q)e_n + h_{n+1}PE_{n+1} \tag{4.2.14}$$

for $n = k-1, \ldots, N-1$.

Similarly, for the interpolant consider the inequality

$$\max_{t_n \le t \le t_{n+1}} \|y(t) - \eta(t)\| \le \max_{t_n \le t \le t_{n+1}} \|y(t) - \zeta(t)\| + \max_{t_n \le t \le t_{n+1}} \|\zeta(t) - \eta(t)\|.$$

By the smoothness of $y(t)$, by hypotheses (h'_3) (see again Definition 3.2.5) and (h'_4), and by (4.2.12), we get

$$\max_{t_n \le t \le t_{n+1}} \|y(t) - \eta(t)\| \le M h_{n+1}^{q+1} + (T + h_{n+1}S) \max_{n - i_n \le s \le n} \|y(t_s) - \eta(t_s)\|$$

$$+h_{n+1}R \max_{t\leq t_{n+1}} \|y(t)-\eta(t)\|.$$

Thus, since there exists a constant $K > 0$ such that

$$\|y(t_s)-\eta(t_s)\| \leq K\|\|\mathbf{y}_s-\boldsymbol{\eta}_s\|\| \quad \text{for all } s,$$

we obtain

$$\max_{t_n\leq t\leq t_{n+1}} \|y(t)-\eta(t)\| \leq Mh_{n+1}^{q+1} + (T+h_{n+1}S)Ke_n + h_{n+1}RE_{n+1} \tag{4.2.15}$$

for $n = k-1, \ldots, N-1$.

With $L = \max\{M, P, Q, R, TK, SK\}$ the inequalities (4.2.14) and (4.2.15) yield

$$e_{n+1} \leq (1+h_{n+1}L)e_n + h_{n+1}LE_{n+1} + h_{n+1}Lh^p \tag{4.2.16}$$

and

$$\max_{t_n\leq t\leq t_{n+1}} \|y(t)-\eta(t)\| \leq (1+h)Le_n + hLE_{n+1} + Lh^{q+1} \tag{4.2.17}$$

for $n = k-1, \ldots, N-1$.

Now assume, without any restriction, that $h \leq h^*$, where $h^* = \min\{1, 1/(2L)\}$. Since both e_n and E_n are monotonic, (4.2.17) implies

$$E_{n+1} \leq (1+h)Le_n + hLE_{n+1} + Lh^{q+1} \tag{4.2.18}$$

and hence

$$E_{n+1} \leq \frac{(1+h)L}{1-hL}e_n + \frac{L}{1-hL}h^{q+1}, \tag{4.2.19}$$

for $n = k-1, \ldots, N-1$. Therefore, with $\Lambda = L + 4L^2$, by substituting (4.2.19) into (4.2.16) we get

$$\begin{aligned} e_{n+1} &\leq \left[1 + h_{n+1}\left(L + \frac{(1+h)L^2}{1-hL}\right)\right] e_n + h_{n+1}\left(L + \frac{L^2}{1-hL}\right) h^{q'} \\ &\leq (1+\Lambda h_{n+1})e_n + h_{n+1}\Lambda h^{q'} \\ &\leq e^{\Lambda h_{n+1}} e_n + h_{n+1}\Lambda h^{q'} \end{aligned} \tag{4.2.20}$$

for $n = k-1, \ldots, N-1$, where $q' = \min\{p, q+1\}$.

Now we are in a position to prove by standard arguments that

$$\begin{aligned} e_n &\leq \mathrm{e}^{\Lambda(t_n-t_{k-1})}e_{k-1} + \left(\sum_{i=k}^{n} \mathrm{e}^{\Lambda(t_n-t_i)}h_i\right)\Lambda h^{q'} \\ &\leq \mathrm{e}^{\Lambda(t_f-t_{k-1})}e_{k-1} + \left(\mathrm{e}^{\Lambda(t_f-t_{k-1})} - 1\right)h^{q'}, \end{aligned}$$

and hence, since the hypotheses (h_2'), (h_3') and (h_4') imply $e_{k-1} = O(h^{q+1})$ (recall that $q \leq p$), the proof is complete. ■

Remark 4.2.6 *In order to extend Theorem 4.2.5 to more general equations with state dependent delay, observe that hypothesis* (h'_1) *was assumed just to guarantee the smoothness of the solution* $z_{n+1}(t)$ *of (4.2.9) and, hence, the order* $p+1$ *for the local error* $|||\mathbf{y}_{n+1}-\mathbf{z}_{n+1}|||$ *in (4.2.13). Because the delay* τ *is constant or time dependent, locating the discontinuities can be easily done a priori with sufficient accuracy. This would not be the case for state dependent delays where we must be satisfied with an approximation of the discontinuities with an accuracy that is related to the accuracy of the approximate solution itself.*

4.3 DDEs with state dependent delay

In this section we enlarge our analysis to the more general equation with state dependent delay

$$\begin{cases} y'(t) = f\Big(t, y(t), y\big(t-\tau(t,y(t))\big)\Big), & t_0 \le t \le t_f, \\ y(t) = \phi(t), \quad t \le t_0, \end{cases} \tag{4.3.1}$$

with no assumptions on $\tau(t, y)$.

Since hypothesis (H_1^*) on the delay τ is not being assumed, overlapping may actually occur for some mesh interval $[t_n, t_{n+1}]$ and therefore, in the local problem

$$\begin{cases} w'_{n+1}(t) = f\Big(t, w_{n+1}(t), x\big(t-\tau(t,w_{n+1}(t))\big)\Big), & t_n \le t \le t_{n+1}, \\ w_{n+1}(t_n) = y_n, \end{cases} \tag{4.3.2}$$

where

$$x(s) = \begin{cases} \phi(s) & \text{for } s \le t_0, \\ \eta(s) & \text{for } t_0 \le s \le t_n, \\ w_{n+1}(s) & \text{for } t_n \le s \le t_{n+1}, \end{cases}$$

the delay function $x(t-\tau(t, w_{n+1}(t)))$ may be partially or totally unknown. Unlike the case of time dependent delays, evaluating the arguments $s = t - \tau(t, w_{n+1}(t))$, where the delay function $x(s)$ must be computed for some $t \in [t_n, t_{n+1}]$, cannot be done a priori and hence, *a fortiori*, the class of interpolants (3.2.5) must be restricted to the case $j_n = 0$. Thus, for the computations in the current interval $[t_n, t_{n+1}]$, we surely avoid using values which are, possibly, still to be computed. Under this constraint on the interpolant, the application of the continuous method (3.2.3), (3.2.5) yields the following equation, to be solved for the extension of $\eta(s)$ on $[t_n, t_{n+1}]$:

$$\begin{aligned} \eta(t_n + \theta h_{n+1}) = {} & \beta_{n,1}(\theta) y_n + \cdots + \beta_{n,1+i_n}(\theta) y_{n-i_n} \\ & + h_{n+1} \Psi(y_n, \ldots, y_{n-i_n}; \theta, g_\eta, \Delta'_n), \quad 0 \le \theta \le 1, \end{aligned} \tag{4.3.3}$$

where

$$g_\eta(t, y) = f\Big(t, y, \eta\big(t-\tau(t,y)\big)\Big). \tag{4.3.4}$$

Recall that, for $s \le t_n$, the term $\eta(s)$ is known.

Since τ depends on the state y, proving the well-posedness of the numerical method is more complicated than for the time dependent delay case considered in Theorem 4.2.3. In fact, in the inequality

$$\|\mathcal{F}(\lambda)(s) - \mathcal{F}(\eta)(s)\| \le \delta_\eta h_{n+1} \sup_{t_{n-i_n} \le t \le t_{n+1},\ y \in \mathbb{R}^d} \|g_\lambda(t,y) - g_\eta(t,y)\|, \quad (4.3.5)$$

which holds for $h_{n+1} \le h_\eta$ sufficiently small, both h_η and the continuity constant δ_η depend continuously on the Lipschitz constant of g_η with respect to y, that is on

$$\frac{\partial}{\partial y} g_\eta(t,y) = \frac{\partial}{\partial y} f\left(t, y, \eta(t - \tau(t,y))\right) = \frac{\partial f}{\partial y} - \frac{\partial f}{\partial x}\eta' \frac{\partial \tau}{\partial y}. \quad (4.3.6)$$

It turns out that h_η and δ_η actually depend on $\eta'(t)$ and, hence, they cannot be uniformly bounded (h_η from below and δ_η from above) in the whole space of polynomials of suitable degree defined in the mesh interval $[t_n, t_{n+1}]$. Therefore, proving existence and uniqueness by means of the contraction mapping principle is not straightforward.

Theorem 4.3.1 (Well-posedness) *Assume that the local problem (4.3.2) possesses a unique solution $w_{n+1}(t)$. Then, for sufficiently small stepsize h_{n+1}, equations (4.3.3) and (4.3.4) admit a unique solution $\eta(t)$ in a suitable polynomial function space.*

Proof. Let us first prove that the solution $\eta(t)$ exists for h_{n+1} sufficiently small. This will be done by the Brouwer fixed point theorem applied to the (continuous) operator $\mathcal{F}$ which transforms the polynomial function $\eta(s)$, $s \in [t_n, t_{n+1}]$, in (4.3.4) into the polynomial function $\mathcal{F}(\eta)(s)$, $s \in [t_n, t_{n+1}]$, defined by the right-hand side of (4.3.3) with

$$\theta = \frac{s - t_n}{h_{n+1}}.$$

To this end, choose $\rho > 0$ and consider the local solution $w_{n+1}(t)$ of (4.3.2), the polynomial $\bar{\eta}(s) = \mathcal{F}(w_{n+1})(s)$ and the compact ball $\mathcal{B}(\bar{\eta}, \rho)$ with center $\bar{\eta}$ and radius ρ in the space of polynomials of the same degree. For any $\eta \in \mathcal{B}(\bar{\eta}, \rho)$ we have

$$\begin{aligned}\|\mathcal{F}(\eta)(s) - \bar{\eta}(s)\| &= \|\mathcal{F}(\eta)(s) - \mathcal{F}(w_{n+1})(s)\| \\ &= h_{n+1}\|\Psi(y_n, \ldots, y_{n-i_n}; \theta, g_\eta, \Delta'_n) \\ &\quad - \Psi(y_n, \ldots, y_{n-i_n}; \theta, g_{w_{n+1}}, \Delta'_n)\|\end{aligned}$$

and, by (3.2.7), for $h_{n+1} \le h_{w_{n+1}}$ sufficiently small,

$$\|\mathcal{F}(\eta)(s) - \bar{\eta}(s)\| \le \delta_{w_{n+1}} h_{n+1} \sup_{t_{n-i_n} \le t \le t_{n+1},\ y \in \mathbb{R}^d} \|g_\eta(t,y) - g_{w_{n+1}}(t,y)\|$$

$$\le \delta_{w_{n+1}} h_{n+1} M \sup_{t_{n-i_n} \le t \le t_{n+1},\ y \in \mathbb{R}^d} \|\eta(t - \tau(t,y)) - w_{n+1}(t - \tau(t,y))\|,$$

where M is the Lipschitz constant of the function $f(t, y, x)$ with respect to x. Therefore, since $\bar{\eta}(s)$ and $w_{n+1}(s)$ are both known and equal to each other for $s \leq t_n$, we have that

$$\begin{aligned}\max_{t_n \leq s \leq t_{n+1}} \|\mathcal{F}(\eta)(s) - \bar{\eta}(s)\| &\leq \delta_{w_{n+1}} h_{n+1} M \max_{t_n \leq s \leq t_{n+1}} \|\eta(s) - w_{n+1}(s)\| \\ &\leq \delta_{w_{n+1}} h_{n+1} M \Big(\max_{t_n \leq s \leq t_{n+1}} \|\eta(s) - \bar{\eta}(s)\| \\ &\quad + \max_{t_n \leq s \leq t_{n+1}} \|\bar{\eta}(s) - w_{n+1}(s)\| \Big).\end{aligned}$$

Since in the interval $[t_n, t_{n+1}]$ the polynomial $\bar{\eta}(s)$ is the continuous numerical solution of an ODE whose solution is $w_{n+1}(s)$, in the light of Remark 3.2.11, it turns out that

$$\max_{t_n \leq s \leq t_{n+1}} \|\mathcal{F}(\eta)(s) - \bar{\eta}(s)\| \leq \delta_{w_{n+1}} h_{n+1} M \left(\rho + O\left(h_{n+1}^r\right)\right),$$

where $r \geq 2$ even if a discontinuity point of order 1 is included in $[t_n, t_{n+1}]$.

Therefore, for $h_{n+1} \leq h^*_{w_{n+1}}$ sufficiently small, $\mathcal{F}(\eta) \in \mathcal{B}(\bar{\eta}, \rho)$ and hence, by Brouwer's theorem, we can conclude that formulae (4.3.3) and (4.3.4) are well defined and that a polynomial solution $\eta(t)$ exists in $[t_n, t_{n+1}]$.

Once we have proved that the approximate solution $\eta(t)$ exists, we may use the same arguments as in the proof of the forthcoming Theorem 4.3.8 to state that also $\|\eta(s) - w_{n+1}(s)\|$ is at least of $O(h_{n+1}^2)$ uniformly in $[t_n, t_{n+1}]$. Therefore, by general arguments on the polynomial approximation of function derivatives, we can claim that $\|\eta'(s) - w'_{n+1}(s)\|$ is at least of $O(h_{n+1})$ uniformly in $[t_n, t_{n+1}]$ (see, for instance, the proof of Theorem 5.1.5). As a consequence, both h_η and the continuity constant δ_η in (4.3.5), which depend continuously on (4.3.6), turn out to be arbitrarily close to $h_{w_{n+1}}$ and $\delta_{w_{n+1}}$, respectively, whenever $\eta(t)$ is a solution of (4.3.3), (4.3.4) for sufficiently small h_{n+1}.

So we can conclude that there exist constants $h > 0$ and $\delta > 0$ such that, for any $\lambda(t)$,

$$\max_{t_n \leq s \leq t_{n+1}} \|\mathcal{F}(\lambda)(s) - \mathcal{F}(\eta)(s)\| \leq \delta h_{n+1} M \max_{t_n \leq s \leq t_{n+1}} \|\lambda(s) - \eta(s)\| \tag{4.3.7}$$

whenever $\eta(t)$ is a solution of (4.3.3), (4.3.4) for any $h_{n+1} \leq h$. This clearly implies that, for h_{n+1} sufficiently small, the approximate solution is unique. ■

Unlike the time dependent case, the operator $\mathcal{F}$ that transforms the polynomial $\eta(t)$ in (4.2.3) into the right-hand side of (4.2.2) is not guaranteed to be a contraction of the space of polynomials of suitable degree defined in the mesh interval $[t_n, t_{n+1}]$, even for sufficiently small stepsizes h_{n+1}. Nevertheless, the inequality (4.3.7) ensures that, for sufficiently small h_{n+1}, the operator $\mathcal{F}$ always contracts the distance $\max_{t_n \leq s \leq t_{n+1}} \|\lambda(s) - \eta(s)\|$ from the approximate solution $\eta(t)$. This is sufficient to guarantee that the standard Picard iteration

$$\eta^{k+1} = \mathcal{F}(\eta^k)$$

is still convergent. Once again, a convenient initial guess $\eta^0(t)$ may be obtained, for instance, by extrapolation from the previous interval.

We recall that, of course, iterations are needed only if overlapping occurs. Otherwise, just one iteration gives the approximate solution $\eta(t)$ in $[t_n, t_{n+1}]$.

Example 4.3.2 In order to clarify the meaning and the use of formulae (4.3.3) and (4.3.4), let us consider the particular case of the BDF formula

$$\begin{aligned} y_{n+1} &= \alpha_{n,1} y_n + \cdots + \alpha_{n,k} y_{n-k+1} \\ &\quad + h_{n+1}\beta_{n,0} f\Big(t_{n+1}, y_{n+1}, \eta\big(t_{n+1} - \tau(t_{n+1}, y_{n+1})\big)\Big), \end{aligned} \tag{4.3.8}$$

with the continuous extension

$$\begin{aligned} \eta(t_n + \theta h_{n+1}) &= \beta_{n,1}(\theta) y_n + \cdots + \beta_{n,1+i_n}(\theta) y_{n-i_n} \\ &\quad + h_{n+1}\beta_{n,0}(\theta) f\Big(t_{n+1}, y_{n+1}, \eta\big(t_{n+1} - \tau(t_{n+1}, y_{n+1})\big)\Big). \end{aligned}$$

The following pseudo-code performs one step of the BDF formula implemented in the $P(EC)^m$ (predictor–(evaluation–corrector)^m) mode.

Algorithm 4.3.3

1 *Predictor:*
 Set

$$y_{n+1}^0 = \eta(t_n).$$

2 *For* $\nu = 0$ *to* $m-1$*:*

- *evaluation:*
  - *compute the argument* $s = t_{n+1} - \tau(t_{n+1}, y_{n+1}^\nu)$,
  - *if* $s \leq t_n$

    then set $X = \eta(s)$,

    else set

$$\theta^\nu = \frac{s - t_n}{h_{n+1}}$$

    and solve for X

$$\begin{aligned} X &= \beta_{n,1}(\theta^\nu) y_n + \ldots + \beta_{n,1+i_n}(\theta^\nu) y_{n-i_n} \\ &\quad + h_{n+1}\beta_{n,0}(\theta^\nu) f(t_{n+1}, y_{n+1}^\nu, X), \end{aligned}$$

    end if;
  - *evaluate* $f(t_{n+1}, y_{n+1}^\nu, X)$*;*
- *correction:*

$$\begin{aligned} y_{n+1}^{\nu+1} &= \alpha_{n,1} y_n + \cdots + \alpha_{n,k} y_{n-k+1} \\ &\quad + h_{n+1} f(t_{n+1}, y_{n+1}^\nu, X). \end{aligned}$$

3 *End.*

Observe that at each evaluation step the procedure might require the solution of a non-linear equation for X. It is not difficult to see that, for sufficiently small stepsize h_{n+1}, such an X is the fixed point of a contraction mapping. For small h_{n+1}, the transformation $y^{\nu}_{n+1} \to y^{\nu+1}_{n+1}$ is also a contraction. So the overall procedure requires, at each integration step, the computation of two nested iterative schemes. This may be avoided by rewriting the BDF method in the form

$$y_{n+1} = \alpha_{n,1} y_n + \cdots + \alpha_{n,k} y_{n-k+1} + h_{n+1}\beta_{n,0} K_{n+1}, \qquad (4.3.9)$$

with the continuous extension

$$\eta(t_n + \theta h_{n+1}) = \beta_{n,1}(\theta) y_n + \cdots + \beta_{n,1+i_n}(\theta) y_{n-i_n} + h_{n+1}\beta_{n,0}(\theta) K_{n+1},$$

where the term

$$K_{n+1} = f\Big(t_{n+1}, y_{n+1}, \eta\big(t_{n+1} - \tau(t_{n+1}, y_{n+1})\big)\Big)$$

depends on y_{n+1} and hence, by (4.3.9), on K_{n+1} itself. As before, it is not possible to state a priori which mesh interval includes the argument s. On the other hand, this may be done for any assigned value of K_{n+1} and hence the following recursive algorithm for the computation of y_{n+1} is well defined.

Algorithm 4.3.4

1 *Starter:*

Set

$$K^0_{n+1} = \begin{cases} f\Big(t_{n+1}, y_n, \eta\big(t_{n+1} - \tau(t_{n+1}, y_n)\big)\Big), & \textit{for } t_{n+1} - \tau(t_{n+1}, y_n) < t_n, \\ f(t_{n+1}, y_n, y_n), & \textit{for } t_{n+1} - \tau(t_{n+1}, y_n) \geq t_n. \end{cases}$$

2 *For* $\nu = 0$ *to* $m-1$*:*

- *set* $z = \alpha_{n,1} y_n + \cdots + \alpha_{n,k} y_{n-k+1} + h_{n+1}\beta_{n,0} K^{\nu}_{n+1}$ *and compute the argument* $s = t_{n+1} - \tau(t_{n+1}, z)$*;*

- *if* $s \leq t_n$

  then set $$K^{\nu+1}_{n+1} = f\big(t_{n+1}, z, \eta(s)\big), \qquad (\star)$$

  else set

$$\begin{aligned} \theta^{\nu} &= \frac{s - t_n}{h_{n+1}}, \\ w &= \beta_{n,1}(\theta^{\nu}) y_n + \cdots + \beta_{n,1+i_n}(\theta^{\nu}) y_{n-i_n} + h_{n+1}\beta_{n,0}(\theta^{\nu}) K^{\nu}_{n+1}, \\ K^{\nu+1}_{n+1} &= f(t_{n+1}, z, w), \end{aligned}$$

  end if.

3 *Advancing:*
Set $\quad y_{n+1} = \alpha_{n,1}y_n + \cdots + \alpha_{n,k}y_{n-k+1} + h_{n+1}\beta_{n,0}K^m_{n+1}$.
4 *End.*

In this case it is not difficult to see that, as long as we are away from a vanishing delay point, the algorithm eventually chooses the option $(\star)$. Therefore, for sufficiently small h_{n+1}, the transformation $K^\nu_{n+1} \to K^{\nu+1}_{n+1}$ is a contraction and hence the iteration converges. On the contrary, in the vicinity of a vanishing delay point the coordinate s approaches t_{n+1} and hence θ^ν approaches 1. In this case the transformation $K^\nu_{n+1} \to K^{\nu+1}_{n+1}$ is not guaranteed to be a contraction. This inconvenience is due to the dependency of the coefficients $\beta_{n,i}$ on θ^ν and, hence, on K^ν_{n+1}. Nevertheless, the iterative scheme is still convergent to the unique numerical solution. $\diamond$

In connection with question (Q_2) in Section 3.3 about the location of discontinuity points, assume for the moment that we are able to compute and include them as mesh points. This assumption allows us to establish the following convergence result.

Theorem 4.3.5 *Consider the state dependent DDE*

$$\begin{cases} y'(t) = f\Big(t, y(t), y\big(t - \tau(t, y(t))\big)\Big), & t_0 \le t \le t_f, \\ y(t) = \phi(t), \quad t \le t_0, \end{cases}$$

where $f(t,y,x)$ is C^p-continuous in $[t_0, t_f] \times \mathbb{R}^d \times \mathbb{R}^d$, the delay $\tau(t,y)$ is C^p-continuous in $[t_0, t_f] \times \mathbb{R}^d$ and $\phi(t)$ is C^p-continuous. Moreover, assume that:

(h_1') *The mesh $\Delta = \{t_0, t_1, \ldots, t_n, \ldots, t_N = t_f\}$ includes the discontinuity points $\xi_1 < \xi_2 < \ldots < \xi_s$ of order $\le p$.*
(h_2') *The ODE method (3.2.3) is consistent of order p, satisfies the stability condition (3.2.14) and, for $k > 1$, is restarted after each discontinuity point $\xi_i, i = 0, 1, \ldots, s$, by a method of order $\ge p - 1$.*
(h_3') *The interpolant (3.2.5), with $j_n = 0$, is consistent of uniform order q.*
(h_4') *For each n, the interval $[t_{n-i_n}, t_{n+1}]$ where the interpolation takes place is included in $[\xi_i, \xi_{i+1}]$ for some index $0 \le i \le s$.*

Then the resulting DDE method has discrete global order and uniform global order $q' = \min\{p, q+1\}$; that is

$$\max_{1 \le n \le N} \|y(t_n) - y_n\| = O(h^{q'})$$

and

$$\max_{t_0 \le t \le t_f} \|y(t) - \eta(t)\| = O(h^{q'}),$$

where $h = \max_{1 \le n \le N} h_n$.

The proof of the theorem is based on the following lemma.

Lemma 4.3.6 *Let $\zeta(t)$ and $\eta(t)$ be the numerical solutions of the initial value problems*

$$\begin{cases} z'_{n+1}(t) = f\Big(t, z_{n+1}(t), u\big(t - \tau(t, z_{n+1}(t))\big)\Big), & t_n \le t \le t_{n+1}, \\ z_{n+1}(t_n) = z_n, \end{cases}$$

and

$$\begin{cases} w'_{n+1}(t) = f\Big(t, w_{n+1}(t), v\big(t - \tau(t, w_{n+1}(t))\big)\Big), & t_n \le t \le t_{n+1}, \\ w_{n+1}(t_n) = w_n, \end{cases}$$

respectively, obtained by the method (3.2.3) with starting values $\mathbf{z}_n = [z_n, z_{n-1}, \ldots, z_{n-k+1}]^T$ and $\mathbf{w}_n = [w_n, w_{n-1}, \ldots, w_{n-k+1}]^T$, and by the interpolant (3.2.5) with nodal values $\tilde{\mathbf{z}}_n = [z_n, z_{n-1}, \ldots, z_{n-i_n}]^T$ and $\tilde{\mathbf{w}}_n = [w_n, w_{n-1}, \ldots, w_{n-i_n}]^T$.

Then there exist a stepsize $h_u > 0$ and constants $P_u > 0$, $Q_u > 0$, $R_u > 0$, $S_u > 0$ and $T > 0$, independent of the mesh Δ, such that, for $h \le h_u$,

$$|||\mathbf{z}_{n+1} - \mathbf{w}_{n+1}||| \le (1 + h_{n+1}Q_u)|||\mathbf{z}_n - \mathbf{w}_n||| + h_{n+1}P_u \max_{t \le t_{n+1}} \|u(t) - v(t)\|,$$

where $\mathbf{z}_{n+1} = [\zeta(t_{n+1}), z_n, \ldots, z_{n-k+2}]^T$, $\mathbf{w}_{n+1} = [\eta(t_{n+1}), w_n, \ldots, w_{n-k+2}]^T$ and $|||\cdot|||$ is the norm on $\mathbb{R}^{kd}$ defined by (3.2.20) and (3.2.21), and

$$\begin{aligned} \max_{t_n \le t \le t_{n+1}} \|\zeta(t) - \eta(t)\| \le (T + h_{n+1}S_u) \max_{n-i_n \le s \le n} \|z_s - w_s\| \\ + h_{n+1}R_u \max_{t \le t_{n+1}} \|u(t) - v(t)\|. \end{aligned}$$

The proof of Lemma 4.3.6 is quite similar to that of Lemma 4.2.4 and hence is omitted. The sole difference is that, for state dependent delay, the parameterized function $f_u(t, y)$ is given by $f(t, y, u(t - \tau(t, y)))$.

Also the proof of Theorem 4.3.5 is omitted since it is obtained from the proof of Theorem 4.2.5 by simply substituting the local problems (4.2.9) and (4.2.10) by

$$\begin{cases} z'_{n+1}(t) = f\Big(t, z_{n+1}(t), y\big(t - \tau(t, z_{n+1}(t))\big)\Big), & t_n \le t \le t_{n+1}, \\ z_{n+1}(t_n) = y(t_n), \end{cases} \tag{4.3.10}$$

and

$$\begin{cases} w'_{n+1}(t) = f\Big(t, w_{n+1}(t), \eta\big(t - \tau(t, w_{n+1}(t))\big)\Big), & t_n \le t \le t_{n+1}, \\ w_{n+1}(t_n) = \eta(t_n). \end{cases}$$

As pointed out in Remark 4.2.6, the crucial condition in the proof of Theorem 4.2.5, and hence of Theorem 4.3.5, is that the approximation of the local problem (4.3.10) has an error of order $p+1$ and this depends on the smoothness of $y(t-\tau(t,z_{n+1}(t)))$ as function of t in the current interval $[t_n,t_{n+1}]$. Essentially, under this condition on the local error of the method, Tavernini [244], Feldstein and Neves [97], Arndt [4] and others proved the convergence of order p for LM and RK methods with suitable continuous extensions. In those papers, how to handle the discontinuities in order to get the order $p+1$ for the local error is deferred to the construction of the method.

A different proof of Theorem 4.3.5, based on the analysis of the defect, was given by Enright and Hayashi [90].

The theorem above applies, in particular, to problems without discontinuity points, that is problems with the initial function $\phi(t)$ fulfilling the splicing conditions, as well as problems with delays such that $t-\tau(t,y(t))\geq t_0$ for all $t\geq t_0$ (see (4.3.11) below).

Example 4.3.7 Let us consider the equation

$$\begin{cases} y'(t)=y(t)-y\big(y(t)\big)+\frac{1}{t}-\log\left(\frac{t}{\log(t)+1}\right), & 1\leq t\leq 5,\\ y(1)=1, \end{cases} \tag{4.3.11}$$

whose solution

$$y(t)=\log(t)+1$$

turns out to be ≥ 1. Since the initial data reduce to the single initial value $y(1)$, the solution $y(t)$ is smooth and no discontinuities are present in $[1,t_f]$. Therefore, Theorem 4.3.5 holds.

Let us integrate the problem by the BDF method (4.3.8) of order 2:

$$y_{n+1}=\tfrac{4}{3}y_n-\tfrac{1}{3}y_{n-1}+\tfrac{2}{3}hf\big(t_{n+1},y_{n+1},\eta\big(t_{n+1}-\tau(t_{n+1},y_{n+1})\big)\big),$$

with constant stepsize h and linear interpolation

$$\eta(\theta h)=(1-\theta)y_n+\theta y_{n+1}.$$

The right-hand side of (4.3.11) being linear with respect to the delay argument $y\big(y(t)\big)$, the equation for X in Algorithm 4.3.3 is solved exactly for each evaluation step of the $P(EC)^m$ mode. Figure 4.1 shows that the method performs quite well to order 2. $\diamond$

Disregarding the location of the discontinuities, an approximate solution accurate to order 2 is obtainable as stated by the following theorem.

Theorem 4.3.8 *If equation (4.3.1) has a smooth solution apart from a finite number of discontinuities of order* 1, *then the standard approach based on any ODE method of order* $p\geq 1$ *and continuous extension of uniform order* $q\geq 1$ *furnishes an approximate solution* $\eta(t)$ *of uniform global order at least* $\min\{p,2\}$ *for any choice of the mesh.*

Proof. The proof is the same as in Theorem 4.3.5 by simply disregarding the location of discontinuities and by taking Remark 3.2.11 into account. In fact, on any integration interval including a discontinuity of order 1 as an inner point, the

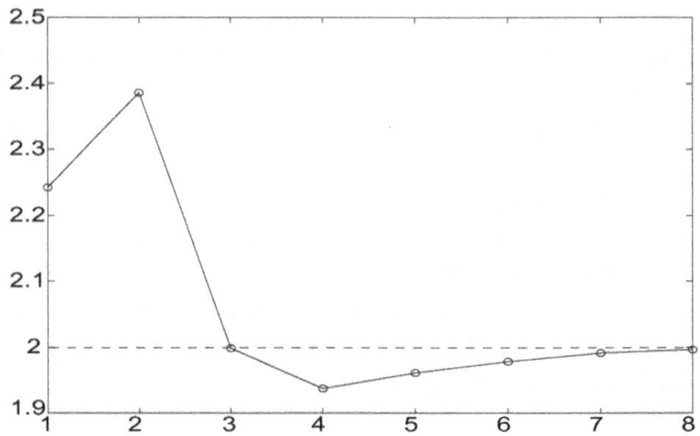

FIG. 4.1. Logarithmic plot of the ratios of consecutive errors for the solution of (4.3.11), obtained by doubling the number of mesh points. The BDF method of order 2 is implemented in the $P(EC)^m$ mode, $m = 3$ and y^0_{n+1} linearly extrapolated from the previous interval.

method performs to order 2 locally. Since there are finitely many such intervals, the whole method gives an approximation that is accurate to order $\min\{p, 2\}$ if $p \geq 1$. ■

Since the approximate solution $\eta(t)$ converges to $y(t)$ as h tends to zero, the set of *computed discontinuities* $\tilde{\xi}_i$, obtained by solving recursively

$$\tilde{\xi}_{i+1} - \tau(\tilde{\xi}_{i+1}, \eta(\tilde{\xi}_{i+1})) = \tilde{\xi}_i, \quad i = 0, 1, \ldots,$$

where $\tilde{\xi}_0 = t_0$, tends to the exact discontinuities ξ_i with an accuracy order that depends on the order of $\eta(t)$ and the multiplicity of ξ_i.

Let us now consider an example with a non-smooth solution.

Example 4.3.9 (Feldstein and Neves [97]) Consider the equation

$$\begin{cases} y'(t) = \frac{1}{2}\frac{1}{\sqrt{t}}y\big(y(t) - \sqrt{2} + 1\big), & 1 \leq t \leq 3, \\ y(t) = 1, \quad t \leq 1. \end{cases}$$

The solution $y(t)$ has a jump discontinuity in $y''(t)$ at $t = 2$ and is given by

$$y(t) = \begin{cases} \sqrt{t}, & 1 \leq t \leq 2, \\ \frac{t}{4} + \frac{1}{2} + (1 - \frac{\sqrt{2}}{2})\sqrt{t}, & 2 \leq t \leq 3. \end{cases}$$

Let us approximate the solution by the following explicit RK method of order 3, with interpolant of order 2 (see Chapter 5):

$$\begin{array}{c|ccc} 0 & 0 & & \\ \frac{1}{3} & \frac{1}{3} & & \\ \frac{2}{3} & 0 & \frac{2}{3} & \\ \hline & \frac{1}{4} & 0 & \frac{3}{4} \end{array} \qquad \begin{aligned} b_1(\theta) &= -\tfrac{3}{4}\theta^2 + \theta, \\ b_2(\theta) &= 0, \\ b_3(\theta) &= \tfrac{3}{4}\theta^2. \end{aligned}$$

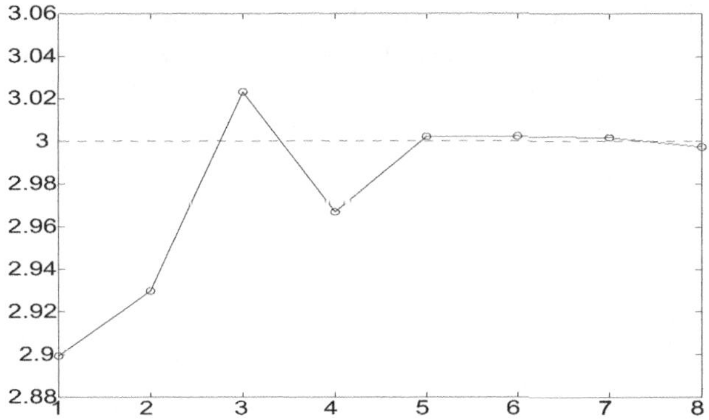

FIG. 4.2. Logarithmic plot of the ratios of consecutive errors for the solution of Example 4.3.9, obtained by doubling the number of mesh points. The discontinuity point $\xi = 2$ is included as a mesh point.

By including the discontinuity point $\xi = 2$ as a mesh point, the method performs to order 3, as stated in Theorem 4.3.5 (see Figure 4.2).

On the contrary, if we assume the discontinuity point to be unknown and we implement the method with equally spaced meshes that do not include the discontinuity, the method still converges but, according to Theorem 4.3.8, the convergence order should decay to 2. In fact, whenever the integration step includes the discontinuity point $\xi = 2$ as an inner point, the solution $y(t)$ is locally C^1-continuous with a Lipschitz continuous derivative. Therefore, the error in the current interval just has order 2.

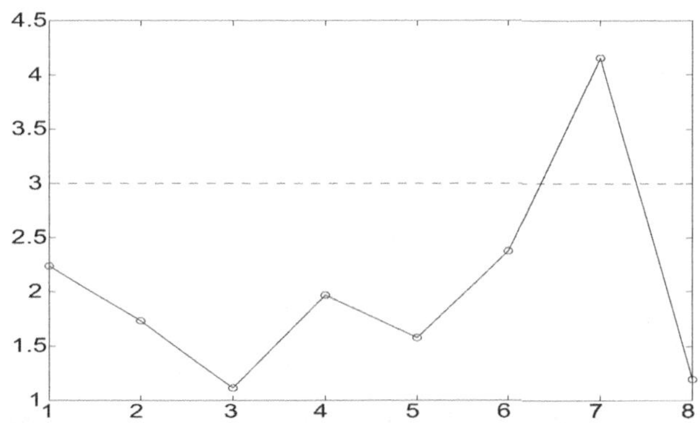

FIG. 4.3. Logarithmic plot of the ratios of consecutive errors for the solution of Example 4.3.9, obtained by doubling the number of mesh points. The equation is integrated in the interval $[1, 3.11111]$ so that, for equally spaced mesh points, the discontinuity $\xi = 2$ moves randomly inside the interval.

However, the analysis based on doubling the number of mesh points does not display any order because, due to the fluctuating position of the discontinuity $\xi = 2$ inside the interval where it lies (possibly very close or even equal to a mesh point), the method occasionally performs better than expected (see Figure 4.3).

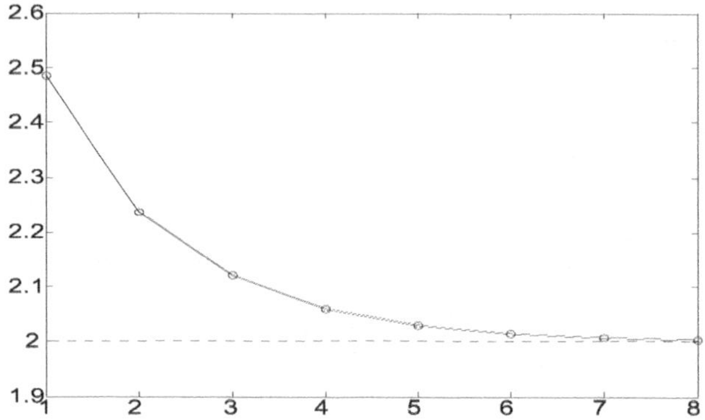

FIG. 4.4. Logarithmic plot of the ratios of consecutive errors for the solution of Example 4.3.9, obtained by doubling the number of mesh points. The discontinuity $\xi = 2$ is forced to stay $h/3$ away from the left end of the interval.

On the other hand, by forcing the discontinuity $\xi = 2$ to stay, for any stepsize, in a fixed relative position inside the interval where it lies, the method performs exactly to order 2, as shown in Figure 4.4.

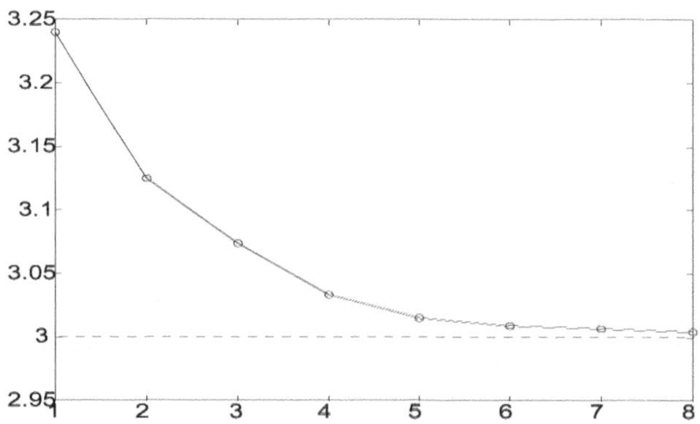

FIG. 4.5. Logarithmic plot of the ratios of consecutive errors for the solution of Example 4.3.9, obtained by doubling the number of mesh points. The discontinuity $\xi = 2$ is forced to stay in the middle of the interval.

It is worth remarking that, for reasons that are not completely investigated, some particular positions of the discontinuity inside the interval are privileged and allow the method to perform again to order 3 (see Figure 4.5). ◇

For equations whose solutions actually possess some discontinuities, the previous example shows that, even for low order schemes, we cannot disregard their location. We can conclude that Theorem 4.3.5 provides strong support for all the strategies based on trying to approximate and include the discontinuities as mesh points.

4.3.1 *Tracking of discontinuities*

Various schemes have been proposed by Feldstein and Neves [97], Paul [216], Willé and Baker [270, 271, 272] and others for the approximate computation of the discontinuities and their inclusion as mesh points. Here we just sketch a possible strategy for their accurate computation.

Assume that the local problem

$$\begin{cases} w'_{n+1}(t) = f\Big(t, w_{n+1}(t), x(t - \tau(t, w_{n+1}(t)))\Big), & t_n \le t \le t_{n+1}, \\ w_{n+1}(t_n) = y_n, \end{cases} \tag{4.3.12}$$

where

$$x(s) = \begin{cases} \phi(s) & \text{for } s \le t_0, \\ \eta(s) & \text{for } t_0 \le s \le t_n, \\ w_{n+1}(s) & \text{for } t_n \le s \le t_{n+1}, \end{cases}$$

has a discontinuity point ξ in $[t_n, t_{n+1}]$ originated by a previously computed discontinuity point $\tilde{\xi}_i$, that is

$$\xi - \tau(\xi, w_{n+1}(\xi)) = \tilde{\xi}_i.$$

Once we have solved (4.3.12) for $w_{n+1}(t)$ in $[t_n, t_{n+1}]$ by the continuous approximation $\eta(t)$, we can check whether or not the approximate deviated argument $t - \tau(t, \eta(t))$ traverses the discontinuity point $\tilde{\xi}_i$ at some point $\tilde{\xi}_{i+1}$ interior to the current interval $[t_n, t_{n+1}]$. If such a point exists, then we must reject the current step along with the approximations $\eta(t)$ and $\tilde{\xi}_{i+1}$, and recompute both of them on a reduced interval $[t_n, t_n + h_{n+1}^{\text{new}}]$. A reasonable guess value for h_{n+1}^{new} is $h_{n+1}^{\text{new}} = \rho(\tilde{\xi}_{i+1} - t_n)$ with the control parameter ρ close to 1. The procedure must be repeated from now on by also rejecting and enlarging intervals that do not contain the computed discontinuity so as to capture it. Otherwise the location of the computed discontinuity is simply deferred to the next interval and the process could be more expensive and even endless (see Figure 3.2 in Section 3.3). Indeed, what we would like to do is determine a stepsize $\bar{h}$ such that the local problem (4.3.12) in $[t_n, t_n + \bar{h}]$ provides an approximation $\bar{\eta}(t)$ for which

$$t_n + \bar{h} - \tau(t_n + \bar{h}, \bar{\eta}(t_n + \bar{h})) = \tilde{\xi}_i.$$

In other words, we are looking for the fixed point of the transformation which associates with the end point of the current interval the solution of the equation $t - \tau(t, \eta(t)) = \tilde{\xi}_i$, $\eta(t)$ being the continuous approximation on *that* interval. If such a point $t_n + \bar{h}$ exists, then it turns out to be the next computed discontinuity $\tilde{\xi}_{i+1}$ and to belong to the mesh as well.

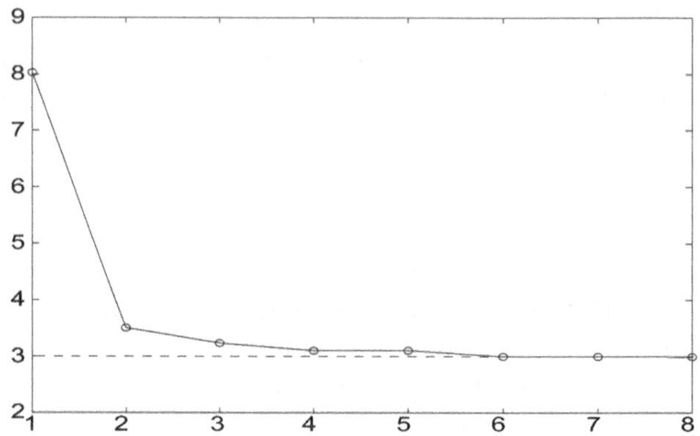

FIG. 4.6. Logarithmic plot of the ratios of consecutive errors for the solution of Example 4.3.9, obtained by doubling the number of mesh points. The computed discontinuity $\tilde{\xi}$ is included as a mesh point.

Since by Theorem 4.3.8 the approximate solution $\eta(t)$ converges to $y(t)$ independently of the mesh, the computed discontinuities $\tilde{\xi}_i$ converge to the exact discontinuities ξ_i and intuitively we can say that, *as the stepsize goes to zero, the method including the computed discontinuities as mesh points approaches the method that includes the exact discontinuities.*

Numerical experiments confirm that, by including the computed discontinuities $\tilde{\xi}_i$ as mesh points, the method eventually performs according to the results of Theorem 4.3.5. For Example 4.3.9, the method based on the strategy described above actually performs to the expected order 3 (see Figure 4.6).

A similar strategy was developed by Feldstein and Neves [97] for higher order methods along with a deep analysis of the required accuracy in the root-finding procedure for the computed discontinuities taking the multiplicity of roots into account.

4.4 DDEs of neutral type

The complexity in the proof of convergence, as well as in the implementation of the standard approach for the NDDE

$$\begin{cases} y'(t) = f\Big(t, y(t), y\big(t-\tau(t,y(t))\big), y'\big(t-\sigma(t,y(t))\big)\Big), & t_0 \le t \le t_f, \\ y(t) = \phi(t), \quad t \le t_0, \end{cases} \tag{4.4.1}$$

lies in possible overlapping, for which the same arguments for the non-neutral case apply. Therefore, in order to shorten the presentation, we will simply consider two extreme cases. On the one hand, we shall consider the method of steps for equations with constant or non-vanishing time dependent delays satisfying (H_1). On the other hand, we shall consider the most general class of equations with state dependent delays, where, in general, no a priori conditions on the delay can be assumed.

4.4.1 *NDDEs with constant or non-vanishing time dependent delay*

As with the non-neutral case, we start the convergence analysis of NDDE methods by analyzing the model equation

$$\begin{cases} y'(t) = f\Big(t, y(t), y\big(t-\tau(t)\big), y'\big(t-\sigma(t)\big)\Big), & t_0 \le t \le t_f, \\ y(t) = \phi(t), \qquad t \le t_0, \end{cases} \tag{4.4.2}$$

where the delays τ and σ satisfy hypothesis (H_1).

In this case, the standard approach may be framed as the *method of steps*, which consists in locating a number of discontinuity points $\xi_1 < \cdots < \xi_s$ in $[t_0, t_f]$, including the principal discontinuity points (see Definition 2.1.4), and solving, for $i = 0, 1, \ldots, s$, the ODEs

$$\begin{cases} z'(t) = f\Big(t, z(t), \eta\big(t-\tau(t)\big), \lambda\big(t-\sigma(t)\big)\Big), & \xi_i \le t \le \xi_{i+1}, \\ z(\xi_i) = \eta(\xi_i), \end{cases}$$

where $\xi_0 = t_0$, $\xi_{s+1} = t_f$, $\eta(t)$ is the interpolant (3.2.5) of the ODE method and $\lambda(t)$ is the continuous approximation of $y'(t)$ given by either

$$\lambda(t) = \eta'(t) \tag{4.4.3}$$

or

$$\lambda(t) = \mathcal{P}\Big(f\Big(\cdot, \eta(\cdot), \eta\big(\cdot - \tau(\cdot)\big), \lambda\big(\cdot - \sigma(\cdot)\big)\Big)\Big)(t), \tag{4.4.4}$$

where, in each mesh interval $[t_k, t_{k+1}]$, $\mathcal{P}$ is an interpolation operator in a suitable polynomial space of degree possibly other than $\deg(\eta')$ and nodes in $[t_{k-i_k}, t_{k+j_k+1}]$.

Note that the presence of two different delays in (4.4.2) simply reflects in a larger proliferation of discontinuities. In fact, under hypothesis (H_1), any discontinuity point ξ_i gives rise to a set of subsequent discontinuities which are the solutions of $t - \tau(t) = \xi_i$ and $t - \sigma(t) = \xi_i$.

The convergence of the method of steps for NDDEs is given by the following theorem, analogous to Theorem 4.1.2. The proof is quite similar and, therefore, omitted.

Theorem 4.4.1 *Consider the NDDE*

$$\begin{cases} y'(t) = f\Big(t, y(t), y\big(t-\tau(t)\big), y'\big(t-\sigma(t)\big)\Big), & t_0 \le t \le t_f, \\ y(t) = \phi(t), \qquad t \le t_0, \end{cases}$$

where $f(t, y, x, w)$ is C^p-continuous in $[t_0, t_f] \times \mathbb{R}^d \times \mathbb{R}^d \times \mathbb{R}^d$, the delays $\tau(t)$ and $\sigma(t)$ are C^p-continuous in $[t_0, t_f]$ and fulfill hypothesis (H_1), and the initial function $\phi(t)$ is C^{p+1}-continuous. Moreover, assume that

(h_1) *The mesh $\Delta = \{t_0, t_1, \ldots, t_n, \ldots, t_N = t_f\}$ includes all the principal discontinuity points and the discontinuity points of order $\le p$ lying in $[t_0, t_f]$, denoted by $\xi_1, \ldots, \xi_s (< t_f)$.*

(h_2) *The ODE method (3.2.3) is consistent of order p, satisfies the stability condition (3.2.14) and, for $k > 1$, is restarted after each discontinuity point $\xi_i, i = 0, 1, \ldots, s$, by a method of order $\geq p - 1$.*

(h_3) *The interpolant (3.2.5) is consistent of uniform order q.*

(h_4) *For each n, the interval $[t_{n-i_n}, t_{n+j_n+1}]$ where the interpolation takes place is included in $[\xi_i, \xi_{i+1}]$ for some index $0 \leq i \leq s$.*

(h_5) *The interpolant (4.4.3) or (4.4.4) has uniform order r.*

Then the resulting DDE method has discrete global order and uniform global order $q' = \min\{p, q+1, r+1\}$; that is

$$\max_{1 \leq n \leq N} \|y(t_n) - y_n\| = O(h^{q'})$$

and

$$\max_{t_0 \leq t \leq t_f} \|y(t) - \eta(t)\| = O(h^{q'}),$$

where $h = \max_{1 \leq n \leq N} h_n$.

As for hypothesis (h_5), observe that by using the option $\lambda(t) = \eta'(t)$ its uniform order of approximation turns out to be $r = q - 1$ and this may reduce the order of the overall method by one. The option (4.4.4) instead allows the preservation of the optimal order $q' = \min\{p, q+1\}$ by using a suitable interpolation process with polynomials of degree $\geq q$.

As for the choice of the function $\lambda(t)$, note that, because $\eta(t - \tau(t))$ and $\lambda(t - \sigma(t))$ are known for any t of the current macro interval $[\xi_i, \xi_{i+1}]$, we have one more option available, i.e.

$$\lambda(t) = f\big(t, \eta(t), \eta(t - \tau(t)), \lambda(t - \sigma(t))\big).$$

In this case the uniform approximation order for $\lambda(t)$ turns out to be the same for $\eta(t)$, that is q, and the global order of the DDE method remains

$$q' = \min\{p, q+1\}.$$

However, this option is not advisable from an implementation point of view in that the computation of its values in the current macro interval $[\xi_i, \xi_{i+1}]$ requires tracing back to the starting interval where the initial function $\phi'(t)$ is defined.

The method of steps for NNDEs, defined by option (4.4.4), can be summarized in the following pseudo-code.

Algorithm 4.4.2

1 *Locate all the principal discontinuities and the jump discontinuities of order $\leq p$*

$$\xi_1, \ldots, \xi_s (< t_f),$$

and set $\xi_0 = t_0$ and $\xi_{s+1} = t_f$.

2 *Solve the equation*

$$\begin{cases} z'(t) = f\Big(t, z(t), \phi\big(t-\tau(t)\big), \phi'\big(t-\sigma(t)\big)\Big), & \xi_0 \le t \le \xi_1, \\ z(\xi_0) = \phi(\xi_0), \end{cases}$$

by the discrete ODE method.

3 *For $i = 1$ to s:*

- *compute and store the continuous extension $\eta(t)$ for $t \in [\xi_{i-1}, \xi_i]$ by using data from the same macro interval $[\xi_{i-1}, \xi_i]$ only;*
- *compute and store the function $\lambda(t)$ for $t \in [\xi_{i-1}, \xi_i]$ by interpolating $f\big(t, \eta(t), u(t-\tau(t)), v(t-\sigma(t))\big)$, at points of the same macro interval $[\xi_{i-1}, \xi_i]$ only, where*

$$u(s) = \begin{cases} \phi(s) & \text{for } s \le t_0, \\ \eta(s) & \text{for } t_0 \le s \le \xi_{i-1}, \end{cases}$$

and

$$v(s) = \begin{cases} \phi'(s) & \text{for } s \le t_0, \\ \lambda(s) & \text{for } t_0 \le s \le \xi_{i-1}; \end{cases}$$

- *solve the equation*

$$\begin{cases} z'(t) = f\Big(t, z(t), u\big(t-\tau(t)\big), v\big(t-\sigma(t)\big)\Big), & \xi_i \le t \le \xi_{i+1}, \\ z(\xi_i) = \eta(\xi_i), \end{cases}$$

by the discrete ODE method.

4 *Plot the solution $\eta(t)$ in $[\xi_0, \xi_{s+1}]$.*

5 *End.*

4.4.2 *NDDEs with state dependent delay*

For the numerical solution of the NDDE (4.4.1), where hypothesis (H_1^*) is not assumed on the delays τ and σ, the $(n+1)$st step of the standard approach consists in solving the local problem

$$\begin{cases} w'_{n+1}(t) = f\Big(t, w_{n+1}(t), x\big(t-\tau(t, w_{n+1}(t))\big), z\big(t-\sigma(t, w_{n+1}(t))\big)\Big), \\ \hfill t_n \le t \le t_{n+1}, \\ w_{n+1}(t_n) = y_n, \end{cases} \tag{4.4.5}$$

where

$$x(s) = \begin{cases} \phi(s) & \text{for } s \le t_0, \\ \eta(s) & \text{for } t_0 \le s \le t_n, \\ w_{n+1}(s) & \text{for } t_n \le s \le t_{n+1}, \end{cases}$$

and

$$z(s) = \begin{cases} \phi'(s) & \text{for } s \le t_0, \\ \lambda(s) & \text{for } t_0 \le s \le t_n, \\ w'_{n+1}(s) & \text{for } t_n \le s \le t_{n+1}, \end{cases}$$

and the interpolants $\eta(s)$ and $\lambda(s)$ are assumed to be known for any $s \le t_n$. On the contrary, when $s = t - \tau(t, w_{n+1}(t)) > t_n$ and/or $s' = t - \sigma(t, w_{n+1}(t)) > t_n$, the interpolant $\eta(s)$ is implicitly defined by

$$\begin{aligned}\eta(t_n + \theta h_{n+1}) = \beta_{n,1}(\theta) y_n + \cdots + \beta_{n,1+i_n}(\theta) y_{n-i_n} \\ + h_{n+1} \Psi(y_n, \ldots, y_{n-i_n}; \theta, g_{\eta,\lambda}, \Delta'_n), \quad 0 \le \theta \le 1, \end{aligned} \quad (4.4.6)$$

with

$$g_{\eta,\lambda}(t, y) = f\Big(t, y, \eta(t - \tau(t, y)), \lambda(t - \sigma(t, y))\Big). \quad (4.4.7)$$

In any case, the continuous approximation $\lambda(t)$ of $y'(t)$ is given by either

$$\lambda(t) = \eta'(t) \quad (4.4.8)$$

or

$$\lambda(t) = \mathcal{P}\Big(f\Big(\cdot, \eta(\cdot), \eta(\,\cdot - \tau(\cdot, \eta(\cdot))), \lambda(\,\cdot - \sigma(\cdot, \eta(\cdot)))\Big)\Big)(t), \quad (4.4.9)$$

where, in each mesh interval $[t_k, t_{k+1}]$, $\mathcal{P}$ is an interpolation operator in a suitable polynomial space of degree possibly other than $\deg(\eta')$, with nodes in $[t_{k-i_k}, t_{k+1}]$.

For both options (4.4.8) and (4.4.9), a well-posedness result analogous to Theorem 4.3.1 may be proved for equations (4.4.6) and (4.4.7) under the condition that the local problem (4.4.5) possesses a unique solution $w_{n+1}(t)$.

In connection with questions (Q_1) and (Q_2) in Section 3.3, we assume that we have the machinery available to solve the possibly implicit equation arising from the numerical approximation of (4.4.5) and to locate the discontinuity point as required in the following theorem, which is analogous to Theorem 4.3.5 for DDEs. The proof is quite similar and, therefore, omitted.

Theorem 4.4.3 *Consider the state dependent NDDE (4.4.1), where $f(t, y, x, w)$ is C^p-continuous in $[t_0, t_f] \times \mathbb{R}^d \times \mathbb{R}^d \times \mathbb{R}^d$, the delays $\tau(t, y)$ and $\sigma(t, y)$ are C^p-continuous in $[t_0, t_f] \times \mathbb{R}^d$ and the initial function $\phi(t)$ is C^p-continuous. Moreover, assume that:*

- (h'_1) *The mesh $\Delta = \{t_0, t_1, \ldots, t_n, \ldots, t_N = t_f\}$ includes the discontinuity points $\xi_1 < \xi_2 < \ldots < \xi_s$ of order $\le p$.*
- (h'_2) *The ODE method (3.2.3) is consistent of order p, satisfies the stability condition (3.2.14) and, for $k > 1$, is restarted after each discontinuity point $\xi_i, i = 0, 1, \ldots, s$, by a method of order $\ge p - 1$.*
- (h'_3) *The interpolant (3.2.5), with $j_n = 0$, is consistent of uniform order q.*
- (h'_4) *For each n, the interval $[t_{n-i_n}, t_{n+1}]$ where the interpolation takes place is included in $[\xi_i, \xi_{i+1}]$ for some index $0 \le i \le s$.*
- (h'_5) *The interpolant (4.4.8) or (4.4.9) has uniform order r.*

Then the resulting DDE method has discrete global order and uniform global order $q' = \min\{p, q+1, r+1\}$*; that is*

$$\max_{1 \le n \le N} \|y(t_n) - y_n\| = O(h^{q'})$$

and

$$\max_{t_0 \le t \le t_f} \|y(t) - \eta(t)\| = O(h^{q'}),$$

where $h = \max_{1 \le n \le N} h_n$.

4.5 Other formulations of DDEs of neutral type

We saw in Chapter 3 that the class of NDDEs is more articulate and that the equations are often stated as, or may be transformed into, systems of DDAEs of the form (3.3.9), (3.3.10) or (3.3.11), where the solution $y(t)$ is coupled with the auxiliary function $z(t)$, $\Phi(t)$ or $\Psi(t)$, respectively.

Whereas the local problems for the step by step integration of (3.3.9) are DAEs for any choice of the stepsize, for non-vanishing delays and sufficiently small stepsizes equations (3.3.10) and (3.3.11) are basically ODEs and may be solved by some continuous ODE method. In fact, for sufficiently small stepsizes, the algebraic components $y(t)$ in (3.3.10) and $\Psi(t)$ in (3.3.11) may be computed separately by recursion after or before the relevant differential components $\Phi(t)$ and $y(t)$, respectively. On the contrary, the components $y(t)$ and $z(t)$ in (3.3.9) are strongly coupled and the system must be solved by some continuous DAE method.

When the neutral equation is given in the form (3.3.10), that is

$$\begin{cases} \Phi'(t) = K\Big(t, \Phi(t), y\big(t - \tau(t)\big)\Big), & t_0 \le t \le t_f, \\ y(t) = \Phi(t) + G\Big(t, y\big(t - \tau(t)\big)\Big), & \\ \Phi(t_0) = \phi(t_0) - G\Big(t_0, \phi\big(t_0 - \tau(t_0)\big)\Big), & \\ y(t) = \phi(t), \quad t \le t_0, & \end{cases} \tag{4.5.1}$$

the method of steps consists in solving, for $i = 0, \ldots, s$, the DDAEs

$$\begin{cases} W'(t) = K\Big(t, W(t), \eta\big(t - \tau(t)\big)\Big), \quad \xi_i \le t \le \xi_{i+1}, \\ z(t) = W(t) + G\Big(t, \eta\big(t - \tau(t)\big)\Big), \\ W(\xi_i) = \eta(\xi_i) - G\Big(\xi_i, \eta\big(\xi_i - \tau(\xi_i)\big)\Big), \\ \eta(t) = \phi(t), \quad t \le t_0. \end{cases}$$

The extension of the approximate continuous solutions $\eta(t)$ and $\lambda(t)$ with nodal approximations y_{n+1} and Φ_{n+1} of $z(t)$ and $W(t)$, respectively, to the current interval $[t_n, t_{n+1}]$ is obtained by solving numerically, for $w_{n+1}(t)$ and $\Lambda_{n+1}(t)$, the local problem

$$\begin{cases} \Lambda'_{n+1}(t) = K\Big(t, \Lambda_{n+1}(t), \eta\big(t-\tau(t)\big)\Big), & t_n \le t \le t_{n+1}, \\ w_{n+1}(t) = \Lambda_{n+1}(t) + G\Big(t, \eta\big(t-\tau(t)\big)\Big), & \\ \Lambda_{n+1}(t_n) = \Phi_n = y_n - G\Big(t_n, \eta\big(t_n - \tau(t_n)\big)\Big). & \end{cases} \tag{4.5.2}$$

Note that, for sufficiently small stepsizes, the differential equation for $\Lambda_{n+1}(t)$ is independent of $w_{n+1}(t)$ and can be solved before the algebraic recursion for $w_{n+1}(t)$ by the continuous ODE method which provides the approximation $\lambda(t)$. The approximation $\eta(t)$ is then extended in $[t_n, t_{n+1}]$ by the algebraic recursion

$$\eta(t) = \lambda(t) + G\Big(t, \eta\big(t-\tau(t)\big)\Big)$$

or, in order to avoid tracing back the solution to $\phi(t)$, by the local approximation

$$\eta(t) = \mathcal{P}\Big(\lambda(\cdot) + G\Big(\cdot, \eta\big(\cdot - \tau(\cdot)\big)\Big)\Big)(t), \tag{4.5.3}$$

where $\mathcal{P}$ is an interpolation operator in a suitable polynomial space and nodes in $[t_{n-i_n}, t_{n+j_n+1}]$.

The resulting method, using option (4.5.3), is summarized in the following pseudo-code.

Algorithm 4.5.1

1 *Locate all the principal discontinuities and the jump discontinuities of order* $\le p$

$$\xi_1, \ldots, \xi_s (< t_f),$$

and set $\xi_0 = t_0$ *and* $\xi_{s+1} = t_f$.

2 *Solve the equation*

$$\begin{cases} W'(t) = K\Big(t, W(t), \phi\big(t-\tau(t)\big)\Big), & \xi_0 \le t \le \xi_1, \\ W(\xi_0) = \phi(\xi_0) - G\Big(\xi_0, \phi\big(\xi_0 - \tau(\xi_0)\big)\Big), & \end{cases}$$

by the discrete ODE method and store the approximate nodal values Φ_n.

3 *Compute and store the corresponding nodal values*

$$y_n = \Phi_n + G\Big(t_n, \phi\big(t_n - \tau(t_n)\big)\Big).$$

4 *For* $i = 1$ *to* s:

- *compute and store the continuous extension* $\lambda(t)$ *of the values* Φ_n, *for* $t \in [\xi_{i-1}, \xi_i]$, *by using data from the same macro interval* $[\xi_{i-1}, \xi_i]$ *only;*

- *compute and store the continuous extension* $\eta(t)$ *of the nodal values* y_n*, for* $t \in [\xi_{i-1}, \xi_i]$*, by interpolating* $\lambda(t) + G\big(t, \eta(t-\tau(t))\big)$ *at points in the same macro interval* $[\xi_{i-1}, \xi_i]$ *only;*

- *solve the equation*

$$\begin{cases} W'(t) = K\Big(t, W(t), \eta\big(t-\tau(t)\big)\Big), & \xi_i \le t \le \xi_{i+1}, \\ W(\xi_i) = \lambda(\xi_i), \end{cases}$$

by the discrete ODE method and store the approximate nodal values Φ_n*;*
- *compute and store the corresponding nodal approximations of the solution*

$$y_n = \Phi_n + G\Big(t_n, \eta\big(t_n - \tau(t_n)\big)\Big).$$

5 *Plot the solution* $\eta(t)$ *in* $[\xi_0, \xi_{s+1}]$.
6 *End.*

The convergence of the method described above is stated in the following theorem proved by Torelli and Vermiglio [250]. The proof follows the same line as the proof of Theorem 4.1.2 and thus is omitted.

Theorem 4.5.2 *Consider the DDAE (4.5.1), where* $K(t,x,y)$ *is* C^p*-continuous in* $[t_0, t_f] \times \mathbb{R}^d \times \mathbb{R}^d$*,* $G(t,x)$ *is* C^p*-continuous in* $[t_0, t_f] \times \mathbb{R}^d$*, the delay* $\tau(t)$ *is* C^p*-continuous in* $[t_0, t_f]$ *and fulfills the hypothesis* (H_1)*, and the initial function* $\phi(t)$ *is* C^p*-continuous. Moreover, assume that:*

(h_1) *The mesh* $\Delta = \{t_0, t_1, \ldots, t_n, \ldots, t_N = t_f\}$ *includes all the principal discontinuity points and the discontinuity points of order* $\le p$ *lying in* $[t_0, t_f]$*, denoted by* $\xi_1, \ldots, \xi_s (< t_f)$.

(h_2) *The ODE method (3.2.3) is consistent of order* p*, satisfies the stability condition (3.2.14) and, for* $k > 1$*, is restarted after each discontinuity point* $\xi_i, i = 0, 1, \ldots, s$*, by a method of order* $\ge p - 1$.

(h_3) *The interpolant (3.2.5) that defines the continuous extension* $\lambda(t)$ *is consistent of uniform order* q.

$(h_3$ *bis)* *The interpolant (4.5.3) of the function* $\lambda(t) + G\big(t, \eta(t-\tau(t))\big)$ *that defines the continuous approximation* $\eta(t)$ *has uniform order* q.

(h_4) *For each* n*, the interval* $[t_{n-i_n}, t_{n+j_n+1}]$ *where the interpolation takes place is included in* $[\xi_i, \xi_{i+1}]$ *for some index* $0 \le i \le s$.

Then the resulting method of steps has discrete global order and uniform global order $q' = \min\{p, q+1\}$*; that is*

$$\max_{1 \le n \le N} \|y(t_n) - y_n\| = O(h^{q'})$$

and

$$\max_{t_0 \le t \le t_f} \|y(t) - \eta(t)\| = O(h^{q'}),$$

where $h = \max_{1 \le n \le N} h_n$.

Similarly, when the neutral equation has the form (3.3.11), that is

$$\begin{cases} y'(t) = F\big(t, y(t), \Psi(t)\big), & t_0 \le t \le t_f, \\ \Psi(t) = H\Big(t, y\big(t-\tau(t)\big), \Psi\big(t-\tau(t)\big)\Big), & t-\tau(t) > t_0, \\ y(t_0) = \phi(t_0), \\ \Psi(t) = G\Big(t, \phi\big(t-\tau(t)\big), \phi'\big(t-\tau(t)\big)\Big), & t-\tau(t) \le t_0, \end{cases} \tag{4.5.4}$$

to be solved for $\Psi(t)$ and $y(t)$, where

$$H(t, y, x) = G\Big(t, y, F\big(t-\tau(t), y, x\big)\Big),$$

the method of steps consists in solving, for $i = 0, \ldots, s$, the DDAEs

$$\begin{cases} z'(t) = F\big(t, z(t), W(t)\big), & \xi_i \le t \le \xi_{i+1}, \\ W(t) = H\Big(t, \eta\big(t-\tau(t)\big), \lambda\big(t-\tau(t)\big)\Big), & t-\tau(t) > t_0, \\ z(t_i) = \eta(t_i), \\ W(t) = G\Big(t, \phi\big(t-\tau(t)\big), \phi'\big(t-\tau(t)\big)\Big), & t-\tau(t) \le t_0. \end{cases}$$

The extension of the approximate continuous solutions $\eta(t)$ and $\lambda(t)$ with nodal approximations y_{n+1} and Ψ_{n+1} of $z(t)$ and $W(t)$, respectively, to the current interval $[t_n, t_{n+1}]$ is obtained by solving numerically, for $w_{n+1}(t)$ and $\Lambda_{n+1}(t)$, the local problem

$$\begin{cases} w'_{n+1}(t) = F\big(t, w_{n+1}(t), \Lambda_{n+1}(t)\big), & t_n \le t \le t_{n+1}, \\ \Lambda_{n+1}(t) = H\Big(t, \eta\big(t-\tau(t)\big), \lambda\big(t-\tau(t)\big)\Big), & t-\tau(t) > t_0, \\ w_{n+1}(t_n) = y_n, \\ \Lambda_{n+1}(t) = G\Big(t, \phi\big(t-\tau(t)\big), \phi'\big(t-\tau(t)\big)\Big), & t-\tau(t) \le t_0. \end{cases}$$

Here, we must first solve the algebraic equation for $\Lambda_{n+1}(t)$ by setting

$$\lambda(t) = \Lambda_{n+1}(t) = G\Big(t, \phi\big(t-\tau(t)\big), \phi'\big(t-\tau(t)\big)\Big) \quad \text{for } t-\tau(t) \le t_0,$$

and, for $t - \tau(t) > t_0$, by the exact recursion

$$\lambda(t) = \Lambda_{n+1}(t) = H\Big(t, \eta\big(t-\tau(t)\big), \lambda\big(t-\tau(t)\big)\Big),$$

or, in order to avoid tracing back to the initial macro interval, by the approximate recursion

$$\lambda(t) = H\Big(t, \eta\big(t-\tau(t)\big), \mathcal{P}\Big(\lambda\big(\cdot - \tau(\cdot)\big)\Big)(t)\Big), \tag{4.5.5}$$

where, in each mesh interval $[t_k, t_{k+1}]$, $\mathcal{P}$ is an interpolation operator in a suitable polynomial space with nodes in $[t_{k-i_k}, t_{k+j_k+1}]$.

Then we have to solve numerically the possibly modified ODE

$$\begin{cases} w'_{n+1}(t) = F\big(t, w_{n+1}(t), \lambda(t)\big), & t_n \le t \le t_{n+1}, \\ w_{n+1}(t_n) = y_n. \end{cases}$$

The resulting method, using option (4.5.5), is summarized in the following pseudo-code.

Algorithm 4.5.3

1 *Locate all the principal discontinuities and the jump discontinuities of order* $\le p$

$$\xi_1, \dots, \xi_s (< t_f),$$

and set $\xi_0 = t_0$ *and* $\xi_{s+1} = t_f$.

2 *Set* $\lambda(t) = G\Big(t, \phi\big(t - \tau(t)\big), \phi'\big(t - \tau(t)\big)\Big)$.

3 *Solve the equation*

$$\begin{cases} z'(t) = F\big(t, z(t), \lambda(t)\big), & \xi_0 \le t \le \xi_1, \\ z(\xi_0) = \phi(\xi_0), \end{cases}$$

by the discrete ODE method and store the approximate nodal values y_n.

4 *Compute and store the continuous extension* $\eta(t)$ *of the nodal values* y_n*, for* $t \in [\xi_0, \xi_1]$*, by using data from the current interval only.*

5 *For* $i = 1$ *to* s*:*

- *compute and store the polynomial* $\mu(t)$*, for* $t \in [\xi_{i-1}, \xi_i]$*, by interpolating* $\lambda(t)$ *at points in the current interval only;*
- *for* $t \in [\xi_i, \xi_{i+1}]$*, set* $\lambda(t) = H\big(t, \eta(t - \tau(t)), \mu(t - \tau(t))\big)$*;*
- *solve the equation*

$$\begin{cases} z'(t) = F\big(t, z(t), \lambda(t)\big), & \xi_i \le t \le \xi_{i+1}, \\ z(\xi_i) = \eta(\xi_i), \end{cases}$$

  by the discrete ODE method and store the approximate nodal values y_n*;*
- *compute and store the continuous extension* $\eta(t)$ *of the nodal values* y_n*, for* $t \in [\xi_i, \xi_{i+1}]$*, by using data from the current interval only.*

5 *Plot the solution* $\eta(t)$ *in* $[\xi_0, \xi_{s+1}]$.

6 *End.*

The convergence of the method described above is stated in the following theorem proved by Bellen, Guglielmi and Zennaro [30]. As with Theorem 4.5.2, the proof follows the same line as the proof of Theorem 4.1.2 and thus is omitted.

Theorem 4.5.4 *Consider the DDAE (4.5.4), where* $F(t, y, x)$ *and* $H(t, y, x)$ *are* C^p*-continuous in* $[t_0, t_f] \times \mathbb{R}^d \times \mathbb{R}^d$*, the delay* $\tau(t)$ *is* C^p*-continuous in* $[t_0, t_f]$ *and fulfills hypothesis* (H_1)*, and the initial function* $\phi(t)$ *is* C^p*-continuous. Moreover, assume that:*

(h_1) *The mesh* $\Delta = \{t_0, t_1, \dots, t_n, \dots, t_N = t_f\}$ *includes all the principal discontinuity points and the discontinuity points of order* $\le p$ *lying in* $[t_0, t_f]$ *denoted by* $\xi_1, \dots, \xi_s (< t_f)$.

(h_2) *The ODE method (3.2.3) is consistent of order* p, *satisfies the stability condition (3.2.14) and, for* $k > 1$, *is restarted after each discontinuity point* $\xi_i, i = 0, 1, \ldots, s$, *by a method of order* $\geq p - 1$.

(h_3) *The interpolant (3.2.5) is consistent of uniform order* q.

(h_3 *bis*) *The interpolant of the function* $\lambda(t - \tau(t))$ *in (4.5.5) has uniform order* q.

(h_4) *For each* n, *the interval* $[t_{n-i_n}, t_{n+j_n+1}]$ *where the interpolation takes place is included in* $[\xi_i, \xi_{i+1}]$ *for some index* $0 \leq i \leq s$.

Then the resulting method of steps has discrete global order and uniform global order $q' = \min\{p, q+1\}$; *that is*

$$\max_{1 \leq n \leq N} \|y(t_n) - y_n\| = O(h^{q'})$$

and

$$\max_{t_0 \leq t \leq t_f} \|y(t) - \eta(t)\| = O(h^{q'}),$$

where $h = \max_{1 \leq n \leq N} h_n$.

From a numerical point of view, the form (4.5.1) is preferable to the standard form (4.4.1), where the right-hand side depends on $y'(t)$, which is discontinuous, and locating the discontinuity points is necessary for the convergence of the method. Instead, the auxiliary function $\Phi(t)$ is of class $C^1[t_0, t_f]$ with Lipschitz continuous derivatives at every discontinuity point. This allows us to implement low order methods without locating the discontinuity points (see Section 4.5.1).

Unlike (4.5.1), the form (4.5.4) does not present any gain in smoothness with respect to the standard form of the neutral equation, since the auxiliary function $\Psi(t)$ is discontinuous at the discontinuity points as well as the derivative of the solution $y'(t)$. Consequently, from a numerical point of view, the location of discontinuities remains important even for low order methods (see again Section 4.5.1).

4.5.1 *Integration of NDDEs with free stepsize*

Since for NDDEs smoothing of the solution does not take place, the right-hand side of (4.4.2) is piecewise continuous and jump discontinuities of order 0 propagate throughout the integration interval $[t_0, t_f]$. Therefore, on the simple basis of Theorem 3.2.8, any convergent numerical method cannot help including all the discontinuities into the mesh. Otherwise the right-hand side of equation (4.4.2) would be discontinuous in any mesh interval $[t_n, t_{n+1}]$, including the discontinuity point, and thus Theorem 3.2.8 could not be used.

On the other hand, it is not difficult to see that, for the ODE $y' = g(t, y)$, any method of order $p \geq 1$ has a local error $O(h)$ on any mesh interval $[t_n, t_{n+1}]$ where the right-hand-side function $g(t, y)$ has a jump discontinuity in t, and a local error $O(h^{p+1})$ elsewhere. Therefore, assuming that the number of discontinuities is finite in $[t_0, t_f]$, we can conclude that any method of order $p \geq 1$ performs to order 1 with free stepsizes.

Dealing with equations of the form (4.5.1) has the advantage that the function $y(t)$ in the right-hand side of the differential equation that defines the auxiliary function $\Phi(t)$ is Lipschitz continuous. On this basis, Torelli and Vermiglio [250] observed that, in view of Remark 3.2.11, any method of order $p \geq 1$ used for the solution of the differential equation (4.5.2) has a local error which is $O(h^2)$ in any mesh interval including a discontinuity point, and $O(h^{p+1})$ elsewhere. Assuming again that the number of discontinuities is finite in $[t_0, t_f]$, they conclude that, for the approximation of the auxiliary function $\Phi(t)$ in (4.5.1), a method of order $p = 1, 2$ performs to order p with free stepsizes, not necessarily including the discontinuity points into the mesh Δ.

Nevertheless, there are still some difficulties with discontinuities for the approximation $\eta(t)$ of the solution $y(t)$. In fact, if the interpolation (4.5.3) is used for the computation of $G\big(t, \eta(t-\tau(t))\big)$ in the recursive formula in (4.5.2), then the lack of smoothness of $y(t)$ at the discontinuity points causes a local loss of accuracy in the interpolation process down to $O(h)$ whenever the delay argument $t - \tau(t)$ crosses one such point. In any case, this unpleasant phenomenon takes place only locally at those mesh points close to the discontinuity points of order ≥ 2, unless all the discontinuities are included in the mesh Δ, in which case no loss of accuracy occurs. The global order of accuracy of $\eta(t)$ at all the other mesh points remains $O(h^2)$. On the other hand, if interpolation is not used and $\eta(t)$ is computed by tracing it back to the initial function $\phi(t)$, not even these local losses of accuracy occur.

Finally, we observe that, for equations of the form (4.5.4), the right-hand-side function of the differential equation that defines the solution $y(t)$ is again discontinuous in any mesh interval $[t_n, t_{n+1}]$ including the discontinuity point. Consequently, once again, the order of convergence would always reduce to 1. In fact, the auxiliary function $\Psi(t)$ defined by the algebraic recursive formula (4.5.4) is discontinuous at the discontinuity points. For this reason, as with the standard formulation (4.4.2), the inclusion of all the discontinuity points is crucial for the accuracy of the method.

Moreover, as happens for the formulation (4.5.1), the use of interpolation for the numerical solution of the algebraic recursive formula causes a local loss of accuracy at those mesh points close to each discontinuity point of order ≥ 2, unless all the discontinuities are included in the mesh Δ. This time, the function to interpolate is discontinuous and, therefore, even convergence is lost, whereas the global error for the approximant of $\Psi(t)$ at all the other mesh points is $O(h)$. Again, if interpolation is not used, then local losses of accuracy do not occur.

All the various occurrences mentioned above are illustrated by the following example.

Example 4.5.5 Let us consider the equation

$$\begin{cases} y'(t) = y(t) + y'(t-1), & 0 \leq t \leq 3, \\ y(t) = 1, \quad t \leq 0, \end{cases} \tag{4.5.6}$$

whose solution is the continuous function

TABLE 4.1. Numerical results for equation (4.5.6) with meshes including $\xi_1 = 1$ and $\xi_2 = 2$.

N	$\max_n \lvert y_n - y(t_n) \rvert$
30	$2.48 \cdot 10^{-1}$
300	$2.69 \cdot 10^{-3}$
3000	$2.71 \cdot 10^{-5}$

TABLE 4.2. Numerical results for equation (4.5.6) without including $\xi_1 = 1$ and $\xi_2 = 2$ in the mesh.

N	$\max_n \lvert y_n - y(t_n) \rvert$
31	$4.01 \cdot 10^{-1}$
301	$2.35 \cdot 10^{-2}$
3001	$2.16 \cdot 10^{-3}$

$$y(t) = \begin{cases} \mathrm{e}^t & \text{for } 0 \le t \le 1, \\ \mathrm{e}^t + (t-1)\mathrm{e}^{t-1} & \text{for } 1 \le t \le 2, \\ \mathrm{e}^t + \mathrm{e}^{t-1} + \frac{1}{2}(t-2)(t+2\mathrm{e})\mathrm{e}^{t-2} & \text{for } 2 \le t \le 3. \end{cases}$$

We apply the Heun method with linear interpolation of order 2 and with constant stepsize $h = 3/N$, where N is the number of steps used to cover $[0, 3]$. In Table 4.1 we show the results obtained by including the discontinuity points $\xi_1 = 1$ and $\xi_2 = 2$ in the mesh, whereas in Table 4.2 we show the results obtained by excluding them from the mesh. The first point $t_0 = 0$ and the final point $t_f = 3$ (which is a discontinuity point) are always included in the mesh.

We denote the continuous numerical solution by $\eta(t)$ and the nodal values by $y_n = \eta(t_n)$.

As expected, in Table 4.1 the maximum uniform global order is $O(h^2)$, whereas in Table 4.2 it is $O(h)$.

Equation (4.5.6) may be rewritten in the form (4.5.1), that is

$$\begin{cases} \Phi'(t) = \Phi(t) + y(t-1), & 0 \le t \le 3, \\ y(t) = \Phi(t) + y(t-1), \\ \Phi(0) = 0, \\ y(t) = 1, & t \le 0. \end{cases} \tag{4.5.7}$$

TABLE 4.3. Numerical results for equation (4.5.7) with meshes including $\xi_1 = 1$ and $\xi_2 = 2$.

N	$\max_n \lvert y_n - y(t_n)\rvert$	$\max_n \lvert \Phi_n - \Phi(t_n)\rvert$
30	$1.90 \cdot 10^{-1}$	$1.57 \cdot 10^{-1}$
300	$2.05 \cdot 10^{-3}$	$1.70 \cdot 10^{-3}$
3000	$2.07 \cdot 10^{-5}$	$1.71 \cdot 10^{-5}$

TABLE 4.4. Numerical results for equation (4.5.7) without including $\xi_1 = 1$ and $\xi_2 = 2$ in the mesh and with interpolation.

N	$\max_n \lvert y_n - y(t_n)\rvert$	$\max_{n\neq n_1,n_2} \lvert y_n - y(t_n)\rvert$	$\max_n \lvert \Phi_n - \Phi(t_n)\rvert$
31	$9.46 \cdot 10^{-2}$	$9.46 \cdot 10^{-2}$	$1.04 \cdot 10^{-1}$
301	$1.69 \cdot 10^{-3}$	$1.30 \cdot 10^{-3}$	$1.22 \cdot 10^{-3}$
3001	$2.83 \cdot 10^{-4}$	$1.34 \cdot 10^{-5}$	$1.25 \cdot 10^{-5}$

The auxiliary function turns out to be

$$\Phi(t) = \begin{cases} \mathrm{e}^t - 1 & \text{for } 0 \le t \le 1, \\ \mathrm{e}^t + (t-2)\mathrm{e}^{t-1} & \text{for } 1 \le t \le 2, \\ \mathrm{e}^t + \frac{1}{2}(t-2)(t+2\mathrm{e}-2)\mathrm{e}^{t-2} & \text{for } 2 \le t \le 3. \end{cases}$$

We again apply the Heun method with linear interpolation of order 2 and with constant stepsize $h = 3/N$ to approximate the auxiliary function $\Phi(t)$ and we use the linear interpolation of nodal values to compute the numerical continuous solution $\eta(t)$ from the recursive formula that defines $y(t)$. We use the same meshes as before and report the relevant results in Tables 4.3 and 4.4. We call n_1 and n_2 the indices of the two mesh points at which the loss of accuracy occurs, due to the interpolation process. The mesh point t_{n_1} is close to $\xi_2 = 2$, whereas $t_{n_2} = t_N = 3$. This is illustrated in Figure 4.7. We denote the nodal approximations of the auxiliary function by Φ_n.

Moreover, in Table 4.5 we show the results obtained by avoiding interpolation of the recursive formula and not including the discontinuities $\xi_1 = 1$ and $\xi_2 = 2$ in the mesh.

All the results are consistent with our expectations.

Finally, we rewrite equation (4.5.6) in the form (4.5.4), that is

TABLE 4.5. Numerical results for equation (4.5.7) without including $\xi_1 = 1$ and $\xi_2 = 2$ in the mesh and without interpolation.

N	$\max_n \|y_n - y(t_n)\|$	$\max_n \|\Phi_n - \Phi(t_n)\|$
31	$1.18 \cdot 10^{-1}$	$1.11 \cdot 10^{-1}$
301	$1.39 \cdot 10^{-3}$	$1.29 \cdot 10^{-3}$
3001	$1.43 \cdot 10^{-5}$	$1.31 \cdot 10^{-5}$

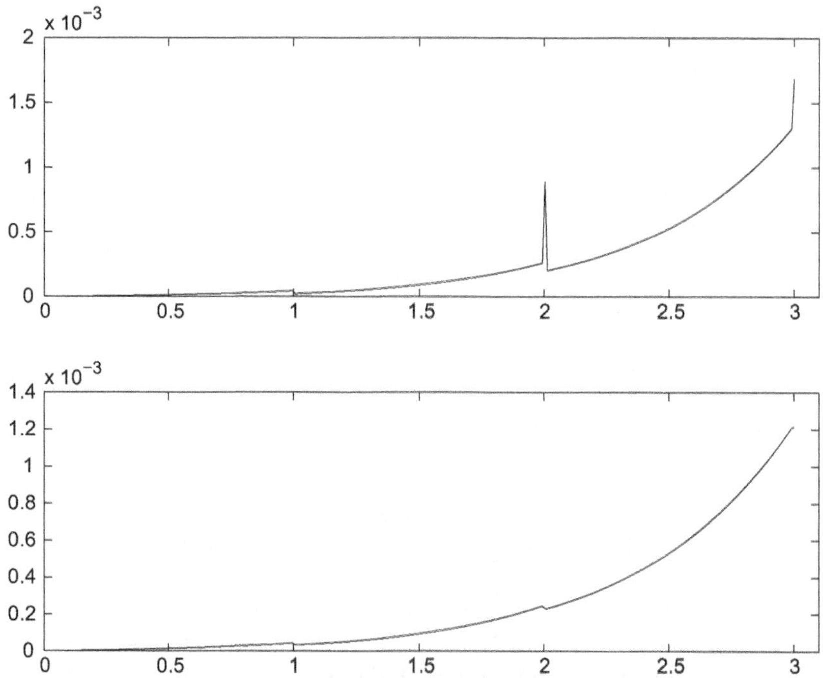

FIG. 4.7. Errors for the approximations of $y(t)$ (top) and $\Phi(t)$ (bottom) with $N = 301$ for equation (4.5.7) by using interpolation (see Table 4.4).

$$\begin{cases} y'(t) = y(t) + \Psi(t), & 0 \le t \le 3, \\ \Psi(t) = y(t-1) + \Psi(t-1), & t > 1, \\ y(0) = 1, \\ \Psi(t) = 0, & 0 \le t \le 1. \end{cases} \tag{4.5.8}$$

The auxiliary function turns out to be

TABLE 4.6. Numerical results for equation (4.5.8) with meshes including $\xi_1 = 1$ and $\xi_2 = 2$.

N	$\max_n \lvert y_n - y(t_n)\rvert$	$\max_n \lvert \Psi_n - \Psi(t_n)\rvert$
30	$1.90 \cdot 10^{-1}$	$3.74 \cdot 10^{-2}$
300	$2.05 \cdot 10^{-3}$	$4.02 \cdot 10^{-4}$
3000	$2.07 \cdot 10^{-5}$	$4.05 \cdot 10^{-6}$

TABLE 4.7. Numerical results for equation (4.5.8) without including $\xi_1 = 1$ and $\xi_2 = 2$ in the mesh and with interpolation.

N	$\max_n \lvert y_n - y(t_n)\rvert$	$\max_n \lvert \Psi_n - \Psi(t_n)\rvert$	$\max_{n\neq n_1,n_2} \lvert \Psi_n - \Psi(t_n)\rvert$
31	$3.18 \cdot 10^{-1}$	$4.67 \cdot 10^{-1}$	$4.74 \cdot 10^{-2}$
301	$2.26 \cdot 10^{-2}$	$4.47 \cdot 10^{-1}$	$4.60 \cdot 10^{-3}$
3001	$2.15 \cdot 10^{-3}$	$4.45 \cdot 10^{-1}$	$4.54 \cdot 10^{-4}$

$$\Psi(t) = \begin{cases} 0 \quad \text{for } 0 \le t \le 1, \\ \mathrm{e}^{t-1} \quad \text{for } 1 \le t \le 2, \\ \mathrm{e}^{t-1} + (t-1)\mathrm{e}^{t-2} \quad \text{for } 2 \le t \le 3. \end{cases}$$

The application of the Heun method with linear interpolation of order 2 and with constant stepsize $h = 3/N$, together with the use of linear interpolation of the nodal values for the recursive formula which defines the auxiliary function $\Psi(t)$, yields the results shown in Tables 4.6 and 4.7. Moreover, in Table 4.8 we show the results obtained by avoiding interpolation for the recursive formula and not including the discontinuities $\xi_1 = 1$ and $\xi_2 = 2$ in the mesh. We denote the nodal approximations of the auxiliary function by Ψ_n.

Again, all the results are consistent with our expectations. In particular, the approximations Ψ_n do not even converge. $\diamond$

4.6 One-step versus multistep methods

In the general approach to the numerical integration of DDEs based on discrete ODE methods endowed with some continuous extension, the two basic options, one-step and multistep methods, have been widely exploited also in combination. All theorems in this chapter refer to both of them and provide the basis for

TABLE 4.8. Numerical results for equation (4.5.8) without including $\xi_1 = 1$ and $\xi_2 = 2$ in the mesh and without interpolation.

N	$\max_n \|y_n - y(t_n)\|$	$\max_n \|\Psi_n - \Psi(t_n)\|$
31	$2.61 \cdot 10^{-1}$	$4.98 \cdot 10^{-2}$
301	$1.39 \cdot 10^{-2}$	$4.63 \cdot 10^{-3}$
3001	$1.25 \cdot 10^{-3}$	$4.54 \cdot 10^{-4}$

designing methods that converge to the solution of DDEs and NDDEs with arbitrarily high order. However, the two options are intrinsically different and it is worth comparing them from the implementative and computational points of view.

In this regard, in 1969 Zwerkina [289] analyzed the two classes of RK and LM methods on the basis of their storage capacity, number of computations and accuracy, and concluded that, to date, the LM were more desirable than RK methods. Along this line are the papers by Zwerkina [286, 287, 288], Linz [187], Tavernini [243], Kemper [169], Cryer [72], Chartes and Stepleman [68], Bock and Schlöder [46], Jackiewicz [152, 154, 157, 158], Jackiewicz and Lo [160], de Gee [101], Baker and Ford [9] and Bocharov, Marchuk and Romanyukha [45].

In the meantime, the growing interest in RK methods, due to their superiority in treating stiff problems and performing stepsize control, has been reflected in development of DDE methods as well. In particular, in the 1970s and early 1980s interest was focused on RK methods combined with Hermite–Birkhoff multistep interpolation. Along this line are the papers by Oberle and Pesh [210], Neves [204], Neves and Feldstein [205], Oppelstrup [212] and the early paper by Stetter [238] who supported the use of Hermite interpolation for the approximation of the retarded argument.

Besides the usual arguments in the analysis of LM versus RK methods for ODEs, new and severe restrictions arise in the implementation of multistep strategies for DDEs and NDDEs when the solution is not smooth because of primary discontinuities. By all the foregoing convergence theorems it turns out that, for a one-step method to achieve its own accuracy order, the solution must be regular on the current interval $[t_n, t_{n+1}]$, and this may be achieved by including the discontinuity points into the mesh Δ. For k-step methods, $k > 1$, regularity of the solution is instead required in the larger interval $[t_{n-k+1}, t_{n+1}]$ between the minimum and maximum nodes involved by the method. This means that, as required by conditions (h_2) or (h_2') throughout the chapter, after each discontinuity point ξ_i the multistep method must be restarted by the usual starting procedure designed for the initial point. Similar inconveniences arise in using multistep interpolation for approximating the delay term $y(\alpha(t))$, which must be regular in the whole interpolation interval $[t_{n-i_n}, t_{n+j_n+1}]$ as well. Therefore, near discontinuities, the choice of the interpolation interval must fluctuate so as to fulfill the hypotheses (h_4) or (h_4'). This, besides being an implementation complication, contradicts the usual request that the interpolant should be evaluated at points as centered as possible with respect to the nodes.

Additional restrictions are imposed by vanishing delays in the proximity of a vanishing delay point where overlapping occurs. Conditions (h_3') on the interpolant were, in fact, requested in order to prevent the deviating argument $t-\tau(t)$ from taking values that are advanced with respect to the current integration interval.

These arguments and the development of *continuous RK* methods became decisive in making them preferable to LM methods and to Hermite–Birkhoff multistep interpolation for solving DDEs and NDDEs. Nevertheless, there are still cases where multistep strategies may be competitive, in particular non-stiff equations with a smooth solution such as the pantograph equation (2.1.1).

Besides some early papers by Feldstein [95], Birkhoff [43], Fox [99], Grafton [105], Cryer and Tavernini [73] and others on low order RK methods, and the papers by Tavernini [242, 244] who developed the general theory of higher order one-step methods for Volterra functional differential equations and for retarded differential equations with vanishing delays, most of the methods proposed since the late 1980s are actually based on RK methods and their continuous extensions.

4.7 Some additional references and codes

The existing literature is very large and several papers are quoted throughout the book. We conclude this chapter by giving a list of additional references, where particular methods that have been proposed for the numerical solution of DDEs can be found. Some of them also contain pseudo-codes and/or computer programs. Bellen and Zennaro [36] presented some algorithms based on a predictor–corrector version of the one-step collocation method at Gaussian points for non-stiff DDEs with time dependent delays. Baker and Paul [12] investigated the convergence of a class of predictor–corrector RK methods for equations with vanishing delay. Corwin and Thompson [70] analyzed the error estimation and stepsize control by continuously embedded RK–Sarafyan methods. Enright and Hayashi [89] analyzed the defect control for continuous RK methods. Paul [216, 219] discussed various phenomena to be tested when designing a robust code based on RK methods. Jackiewicz and Lo [161] used a fully implicit RK method for neutral functional differential equations. Other software-oriented papers include Al Mutib [1], Weiner and Strehmel [266], Willé and Baker [271], Butcher [63], Baker and Paul [11], Baker, Butcher and Paul [8], Hayashi [88] and Shampine and Thompson [229].

Finally, we quote the set of DDEs collected by Paul [217] and the set of very challenging stiff problems collected and made available by Guglielmi and Hairer (see references in [112]) for testing and comparing numerical algorithms.

Although for decades engineers had been solving DDEs by combining an integrator for ODEs with an interpolation formula to reconstruct the history of the solution from values stored in an array, Tavernini [245, 246] was the first to produce and make a package available for its automated numerical solution, named CTMS (continuous-time model simulation), that was based on a firm mathematical foundation.

Since the pioneering code by Tavernini, many DDE codes have been produced in the last decades for the numerical integration of functional differential equations. Some of them are devoted to the restricted class of DDEs with constant delays, others to larger classes of equations including state dependent delays, neutral equations, singularly perturbed problems, delay differential algebraic equations of any index and also Volterra integro-differential equations. Among them we quote:

- ARCHI by Paul [218]. This code is based on the RK formulae of Dormand and Prince extended by fifth-order Hermite interpolation. It tracks the discontinuities and includes them as mesh points.
- CTMS by Tavernini [245]. This is a package providing a unified numerical tool for DDEs, hysteresis problems, impulsive systems and hybrid systems. It is based on LM methods extended by Hermite interpolation. The package is described quite extensively in the book by Tavernini [246].
- DDE23 by Shampine and Thompson [230]. This uses the imbedded RK pair of order 2(3) with continuous extension based on cubic Hermite interpolation for solving DDEs with many constant delays.
- DDE-STRIDE by Butcher [63] (see also Baker, Butcher and Paul [8]). This is an adaptation of STRIDE to stiff DDEs with several state dependent delays and NDDEs.
- DDVERK by Enright and Hayashi [89]. This is based on the explicit RK method and performs stepsize control by computing the residual. The code is suitable for non-stiff DDEs with possible vanishing state dependent delays.
- DELH by Weiner and Strehmel [266]. This uses a Rosenbrock type method for stiff/non-stiff DDEs with a single constant delay.
- DESOL by Willé and Baker [271]. This is based on LM methods with particular attention to the presence of discontinuities.
- DIFSUB-DDE by Bocharov, Marchuk and Romanyukha [45]. This is an extension of DIFSUB (the code by Gear based on BDF) for DDEs with constant delays.
- DMRODE (Algorithm No. 497 of ACM) by Neves [204, 205]. This is based on Merson's variable step method of order 4 with two-point Hermite interpolation for solving DDEs with state dependent delays and Volterra integro-differential equations.
- DRLAG6 by Corwin, Sarafyan and Thompson [69]. This uses methods based on continuously embedded sixth-order methods for state dependent DDEs.
- RADAR5 by Guglielmi and Hairer [112]. This uses the Radau collocation method for solving stiff DDEs including differential algebraic equations and

neutral problems. It allows steps larger than delays and performs stepsize control by combining the discrete and continuous local error estimation.

- RETARD by Hairer, Wanner and Norsett [117]. This is a simple modification of the code DOPRI5 for DDEs with non-vanishing delay.
- SNDDELM by Jackiewicz and Lo [160]. This is a variable step, variable order code for the numerical integration of neutral DDEs with state dependent delay based on Adams methods of order ranging from 1 to 12.

For the sake of completeness, we quote the Time-Delay System Toolbox, a general purpose collection of algorithms developed by Kim et al. [170] for time-domain analysis of linear systems and numerical simulation of more general linear and non-linear systems with discrete and distributed, possibly time dependent, bounded delays. The numerical integration is based on a particular technique presented in Kim and Pimenov [171].

5

CONTINUOUS RUNGE–KUTTA METHODS FOR ODES

In the 1980s and early 1990s more and more attention was paid to the study of continuous extensions of numerical methods. The aim was to enhance and refine the ODE solvers and to make them suitable for providing dense output and for treating more general problems, such as systems with driving equations, discontinuous equations, waveform relaxation, Volterra functional differential equations and, in particular, delay differential equations (DDEs).

Owing to the theoretical and practical disadvantages of multistep methods and multistep interpolants for DDEs pointed out in Chapter 4, throughout the rest of the book we shall consider only one-step methods and, in particular, Runge–Kutta (RK) methods.

It is well known that, given a mesh $\Delta = \{t_0, t_1, \ldots, t_n, \ldots, t_N = t_f\}$, a ν-stage RK method for the numerical solution of the ODE

$$\begin{cases} y'(t) = g\big(t, y(t)\big), & t_0 \le t \le t_f, \\ y(t_0) = y_0, \end{cases} \tag{5.0.1}$$

has the form (in the so-called Y *notation*)

$$Y_{n+1}^i = y_n + h_{n+1} \sum_{j=1}^{\nu} a_{ij} g(t_{n+1}^j, Y_{n+1}^j), \quad i = 1, \ldots, \nu, \tag{5.0.2}$$

$$y_{n+1} = y_n + h_{n+1} \sum_{i=1}^{\nu} b_i g(t_{n+1}^i, Y_{n+1}^i), \tag{5.0.3}$$

where $t_{n+1}^i = t_n + c_i h_{n+1}, c_i = \sum_{j=1}^{\nu} a_{ij}, i = 1, \ldots, \nu$, $h_{n+1} = t_{n+1} - t_n$ and ν is referred to as the number of *stages*. The b_i's are called *weights* of the quadrature formula (5.0.3) and the c_i's are called *abscissae* and, for most of common methods, they belong to $[0,1]$. Since the RK method (5.0.2), (5.0.3) is characterized by the weights b_i and the matrix coefficients $A = [a_{ij}]_{i,j=1}^{\nu}$, it will be denoted by (A, b). It is worth observing that in many papers and books the RK formulae are written in an equivalent different form, the so-called K *notation*. So the RK method (5.0.2), (5.0.3) takes the form

$$K_{n+1}^i = g\Big(t_{n+1}^i, y_n + h_{n+1} \sum_{j=1}^{\nu} a_{ij} K_{n+1}^j\Big), \quad i = 1, \ldots, \nu,$$

$$y_{n+1} = y_n + h_{n+1} \sum_{i=1}^{\nu} b_i K_{n+1}^i.$$

Note that the K notation is obtained from formulae (5.0.2), (5.0.3) by setting

$$K_{n+1}^i = g(t_{n+1}^i, Y_{n+1}^i), \quad i = 1, \ldots, \nu.$$

Although in developing and implementing RK methods for ODEs the two notations are basically equivalent, in the application of RK methods to DDEs it will often be preferable to adopt the K notation.

The computational complexity of the method is mainly determined by the number of stages and by the form of the coefficient matrix A. It is well known that when the matrix A is lower triangular with zero diagonal elements, the method is called *explicit* and the computational cost is lower, whereas when the matrix A is full, the method is called *implicit* and the computational cost is higher.

In this chapter we investigate deeply the one-step interpolants of the RK method (5.0.2), (5.0.3). However, in the literature there are papers where, in view of high order accuracy and stability needs, some RK methods have been extended with multistep interpolants (see, for example, Tsitouras and Papageorgiou [252], In 't Hout [133], Vermiglio [258] and Vermiglio and Zennaro [259]).

5.1 Continuous extensions of RK methods

According to Definition 3.2.2, the one-step interpolants of the RK method (5.0.2), (5.0.3) are constructed step by step by making use of information from the underlying mesh interval $[t_n, t_{n+1}]$ only, possibly by including some additional stages, that is by some extra evaluations of the right-hand-side function $g(t, y)$ in (5.0.1).

Interpolants constructed using no extra stages are called *interpolants of the first class* and the resulting continuous extension $\eta(t)$ is defined, in each subinterval of the mesh Δ, by a one-step continuous quadrature rule of the form

$$\eta(t_n + \theta h_{n+1}) = y_n + h_{n+1} \sum_{i=1}^{\nu} b_i(\theta) g(t_{n+1}^i, Y_{n+1}^i), \quad 0 \le \theta \le 1, \tag{5.1.1}$$

or, in the K notation,

$$\eta(t_n + \theta h_{n+1}) = y_n + h_{n+1} \sum_{i=1}^{\nu} b_i(\theta) K_{n+1}^i, \quad 0 \le \theta \le 1,$$

where the $b_i(\theta)$'s are polynomials of suitable degree $\le \delta$ satisfying

$$b_i(0) = 0 \text{ and } b_i(1) = b_i, \quad i = 1, \ldots, \nu, \tag{5.1.2}$$

so as to fulfil the continuity conditions (3.2.6).

Interpolants constructed by means of additional stages are called *interpolants of the second class* and the continuous extension is given by

$$\eta(t_n + \theta h_{n+1}) = y_n + h_{n+1} \sum_{i=1}^{s} b_i(\theta) g(t_{n+1}^i, Y_{n+1}^i), \quad 0 \le \theta \le 1, \tag{5.1.3}$$

or, in the K notation, by

$$\eta(t_n + \theta h_{n+1}) = y_n + h_{n+1} \sum_{i=1}^{s} b_i(\theta) K_{n+1}^i, \quad 0 \le \theta \le 1, \tag{5.1.4}$$

where the $b_i(\theta)$'s are again polynomials of suitable degree $\le \delta$ satisfying the continuity conditions

$$b_i(0) = 0, \quad i = 1, \ldots, s;$$

$$b_i(1) = b_i, \quad i = 1, \ldots, \nu; \qquad b_i(1) = 0, \quad i = \nu + 1, \ldots, s. \tag{5.1.5}$$

The additional $s - \nu$ stages are given by

$$Y_{n+1}^i = y_n + h_{n+1} \sum_{j=1}^{s} a_{ij} g(t_{n+1}^j, Y_{n+1}^j), \quad i = \nu + 1, \ldots, s, \tag{5.1.6}$$

or, in the K notation, by

$$K_{n+1}^i = g\Big(t_{n+1}^i, y_n + h_{n+1} \sum_{j=1}^{s} a_{ij} K_{n+1}^j\Big), \quad i = \nu + 1, \ldots, s,$$

so that the original coefficient matrix $A = [a_{ij}]_{i,j=1}^{\nu}$ is embedded into the block lower triangular matrix

$$A' = \begin{bmatrix} A & 0 \\ [a_{ij}]_{i=\nu+1,j=1}^{s,\nu} & [a_{ij}]_{i,j=\nu+1}^{s} \end{bmatrix}. \tag{5.1.7}$$

The overall *continuous Runge–Kutta* (CRK) methods (5.0.2), (5.0.3), (5.1.1) and (5.0.2), (5.0.3), (5.1.6), (5.1.3), denoted by $\big(A, b(\theta)\big)$ and $\big(A', b(\theta)\big)$, respectively, are the continuous extensions of the RK method (A, b) and δ will be referred to as the *degree of the interpolant.* In contrast, the method (A, b) will be called the *underlying (discrete)* RK method.

It is worth remarking that, in general, $\eta(t_n + c_i h_{n+1}) \ne Y_{n+1}^i$. Nevertheless, equality holds for every right-hand side $g(t, y)$ whenever $b_i(c_j) = a_{ji}$, as appears evident by comparing (5.1.3) and (5.1.6). So we have

$$\{\eta(t_n + c_i h_{n+1}) = Y_{n+1}^i \quad \forall i\} \Longleftrightarrow \{b_i(c_j) = a_{ji} \quad \forall i, j\}. \tag{5.1.8}$$

An interpolant, either of the first or second class, determines a matrix B, whose elements are $b_{ij} = b_j(c_i)$.

Definition 5.1.1 *A CRK method is called* natural *if $A = B$ ($A' = B$).*

In connection with the general form of discrete and continuous ODE methods (3.2.3) and (3.2.5), the increment functions Φ and Ψ of the discrete and continuous RK methods take the form

$$\Phi(y_n; g, t_n, t_{n+1}) = \sum_{i=1}^{\nu} b_i g(t_{n+1}^i, Y_{n+1}^i)$$

and

$$\Psi(y_n; \theta, g, t_n, t_{n+1}) = \sum_{i=1}^{s} b_i(\theta) g(t_{n+1}^i, Y_{n+1}^i), \tag{5.1.9}$$

respectively.

Proposition 5.1.2 *The increment functions Φ and Ψ of an RK method (5.0.2), (5.0.3) and its interpolants (5.1.1) and (5.1.3) satisfy the continuity conditions (3.2.4) and (3.2.7) with respect to the functional argument g.*

Proof. It is sufficient to prove the result for the interpolant (5.1.3), since it includes the interpolant (5.1.1) for $s = \nu$ and the discrete formula (5.0.3) for $\theta = 1$.

Thus consider a right-hand-side function $g(t, y)$ and an arbitrary perturbation $\tilde{g}(t, y)$. Let Y_{n+1}^i and $\tilde{Y}_{n+1}^i$, $i = 1, \ldots, s$, be the corresponding stage values obtained by applying formulae (5.0.2), (5.1.6). Then (5.1.9) yields

$$\begin{aligned}
&\|\Psi(y_n; \theta, \tilde{g}, t_n, t_{n+1}) - \Psi(y_n; \theta, g, t_n, t_{n+1})\| \\
&\quad \leq \beta \max_{1 \leq i \leq s} \|\tilde{g}(t_{n+1}^i, \tilde{Y}_{n+1}^i) - g(t_{n+1}^i, Y_{n+1}^i)\|,
\end{aligned} \tag{5.1.10}$$

where

$$\beta = \max_{0 \leq \theta \leq 1} \sum_{i=1}^{s} |b_i(\theta)|.$$

Now, if L_g is the Lipschitz constant of g, then we have

$$\begin{aligned}
&\|\tilde{g}(t_{n+1}^i, \tilde{Y}_{n+1}^i) - g(t_{n+1}^i, Y_{n+1}^i)\| \\
&\quad \leq \|\tilde{g}(t_{n+1}^i, \tilde{Y}_{n+1}^i) - g(t_{n+1}^i, \tilde{Y}_{n+1}^i)\| + \|g(t_{n+1}^i, \tilde{Y}_{n+1}^i) - g(t_{n+1}^i, Y_{n+1}^i)\| \\
&\quad \leq \|\tilde{g}(t_{n+1}^i, \tilde{Y}_{n+1}^i) - g(t_{n+1}^i, \tilde{Y}_{n+1}^i)\| \\
&\qquad + h_{n+1} L_g \sum_{j=1}^{s} |a_{ij}| \cdot \|\tilde{g}(t_{n+1}^j, \tilde{Y}_{n+1}^j) - g(t_{n+1}^j, Y_{n+1}^j)\|.
\end{aligned}$$

Therefore, if $h_{n+1} \leq h_g$ for some $h_g < \left(L_g \max_{1 \leq i \leq s} \sum_{j=1}^{s} |a_{ij}|\right)^{-1}$, then by (5.1.10) we can easily conclude that

$$\|\Psi(y_n; \theta, \tilde{g}, t_n, t_{n+1}) - \Psi(y_n; \theta, g, t_n, t_{n+1})\|$$

$$
\begin{aligned}
&\leq \delta_g \max_{1\leq i\leq s} \|\tilde{g}(t_{n+1}^i, \tilde{Y}_{n+1}^i) - g(t_{n+1}^i, \tilde{Y}_{n+1}^i)\| \\
&\leq \delta_g \sup_{t_n\leq t\leq t_{n+1},\ y\in\mathbb{R}^d} \|\tilde{g}(t,y) - g(t,y)\|,
\end{aligned} \tag{5.1.11}
$$

where

$$
\delta_g = \frac{\beta}{1 - h_g L_g \max\limits_{1\leq i\leq s} \sum_{j=1}^{s} |a_{ij}|}.
$$

Clearly, both h_g and δ_g can be chosen as continuous functions of L_g. In particular, δ_g can even be assumed to be constant, whereas h_g can be chosen to be proportional to L_g^{-1}. ■

According to Definition 3.2.5 for the discrete and continuous order of a general ODE method, we have the following definition for the class of RK methods and their interpolants.

Definition 5.1.3 *We say that the RK method (5.0.2), (5.0.3) is* consistent of order *(or, equivalently, has* order*) p if $p \geq 1$ is the largest integer such that, for all C^p-continuous right-hand-side functions $g(t,y)$ in (5.0.1) and for all mesh points, we have that*

$$
\|z_{n+1}(t_{n+1}) - y_{n+1}\| = O(h_{n+1}^{p+1}),
$$

uniformly with respect to y_n^* in any bounded subset of $\mathbb{R}^d$ and to $n = 0, \ldots, N-1$, where $z_{n+1}(t)$ is the local solution to the local problem

$$
\begin{cases} z_{n+1}'(t) = g\big(t, z_{n+1}(t)\big), & t_n \leq t \leq t_{n+1}, \\ z_{n+1}(t_n) = y_n^*. \end{cases} \tag{5.1.12}
$$

We say that the interpolant (5.1.1) or (5.1.3) is consistent of uniform order *(or, equivalently, has* uniform order*) q if $q \geq 1$ is the largest integer such that, for all C^q-continuous right-hand side functions $g(t,y)$ and for all mesh points, we have that*

$$
\max_{t_n\leq t\leq t_{n+1}} \|z_{n+1}(t) - \eta(t)\| = O(h_{n+1}^{q+1}).
$$

The convergence results on general discrete and continuous ODE methods have been stated and proved in Theorem 3.2.8. The following theorem summarizes those results for discrete and continuous RK methods.

Theorem 5.1.4 *Let the RK method (5.0.2), (5.0.3) be consistent of order p and let the right-hand-side function $g(t,y)$ in (5.0.1) be C^p-continuous. Then the method is convergent of order (or, equivalently, has global order) p on any bounded interval $[t_0, t_f]$, that is*

$$
\max_{1\leq n\leq N} \|y(t_n) - y_n\| = O(h^p), \tag{5.1.13}
$$

where $h = \max_{1\leq n\leq N} h_n$.

If the interpolant (5.1.1) or (5.1.3) has uniform order q, then the CRK method (5.0.2), (5.0.3), (5.1.1) or (5.0.2), (5.0.3), (5.1.6), (5.1.3) is uniformly convergent of order (or, equivalently, has uniform global order) $q' = \min\{p, q+1\}$; that is

$$\max_{t_0 \le t \le t_f} \|y(t) - \eta(t)\| = O(h^{q'}). \tag{5.1.14}$$

As we did for general continuous ODE methods (see Remarks 3.2.7 and 3.2.9), we shall often refer to the order of consistency and to the order of convergence of the RK method (5.0.2), (5.0.3) as the *discrete order* and the *discrete global order* of the CRK method (5.0.2), (5.0.3), (5.1.1) or (5.0.2), (5.0.3), (5.1.6), (5.1.3).

The following theorem provides additional results on the derivatives of the continuous extension.

Theorem 5.1.5 *If, in addition to the hypotheses of Theorem 5.1.4, the interpolant is a piecewise polynomial of degree $\delta \ge q$ and the right-hand-side function $g(t, y)$ in (5.0.1) is $C^{\max\{\delta,p\}}$-continuous, then the following convergence, boundedness and unboundedness estimates hold for the derivatives of the global error function:*

$$\max_{t_0 \le t \le t_f} \|y^{(j)}(t) - \eta^{(j)}(t)\| = O(h^{q+1-j}), \quad j = 1, \dots, \delta, \tag{5.1.15}$$

where the derivatives of $\eta(t)$ at the mesh points are meant in the left/right sense.

Proof. Let us first prove, by induction, that, on each interval $[t_n, t_{n+1}]$, the derivatives of the local error function satisfy the following estimates:

$$\max_{t_n \le t \le t_{n+1}} \|z_{n+1}^{(j)}(t) - \eta^{(j)}(t)\| = O(h_{n+1}^{q+1-j}), \quad j = 0, \dots, \delta, \tag{5.1.16}$$

uniformly with respect to $n = 0, \dots, N-1$. By the definition of the interpolant of uniform order q, the assertion is true for $j = 0$. Assume (5.1.16) for $j (< \delta)$ and prove it for $j+1$. To this end, let $p(t)$ be the Taylor expansion of degree $\delta - j$ of the function $z_{n+1}^{(j)}(t)$ at the point t_n. Since $z_{n+1}(t)$ is $C^{\delta+1}$-continuous, it holds that

$$\max_{t_n \le t \le t_{n+1}} \|z_{n+1}^{(j)}(t) - p(t)\| = O(h_{n+1}^{\delta+1-j}) \tag{5.1.17}$$

and, consequently,

$$\max_{t_n \le t \le t_{n+1}} \|z_{n+1}^{(j+1)}(t) - p'(t)\| = O(h_{n+1}^{\delta-j}), \tag{5.1.18}$$

$p'(t)$ being the Taylor expansion of $z_{n+1}^{(j+1)}(t)$. All these estimates are uniform with respect to $n = 0, \dots, N-1$. By inductive hypothesis and by (5.1.17), we have

$$\max_{t_n \le t \le t_{n+1}} \|\eta^{(j)}(t) - p(t)\| \le \max_{t_n \le t \le t_{n+1}} \|\eta^{(j)}(t) - z_{n+1}^{(j)}(t)\|$$

$$+ \max_{t_n \le t \le t_{n+1}} \|z_{n+1}^{(j)}(t) - p(t)\| = O(h_{n+1}^{q+1-j}),$$

uniformly with respect to $n = 0, \ldots, N-1$.

It can be easily seen that the derivative operator $\frac{\mathrm{d}}{\mathrm{d}t}$, acting on the space of polynomials $\Pi_{\delta-j}([t_n, t_{n+1}])$, is bounded and that its norm is $C_{\delta-j}/h_{n+1}$ for a suitable constant $C_{\delta-j}$, independent of the stepsize h_{n+1}. Therefore,

$$\max_{t_n \le t \le t_{n+1}} \|\eta^{(j+1)}(t) - p'(t)\| \le \frac{C_{\delta-j}}{h_{n+1}} \max_{t_n \le t \le t_{n+1}} \|\eta^{(j)}(t) - p(t)\| = O(h_{n+1}^{q-j}).$$

Then, by (5.1.18),

$$\max_{t_n \le t \le t_{n+1}} \|\eta^{(j+1)}(t) - z_{n+1}^{(j+1)}(t)\| \le \max_{t_n \le t \le t_{n+1}} \|\eta^{(j+1)}(t) - p'(t)\|$$
$$+ \max_{t_n \le t \le t_{n+1}} \|p'(t) - z_{n+1}^{(j+1)}(t)\| = O(h_{n+1}^{q-j})$$

and, hence, (5.1.16) is proved for $j+1$ too.

To conclude the proof, it is sufficient to observe that

$$\max_{t_n \le t \le t_{n+1}} \|y^{(j)}(t) - \eta^{(j)}(t)\| \le \max_{t_n \le t \le t_{n+1}} \|y^{(j)}(t) - z_{n+1}^{(j)}(t)\|$$
$$+ \max_{t_n \le t \le t_{n+1}} \|z_{n+1}^{(j)}(t) - \eta^{(j)}(t)\|,$$

where, by the C^δ-continuity of $g(t,y)$ and by the global order p of the discrete method (see (5.1.13)), there exists $M > 0$, independent of $n = 0, \ldots, N-1$, such that

$$\max_{t_n \le t \le t_{n+1}} \|y^{(j)}(t) - z_{n+1}^{(j)}(t)\| \le M \max_{t_n \le t \le t_{n+1}} \|y(t) - z_{n+1}(t)\| = O(h^p).$$

Since $p \ge q$, (5.1.15) is proved. ■

The estimates (5.1.14) and (5.1.15) show that the first derivative retains the global uniform order of the interpolant if and only if the interpolant has the maximum attainable uniform order p. It is also evident that, in order to get the uniform order q, the interpolant must be of degree $\delta \ge q$. On the other hand, polynomials of degree $\delta > q$ are unnecessary, as shown by the following theorem.

Theorem 5.1.6 *Assume that the RK method (5.0.2), (5.0.3) has a continuous extension $\eta(t)$ of order q and degree $\delta > q$. Then there exists another continuous extension $\tilde{\eta}(t)$ of order q whose degree is also q.*

Proof. Let

$$\eta(t_n + \theta h_{n+1}) = y_n + h_{n+1} \sum_{i=1}^{\nu} b_i(\theta) K_{n+1}^i \tag{5.1.19}$$

be an interpolant of order q and degree $\delta > q$ of the RK method (A, b) in the interval $[t_n, t_{n+1}]$, and let $\tilde{\eta}(t)$ be the polynomial of degree q which interpolates

$\eta(t)$ at the nodes $t_n + \tilde{c}_j h_{n+1}$, $j = 1, \ldots, q+1$, for suitable abscissae $\tilde{c}_1 = 0 < \tilde{c}_2 < \cdots < \tilde{c}_{q+1} = 1$.

The Lagrange interpolation formula with polynomial coefficients

$$\ell_j(\theta) = \prod_{k=1,k\neq j}^{q+1} \frac{\theta - \tilde{c}_k}{\tilde{c}_j - \tilde{c}_k}$$

applied to the function (5.1.19) yields

$$\begin{aligned}
\tilde{\eta}(t_n + \theta h_{n+1}) &= \sum_{j=1}^{q+1} \ell_j(\theta)\eta(t_n + \tilde{c}_j h_{n+1}) \\
&= \sum_{j=1}^{q+1} \ell_j(\theta)\left(y_n + h_{n+1}\sum_{i=1}^{\nu} b_i(\tilde{c}_j)K^i_{n+1}\right) \\
&= y_n + h_{n+1}\sum_{i=1}^{\nu}\sum_{j=1}^{q+1} \ell_j(\theta)b_i(\tilde{c}_j)K^i_{n+1} \\
&= y_n + h_{n+1}\sum_{i=1}^{\nu} \tilde{b}_i(\theta)K^i_{n+1},
\end{aligned}$$

where $\tilde{b}_i(\theta) = \sum_{j=1}^{q+1} \ell_j(\theta)b_i(\tilde{c}_j)$ is a polynomial of degree q.

Since $\tilde{b}_i(0) = b_i(0)$ and $\tilde{b}_i(1) = b_i(1)$, $i = 1, \ldots, \nu$, the polynomial $\tilde{\eta}(t_n + \theta h_{n+1})$ is an interpolant of the discrete method (A, b) of degree q. In order to prove that the interpolant $\tilde{\eta}(t)$ has uniform order q, it is sufficient to prove that it agrees with $\eta(t)$ to the same order q. This is immediately obtained from (5.1.16) for $j = q+1$ and from the componentwise error interpolation formula

$$\begin{aligned}
&\tilde{\eta}_k(t_n + \theta h_{n+1}) - \eta_k(t_n + \theta h_{n+1}) \\
&\quad = \frac{\eta_k^{(q+1)}(t_n + \xi_k(\theta)h_{n+1})}{(q+1)!}(\theta - \tilde{c}_1)\ldots(\theta - \tilde{c}_{q+1})h_{n+1}^{q+1},
\end{aligned}$$

where $\xi_k(\theta) \in (0, 1)$, $k = 1, \ldots, d$. ■

Remark 5.1.7 *By Theorems 5.1.5 and 5.1.6 we may observe that, not only is the employment of interpolants of degree higher than q unnecessary, but interpolants of degree $\delta > q+1$ are even dangerous in that the derivatives of order k, with $q+2 \leq k \leq \delta$, may diverge as $h \to 0$. For these reasons we shall assume that continuous extensions of order q will be always made by interpolants of degree $\delta = q$.*

In the context of interpolants of the first and the second class, two questions are to be answered.

Question 1 What is the maximum uniform order an RK method of order p can achieve by means of an interpolant of the first class?

Question 2 For an RK method of discrete order p and continuous extension of uniform order $q < p$, what is the number of extra stages necessary to raise the uniform order to $p - 1$ or even p?

We shall answer both questions in the forthcoming sections. So far, we can give the following upper bound to the uniform order of an interpolant (for both classes).

Theorem 5.1.8 *Assume that the RK method (5.0.2), (5.0.3) has a continuous extension $\eta(t)$ given by (5.1.3). Then its uniform order q cannot exceed s^*, the number of distinct abscissae of the extended RK method represented by (5.1.7).*

Proof. The result is obvious after observing that formula (5.1.3) is a continuous quadrature rule based exactly on s^* distinct abscissae. ■

5.2 Interpolants of the first class

A general analysis of the uniform order for the continuous extension (5.1.1) is based on the property that, for any $0 < \theta \leq 1$, it can be viewed as the discrete method $(\frac{A}{\theta}, \frac{b(\theta)}{\theta})$ with stepsize θh_{n+1}. So, we immediately get the uniform order conditions for the polynomials $b_i(\theta)$ from the well-known order conditions of the RK methods. The conditions up to order $p = 4$ are shown in Table 5.1.

In order to answer Question 1, each method has to be analyzed individually by checking the order conditions. In general, we can give only a partial answer by means of the following theorem, the proof of which does not directly involve the order conditions.

Theorem 5.2.1 *Every RK method (5.0.2), (5.0.3) of order $p \geq 1$ has a continuous extension $\eta(t)$ of order (and degree) $q = 1, \ldots, \lfloor \frac{p+1}{2} \rfloor$.*

Proof. Let $1 \leq q \leq p$. Then define the polynomials $b_i(\theta)$, $i = 1, \ldots, \nu$, by imposing the following conditions:

$$b_i(0) = 0 \quad \text{and} \quad \int_0^1 \theta^r b_i'(\theta) \mathrm{d}\theta = b_i c_i^r, \ r = 0, \ldots, q-1. \tag{5.2.1}$$

Since the above integral condition for $r = 0$ is equivalent to $b_i(1) = b_i$, such polynomials satisfy the continuity conditions (5.1.2) and, hence, determine an interpolant $\eta(t)$ of the form (5.1.1).

Now consider the following system of ODEs, enlarged from the local problem (5.1.12):

$$\begin{cases} z_{n+1}'(t) = g\big(t, z_{n+1}(t)\big), \\ z_{n+1}(t_n) = y_n^*, \\ w_1'(t) = (t - t_n) g\big(t, z_{n+1}(t)\big), \\ w_1(t_n) = 0, \\ \quad \vdots \\ w_{q-1}'(t) = (t - t_n)^{q-1} g\big(t, z_{n+1}(t)\big), \\ w_{q-1}(t_n) = 0. \end{cases}$$

TABLE 5.1. Order conditions for continuous RK methods.

Order	Conditions
1	$\sum_{i=1}^{\nu} b_i(\theta) = \theta$
2	$\sum_{i=1}^{\nu} b_i(\theta)c_i = \frac{1}{2}\theta^2$
3	$\sum_{i=1}^{\nu} b_i(\theta)c_i^2 = \frac{1}{3}\theta^3$ $\sum_{i,j=1}^{\nu} b_i(\theta)a_{ij}c_j = \frac{1}{6}\theta^3$
4	$\sum_{i=1}^{\nu} b_i(\theta)c_i^3 = \frac{1}{4}\theta^4$ $\sum_{i,j=1}^{\nu} b_i(\theta)c_i a_{ij}c_j = \frac{1}{8}\theta^4$ $\sum_{i,j=1}^{\nu} b_i(\theta)a_{ij}c_j^2 = \frac{1}{12}\theta^4$ $\sum_{i,j,k=1}^{\nu} b_i(\theta)a_{ij}a_{jk}c_k = \frac{1}{24}\theta^4$

Let $(y_{n+1}, \bar{w}_1, \dots, \bar{w}_{q-1})$ be the numerical solution of the above system at the mesh point $t_{n+1} = t_n + h_{n+1}$. By (5.2.1) for $r = 0$ we have

$$y_{n+1} = y_n + h_{n+1} \sum_{i=1}^{\nu} b_i g(t_{n+1}^i, Y_{n+1}^i) = y_n + h_{n+1} \int_0^1 \eta'(t_n + \theta h_{n+1}) \mathrm{d}\theta,$$

whereas, as is immediately seen, by (5.2.1) for $r = 1, \dots, q-1$ we have

$$\bar{w}_r = h_{n+1}^{r+1} \sum_{i=1}^{\nu} b_i c_i^r g(t_{n+1}^i, Y_{n+1}^i) = h_{n+1}^{r+1} \int_0^1 \theta^r \eta'(t_n + \theta h_{n+1}) \mathrm{d}\theta.$$

Thus, since

$$z_{n+1}(t_{n+1}) = y_n + \int_{t_n}^{t_{n+1}} z_{n+1}'(t) \mathrm{d}t = y_n + h_{n+1} \int_0^1 z_{n+1}'(t_n + \theta h_{n+1}) \mathrm{d}\theta$$

and

$$w_r(t_{n+1}) = \int_{t_n}^{t_{n+1}} (t - t_n)^r z_{n+1}'(t) \mathrm{d}t = h_{n+1}^{r+1} \int_0^1 \theta^r z_{n+1}'(t_n + \theta h_{n+1}) \mathrm{d}\theta,$$

and since the RK method has order p, it holds that

$$\left\| \int_0^1 \theta^r [z'_{n+1}(t_n + \theta h_{n+1}) - \eta'(t_n + \theta h_{n+1})] \mathrm{d}\theta \right\| = O(h_{n+1}^{p-r}) \qquad (5.2.2)$$

for $r = 0, \dots, q-1$.

It is known that, for the error function, the expansion

$$z_{n+1}(t_n + \theta h_{n+1}) - \eta(t_n + \theta h_{n+1}) = \sum_{k=1}^{p} e_k(\theta) h_{n+1}^k + O(h_{n+1}^{p+1})$$

holds, where $e_k(\theta)$, $k = 1, \dots, p$, are d-vector polynomials independent of the stepsize h_{n+1} such that $e_k(0) = 0$. For $k \le q$ their degree is $\le q$, whereas for $k > q$ their degree is $\le k$. Consequently, by differentiating and proceeding as in the proof of Theorem 5.1.5, we obtain

$$z'_{n+1}(t_n + \theta h_{n+1}) - \eta'(t_n + \theta h_{n+1}) = \sum_{k=1}^{p} e'_k(\theta) h_{n+1}^{k-1} + O(h_{n+1}^{p}).$$

Therefore, (5.2.2) implies

$$\int_0^1 \theta^r e'_k(\theta) \mathrm{d}\theta = 0 \qquad \text{for } r + k \le p.$$

Since $r = 0, \dots, q-1$ and $e'_k(\theta)$ is a polynomial of degree $\le q-1$ for all $k \le q$, it follows that

$$e_k(\theta) \equiv 0 \qquad \text{for } 1 \le k \le \tilde{q} = \min\{q, p-q+1\}.$$

This means that the interpolant $\eta(t)$ has uniform order $\tilde{q}$.

In order to complete the proof, it is sufficient to observe that $\tilde{q}$ is maximized by the choice $q = \lfloor \frac{p+1}{2} \rfloor$ and that, for any $q \le \lfloor \frac{p+1}{2} \rfloor$, we have $\tilde{q} = q$. ■

Theorem 5.2.2 *If an RK method (5.0.2), (5.0.3) has a continuous extension $\eta(t)$ of order (and degree) $q \ge 2$, then it also has another continuous extension $\tilde{\eta}(t)$ of order (and degree) $\tilde{q}$ for each $\tilde{q} \le q-1$.*

Proof. Consider the polynomial $\tilde{\eta}(t)$ of degree $\tilde{q}$ that interpolates $\eta(t)$ at the nodes $t_n + \tilde{c}_j h_{n+1}$, $j = 1, \dots, \tilde{q}+1$, for suitable abscissae $\tilde{c}_1 = 0 < \tilde{c}_2 < \cdots < \tilde{c}_{\tilde{q}+1} = 1$.

Then the proof proceeds as for Theorem 5.1.6. ■

In conclusion, we can answer Question 1 by saying that, in general, only interpolants up to order $\lfloor \frac{p+1}{2} \rfloor$ are assured to exist. On the other hand, it might well be that the maximum uniform order reachable by means of interpolants of the first class is actually $> \lfloor \frac{p+1}{2} \rfloor$ and, possibly, even $= p$.

Definition 5.2.3 *We say that an RK method (5.0.2), (5.0.3) of discrete order p is* superconvergent *if the maximum uniform order q reachable by means of interpolants of the first class is $\le p-1$.*

In other words, *superconvergence* is attained at the end-point of the step-interval with respect to the maximum uniform accuracy order q. Of course, it might well be that the interpolant attains a higher order $p' > q$, not necessarily equal to the discrete order p, also at some additional points inside the step-interval. They will be called *inner superconvergence points.*

5.2.1 *Collocation methods*

A particular class of continuous RK methods that has been studied extensively are the *one-step collocation* methods. They were originally introduced by Guillou and Soulé [113] and developed since the early 1970s by Wright [274], Hulme [142], de Boor and Swartz [48] and many others for both initial value and boundary value problems. For an extensive list of references and the historical development of collocation methods, see Brunner [55]. However, the interest in piecewise collocation was due to the simplicity in determining the order of convergence and superconvergence via the non-linear variation-of-constants formula and to the optimal stability properties as discrete methods, rather than to its intrinsically continuous nature.

The one-step collocation method can be defined as follows. Chose ν distinct abscissae $c_1, \ldots, c_\nu \in [0,1]$ and, in each mesh interval $[t_n, t_{n+1}]$, compute the polynomial $\eta(t)$ of degree $\leq \nu$ satisfying

$$\eta'(t_{n+1}^i) = g\big(t_{n+1}^i, \eta(t_{n+1}^i)\big), \quad i = 1, \ldots, \nu, \quad \eta(t_n) = y_n.$$

It is easy to check that such methods can be rewritten as a continuous implicit RK method (5.0.2), (5.1.1), where

$$a_{ij} = \int_0^{c_i} \ell_j(\xi) \mathrm{d}\xi, \quad i,j = 1, \ldots, \nu,$$

$$b_i(\theta) = \int_0^{\theta} \ell_i(\xi) \mathrm{d}\xi, \quad i = 1, \ldots, \nu,$$

$\ell_i(\xi)$ being the Lagrange polynomial coefficient $\prod_{k=1,k\neq i}^{\nu} \frac{\xi - c_k}{c_i - c_k}$.

In particular, we have $b_i(c_j) = a_{ji}$ and, therefore, any collocation method is a natural CRK method.

It is known that, for any choice of the abscissae $c_1, \ldots, c_\nu \in [0,1]$, the collocation method has order $p \geq \nu$ and the uniform order of the interpolant (5.1.1) is $q = \nu$ (see, for example, Nørsett and Wanner [209]). Consequently, by Theorem 5.1.4, the collocation method is a continuous RK method of global uniform order $q' = \nu$ (if $p = \nu$) or $q' = \nu + 1$ (if $p > \nu$). In this sense the collocation method is optimal in that it achieves the maximum attainable uniform order for the given number of stages. In particular, if the abscissae are the shifted roots of the Legendre orthogonal polynomial of degree ν, then the method has order $p = 2\nu$. This is the most famous example of superconvergence.

The following are some examples of superconvergent collocation methods.

- Gaussian methods: discrete order $p = 2\nu$, uniform order $q = \nu$, global uniform order $q' = \min\{p, q+1\}$;

$\nu = 1$ (*midpoint rule*): $p = 2, \quad q = 1, \quad q' = 2;$

$$\begin{array}{c|c} \frac{1}{2} & \frac{1}{2} \\ \hline & 1 \end{array} \qquad b_1(\theta) = \theta;$$

$\nu = 2$ (*Hammer–Hollingsworth* method): $p = 4, \quad q = 2, \quad q' = 3;$

$$\begin{array}{c|cc} \frac{1}{2} - \frac{\sqrt{3}}{6} & \frac{1}{4} & \frac{1}{4} - \frac{\sqrt{3}}{6} \\ \frac{1}{2} + \frac{\sqrt{3}}{6} & \frac{1}{4} + \frac{\sqrt{3}}{6} & \frac{1}{4} \\ \hline & \frac{1}{2} & \frac{1}{2} \end{array} \qquad \begin{array}{l} b_1(\theta) = -\frac{\sqrt{3}}{2}\theta(\theta - 1 - \frac{\sqrt{3}}{3}), \\ b_2(\theta) = \frac{\sqrt{3}}{2}\theta(\theta - 1 + \frac{\sqrt{3}}{3}). \end{array}$$

- Radau IIA methods: discrete order $p = 2\nu - 1$, uniform order $q = \nu$, global uniform order $q' = \min\{p, q+1\}$;

$\nu = 1$ (*backward Euler* method): $p = 1, \quad q = 1, \quad q' = 1;$

$$\begin{array}{c|c} 1 & 1 \\ \hline & 1 \end{array} \qquad b_1(\theta) = \theta;$$

$\nu = 2$ (*Ehle* method): $p = 3, \quad q = 2, \quad q' = 3;$

$$\begin{array}{c|cc} \frac{1}{3} & \frac{5}{12} & -\frac{1}{12} \\ 1 & \frac{3}{4} & \frac{1}{4} \\ \hline & \frac{3}{4} & \frac{1}{4} \end{array} \qquad \begin{array}{l} b_1(\theta) = -\frac{3}{4}\theta(\theta - 2), \\ b_2(\theta) = \frac{3}{4}\theta(\theta - \frac{2}{3}). \end{array}$$

- Lobatto IIIA methods: discrete order $p = 2\nu - 2$, uniform order $q = \nu$, global uniform order $q' = \min\{p, q+1\}$;

$\nu = 2$ (*trapezoidal rule*): $p = 2, \quad q = 2, \quad q' = 2;$

$$\begin{array}{c|cc} 0 & 0 & 0 \\ 1 & \frac{1}{2} & \frac{1}{2} \\ \hline & \frac{1}{2} & \frac{1}{2} \end{array} \qquad \begin{array}{l} b_1(\theta) = -\frac{1}{2}\theta(\theta - 2), \\ b_2(\theta) = \frac{1}{2}\theta^2; \end{array}$$

$\nu = 3$ (*Ehle* method): $p = 4, \quad q = 3, \quad q' = 4;$

$$\begin{array}{c|ccc} 0 & 0 & 0 & 0 \\ \frac{1}{2} & \frac{5}{24} & \frac{1}{3} & -\frac{1}{24} \\ 1 & \frac{1}{6} & \frac{2}{3} & \frac{1}{6} \\ \hline & \frac{1}{6} & \frac{2}{3} & \frac{1}{6} \end{array} \qquad \begin{aligned} b_1(\theta) &= 2\theta(\tfrac{1}{3}\theta^2 - \tfrac{3}{4}\theta + \tfrac{1}{2}), \\ b_2(\theta) &= -4\theta^2(\tfrac{1}{3}\theta - \tfrac{1}{2}), \\ b_3(\theta) &= 2\theta^2(\tfrac{1}{3}\theta - \tfrac{1}{4}). \end{aligned}$$

5.2.2 *Natural continuous extensions*

In this section we introduce particular kinds of interpolants of the first class that were developed by Zennaro [279]. They are characterized by an additional *asymptotic orthogonality condition*, which allows preservation of the superconvergence of RK methods as applied to particular functional differential equations including DDEs and NDDEs.

Definition 5.2.4 *We say that the interpolant $\eta(t)$ in (5.1.1) of order (and degree) q is a* natural continuous extension *(NCE) of the RK method (5.0.2), (5.0.3) of order p if the polynomials $b_i(\theta), i = 1, \ldots, \nu$, are such that $\eta(t)$ satisfies the additional asymptotic orthogonality condition*

$$\left\| \int_{t_n}^{t_{n+1}} G(t)[z'_{n+1}(t) - \eta'(t)]\mathrm{d}t \right\| = O(h_{n+1}^{p+1}) \tag{5.2.3}$$

for every sufficiently smooth matrix-valued function G, uniformly with respect to $n = 0, \ldots, N-1$, where $z_{n+1}(t)$ is the solution to the local problem (5.1.12).

It is easy to see that the NCEs necessarily also satisfy

$$\left\| \int_{t_n}^{t_{n+1}} G(t)[z_{n+1}(t) - \eta(t)]\mathrm{d}t \right\| = O(h_{n+1}^{p+1}) \tag{5.2.4}$$

for every sufficiently smooth matrix-valued function G.

Moreover, by (5.2.3) and (5.2.4), we easily get

$$\max_{0 \le n \le N-1} \left\| \int_{t_n}^{t_{n+1}} G(x)[y^{(j)}(x) - \eta^{(j)}(x)]\mathrm{d}x \right\| = O(h^{p+1}), \quad j = 0, 1, \tag{5.2.5}$$

for every sufficiently smooth matrix-valued function G, where $h = \max_{1 \le n \le N} h_n$.

With respect to the results of Theorems 5.1.4 and 5.1.5, condition (5.2.5) states the additional global convergence property of NCEs. It is worth remarking that, for any one-step collocation method, the collocation polynomial is an NCE of degree $q = \nu$ (see [279]).

With regard to the existence of NCEs and their attainable uniform order, the following theorems hold.

Theorem 5.2.5 *If the interpolant $\eta(t)$ in (5.1.1) of order (and degree) q is an NCE of the RK method (5.0.2), (5.0.3) of order p, then $q \ge \lfloor \frac{p+1}{2} \rfloor$.*

Proof. Let us consider the following matrix-valued function

$$G_r(t) = (t - t_n)^r I = \theta^r h_{n+1}^r I, \quad r = 0, \ldots, p-1.$$

Since (5.2.3) holds, we get

$$\left\| \int_0^1 \theta^r [z'_{n+1}(t_n + \theta h_{n+1}) - \eta'(t_n + \theta h_{n+1})] \mathrm{d}\theta \right\| = O(h_{n+1}^{p-r})$$

for $r = 0, \ldots, p-1$. Therefore, as in the proof of Theorem 5.2.1, we obtain

$$\int_0^1 \theta^r e'_k(\theta) \mathrm{d}\theta = 0 \quad \text{for } r + k \leq p. \tag{5.2.6}$$

For every $k = 1, \ldots, \lfloor \frac{p+1}{2} \rfloor$, (5.2.6) holds for all $r = 0, \ldots, p - \lfloor \frac{p+1}{2} \rfloor$. Since $e'_k(\theta)$ is a polynomial of degree $\leq \lfloor \frac{p+1}{2} \rfloor - 1 \leq p - \lfloor \frac{p+1}{2} \rfloor$ for all $k \leq \lfloor \frac{p+1}{2} \rfloor$, it follows that

$$e_k(\theta) \equiv 0, \quad k = 1, \ldots, \left\lfloor \frac{p+1}{2} \right\rfloor.$$

Hence, the uniform order q of the interpolant $\eta(t)$ must be $\geq \lfloor \frac{p+1}{2} \rfloor$. ■

Theorem 5.2.6 *For every RK method (5.0.2), (5.0.3) of order p, formula (5.2.1) furnishes an NCE $\eta(t)$ of minimal order (and degree) $q = \lfloor \frac{p+1}{2} \rfloor$.*

Proof. Formula (5.2.1) defines an interpolant $\eta(t)$ of order $q = \lfloor \frac{p+1}{2} \rfloor$ (see the proof of Theorem 5.2.1). It remains to prove condition (5.2.3). To this end consider a smooth matrix-valued function G and its Taylor expansion

$$G(t_n + \theta h_{n+1}) = G(t_n) + \theta h_{n+1} G'(t_n) + \cdots + \frac{(\theta h_{n+1})^{q-1}}{(q-1)!} G^{(q-1)}(t_n) + O(h_{n+1}^q).$$

Then we have

$$\begin{aligned}
&\left\| \int_{t_n}^{t_{n+1}} G(t)[z'_{n+1}(t) - \eta'(t)] \mathrm{d}t \right\| \\
&= h_{n+1} \left\| \int_0^1 G(t_n + \theta h_{n+1})[z'_{n+1}(t_n + \theta h_{n+1}) - \eta'(t_n + \theta h_{n+1})] \mathrm{d}\theta \right\| \\
&\leq \sum_{r=0}^{q-1} \frac{h_{n+1}^{r+1}}{r!} \left\| G^{(r)}(t_n) \int_0^1 \theta^r [z'_{n+1}(t_n + \theta h_{n+1}) - \eta'(t_n + \theta h_{n+1})] \mathrm{d}\theta \right\| \\
&\quad + O(h_{n+1}^{q+1}) \left\| \int_0^1 [z'_{n+1}(t_n + \theta h_{n+1}) - \eta'(t_n + \theta h_{n+1})] \mathrm{d}\theta \right\|.
\end{aligned}$$

By (5.2.2), the first term is $O(h_{n+1}^{p+1})$. On the other hand, $\eta(t)$ being an interpolant of uniform order q, the second term is $O(h_{n+1}^{2q+1})$. Since $2q + 1 \geq p + 1$, the theorem is proved. ■

Note that the NCE of an RK method is not unique and that formula (5.2.1) simply provides a rule for obtaining one of them. In general, one can use the order conditions of Table 5.1 along with the orthogonality condition (5.2.3).

The following are the NCEs of some explicit RK (ERK) methods.

- One-stage ERK method of order $p = 1$ (Euler method):
  - $q = \lfloor \frac{p+1}{2} \rfloor = \nu = p = 1$

$$b_1(\theta) = \theta.$$

- Two-stage ERK methods of order $p = 2$:
  - $q = \lfloor \frac{p+1}{2} \rfloor = 1$

$$b_i(\theta) = b_i\theta, \quad i = 1, 2;$$

  - $q = \nu = p = 2$

$$\begin{aligned} b_1(\theta) &= (b_1 - 1)\theta^2 + \theta, \\ b_2(\theta) &= b_2\theta^2. \end{aligned}$$

- Three-stage ERK methods of order $p = 3$ (with c_2, $c_3 \neq 0$):
  - $q = \lfloor \frac{p+1}{2} \rfloor = 2$

$$b_i(\theta) = w_i\theta^2 + (b_i - w_i)\theta, \quad i = 1, 2, 3,$$

where

$$\begin{aligned} w_1 &= -\tfrac{1}{2c_3} - (c_3 - c_2)\lambda, \\ w_2 &= c_3\lambda, \\ w_3 &= \tfrac{1}{2c_3} - c_2\lambda, \end{aligned}$$

and $\lambda \in \mathbb{R}$. Here we have a one-parameter family of NCEs, that is an example of non-uniqueness of NCEs of minimal order q. In particular, the polynomial furnished by the conditions (5.2.1) is

$$b_i(\theta) = 3(2c_i - 1)b_i\theta^2 + 2(2 - 3c_i)b_i\theta, \quad i = 1, 2, 3,$$

which corresponds to the value $\lambda = \frac{3(2c_2-1)}{c_3}b_2$;

  - $q = \nu = p = 3$
    there are no NCEs of order $q = 3$.

- Four-stage ERK methods of order $p = 4$:
  - $q = \lfloor \frac{p+1}{2} \rfloor = 2$

$$b_i(\theta) = 3(2c_i - 1)b_i\theta^2 + 2(2 - 3c_i)b_i\theta, \quad i = 1, 2, 3, 4;$$

  - $q = 3$

$$\begin{aligned} b_1(\theta) &= 2(1 - 4b_1)\theta^3 + 3(3b_1 - 1)\theta^2 + \theta, \\ b_i(\theta) &= 4(3c_i - 2)b_i\theta^3 + 3(3 - 4c_i)b_i\theta^2, \quad i = 2, 3, 4; \end{aligned}$$

* $q = \nu = p = 4$
  there are no NCEs of order $q = 4$.

Example 5.2.7 Consider the test equation

$$\begin{cases} y'(t) = -y(t), & t \geq t_0, \\ y(0) = 1, \end{cases} \tag{5.2.7}$$

and compute the approximate solution in the interval $[0, h]$ using the following well-known ERK method of order $p = 4$ (the classical four-stage RK method), interpolated by its unique NCE of order $q = 2$:

$$\begin{array}{c|cccc} 0 & & & & \\ \frac{1}{2} & \frac{1}{2} & & & \\ \frac{1}{2} & 0 & \frac{1}{2} & & \\ 1 & 0 & 0 & 1 & \\ \hline & \frac{1}{6} & \frac{1}{3} & \frac{1}{3} & \frac{1}{6} \end{array} \qquad \begin{aligned} b_1(\theta) &= (-\tfrac{1}{2}\theta + \tfrac{2}{3})\theta, \\ b_2(\theta) &= \tfrac{1}{3}\theta, \\ b_3(\theta) &= \tfrac{1}{3}\theta, \\ b_4(\theta) &= (\tfrac{1}{2}\theta - \tfrac{1}{3})\theta. \end{aligned}$$

Since the integration is made on the interval $[0, h]$, the expected nodal error $|y_1 - y(h)|$ is $O(h^5)$ and the uniform error $\max_{0 \leq \theta \leq 1} |\eta(\theta h) - y(\theta h)|$ is $O(h^3)$. This means that the order conditions of Table 5.1 are fulfilled up to order 4 at $\theta = 1$, and up to order 2 for all $\theta \in [0, 1]$. On the other hand, it is easy to verify that, for $\theta = \frac{1}{2}$, the conditions of order 3 are fulfilled as well, and this is the sole inner superconvergence point, where the error is $O(h^4)$. Table 5.2 illustrates these occurrences.

For the NCE of order $q = 3$ given by

$$\begin{aligned} b_1(\theta) &= (\tfrac{2}{3}\theta^2 - \tfrac{3}{2}\theta + 1)\theta, \\ b_2(\theta) &= (-\tfrac{2}{3}\theta + 1)\theta^2, \\ b_3(\theta) &= (-\tfrac{2}{3}\theta + 1)\theta^2, \\ b_4(\theta) &= (\tfrac{2}{3}\theta - \tfrac{1}{2})\theta^2, \end{aligned}$$

the uniform error is $O(h^4)$ and no inner superconvergence points exist. In fact, the first of the conditions of order 4 in Table 5.1 reads $\theta^2(\theta^2 - 2\theta + 1) = 0$, which is satisfied only for $\theta = 0$ and $\theta = 1$. Numerical experiments are shown in Table 5.3.

The continuous errors for $q = 2$ and $q = 3$ are plotted in Figure 5.1. ◇

The following are the NCEs of general ν-stage RK methods (also implicit) of order $p \geq \nu$ up to $\nu = 3$ with distinct abscissae $c_1, \ldots, c_\nu$.

- One-stage RK methods of order $p = 1, 2$:
  the discrete weight is

$$b_1 = 1;$$

TABLE 5.2. Numerical results for the NCE of order $q = 2$ of the ERK method of order $p = 4$ applied to equation (5.2.7).

h	$\lvert y_1 - y(h)\rvert$	$\max_{0\le\theta\le1} \lvert\eta(\theta h) - y(\theta h)\rvert$	$\lvert\eta(\frac{1}{2}h) - y(\frac{1}{2}h)\rvert$
1	$9.95 \cdot 10^{-3}$	$2.12 \cdot 10^{-2}$	$1.33 \cdot 10^{-2}$
0.1	$8.47 \cdot 10^{-8}$	$9.25 \cdot 10^{-6}$	$1.30 \cdot 10^{-6}$
0.01	$8.35 \cdot 10^{-13}$	$8.14 \cdot 10^{-9}$	$1.30 \cdot 10^{-10}$

TABLE 5.3. Numerical results for the NCE of order $q = 3$ of the ERK method of order $p = 4$ applied to equation (5.2.7).

h	$\lvert y_1 - y(h)\rvert$	$\max_{0\le\theta\le1} \lvert\eta(\theta h) - y(\theta h)\rvert$
1	$9.95 \cdot 10^{-3}$	$1.57 \cdot 10^{-2}$
0.1	$8.47 \cdot 10^{-8}$	$1.46 \cdot 10^{-6}$
0.01	$8.35 \cdot 10^{-13}$	$1.45 \cdot 10^{-10}$

* $q = \lfloor \frac{p+1}{2} \rfloor = \nu = 1$

$$b_1(\theta) = \theta.$$

- Two-stage RK methods of order $p = 2, 3, 4$:
  the discrete weights are

$$b_1 = \frac{2c_2 - 1}{2(c_2 - c_1)},$$
$$b_2 = \frac{1 - 2c_1}{2(c_2 - c_1)};$$

* $q = \lfloor \frac{p+1}{2} \rfloor = 1$ (only for $p = 2$)

$$b_1(\theta) = b_1\theta,$$
$$b_2(\theta) = b_2\theta;$$

* $q = \nu = 2$

$$b_1(\theta) = -\frac{1}{2(c_2 - c_1)}\theta^2 + \frac{c_2}{c_2 - c_1}\theta,$$
$$b_2(\theta) = \frac{1}{2(c_2 - c_1)}\theta^2 - \frac{c_1}{c_2 - c_1}\theta.$$

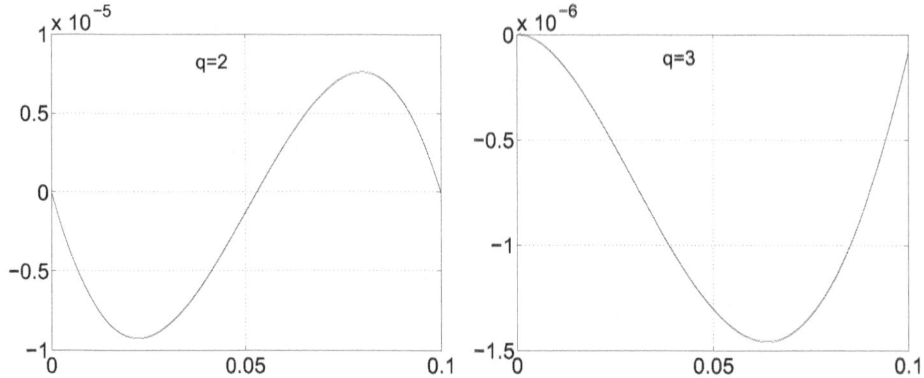

FIG. 5.1. Continuous errors of the NCEs of order $q = 2, 3$ of the ERK method of order $p = 4$ applied to equation (5.2.7) in the interval $[0, 0.1]$.

- Three-stage RK methods of order $p = 3, 4, 5, 6$:
  the discrete weights are

$$\begin{aligned} b_1 &= \tfrac{6c_2c_3 - 3(c_2+c_3)+2}{6(c_2-c_1)(c_3-c_1)}, \\ b_2 &= \tfrac{-6c_1c_3 + 3(c_1+c_3)-2}{6(c_2-c_1)(c_3-c_2)}, \\ b_3 &= \tfrac{6c_1c_2 - 3(c_1+c_2)+2}{6(c_3-c_1)(c_3-c_2)}; \end{aligned}$$

  * $q = \lfloor \frac{p+1}{2} \rfloor = 2$ (only for $p = 3$)

$$b_i(\theta) = w_i\theta^2 + (b_i - w_i)\theta, \qquad i = 1, 2, 3,$$

where

$$\begin{aligned} w_1 &= -\tfrac{1}{2(c_3-c_1)} - (c_3 - c_2)\lambda, \\ w_2 &= (c_3 - c_1)\lambda, \\ w_3 &= \tfrac{1}{2(c_3-c_1)} - (c_2 - c_1)\lambda, \end{aligned}$$

and $\lambda \in \mathbb{R}$. Here we have a one-parameter family of NCEs.

For $p = 4$ there exists a unique NCE of order $q = 2$, the one furnished by the conditions (5.2.1), that is

$$b_i(\theta) = 3(2c_i - 1)b_i\theta^2 + 2(2 - 3c_i)b_i\theta, \qquad i = 1, 2, 3,$$

which corresponds to the value $\lambda = \frac{(6c_1c_3 - 3(c_1+c_3)+2)(1-2c_2)}{2(c_3-c_1)(c_2-c_1)(c_3-c_2)}$;

$* \; q = \nu = 3$

$$\begin{aligned}
b_1(\theta) &= \tfrac{1}{3(c_2-c_1)(c_3-c_1)}\theta^3 - \tfrac{c_2+c_3}{2(c_2-c_1)(c_3-c_1)}\theta^2 + \tfrac{c_2c_3}{(c_2-c_1)(c_3-c_1)}\theta, \\
b_2(\theta) &= -\tfrac{1}{3(c_2-c_1)(c_3-c_2)}\theta^3 + \tfrac{c_1+c_3}{2(c_2-c_1)(c_3-c_2)}\theta^2 - \tfrac{c_1c_3}{(c_2-c_1)(c_3-c_2)}\theta, \\
b_3(\theta) &= \tfrac{1}{3(c_3-c_1)(c_3-c_2)}\theta^3 - \tfrac{c_1+c_2}{2(c_2-c_1)(c_3-c_1)}\theta^2 + \tfrac{c_1c_2}{(c_3-c_1)(c_3-c_2)}\theta;
\end{aligned}$$

it exists if and only if the RK method satisfies the simplifying assumptions (see, for example, [62])

$$\sum_{j=1}^{3} a_{ij}c_j = \frac{1}{2}c_i^2, \qquad i = 1, 2, 3,$$

which always hold for $p = 5, 6$.

Bellen and Zennaro [37] proved the existence of another NCE of minimal degree $\lfloor \frac{p+1}{2} \rfloor$, which may or may not coincide with the one furnished by the conditions (5.2.1) and which, in any case, attains some useful *stability* properties (see Section 8.4.1). The result is given by the following theorem, the rather technical proof of which is omitted.

Theorem 5.2.8 *Every RK method (5.0.2), (5.0.3) of order $p \geq 3$ has an NCE $\eta(t)$ of minimal order (and degree) $q = \lfloor \frac{p+1}{2} \rfloor$, whose continuous weights satisfy the continuity conditions (5.1.5) and, for $i = 1, \ldots, \nu$,*

$$\int_0^1 \theta^r b_i(\theta)\mathrm{d}\theta = \sum_{k=1}^{r} \lambda_{rk} \sum_{j=1}^{\nu} b_j (A^k)_{ji}, \quad r = 1, \ldots, q-1, \tag{5.2.8}$$

where λ_{rk} are defined by the recurrence relation

$$\lambda_{r1} = 1 \quad \textit{and} \quad \lambda_{r+1,k+1} = -r\lambda_{rk} \quad \forall r, k \geq 1. \tag{5.2.9}$$

In general, the NCEs of minimal degree determined by conditions (5.2.1) and (5.2.8) are different from each other. The next theorem, thee proof of which is again rather technical, characterizes the case in which equality holds.

Theorem 5.2.9 *The NCEs of minimal order (and degree) $q = \lfloor \frac{p+1}{2} \rfloor$ determined by conditions (5.2.1) and (5.2.8) are equal to each other if and only if the RK method satisfies the simplifying assumptions (see, for example, [62])*

$$\sum_{i=1}^{\nu} b_i c_i^{r-1} a_{ij} = \frac{1}{r}(1 - c_j^r)b_j, \quad j = 1, \ldots, \nu \quad \textit{and} \quad r = 1, \ldots, q-1.$$

Some further theoretical results on NCEs, which are beyond the scope of this book, may be found in Zennaro [279, 281].

5.2.3 *An application of the NCEs*

In this section we illustrate the application of the NCEs to the numerical solution of an IVP with a *driving equation*. As we shall see in Chapter 6, it includes the application to DDEs and NDDEs.

Consider an interval $[a, b]$ and an integer q, $\lfloor \frac{p+1}{2} \rfloor \leq q \leq p$, and assume that, to any mesh $\Delta = \{t_0 = a, t_1, \ldots, t_n, \ldots, t_N = b\}$, two piecewise C^{p+1}-continuous functions $u_\Delta(t)$ and $v_\Delta(t)$ are associated which fulfil the following properties:

$$\max_{0 \leq n \leq N} \|u_\Delta(t_n) - v_\Delta(t_n)\| = O(h^p),$$

$$\max_{a \leq t \leq b} \|u_\Delta(t) - v_\Delta(t)\| = O(h^{q'}), \tag{5.2.10}$$

where $q' = \min\{q+1, p\}$,

$$\max_{a \leq t \leq b} \|u_\Delta^{(j)}(t) - v_\Delta^{(j)}(t)\| = O(h^{q-j+1}), \quad j = 1, \ldots, q, \tag{5.2.11}$$

$$\max_{0 \leq n \leq N-1} \left\| \int_{t_n}^{t_{n+1}} G(t)[u_\Delta^{(j)}(t) - v_\Delta^{(j)}(t)]\mathrm{d}t \right\| = O(h^{p+1}), \quad j = 0, 1, \tag{5.2.12}$$

for every sufficiently smooth matrix-valued function G.

Moreover, assume that all the derivatives of $u_\Delta(t)$ and $v_\Delta(t)$ are uniformly bounded as $h \to 0$.

Then consider another interval $[\tilde{a}, \tilde{b}]$ and, for any mesh Δ in $[a, b]$, consider the following two IVPs for ODEs with the same initial condition but with the right-hand side depending on $u_\Delta(t)$ and $v_\Delta(t)$, respectively:

$$\begin{cases} z'(x) = f\Big(x, z(x), u_\Delta(\alpha(x)), u'_\Delta(\alpha(x))\Big), & \tilde{a} \leq x \leq \tilde{b}, \\ z(\tilde{a}) = z_a, \end{cases} \tag{5.2.13}$$

$$\begin{cases} w'(x) = f\Big(x, w(x), v_\Delta(\alpha(x)), v'_\Delta(\alpha(x))\Big), & \tilde{a} \leq x \leq \tilde{b}, \\ w(\tilde{a}) = z_a, \end{cases} \tag{5.2.14}$$

where $\alpha(x)$ is one-to-one between the intervals $[\tilde{a}, \tilde{b}]$ and $[a = \alpha(\tilde{a}), b = \alpha(\tilde{b})]$. Moreover, denote by $\beta(t)$ the inverse of $\alpha(x)$ and assume that the functions $f(x, z, y, w)$, $\alpha(x)$ and $\beta(t)$ are sufficiently smooth. Finally, let $\tilde{\Delta} = \{x_0 = \tilde{a}, x_1, \ldots, x_n, \ldots, x_N = \tilde{b}\}$ be the corresponding mesh of $[\tilde{a}, \tilde{b}]$ such that $x_n = \beta(t_n)$, $n = 0, 1, \ldots, N$.

Then we have the following result.

Theorem 5.2.10 *The solutions $z(x)$ and $w(x)$ of the ODEs (5.2.13) and (5.2.14) satisfy, in the interval $[\tilde{a}, \tilde{b}]$ and with respect to the mesh $\tilde{\Delta}$, the same* *properties satisfied by* $u_\Delta(t)$ *and* $v_\Delta(t)$ *in the interval* $[a, b]$ *with respect to the mesh* Δ*, that is:*

$$\max_{0\le n\le N} \|z(x_n) - w(x_n)\| = O(h^p), \tag{5.2.15}$$

$$\max_{\tilde a\le x\le \tilde b} \|z(x) - w(x)\| = O(h^{q'}), \tag{5.2.16}$$

where $q' = \min\{q+1, p\}$,

$$\max_{\tilde a\le x\le \tilde b} \|z^{(j)}(x) - w^{(j)}(x)\| = O(h^{q-j+1}), \quad j = 1, \dots, q, \tag{5.2.17}$$

$$\max_{0\le n\le N-1} \left\| \int_{x_n}^{x_{n+1}} G(x)[z^{(j)}(x) - w^{(j)}(x)]\mathrm{d}x \right\| = O(h^{p+1}), \quad j = 0, 1, \tag{5.2.18}$$

for every sufficiently smooth matrix-valued function G.

Moreover, all the derivatives of $z(x)$ *and* $w(x)$ *are uniformly bounded as* $h \to 0$.

Proof. By (5.2.10) and (5.2.11) for $j = 1$ we have

$$\max_{\tilde a\le x\le \tilde b} \|u_\Delta^{(j)}(\alpha(x)) - v_\Delta^{(j)}(\alpha(x))\| = O(h^q), \quad j = 0, 1,$$

which yields

$$\max_{\tilde a\le x\le \tilde b} \|f\Big(x, z, u_\Delta(\alpha(x)), u'_\Delta(\alpha(x))\Big) - f\Big(x, z, v_\Delta(\alpha(x)), v'_\Delta(\alpha(x))\Big)\| = O(h^q)$$

uniformly with respect to z in any bounded set of $\mathbb{R}^d$.

Therefore, standard arguments allow us to conclude with (5.2.17) for $j = 1$.

Observe that (5.2.18) for $j = 0$ is easily obtained from (5.2.15) and (5.2.18) for $j = 1$ by using integration by parts and that (5.2.16) and (5.2.17) for $j = 2, \dots, q$ are obtained from (5.2.15) and (5.2.17) for $j = 1$. Here we have to use (5.2.10) and (5.2.11) along with successive differentiation of (5.2.13) and (5.2.14). Moreover, the uniform boundedness of $z(x)$ and $w(x)$ as $h \to 0$ follows from the uniform boundedness of $u_\Delta(t)$ and $v_\Delta(t)$.

Therefore, we are left to prove (5.2.15) and (5.2.18) for $j = 1$.

To proceed with the proof, for each $n = 0, \dots, N-1$, we consider the auxiliary problem

$$\begin{cases} w'_{n+1}(x) = f\Big(x, w_{n+1}(x), v_\Delta(\alpha(x)), v'_\Delta(\alpha(x))\Big), & x_n \le x \le x_{n+1}, \\ w_{n+1}(x_n) = z(x_n). \end{cases} \tag{5.2.19}$$

By the smoothness of $f(x, z, y, w)$, we have that, for all meshes $\tilde\Delta$ with $\tilde h = \max_{0\le n\le N-1} |x_{n+1} - x_n|$ sufficiently small, there exists a constant $C > 0$ such that

$$\|w(x_{n+1}) - w_{n+1}(x_{n+1})\| \le \mathrm{e}^{C\tilde h}\|z(x_n) - w(x_n)\|. \tag{5.2.20}$$

By applying the Gröbner–Alekseev non-linear variation-of-constants formula to (5.2.19) in the step $[x_n, x_{n+1}]$, we get

$$w_{n+1}(x) - z(x) = \int_{x_n}^{x} K(x,s)\left[f\Big(s, z(s), v_\Delta(\alpha(s)), v'_\Delta(\alpha(s))\Big) - z'(s)\right] \mathrm{d}s,$$

where $K(x,s)$ is a suitable variational matrix depending on the functions z and v_Δ such that $K(x,x) \equiv I$, the identity matrix. By differentiating we also get

$$\begin{aligned} w'_{n+1}(x) - z'(x) = & \int_{x_n}^{x} \frac{\partial K}{\partial x}(x,s)\left[f\Big(s, z(s), v_\Delta(\alpha(s)), v'_\Delta(\alpha(s))\Big) - z'(s)\right]\mathrm{d}s \\ & + f\Big(x, z(x), v_\Delta(\alpha(x)), v'_\Delta(\alpha(x))\Big) - z'(x). \end{aligned}$$

Then, after a small calculation, multiplication by $G(x)$ yields

$$\begin{aligned} & \int_{x_n}^{x_{n+1}} G(x)\left[w'_{n+1}(x) - z'(x)\right]\mathrm{d}x \\ & \quad = \int_{x_n}^{x_{n+1}} L(s)\left[f\Big(s, z(s), v_\Delta(\alpha(s)), v'_\Delta(\alpha(s))\Big) - z'(s)\right]\mathrm{d}s, \quad (5.2.21) \end{aligned}$$

where

$$L(s) = G(s) + \int_{s}^{x_{n+1}} G(\xi)\frac{\partial K}{\partial x}(\xi, s)\mathrm{d}\xi.$$

By using (5.2.13), by subtracting and adding the term

$$L(s) f(s, z(s), u_\Delta(\alpha(s)), v'_\Delta(\alpha(s)))$$

in the argument of the right-hand-side integral in (5.2.21), and by setting

$$\Gamma_y(s) = \int_0^1 \frac{\partial f}{\partial y}\Big(s, z(s), u_\Delta(\alpha(s)) + \omega\big(v_\Delta(\alpha(s)) - u_\Delta(\alpha(s))\big), v'_\Delta(\alpha(s))\Big)\mathrm{d}\omega$$

and

$$\Gamma_{y'}(s) = \int_0^1 \frac{\partial f}{\partial y'}\Big(s, z(s), u_\Delta(\alpha(s)), u'_\Delta(\alpha(s)) + \omega\big(u'_\Delta(\alpha(s)) - v'_\Delta(\alpha(s))\big)\Big)\mathrm{d}\omega,$$

we get

$$\begin{aligned} \int_{x_n}^{x_{n+1}} G(x)\left[w'_{n+1}(x) - z'(x)\right]\mathrm{d}x = & \int_{x_n}^{x_{n+1}} L(s)\Big[\Gamma_y(s)\Big(v_\Delta(\alpha(s)) - u_\Delta(\alpha(s))\Big) \\ & + \Gamma_{y'}(s)\Big(v'_\Delta(\alpha(s)) - u'_\Delta(\alpha(s))\Big)\Big]\mathrm{d}s. \end{aligned}$$

With the change of variable $s = \beta(t)$, this equality becomes

$$\int_{x_n}^{x_{n+1}} G(x)\left[w'_{n+1}(x) - z'(x)\right]\mathrm{d}x = \int_{t_n}^{t_{n+1}} L(\beta(t))\beta'(t)\Big[\Gamma_y(\beta(t))\left(v_\Delta(t) - u_\Delta(t)\right)$$

$$+\Gamma_{y'}(\beta(t))\left(v'_\Delta(t) - u'_\Delta(t)\right)\Big]\mathrm{d}t.$$

Therefore, by the asymptotic orthogonality conditions (5.2.12), we can conclude that

$$\max_{0\le n\le N-1}\left\|\int_{x_n}^{x_{n+1}} G(x)[w'_{n+1}(x) - z'(x)]\mathrm{d}x\right\| = O(h^{p+1}). \tag{5.2.22}$$

Observe that, in order to prove (5.2.22), we use the fact that the matrices $\Gamma_y(s)$, $\Gamma_{y'}(s)$ and $L(s)$ are uniformly bounded together with all their derivatives as $h \to 0$. Indeed, since they depend on the functions $u_\Delta(t)$, $v_\Delta(t)$ and $z(x)$, they also depend on h. However, since all the derivatives of $u_\Delta(t)$, $v_\Delta(t)$ and $z(x)$ are uniformly bounded as $h \to 0$, by the smoothness of $\alpha(x)$ and $f(x, z, y, w)$, $\Gamma_y(s)$, $\Gamma_{y'}(s)$ and $L(s)$ also inherit the same property.

In particular, for $G(x) \equiv I$, the identity matrix, the estimate (5.2.22) yields

$$\max_{0\le n\le N-1}\|w_{n+1}(x_{n+1}) - z(x_{n+1})\| = O(h^{p+1}).$$

Therefore, by using (5.2.20), we get the recurrence relation

$$\|z(x_{n+1}) - w(x_{n+1})\| \le \mathrm{e}^{C\tilde{h}}\|z(x_n) - w(x_n)\| + O(h^{p+1}),$$

with $\|z(x_0) - w(x_0)\| = 0$. Since $\tilde{h} = O(h)$, such a recurrence relation implies (5.2.15).

In order to conclude the proof, observe that, by comparing (5.2.14) and (5.2.19), (5.2.15) yields

$$\max_{0\le n\le N-1}\ \max_{x_n\le x\le x_{n+1}}\|w'(x) - w'_{n+1}(x)\| = O(h^p).$$

Finally, this last estimate together with (5.2.22) implies (5.2.18) for $j = 1$. ■

Besides the forthcoming application to DDEs and NDDEs, Theorem 5.2.10 may be applied directly to the numerical solution of the ODE

$$\begin{cases} z'(x) = f\Big(x, z(x), y\big(\alpha(x)\big), y'\big(\alpha(x)\big)\Big), & x_0 \le x \le x_f, \\ z(x_0) = z_0, \end{cases} \tag{5.2.23}$$

where $y(t)$ is the solution of the *driving equation*

$$\begin{cases} y'(t) = g\big(t, y(t)\big), & t_0 \le t \le t_f, \\ y(t_0) = y_0. \end{cases} \tag{5.2.24}$$

In order to solve (5.2.23) numerically, we first solve (5.2.24) on a given mesh $\Delta = \{t_0, t_1, \ldots, t_n, \ldots, t_N = t_f\}$ by an RK method of order p with an NCE $\eta(t)$ of order q. Then we solve the perturbed ODE

$$\begin{cases} w'(x) = f\Big(x, w(x), \eta\big(\alpha(x)\big), \eta'\big(\alpha(x)\big)\Big), & x_0 \le x \le x_f, \\ w(x_0) = z_0. \end{cases} \tag{5.2.25}$$

Since the NCE $\eta(t)$ satisfies conditions (5.1.14), (5.1.15) and (5.2.5), Theorem 5.2.10 applies with $[a, b] = [t_0, t_f]$, $[\tilde{a}, \tilde{b}] = [x_0, x_f]$, $u_\Delta(t) = y(t)$ and

$v_\Delta(t) = \eta(t)$. In particular, (5.2.15) holds for the points of the corresponding mesh $\tilde{\Delta}$. Therefore, for the numerical solution of (5.2.25), any discrete RK method of order p across the mesh $\tilde{\Delta}$ preserves the nodal order p, despite the possibly lower uniform accuracy order q of the employed NCE $\eta(t)$.

5.3 Interpolants of the second class

To the authors' knowledge, the first method for constructing high order continuous extensions of RK schemes is due to Sarafyan [223]. His technique is based on the use of suitable linear combinations of the stage values aimed to provide high order approximations of $y(t)$ at some interior points. Then he constructs the Hermite–Birkhoff interpolant at these points and, possibly, at the extremes of the mesh interval by some extra evaluations of $g(t, y)$. In the original paper [223], the author proposes a set of eight-stage RK methods of order 6 endowed with a continuous extension of uniform order 4. He subsequently developed several families of RK methods of high order, say p, with interpolation of order p or $p-1$ (see [224] and [225]). The following scheme is a ten-stage method of order 6 with a continuous approximant of order 5 which forms the basis for the code DKLAG6 developed by Corwin, Sarafyan and Thompson [69]:

$$y_{n+1} = y_n + h_{n+1} \sum_{i=1}^{10} b_i K_{n+1}^i, \tag{5.3.1}$$

with Butcher's tableau

$$\begin{array}{c|cccccccccc}
0 & & & & & & & & & & \\
\frac{1}{9} & \frac{1}{9} & & & & & & & & & \\
\frac{1}{6} & \frac{1}{24} & \frac{3}{24} & & & & & & & & \\
\frac{1}{3} & \frac{1}{6} & -\frac{3}{6} & \frac{4}{6} & & & & & & & \\
\frac{1}{2} & \frac{1}{8} & 0 & 0 & \frac{3}{8} & & & & & & \\
\frac{2}{3} & -\frac{8}{21} & \frac{6}{21} & \frac{24}{21} & -\frac{24}{21} & \frac{16}{21} & & & & & \\
\frac{5}{6} & \frac{65}{456} & \frac{15}{456} & -\frac{80}{456} & \frac{330}{456} & -\frac{160}{456} & \frac{210}{456} & & & & \\
1 & \frac{35}{322} & -\frac{81}{322} & \frac{240}{322} & -\frac{360}{322} & \frac{680}{322} & -\frac{420}{322} & \frac{228}{322} & & & \\
1 & \frac{161}{3000} & 0 & \frac{684}{3000} & \frac{315}{3000} & \frac{680}{3000} & \frac{315}{3000} & \frac{684}{3000} & \frac{161}{3000} & & \\
\frac{2}{3} & \frac{7}{135} & 0 & \frac{32}{135} & \frac{12}{135} & \frac{32}{135} & \frac{7}{135} & 0 & 0 & 0 & \\
\hline
 & \frac{161}{300} & 0 & \frac{57}{250} & \frac{21}{200} & \frac{17}{75} & \frac{21}{200} & \frac{57}{250} & \frac{161}{300} & 0 & 0
\end{array}$$

and interpolant

$$\eta(t_n + \theta h_{n+1}) = y_n + h_{n+1} \sum_{i=1}^{10} b_i(\theta) K_{n+1}^i, \quad 0 \le \theta < 1,$$

where

$$
\begin{aligned}
b_1(\theta) &= \theta - \tfrac{2427}{500}\theta^2 + \tfrac{3586}{375}\theta^3 - \tfrac{1659}{200}\theta^4 + \tfrac{8864}{3375}\theta^5, \\
b_2(\theta) &= 0, \\
b_3(\theta) &= \tfrac{708}{125}\theta^2 - \tfrac{1977}{125}\theta^3 + \tfrac{789}{50}\theta^4 - \tfrac{6028}{1125}\theta^5, \\
b_4(\theta) &= \tfrac{87}{50}\theta^2 - \tfrac{243}{50}\theta^3 + \tfrac{201}{40}\theta^4 - \tfrac{2026}{1125}\theta^5, \\
b_5(\theta) &= -\tfrac{88}{25}\theta^2 + \tfrac{1226}{75}\theta^3 - 21\theta^4 + \tfrac{5584}{675}\theta^5, \\
b_6(\theta) &= -\tfrac{21}{100}\theta^2 + \tfrac{21}{25}\theta^3 - \tfrac{21}{40}\theta^4 - \tfrac{49}{900}\theta^5, \\
b_7(\theta) &= \tfrac{228}{125}\theta^2 - \tfrac{1197}{125}\theta^3 + \tfrac{741}{50}\theta^4 - \tfrac{38608}{1125}\theta^5, \\
b_8(\theta) &= -\tfrac{161}{250}\theta^2 + \tfrac{1127}{750}\theta^3 - \tfrac{161}{200}\theta^4 - \tfrac{161}{3775}\theta^5, \\
b_9(\theta) &= 2\theta^3 - 5\theta^4 + \tfrac{83}{27}\theta^5, \\
b_{10}(\theta) &= \tfrac{1}{4}\theta^5.
\end{aligned}
$$

Observe that the continuous extension does not satisfy the continuity condition (5.1.2) for $\theta = 1$ and, hence, it does not meet the nodal approximation (5.3.1).

Similar techniques were developed later by Horn [130], who obtained extensions of order 4 and 5 for the classical RKF4(5) embedded pair. Also in this case the method of order 5 lacks the continuity condition (5.1.2) at the nodal point with respect to the underlying discrete method.

In this section we consider some general procedures for iterated *uniform corrections* that, by starting from an interpolant $\eta_{(0)}(t)$ of uniform order (and degree) q, provide a sequence of interpolants $\eta_{(1)}(t), \eta_{(2)}(t), \ldots$, of increasing order $q+1, q+2, \ldots$ up to order p of the underlying RK method.

In each interval $[t_n, t_{n+1}]$, the starting polynomial $\eta_{(0)}(t)$ of order q is assumed to be either an interpolant of the first class or a polynomial obtained by using also the value $g(t_{n+1}, y_{n+1})$, whenever available for free from the next step. This is the case of those RK methods whose coefficient matrix A has all zero entries in the first row (for example, all the explicit methods). In this context, this latter type of interpolants will still be called interpolants of the first class.

Several procedures have been proposed for doing this. They differ from one another in the cost, that is the number of extra stages per correction, and in the properties required for increased order polynomials.

5.3.1 *A first uniform correction procedure*

The procedure we are going to describe was proposed by Enright et al. [93] for ERK methods. It is based on the Hermite–Birkhoff interpolation and was called the *bootstrapping process*. Although the authors presented their theory to provide uniform high order approximations of embedded RK pairs, here we show how it may work in general to raise the uniform order of interpolants of any RK method.

Let us consider the local problem

$$
\begin{cases} z_{n+1}'(t) = g\big(t, z_{n+1}(t)\big), & t_n \le t \le t_{n+1}, \\ z_{n+1}(t_n) = y_n, \end{cases} \tag{5.3.2}
$$

and assume that the RK method has order p. The general step of the procedure, which leads from the interpolant $u(t) = \eta_{(r-q)}(t)$ of order r, $q \le r < p$, to the interpolant $v(t) = \eta_{(r+1-q)}(t)$ of order $r+1$, is as follows.

The interpolant $u(t_n + \theta h_{n+1})$, $\theta \in [0,1]$, uniformly approximates $z_{n+1}(t_n + \theta h_{n+1})$ in $[t_n, t_{n+1}]$ with local error $O(h_{n+1}^{r+1})$. Then $u'(t_n+\theta h_{n+1})$ and $g(t, u(t_n+\theta h_{n+1}))$ are approximations of uniform order $r-1$ and r (local errors $O(h_{n+1}^{r})$ and $O(h_{n+1}^{r+1})$), respectively, for $z'_{n+1}(t_n+\theta h_{n+1})$.

Let $t_n + \xi_i h_{n+1}$, $i = 1, \ldots, k$, and $t_n + \zeta_i h_{n+1}$, $i = 1, \ldots, k'$, be the points, if any, of inner superconvergence for $u(t)$ and $u'(t)$, that is

$$\|z_{n+1}(t_n + \xi_i h_{n+1}) - u(t_n + \xi_i h_{n+1})\| = O(h_{n+1}^{r+2}) \tag{5.3.3}$$

and

$$\|z'_{n+1}(t_n + \zeta_i h_{n+1}) - u'(t_n + \zeta_i h_{n+1})\| = O(h_{n+1}^{r+1}). \tag{5.3.4}$$

Now, for a polynomial $P(\theta)$ of degree $r+1$ in $[0,1]$, consider the set of $r+2$ Hermite–Birkhoff interpolation conditions on $P(0)$, $P(1)$, $P'(0)$, $P'(1)$, $P(\xi_i)$, $i = 1, \ldots, k$ (possibly $k = 0$), $P'(\zeta_i)$, $i = 1, \ldots, k'$ (possibly $k' = 0$) and $P'(\theta_i)$, $i = 1, \ldots, k'' = r - 2 - k - k'$ (possibly $k'' = 0$). The points ξ_i, ζ_i and θ_i, if any, are assumed to be such that the $r+2$ conditions are linearly independent, so that the polynomial $P(\theta)$ is uniquely determined.

Note that the sole conditions on $P(0)$, $P(1)$, $P'(0)$ and $P'(1)$ determine a third-degree Hermite interpolant. Therefore, since the values $g(t_n, y_n)$ and $g(t_{n+1}, y_{n+1})$ are known, an interpolant of uniform order 3 is already available. Hence the present procedure is relevant to the case $3 \le q \le r < p$.

Since, for a given function, the polynomial defined by the $r+2$ conditions above is accurate to order $r+1$ in the interval $[t_n, t_{n+1}]$ (local error $O(h_{n+1}^{r+2})$), it preserves this order even if the values of the function to be interpolated and those of its derivative are perturbed by terms of magnitude $O(h_{n+1}^{r+2})$ and $O(h_{n+1}^{r+1})$, respectively.

Thus the (uniquely solvable) system

(i) $v(t_n) = y_n, \quad v(t_{n+1}) = y_{n+1},$

(ii) $v'(t_n) = g(t_n, y_n), \quad v'(t_{n+1}) = g(t_{n+1}, y_{n+1}),$

(iii) $v(t_n + \xi_i h_{n+1}) = u(t_n + \xi_i h_{n+1}), \quad i = 1, \ldots, k,$

(iv) $v'(t_n + \zeta_i h_{n+1}) = u'(t_n + \zeta_i h_{n+1}), \quad i = 1, \ldots, k',$

(v) $v'(t_n + \theta_i h_{n+1}) = g(t_n + \theta_i h_{n+1}, u(t_n + \theta_i h_{n+1})), \quad i = 1, \ldots, k'',$

provides an interpolant $v(t) = \eta_{(r+1-q)}(t)$ of uniform order $r+1$, whose form is

$$\begin{aligned} v(t_n + \theta h_{n+1}) = {} & d_1(\theta) y_n + d_2(\theta) y_{n+1} \\ & + d_3(\theta) h_{n+1} g(t_n, y_n) + d_4(\theta) h_{n+1} g(t_{n+1}, y_{n+1}) \\ & + \sum_{i=1}^{k} d_{4+i}(\theta) u(t_n + \xi_i h_{n+1}) + \sum_{i=1}^{k'} d_{4+k+i}(\theta) h_{n+1} u'(t_n + \zeta_i h_{n+1}) \end{aligned}$$

TABLE 5.4. Butcher's tableau for the RKV6 method.

$$
\begin{array}{c|cccccccc}
0 & & & & & & & & \\
\frac{1}{6} & \frac{1}{6} & & & & & & & \\
\frac{4}{15} & \frac{4}{75} & \frac{16}{75} & & & & & & \\
\frac{2}{3} & \frac{5}{6} & -\frac{8}{3} & \frac{5}{2} & & & & & \\
\frac{5}{6} & -\frac{165}{64} & \frac{55}{6} & -\frac{425}{64} & \frac{85}{96} & & & & \\
1 & \frac{12}{5} & -8 & \frac{4015}{612} & -\frac{11}{36} & \frac{88}{255} & & & \\
\frac{1}{15} & -\frac{8263}{15000} & \frac{124}{75} & -\frac{643}{680} & -\frac{81}{250} & \frac{2484}{10625} & 0 & & \\
1 & \frac{3501}{1720} & -\frac{300}{43} & \frac{297275}{52632} & -\frac{319}{2322} & \frac{24068}{84065} & 0 & \frac{3850}{26703} & \\
\hline
 & \frac{3}{40} & 0 & \frac{875}{2244} & \frac{23}{72} & \frac{264}{1955} & 0 & \frac{125}{11592} & \frac{43}{616}
\end{array}
$$

$$
+\sum_{i=1}^{k''} d_{4+k+k'+i}(\theta) h_{n+1} g\big(t_n + \theta_i h_{n+1}, u(t_n + \theta_i h_{n+1})\big),
$$

where the polynomials $d_i(\theta)$, $i = 1, \ldots, r+2$, have degree $r+1$.

Since conditions (i), (ii), (iii) and (iv) do not involve additional function evaluations, in passing from order r to $r+1$, the procedure requires k'' new evaluations of the right-hand side function $g(t, y)$ in (v). This results in k'' explicit extra stages of the form (5.1.6).

In general, the interpolant $u(t)$ and its derivative $u'(t)$ may or may not possess inner superconvergence points. However, if superconvergence points ξ_i exist for $u(t)$ (see (5.3.3)), it is recommended to use them also among the k'' conditions in (v), even if they are not used in (iii). In fact, this choice makes them also superconvergence points for $v'(t)$ (see (5.3.4)) so that, in turn, they can be exploited again for the next correction in (iv).

Summarizing, if the RK method of order p is superconvergent and possesses an interpolant of the first class of order q, then $p-q$ corrections are needed in order to construct an interpolant $\eta_{(p-q)}(t)$ of uniform order p. The total number $s-\nu$ of additional stages in (5.1.6) is then given by

$$
s - \nu \leq \sum_{r=q}^{p-1} (r-2) = \frac{(p-3)(p-2) - (q-3)(q-2)}{2}.
$$

The inequality is strict whenever some inner superconvergence points exist to be exploited for one or more corrections in (iii) and/or (iv). On the contrary, the equality holds whenever $k = k' = 0$ for all corrections.

Finally, note that the use of conditions (ii) makes the interpolant globally C^1-continuous.

Example 5.3.1 We report from Enright et al. [93] the procedure applied to the eight-stage method of order 6 of the embedded pair RKV(5,6), which we denote by RKV6. This method is displayed in Table 5.4.

Observe that the sixth stage K^6_{n+1} is not used by the RKV6 method. Indeed, it is used by the embedded approximation of lower order 5, whose weights are not displayed in Table 5.4.

By setting

$$K^9_{n+1} = K^1_{n+2} = g(t_{n+1}, y_{n+1}), \tag{5.3.5}$$

which is free from the next step, we construct the following interpolant of the first class of uniform order (and degree) 4:

$$\eta_{(0)}(t_n + \theta h_{n+1}) = y_n + h_{n+1} \sum_{i=1}^{9} b_i(\theta) K^i_{n+1}, \quad 0 \le \theta < 1,$$

where

$$\begin{aligned}
b_1(\theta) &= -\tfrac{1}{480}\theta(643\theta^3 - 1694\theta^2 + 1495\theta - 480),\\
b_2(\theta) &= 0,\\
b_3(\theta) &= \tfrac{125}{53856}\theta^2(1193\theta^2 - 2722\theta + 1697),\\
b_4(\theta) &= -\tfrac{1}{432}\theta^2(707\theta^2 - 1138\theta + 293),\\
b_5(\theta) &= -\tfrac{4}{1955}\theta^2(643\theta^2 - 1154\theta + 445),\\
b_6(\theta) &= \tfrac{17}{44}\theta^2(\theta - 1)^2,\\
b_7(\theta) &= \tfrac{125}{34776}\theta^2(22\theta^2 - 50\theta + 31),\\
b_8(\theta) &= -\tfrac{43}{1848}\theta^2(62\theta^2 - 118\theta + 53),\\
b_9(\theta) &= \tfrac{1}{2}\theta^2(5\theta - 3)(\theta - 1).
\end{aligned}$$

Now it turns out that the point $t_n + \frac{1}{2}h_{n+1}$ is the only inner superconvergence point of order 5 for the interpolant $\eta_{(0)}(t)$. The corresponding vector of weights is

$$b\left(\tfrac{1}{2}\right) = \left[\tfrac{121}{1536}, 0, \tfrac{317125}{861696}, \tfrac{397}{6912}, -\tfrac{1}{68}, \tfrac{17}{704}, \tfrac{125}{12096}, -\tfrac{817}{14784}, \tfrac{1}{32}\right]^T.$$

Enright et al. [93] observe that there exists a one-parameter family of interpolants of uniform order 4 for which $t_n + \frac{1}{2}h_{n+1}$ is an inner superconvergence point of order 5. However, they made the choice displayed above because it minimizes some estimate of the local error at $t_n + \frac{1}{2}h_{n+1}$.

So we can obtain an interpolant $\eta_{(1)}(t)$ of uniform order 5 by solving the following set of conditions:

(i) $\eta_{(1)}(t_n) = y_n, \quad \eta_{(1)}(t_{n+1}) = y_{n+1}$,
(ii) $\eta'_{(1)}(t_n) = g(t_n, y_n), \quad \eta'_{(1)}(t_{n+1}) = g(t_{n+1}, y_{n+1})$,
(iii) $\eta_{(1)}(t_n + \frac{1}{2}h_{n+1}) = \eta_{(0)}(t_n + \frac{1}{2}h_{n+1})$,
(v) $\eta'_{(1)}(t_n + \frac{1}{2}h_{n+1}) = g\big(t_n + \frac{1}{2}h_{n+1}, \eta_{(0)}(t_n + \frac{1}{2}h_{n+1})\big)$.

We obtain

$$\begin{aligned}
\eta_{(1)}(t_n + \theta h_{n+1}) = {} & d_1(\theta) y_n + d_2(\theta) y_{n+1} + d_3(\theta) h_{n+1} g(t_n, y_n)\\
& + d_4(\theta) h_{n+1} g(t_{n+1}, y_{n+1}) + d_5(\theta)\eta_{(0)}(t_n + \tfrac{1}{2}h_{n+1})\\
& + d_6(\theta) h_{n+1} g\big(t_n + \tfrac{1}{2}h_{n+1}, \eta_{(0)}(t_n + \tfrac{1}{2}h_{n+1})\big),
\end{aligned}$$

where

$$\begin{aligned}
d_1(\theta) &= 4(\theta-1)^2(\theta-\tfrac{1}{2})^2(6\theta+1),\\
d_2(\theta) &= 4\theta^2(\theta-1)^2(7-6\theta),\\
d_3(\theta) &= 4\theta(\theta-1)^2(\theta-\tfrac{1}{2})^2,\\
d_4(\theta) &= 4\theta^2(\theta-1)(\theta-\tfrac{1}{2})^2,\\
d_5(\theta) &= 16\theta^2(\theta-1)^2,\\
d_6(\theta) &= 16\theta^2(\theta-1)^2(\theta-\tfrac{1}{2}).
\end{aligned}$$

As observed when describing the general procedure, it turns out that the derivative of the interpolant $\eta'_{(1)}(t)$, whose uniform order is 4, has an inner superconvergence point of order 5 at $t_n + \frac{1}{2}h_{n+1}$. No other inner superconvergences occur. Thus we need two more extra stages in order to construct an interpolant $\eta_{(2)}(t)$ of uniform order 6. In [93] the authors choose the extra abscissae $t_n + \frac{9}{10}h_{n+1}$ and $t_n + \frac{19}{20}h_{n+1}$, and the corresponding values are

$$\begin{aligned}
\eta_{(1)}\left(t_n + \tfrac{9}{10}h_{n+1}\right) = {}& \tfrac{128}{3125}y_n + \tfrac{2592}{3125}y_{n+1} + \tfrac{18}{3125}h_{n+1}g(t_n, y_n)\\
&-\tfrac{162}{3125}h_{n+1}g(t_{n+1}, y_{n+1}) + \tfrac{81}{625}\eta_{(0)}(t_n + \tfrac{1}{2}h_{n+1})\\
&+\tfrac{162}{3125}h_{n+1}g\big(t_n + \tfrac{1}{2}h_{n+1}, \eta_{(0)}(t_n + \tfrac{1}{2}h_{n+1})\big)
\end{aligned}$$

and

$$\begin{aligned}
\eta_{(1)}\left(t_n + \tfrac{19}{20}h_{n+1}\right) = {}& \tfrac{5427}{400000}y_n + \tfrac{380133}{400000}y_{n+1} + \tfrac{1539}{400000}h_{n+1}g(t_n, y_n)\\
&-\tfrac{29241}{800000}h_{n+1}g(t_{n+1}, y_{n+1}) + \tfrac{361}{10000}\eta_{(0)}(t_n + \tfrac{1}{2}h_{n+1})\\
&+\tfrac{3249}{200000}h_{n+1}g\big(t_n + \tfrac{1}{2}h_{n+1}, \eta_{(0)}(t_n + \tfrac{1}{2}h_{n+1})\big).
\end{aligned}$$

So we solve the set of conditions

(i) $\eta_{(2)}(t_n) = y_n, \quad \eta_{(2)}(t_{n+1}) = y_{n+1}$,

(ii) $\eta'_{(2)}(t_n) = g(t_n, y_n), \quad \eta'_{(2)}(t_{n+1}) = g(t_{n+1}, y_{n+1})$,

(iv) $\eta'_{(2)}(t_n + \frac{1}{2}h_{n+1}) = \eta'_{(1)}(t_n + \frac{1}{2}h_{n+1}) = g\big(t_n + \frac{1}{2}h_{n+1}, \eta_{(0)}(t_n + \frac{1}{2}h_{n+1})\big)$,

(v) $\eta'_{(2)}(t_n + \frac{9}{10}h_{n+1}) = g\big(t_n + \frac{9}{10}h_{n+1}, \eta_{(1)}(t_n + \frac{9}{10}h_{n+1})\big)$,
$\eta'_{(2)}(t_n + \frac{19}{20}h_{n+1}) = g\big(t_n + \frac{19}{20}h_{n+1}, \eta_{(1)}(t_n + \frac{19}{20}h_{n+1})\big)$,

and obtain

$$\begin{aligned}
\eta_{(2)}(t_n + \theta h_{n+1}) = {}& d_1(\theta)y_n + d_2(\theta)y_{n+1} + d_3(\theta)h_{n+1}g(t_n, y_n)\\
&+d_4(\theta)h_{n+1}g(t_{n+1}, y_{n+1}) + d_5(\theta)h_{n+1}\eta'_{(1)}(t_n + \tfrac{1}{2}h_{n+1})\\
&+d_6(\theta)h_{n+1}g\big(t_n + \tfrac{9}{10}h_{n+1}, \eta_{(1)}(t_n + \tfrac{9}{10}h_{n+1})\big)\\
&+d_7(\theta)h_{n+1}g\big(t_n + \tfrac{19}{20}h_{n+1}, \eta_{(1)}(t_n + \tfrac{19}{20}h_{n+1})\big),
\end{aligned}$$

where

$$\begin{aligned}
d_1(\theta) &= -\tfrac{1}{17}(\theta-1)^2(400\theta^4 - 808\theta^3 + 462\theta^2 - 34\theta - 17),\\
d_2(\theta) &= \tfrac{1}{17}\theta^2(400\theta^4 - 1608\theta^3 + 2478\theta^2 - 1766\theta - 513),\\
d_3(\theta) &= -\tfrac{1}{8721}\theta(\theta-1)^2(302000\theta^3 - 650840\theta^2 + 438060\theta - 8721),\\
d_4(\theta) &= \tfrac{1}{51}\theta^2(\theta-1)(3800\theta^3 - 7396\theta^2 + 4160\theta - 513),\\
d_5(\theta) &= -\tfrac{1}{918}\theta^2(\theta-1)^2(184000\theta^2 - 33088\theta + 14877),\\
d_6(\theta) &= \tfrac{125}{918}\theta^2(\theta-1)^2(800\theta^2 - 800\theta + 57),\\
d_7(\theta) &= -\tfrac{8000}{8721}\theta^2(\theta-1)^2(200\theta^2 - 200\theta + 27).
\end{aligned}$$

We can conclude that three extra stages, namely

$$\begin{aligned}
K_{n+1}^{10} &= g\big(t_n + \tfrac{1}{2}h_{n+1}, \eta_{(0)}(t_n + \tfrac{1}{2}h_{n+1})\big),\\
K_{n+1}^{11} &= g\big(t_n + \tfrac{9}{10}h_{n+1}, \eta_{(1)}(t_n + \tfrac{9}{10}h_{n+1})\big),\\
K_{n+1}^{12} &= g\big(t_n + \tfrac{19}{20}h_{n+1}, \eta_{(1)}(t_n + \tfrac{19}{20}h_{n+1})\big)
\end{aligned}$$

(plus the free extra stage (5.3.5)), are needed in order to obtain a continuous RK method of uniform order 6. Of course, the overall continuous method may be written in the form (5.0.2), (5.1.6), (5.1.3). ◇

5.3.2 *Explicit uniform corrections of NCEs*

If the starting polynomial $\eta_{(0)}(t)$ is an NCE of an ERK method, then the iterated correction procedure can be carried out in the space of polynomials still satisfying the asymptotic orthogonality conditions (5.2.3). This procedure was first developed by Zennaro [277] for collocation methods and later extended to all RK methods in [279], also in view of applications to DDEs and NDDEs. Here we present an improved version that exploits also the free extra stage $g(t_{n+1}, y_{n+1})$.

Let $u(t) = \eta_{(r-q)}(t)$ be an interpolant of order (and degree) r, $q \le r < p$, satisfying the asymptotic orthogonality condition

$$\left\| \int_{t_n}^{t_{n+1}} G(t)[z'_{n+1}(t) - u'(t)]\mathrm{d}t \right\| = O(h_{n+1}^{p+1}) \tag{5.3.6}$$

for every sufficiently smooth matrix-valued function G, where $z_{n+1}(t)$ is the solution to the local problem (5.3.2).

Moreover, as in Section 5.3.1, let $t_n + \xi_i h_{n+1}$, $i = 1, \ldots, k$, and $t_n + \zeta_i h_{n+1}$, $i = 1, \ldots, k'$, be the points, if any, of inner superconvergence for $u(t)$ and $u'(t)$, that is

$$\|z_{n+1}(t_n + \xi_i h_{n+1}) - u(t_n + \xi_i h_{n+1})\| = O(h_{n+1}^{r+2}) \tag{5.3.7}$$

and

$$\|z'_{n+1}(t_n + \zeta_i h_{n+1}) - u'(t_n + \zeta_i h_{n+1})\| = O(h_{n+1}^{r+1}). \tag{5.3.8}$$

Now, for a polynomial $Q(\theta)$ of degree $r+1$ in $[0,1]$, consider the set of $r+2$ linear conditions formed by the Hermite–Birkhoff interpolation conditions on $Q(0)$, $Q(1)$, $Q'(0)$, $Q'(1)$, $Q(\xi_i)$, $i = 1, \ldots, k$ (possibly $k = 0$), $Q'(\zeta_i)$, $i = 1, \ldots, k'$ (possibly $k' = 0$) and $Q'(\theta_i)$, $i = 1, \ldots, k'' = 2r - 1 - p - k - k'$ (possibly $k'' = 0$), and by the integral conditions obtained by assigning the values of $\int_0^1 \theta^s Q'(\theta)\mathrm{d}\theta$,

$s = 1, \ldots, p-r-1$. The points ξ_i, ζ_i and θ_i, if any, are assumed to be such that the $r+2$ conditions above are linearly independent, so that the polynomial $Q(\theta)$ is uniquely determined.

Note that if $p = 2r$, then the sole conditions on $Q(0)$, $Q(1)$, $Q'(0)$ and $\int_0^1 \theta^s Q'(\theta)\mathrm{d}\theta$, $s = 1, \ldots, p-r-1$, already form a set of $r+2$ linearly independent conditions in the space of polynomials of degree $r+1$. Moreover, if $p = 2r-1$, then by also using the condition on $Q'(1)$, a set of $r+2$ linearly independent conditions is obtained too. We can conclude that interpolants of order $1 + \lfloor \frac{p+1}{2} \rfloor$ are already available for explicit methods and that, consequently, both the present procedure and the previous bootstrapping process are relevant to the case $3 \leq 1 + \lfloor \frac{p+1}{2} \rfloor \leq q \leq r < p$.

Since, for a given function, the polynomial defined by the $r+2$ conditions above is accurate to order $r+1$ in the interval $[t_n, t_{n+1}]$ (local error $O(h_{n+1}^{r+2})$), it preserves this order even if the values of the function to be interpolated and those of its derivative are perturbed by terms of magnitude $O(h_{n+1}^{r+2})$ and $O(h_{n+1}^{r+1})$, respectively.

Thus the (uniquely solvable) system

$$
\begin{aligned}
&\text{(i)}\ \ v(t_n) = y_n, \quad v(t_{n+1}) = y_{n+1}, \\
&\text{(ii)}\ \ v'(t_n) = g(t_n, y_n), \quad v'(t_{n+1}) = g(t_{n+1}, y_{n+1}), \\
&\text{(iii)}\ \ v(t_n + \xi_i h_{n+1}) = u(t_n + \xi_i h_{n+1}), \quad i = 1, \ldots, k, \\
&\text{(iv)}\ \ v'(t_n + \zeta_i h_{n+1}) = u'(t_n + \zeta_i h_{n+1}), \quad i = 1, \ldots, k', \\
&\text{(v)}\ \ v'(t_n + \theta_i h_{n+1}) = g\big(t_n + \theta_i h_{n+1}, u(t_n + \theta_i h_{n+1})\big), \quad i = 1, \ldots, k'', \\
&\text{(vi)}\ \ \textstyle\int_{t_n}^{t_{n+1}} (t - t_n)^i v'(t)\mathrm{d}t = \int_{t_n}^{t_{n+1}} (t - t_n)^i u'(t)\mathrm{d}t, \quad i = 1, \ldots, p-r-1,
\end{aligned}
$$

provides an interpolant $v(t) = \eta_{(r+1-q)}(t)$ of uniform order $r+1$, whose form is

$$
\begin{aligned}
v(t_n + \theta h_{n+1}) = {} & d_1(\theta) y_n + d_2(\theta) y_{n+1} \\
& + d_3(\theta) h_{n+1} g(t_n, y_n) + d_4(\theta) h_{n+1} g(t_{n+1}, y_{n+1}) \\
& + \sum_{i=1}^{k} d_{4+i}(\theta) u(t_n + \xi_i h_{n+1}) + \sum_{i=1}^{k'} d_{4+k+i}(\theta) h_{n+1} u'(t_n + \zeta_i h_{n+1}) \\
& + \sum_{i=1}^{k''} d_{4+k+k'+i}(\theta) h_{n+1} g\big(t_n + \theta_i h_{n+1}, u(t_n + \theta_i h_{n+1})\big) \\
& + \sum_{i=1}^{p-r-1} d_{2r+3-p+i}(\theta) \int_{t_n}^{t_{n+1}} \left(\frac{t - t_n}{h_{n+1}}\right)^i u'(t)\mathrm{d}t,
\end{aligned}
$$

where the polynomials $d_i(\theta)$, $i = 1, \ldots, r+2$, have degree $r+1$. Moreover, it is immediately seen that $v(t)$ preserves the asymptotic orthogonality condition

$$
\left\| \int_{t_n}^{t_{n+1}} G(t)[z'_{n+1}(t) - v'(t)]\mathrm{d}t \right\| = O(h_{n+1}^{p+1})
$$

for every sufficiently smooth matrix-valued function G.

Since conditions (i), (ii), (iii), (iv) and (vi) do not involve additional function evaluations, in passing from order r to $r+1$, the procedure requires k'' new evaluations of the right-hand-side function $g(t,y)$ in (v). This results in k'' explicit extra stages of the form (5.1.6). Obviously, one would like to exploit as many integral conditions of the form (vi) as possible. By using (5.3.6) it can be proved that $p-r-1$ is the largest value of s for which the use of $u'(t)$ instead of $z'_{n+1}(t)$ causes a perturbation of order $\geq r+1$.

Assume that some inner superconvergence points ξ_i in (iii) and/or other superconvergence points leading to further interpolation conditions which are linearly dependent of the others, exist for $u(t)$. Then it is recommended to use all of them for conditions (v). In fact, this choice makes them superconvergence points for $v'(t)$ which, in turn, could be exploited for the next correction in (iv).

Summarizing, if the RK method of order p is superconvergent and possesses an NCE $\eta_{(0)}(t)$ of order q, then $p-q$ corrections are needed in order to construct an NCE $\eta_{(p-q)}(t)$ of uniform order p. The total number $s-\nu$ of additional stages in (5.1.6) is then given by

$$s-\nu \leq \sum_{r=q}^{p-1}(2r-1-p) = (p-q)(q-2).$$

The inequality is strict whenever some inner superconvergence points exist to be exploited for one or more corrections in (iii) and/or (iv). On the contrary, the equality holds whenever $k = k' = 0$ for all corrections.

In principle, the present procedure is more convenient than the bootstrapping process of the previous section because the former also uses the asymptotic orthogonality condition (5.3.6), which is not done by the latter. Nevertheless, given the standard Hermite interpolation conditions at the points t_n and t_{n+1}, it often happens that the asymptotic orthogonality condition (5.3.6) is equivalent to the existence of one or more inner superconvergence points of the type (5.3.7) and/or (5.3.8). As a consequence, we must choose to use either conditions of the type (vi) or (the equivalent) conditions of the type (iii) and/or (iv). In conclusion, it turns out that, in these cases, the number of free conditions to be exploited is not actually larger, making the two procedures equivalent from the point of view of computational cost.

For example, consider the starting interpolant $\eta_{(0)}(t)$ in Example 5.3.1. It may be seen that it also satisfies the asymptotic orthogonality condition (5.3.6). Therefore, since in this case $p-r-1=1$, in the first correction we could exploit the corresponding condition

$$\int_{t_n}^{t_{n+1}}(t-t_n)v'(t)\mathrm{d}t = \int_{t_n}^{t_{n+1}}(t-t_n)u'(t)\mathrm{d}t$$

of the type (vi). Unfortunately, given the Hermite interpolation conditions (i) and (ii) in the space of polynomials of degree 5, this integral condition is equivalent to the interpolation condition at the inner superconvergence point $t_n + \frac{1}{2}h_{n+1}$ in (iii).

5.3.3 *Implicit uniform corrections of NCEs*

For implicit RK methods, Bellen and Zennaro [37] modified the procedure in [279] by involving implicit extra stages of collocation type, starting from an NCE $\eta_{(0)}(t)$. Here we present an improved version.

First of all we observe that the procedure is justified by the need to preserve the stability properties of the underlying discrete RK method (see Section 8.4.1). This makes it necessary to use exclusively implicit extra stages. Therefore, unless they are already part of the original set of ν stages of the discrete RK method, the interpolation conditions for the derivative at the points t_n and t_{n+1} (see (ii) in the previous section) are not to be considered.

Let $u(t) = \eta_{(r-q)}(t)$ again be an interpolant of order (and degree) r, $1 \le q \le r < p$, satisfying the asymptotic orthogonality condition

$$\left\| \int_{t_n}^{t_{n+1}} G(t)[z'_{n+1}(t) - u'(t)]\mathrm{d}t \right\| = O(h_{n+1}^{p+1})$$

for every sufficiently smooth matrix-valued function G, where $z_{n+1}(t)$ is the solution to the local problem (5.3.2).

Moreover, let $t_n + \xi_i h_{n+1}$, $i = 1, \dots, k$, and $t_n + \zeta_i h_{n+1}$, $i = 1, \dots, k'$, be the points, if any, of inner superconvergence for $u(t)$ and $u'(t)$ satisfying (5.3.7) and (5.3.8), respectively. Note that the points $t_n + \zeta_j h_{n+1}$, if any, such that the collocation-like condition

$$u'(t_n + \zeta_j h_{n+1}) = g\big(t_n + \zeta_j h_{n+1}, u(t_n + \zeta_j h_{n+1})\big) \tag{5.3.9}$$

holds, also satisfy condition (5.3.8).

For a polynomial $Q(\theta)$ of degree $r + 1$ in $[0, 1]$, consider the set of $r + 2$ linear conditions formed by the Hermite–Birkhoff interpolation conditions on $Q(0)$, $Q(1)$, $Q(\xi_i)$, $i = 1, \dots, k$ (possibly $k = 0$), $Q'(\zeta_i)$, $i = 1, \dots, k'$ (possibly $k' = 0$) and $Q'(\theta_i)$, $i = 1, \dots, k'' = 2r + 1 - p - k - k'$ (possibly $k'' = 0$), and by the integral conditions obtained by assigning the values of $\int_0^1 \theta^s Q'(\theta)\mathrm{d}\theta$, $s = 1, \dots, p - r - 1$.

As in Section 5.3.2, the points θ_i, ζ_i and ξ_i are assumed to be such that the above set of $r+2$ conditions is linearly independent, so that the polynomial $Q(\theta)$ is uniquely determined.

Again, for a given function, the polynomial defined by the $r + 2$ conditions above is accurate to order $r+1$ in the interval $[t_n, t_{n+1}]$ (local error $O(h_{n+1}^{r+2})$) and, thus, it preserves this order even if the values of the function to be interpolated and those of its derivative are perturbed by terms of magnitude $O(h_{n+1}^{r+2})$ and $O(h_{n+1}^{r+1})$, respectively.

Thus the (uniquely solvable) system

(i) $v(t_n) = y_n, \quad v(t_{n+1}) = y_{n+1},$
(ii) $v(t_n + \xi_i h_{n+1}) = u(t_n + \xi_i h_{n+1}), \quad i = 1, \dots, k,$
(iii) $v'(t_n + \zeta_i h_{n+1}) = u'(t_n + \zeta_i h_{n+1}), \quad i = 1, \dots, k',$

(iv) $v'(t_n + \theta_i h_{n+1}) = g\big(t_n + \theta_i h_{n+1}, v(t_n + \theta_i h_{n+1})\big), \quad i = 1, \ldots, k''$,
(v) $\int_{t_n}^{t_{n+1}} (t - t_n)^i v'(t) \mathrm{d}t = \int_{t_n}^{t_{n+1}} (t - t_n)^i u'(t) \mathrm{d}t, \quad i = 1, \ldots, p - r - 1$,

provides an interpolant $v(t) = \eta_{(r+1-q)}(t)$ of uniform order $r + 1$, whose form is

$$\begin{aligned} v(t_n + \theta h_{n+1}) &= d_1(\theta) y_n + d_2(\theta) y_{n+1} \\ &\quad + \sum_{i=1}^{k} d_{2+i}(\theta) u(t_n + \xi_i h_{n+1}) + \sum_{i=1}^{k'} d_{2+k+i}(\theta) h_{n+1} u'(t_n + \zeta_i h_{n+1}) \\ &\quad + \sum_{i=1}^{k''} d_{2+k+k'+i}(\theta) h_{n+1} g\big(t_n + \theta_i h_{n+1}, u(t_n + \theta_i h_{n+1})\big) \\ &\quad + \sum_{i=1}^{p-r-1} d_{2r+3-p+i}(\theta) \int_{t_n}^{t_{n+1}} \left(\frac{t - t_n}{h_{n+1}} \right)^i u'(t) \mathrm{d}t, \end{aligned}$$

where the polynomials $d_i(\theta)$, $i = 1, \ldots, r + 2$, have degree $r + 1$. Moreover, $v(t)$ preserves the asymptotic orthogonality condition

$$\left\| \int_{t_n}^{t_{n+1}} G(t) [z'_{n+1}(t) - v'(t)] \mathrm{d}t \right\| = O(h_{n+1}^{p+1})$$

for every sufficiently smooth matrix-valued function G.

Conditions (i), (ii), (iii) and (v) do not involve additional function evaluations and, in passing from order r to $r+1$, the remaining collocation-like conditions (iv) lead to k'' implicit extra stages of the form (5.1.6), which guarantee the existence of the points ζ_j, $j = 1, \ldots, k''$, where the polynomial $v(t)$ satisfies conditions (5.3.9). In turn, these points are to be exploited for the next correction in (iii).

Summarizing, if the RK method of order p is superconvergent and possesses an NCE $\eta_{(0)}(t)$ of order q, then $p - q$ corrections are needed in order to construct an NCE $\eta_{(p-q)}(t)$ of uniform order p. If k'_1 is the number of interpolation conditions of type (iii) to be used for the first correction, then it is easy to see that the total number $s - \nu$ of additional stages in (5.1.6) is given by

$$\begin{aligned} s - \nu &\le \frac{(p - q)(q + 1)}{2} \quad \text{if } p - q \text{ is even}, \\ s - \nu &\le \frac{(p - q + 1) q}{2} - k'_1 \quad \text{if } p - q \text{ is odd}. \end{aligned}$$

Note that $k'_1 > 0$ results in a lower cost of the overall procedure only if the total number of corrections is odd.

In particular, if the procedure starts from an NCE of minimal order $q = \lfloor \frac{p+1}{2} \rfloor$, the total number of extra stages is

$$\begin{aligned} s - \nu &\le \frac{q(q + 1)}{2} \quad \text{if } p \text{ is even and } q \text{ is even}, \\ s - \nu &\le \frac{q(q + 1)}{2} - k'_1 \quad \text{if } p \text{ is even and } q \text{ is odd}, \end{aligned}$$

$$s-\nu \leq \frac{q(q+2)}{2} - k_1' \quad \text{if } p \text{ is odd and } q \text{ is even},$$

$$s-\nu \leq \frac{(q+1)^2}{2} \quad \text{if } p \text{ is odd and } q \text{ is odd}.$$

The inequalities are strict whenever some inner superconvergence points exist to be exploited for one or more corrections in (ii). On the contrary, equalities generally hold whenever $k=0$ for all corrections.

Example 5.3.2 Let us consider a superconvergent collocation method of order $p>\nu$ (see Section 5.2.1) and, for simplicity, assume that $k=0$ for all corrections. We know that the collocation polynomial is an NCE of order $q=\nu$ and that it is the interpolant $\eta_{(0)}(t)$ of the first class of maximum attainable order. Since collocation methods are implicit, it makes sense to use the implicit uniform correction procedure.

By definition of the method, it turns out that the collocation polynomial $\eta_{(0)}(t)$ satisfies conditions (5.3.9) for $\zeta_j = c_j$, $j=1,\ldots,\nu$. Therefore, there are exactly ν conditions of type (iii) which are candidates to be exploited for the first correction. On the other hand, it is possible to see that only $2\nu - p$ of such conditions are linearly independent of the other conditions (i) and (v). So we have $k_1' = 2\nu - p$ and, thus, the total number $s-\nu$ of additional stages needed to construct an NCE $\eta_{(p-\nu)}(t)$ of uniform order p is given by

$$s-\nu = \frac{(p-\nu)(\nu+1)}{2} \quad \text{if } p-\nu \text{ is even},$$

$$s-\nu = \frac{(p-\nu)(\nu+2)-\nu}{2} \quad \text{if } p-\nu \text{ is odd}.$$

In particular, for the Gaussian collocation ($p=2\nu$) we have $k_1'=0$ and both the above formulae become

$$s-\nu = \frac{\nu(\nu+1)}{2}. \ \diamond$$

5.3.4 *Some further references*

We conclude this section by mentioning some further papers, mainly devoted to finding interpolants of the second class of embedded pairs of RK methods.

Shampine [227, 228], Gladwell et al. [102], Tsitouras and Papageorgiou [251], Bogacki and Shampine [47] and Calvo, Montijano and Randez [64, 65] presented various interpolation procedures for well-known RK pairs due to Fehlberg [94] and to Dormand and Prince [76].

Calvo, Montijano and Randez [66], Sharp and Verner [231] and Papageorgiou, Simos and Tsitouras [215] proposed new classes of embedded RK pairs endowed with high order interpolants.

5.4 Direct construction of continuous RK methods

So far we have considered continuous extensions of a priori given discrete RK methods. In this section we present the other philosophy of constructing directly a CRK method, without necessarily starting from a given discrete formula.

As pointed out in Section 5.2, a general analysis of the uniform order for the continuous extensions (5.1.1) is based on the property that, for any $0 < \theta \leq 1$, it can be viewed as a discrete method $(\frac{A}{\theta}, \frac{b(\theta)}{\theta})$ with scaled stepsize θh_{n+1}. Horn [130] called it the *scaled RK method.* So we immediately get the uniform order conditions for the parameters c_i and a_{ij} and for the polynomials $b_i(\theta)$ from the well-known order conditions of the RK methods. The conditions up to order $p = 4$ have already been given in Table 5.1.

Let $N(p)$ and $CN(q)$ be the minimum number of stages for which there exist RK methods of (discrete) order p and CRK methods of uniform order q, respectively. Similarly, let $EN(p)$ and $CEN(q)$ be the same quantities restricted to the class of ERK methods and continuous ERK (CERK) methods.

In the general case, it is well known that

$$N(p) = \left\lfloor \frac{p+1}{2} \right\rfloor \quad \text{and} \quad CN(q) = q$$

and that these optimal bounds are attained, for instance, by collocation methods (see Section 5.2.1).

For ERK methods the theory was mainly developed by Butcher and the results can be found in his book [62].

For CERK methods the theory was developed much later by Owren and Zennaro [213, 214], Santo [222], Verner [260] and Verner and Zennaro [261, 262]. The results are often obtained by making somewhat sophisticated analyses of the continuous order conditions. This goes beyond the scope of this book and, hence, the interested reader is referred directly to the foregoing references.

All the order barriers for explicit methods are summarized in Tables 5.5 and 5.6.

We now concentrate on CERK methods with a minimum number of stages $CEN(q)$. It easily turns out that, for given $q \geq 2$, a whole family of such methods exists, which depends on a certain number of parameters. So the parameters can be selected in order to guarantee some nice properties of the method, such as minimization of a suitable estimate of the local error constant and maximization of the absolute stability region of the underlying discrete method (see Section 8.2.1).

Another nice characteristic of some CERK methods is the FSAL (*first same as last*) property. The FSAL property means that the last stage can be reused as the first stage $K^1_{n+1} = g(t_{n+1}, y_{n+1})$ of the next step. This implies that the actual cost of the method is reduced by one function evaluation per step. Of course, because the method is explicit, the reusable stage can be involved only for computation of the interpolant $\eta(t_n + \theta h_{n+1})$ for $\theta \neq 1$ and not for computation of $y_{n+1} = \eta(t_{n+1})$. To this purpose, see also the use of the stage $K^1_{n+1} = g(t_{n+1}, y_{n+1})$ in Sections 5.3.1 and 5.3.2.

In Tables 5.7, 5.8 and 5.9 we report from [214] three examples of CERK methods with a minimum number of stages $CEN(q)$ for uniform orders $q = 3, 4$ and 5, respectively. Such methods satisfy the FSAL property and minimize a suitable measure of the local error constant.

TABLE 5.5. The minimum number of stages necessary for an ERK method to attain the discrete order p.

p	$EN(p)$
1	1
2	2
3	3
4	4
5	6
6	7
7	9
8	11
$r \geq 9$	$\geq r+3$

TABLE 5.6. The minimum number of stages necessary for a CERK method to attain the uniform order q.

q	$CEN(q)$
1	1
2	2
3	4
4	6
5	8
6	11
$r \geq 7$	$\geq 2r-2$

We conclude this section by mentioning the methodology proposed by Verner [260], which is a sort of mixture between the idea behind the procedures surveyed in Section 5.3 and the direct construction of CRK methods. In fact, although his goal still is to upgrade the uniform order of a given discrete RK method (more specifically, of a given embedded pair of RK methods), his approach does not aim to construct a sequence of interpolants of increasing order up to the maximum order p but, rather, to determine a sequence of interpolants whose last element is of appropriate order p or $p-1$, trying to minimize the total number of extra stages.

TABLE 5.7. Butcher's tableau and continuous weights of a CERK method of order 3 with four stages.

$$
\begin{array}{c|cccc}
0 & & & & \\
\frac{12}{23} & \frac{12}{23} & & & \\
\frac{4}{5} & -\frac{68}{375} & \frac{368}{375} & & \\
1 & \frac{31}{144} & \frac{529}{1152} & \frac{125}{384} & \\
\hline
 & b_1(\theta) & b_2(\theta) & b_3(\theta) & b_4(\theta)
\end{array}
$$

$$
\begin{aligned}
b_1(\theta) &= \tfrac{41}{72}\theta^3 - \tfrac{65}{48}\theta^2 + \theta, \\
b_2(\theta) &= -\tfrac{529}{576}\theta^3 + \tfrac{529}{384}\theta^2, \\
b_3(\theta) &= -\tfrac{125}{192}\theta^3 + \tfrac{125}{128}\theta^2, \\
b_4(\theta) &= \theta^3 - \theta^2.
\end{aligned}
$$

TABLE 5.8. Butcher's tableau and continuous weights of a CERK method of order 4 with six stages.

$$
\begin{array}{c|cccccc}
0 & & & & & & \\
\frac{1}{6} & \frac{1}{6} & & & & & \\
\frac{11}{37} & \frac{44}{1369} & \frac{363}{1369} & & & & \\
\frac{11}{17} & \frac{3388}{4913} & -\frac{8349}{4913} & \frac{8140}{4913} & & & \\
\frac{13}{15} & -\frac{36764}{408375} & \frac{767}{1125} & -\frac{32708}{136125} & \frac{210392}{408375} & & \\
1 & \frac{1697}{18876} & 0 & \frac{50653}{116160} & \frac{299693}{1626240} & \frac{3375}{11648} & \\
\hline
 & b_1(\theta) & b_2(\theta) & b_3(\theta) & b_4(\theta) & b_5(\theta) & b_6(\theta)
\end{array}
$$

$$
\begin{aligned}
b_1(\theta) &= -\tfrac{866577}{824252}\theta^4 + \tfrac{1806901}{618189}\theta^3 - \tfrac{104217}{37466}\theta^2 + \theta, \\
b_2(\theta) &= 0, \\
b_3(\theta) &= \tfrac{12308679}{5072320}\theta^4 - \tfrac{2178079}{380424}\theta^3 + \tfrac{861101}{230560}\theta^2, \\
b_4(\theta) &= -\tfrac{7816583}{10144640}\theta^4 + \tfrac{6244423}{5325936}\theta^3 - \tfrac{63869}{293440}\theta^2, \\
b_5(\theta) &= -\tfrac{624375}{217984}\theta^4 + \tfrac{982125}{190736}\theta^3 - \tfrac{1522125}{762944}\theta^2, \\
b_6(\theta) &= \tfrac{296}{131}\theta^4 - \tfrac{461}{131}\theta^3 + \tfrac{165}{131}\theta^2.
\end{aligned}
$$

5.5 An application to the local error estimate

As is well known, there are mainly two approaches for the implementation with variable stepsize of a (discrete) RK method of order p, the aim of which is to obtain a numerical solution whose maximum global error,

$$\max_{1 \le n \le N} \|y(t_n) - y_n\|,$$

TABLE 5.9. Butcher's tableau and continuous weights of a CERK method of order 5 with eight stages.

c								
0								
$\frac{1}{6}$	$\frac{1}{6}$							
$\frac{1}{4}$	$\frac{1}{16}$	$\frac{3}{16}$						
$\frac{1}{2}$	$\frac{1}{4}$	$-\frac{3}{4}$	1					
$\frac{1}{2}$	$-\frac{3}{4}$	$\frac{15}{4}$	-3	$\frac{1}{2}$				
$\frac{9}{14}$	$\frac{369}{1372}$	$-\frac{243}{343}$	$\frac{297}{343}$	$\frac{1485}{9604}$	$\frac{297}{4802}$			
$\frac{7}{8}$	$-\frac{133}{4512}$	$\frac{1113}{6016}$	$\frac{7945}{16544}$	$-\frac{12845}{24064}$	$-\frac{315}{24064}$	$\frac{156065}{198528}$		
1	$\frac{83}{945}$	0	$\frac{248}{825}$	$\frac{41}{180}$	$\frac{1}{36}$	$\frac{2401}{38610}$	$\frac{6016}{20475}$	
	$b_1(\theta)$	$b_2(\theta)$	$b_3(\theta)$	$b_4(\theta)$	$b_5(\theta)$	$b_6(\theta)$	$b_7(\theta)$	$b_8(\theta)$

$$
\begin{aligned}
b_1(\theta) &= \tfrac{596}{315}\theta^5 - \tfrac{4969}{819}\theta^4 + \tfrac{17893}{2457}\theta^3 - \tfrac{3292}{819}\theta^2 + \theta,\\
b_2(\theta) &= 0,\\
b_3(\theta) &= -\tfrac{1984}{275}\theta^5 + \tfrac{1344}{65}\theta^4 - \tfrac{43568}{2145}\theta^3 + \tfrac{5112}{715}\theta^2,\\
b_4(\theta) &= \tfrac{118}{15}\theta^5 - \tfrac{1465}{78}\theta^4 + \tfrac{3161}{234}\theta^3 - \tfrac{123}{52}\theta^2,\\
b_5(\theta) &= 2\theta^5 - \tfrac{413}{78}\theta^4 + \tfrac{1061}{234}\theta^3 - \tfrac{63}{52}\theta^2,\\
b_6(\theta) &= -\tfrac{9604}{6435}\theta^5 + \tfrac{2401}{1521}\theta^4 + \tfrac{60025}{50193}\theta^3 - \tfrac{40817}{33462}\theta^2,\\
b_7(\theta) &= -\tfrac{48128}{6825}\theta^5 + \tfrac{96256}{5915}\theta^4 - \tfrac{637696}{53235}\theta^3 + \tfrac{18048}{5915}\theta^2,\\
b_8(\theta) &= 4\theta^5 - \tfrac{109}{13}\theta^4 + \tfrac{75}{13}\theta^3 - \tfrac{18}{13}\theta^2.
\end{aligned}
$$

shows a rough proportionality with respect to a user-supplied parameter TOL, called the *tolerance*. The concept of *tolerance proportionality* has been precisely defined and investigated by Stetter [239, 240].

The first approach consists in finding an estimate $\tilde{\sigma}_{n+1}$ of the *local error per unit step*,

$$\sigma_{n+1} = \|z_{n+1}(t_{n+1}) - y_{n+1}\|/h_{n+1},$$

at the end-point t_{n+1} along with a choice of the stepsize h_{n+1} that keeps $\tilde{e}_{n+1}$ below the tolerance TOL.

In turn, there are two main ways of estimating the local error. The more general one, applicable to all RK methods, consists in using *Richardson extrapolation* to obtain another RK scheme of one order higher $p+1$, whose approximation $\tilde{y}_{n+1}$ at the end-point t_{n+1} is used to set

$$\tilde{\sigma}_{n+1} = \|\tilde{y}_{n+1} - y_{n+1}\|/h_{n+1}. \tag{5.5.1}$$

This technique is usually quite expensive, in that it requires computing the RK formula three times for each integration step (if halving of the stepsize is used for extrapolation). However, in the case of explicit methods, it is possible to avoid

Richardson extrapolation if suitable *embedded pairs* of methods, such as *Runge–Kutta–Fehlberg* methods, are used (see Fehlberg [94]). The idea behind the pair is that, with a slightly bigger number of stages, it is possible to construct two different approximations y_{n+1} and $\tilde{y}_{n+1}$ of order p and $p+1$, respectively, and then to use (5.5.1).

The second approach, more suitable for explicit methods, consists in considering a pair in which another approximation $\bar{y}_{n+1}$ of one order lower $p-1$ is coupled to y_{n+1} of order p. We quote the famous pairs by Dormand and Prince [76]. Then the estimate

$$\bar{\sigma}_{n+1} = \|\bar{y}_{n+1} - y_{n+1}\|$$

of the local error $\|z_{n+1}(t_{n+1}) - \bar{y}_{n+1}\|$ of the lower order method is substituted for the estimate $\tilde{\sigma}_{n+1}$ of the local error per unit step and *local extrapolation* is used by choosing the stepsize that keeps $\bar{\sigma}_{n+1}$ below the tolerance TOL.

A quick comparison between the two strategies reveals that the second one is more efficient, since it advances with the more accurate method, even if the tolerance proportionality is generally better in the first strategy.

The availability of interpolants has also suggested alternative strategies for stepsize control. For RK methods of order p that possess a continuous extension $\eta(t)$ of maximum uniform order p, Enright [85, 86] proposed monitoring the magnitude of the *defect*

$$\delta(t) = g(t, \eta(t)) - \eta'(t).$$

In fact, by using the Gröbner–Alekseev non-linear variation-of-constants formula, we get

$$z_{n+1}(t) - \eta(t) = \int_{t_n}^{t} K(t, x)[g(x, \eta(x)) - \eta'(x)]\mathrm{d}x,$$

where $K(t,x)$ is a suitable variational matrix that depends on the interpolant $\eta(t)$ such that $K(x,x) \equiv I$, the identity matrix. Therefore, for $h_{n+1} \leq h_0$ sufficiently small, there exists a constant $K > 0$, independent of the particular RK method, such that

$$\|z_{n+1}(t_{n+1}) - y_{n+1}\| \leq \max_{t_n \leq t \leq t_{n+1}} \|z_{n+1}(t) - \eta(t)\| \leq h_{n+1} K \max_{t_n \leq t \leq t_{n+1}} \|\delta(t)\|.$$

Thus we can conclude that controlling the defect implies controlling the local error per unit step σ_{n+1}, and even the *uniform local error*

$$\Sigma_{n+1} = \max_{t_n \leq t \leq t_{n+1}} \|z_{n+1}(t) - \eta(t)\|.$$

Clearly, it is not efficient to try to compute exactly the quantity

$$\max_{t_n \leq t \leq t_{n+1}} \|\delta(t)\|.$$

However, under suitable conditions on the interpolant $\eta(t)$, there exists a problem-independent $\theta^* \in [0,1]$ such that $\|\delta(t_n + \theta^* h_{n+1})\|$ is a good approximation to it. Hence, the final procedure consists in choosing the stepsize that keeps $\|\delta(t_n + \theta^* h_{n+1})\|$ below the tolerance TOL.

For a deeper analysis of all the previous approaches, we refer the reader to the survey paper by Enright et al. [91].

The final strategy we consider for estimation of the local error is due to Zennaro [277, 279] and makes use of a continuous extension $\eta(t)$ of maximum uniform order p. It is based on the observation that an approximation $\tilde{y}_{n+1}$ of higher order $p+1$ may be obtained by applying a quadrature formula, exact for polynomials of degree $\leq p$, to the integral

$$y_n + \int_{t_n}^{t_{n+1}} g(x, \eta(x)) \mathrm{d}x.$$

The higher order approximation has the form

$$\tilde{y}_{n+1} = y_n + h_{n+1} \sum_{i=1}^{k} \beta_i g\big(t_n + \pi_i h_{n+1}, \eta(t_n + \pi_i h_{n+1})\big) \tag{5.5.2}$$

for suitable abscissae $\pi_1, \ldots, \pi_k$, requiring the computation of some additional stages. Of course, it is convenient to use a superconvergent quadrature formulae, such as the Gaussian formulae, in order to minimize the number of extra stages. Moreover, for implicit methods, it may happen that points satisfying the collocation-like conditions (5.3.9) (where $u(t) = \eta(t)$) exist. In this case it could be beneficial to include such points among the abscissae of the quadrature formula.

We also mention that, for implicit methods, Bellen and Zennaro [37] proposed modifying the quadrature formula (5.5.2) to the implicit form

$$\tilde{y}_{n+1} = y_n + h_{n+1} \sum_{i=1}^{k-1} \beta_i g\big(t_n + \pi_i h_{n+1}, \eta(t_n + \pi_i h_{n+1})\big) + h_{n+1}\beta_k g(t_{n+1}, \tilde{y}_{n+1}),$$

which is more suitable to providing an approximation with good stability properties.

In any case, formula (5.5.1) may be applied and used as for Runge–Kutta–Fehlberg methods.

6

RUNGE–KUTTA METHODS FOR DDES

In this chapter we develop the standard approach, introduced in Section 3.3 and analyzed in Chapter 4, based on the class of CRK methods discussed in Chapter 5.

Recall that the standard approach for solving the DDE

$$\begin{cases} y'(t) = f\Big(t, y(t), y\big(t - \tau(t, y(t))\big)\Big), & t_0 \le t \le t_f, \\ y(t) = \phi(t), \quad t \le t_0, \end{cases} \tag{6.0.1}$$

consists in solving step by step the local problems

$$\begin{cases} w'_{n+1}(t) = f\Big(t, w_{n+1}(t), x\big(t - \tau(t, w_{n+1}(t))\big)\Big), & t_n \le t \le t_{n+1}, \\ w_{n+1}(t_n) = y_n, \end{cases}$$

where

$$x(s) = \begin{cases} \phi(s) & \text{for } s \le t_0, \\ \eta(s) & \text{for } t_0 \le s \le t_n, \\ w_{n+1}(s) & \text{for } t_n \le s \le t_{n+1}, \end{cases}$$

and $\eta(s)$ is the continuous approximate solution computed by the method itself up to t_n. The continuous RK method $\big(A, b(\theta)\big)$ being given by (5.0.2) and (5.1.1) or, for interpolants based on extra stages, by (5.0.2), (5.1.3) and (5.1.6), the overall method for DDEs turns out to be

$$Y^i_{n+1} = y_n + h_{n+1} \sum_{j=1}^{s} a_{ij} f\Big(t^j_{n+1}, Y^j_{n+1}, \eta\big(t^j_{n+1} - \tau(t^j_{n+1}, Y^j_{n+1})\big)\Big), \quad i = 1, \ldots, s, \tag{6.0.2}$$

$$\eta(t_n + \theta h_{n+1}) = y_n + h_{n+1} \sum_{i=1}^{s} b_i(\theta) f\Big(t^i_{n+1}, Y^i_{n+1}, \eta\big(t^i_{n+1} - \tau(t^i_{n+1}, Y^i_{n+1})\big)\Big),$$

$$0 \le \theta \le 1. \tag{6.0.3}$$

This method will be called the *RK method for DDEs* or, in short, the *DDE method*. Moreover, we shall say that $\big(A, b\big)$ is the *underlying discrete RK method*, whereas $\big(A, b(\theta)\big)$ is the *underlying interpolant*. The pair formed by the underlying discrete RK method and by the underlying interpolant is called the *underlying continuous RK method*.

We describe this class of methods starting from the most general state dependent DDE (6.0.1) and then we specialize the procedure and the related convergence results to DDEs with constant delay, non-vanishing time dependent delay and vanishing delay. We conclude by illustrating the application of the procedure to the numerical solution of NDDEs.

6.1 The general delay case

In this section we consider the most general formulation (6.0.2), (6.0.3) of the RK method for the state dependent DDE (6.0.1).

For simplicity, we have put both formula (5.0.2) for the stages of the discrete RK method and formula (5.1.6) for the possible additional stages of the interpolant together in (6.0.2).

Note that the use of RK methods with an abscissa $c_i > 1$ could lead to an advanced deviated argument $t_{n+1}^i - \tau(t_{n+1}^i, Y_{n+1}^i) > t_{n+1}$, where the continuous extension $x(s)$ still has to be computed. Therefore, we make the following assumption.

Assumption 6.1.1 *The RK method for DDEs (6.0.2), (6.0.3) is such that the abscissae satisfy the constraint* $0 \le c_i \le 1$, $i = 1, \ldots, s$.

If overlapping occurs, that is if, for some index i, the argument $t_{n+1}^i - \tau(t_{n+1}^i, Y_{n+1}^i)$ of $\eta(s)$ lies in the current interval $[t_n, t_{n+1}]$, then the *spurious stage* $\tilde{Y}_{n+1}^i = \eta(t_{n+1}^i - \tau(t_{n+1}^i, Y_{n+1}^i))$ is given by formula (6.0.3) itself for

$$\theta = \theta_{n+1}^i = c_i - \frac{\tau(t_{n+1}^i, Y_{n+1}^i)}{h_{n+1}}.$$

Otherwise it is given by the interpolant $\eta(t)$ as computed in the past. In any case, in the mesh interval $[t_n, t_{n+1}]$ the method takes the form (in the Y notation)

$$\eta(t_n + \theta h_{n+1}) = y_n + h_{n+1} \sum_{i=1}^{s} b_i(\theta) f(t_{n+1}^i, Y_{n+1}^i, \tilde{Y}_{n+1}^i), \quad 0 \le \theta \le 1, \tag{6.1.1}$$

$$Y_{n+1}^i = y_n + h_{n+1} \sum_{j=1}^{s} a_{ij} f(t_{n+1}^j, Y_{n+1}^j, \tilde{Y}_{n+1}^j), \quad i = 1, \ldots, s, \tag{6.1.2}$$

where the spurious stages $\tilde{Y}_{n+1}^i$ are given by

$$\tilde{Y}_{n+1}^i = y_n + h_{n+1} \sum_{j=1}^{s} b_j(\theta_{n+1}^i) f(t_{n+1}^j, Y_{n+1}^j, \tilde{Y}_{n+1}^j) \tag{6.1.3}$$

if $t_{n+1}^i - \tau(t_{n+1}^i, Y_{n+1}^i) > t_n$, and by

$$\tilde{Y}_{n+1}^i = \eta\big(t_{n+1}^i - \tau(t_{n+1}^i, Y_{n+1}^i)\big) \tag{6.1.4}$$

otherwise.

Note that, whereas the system (6.1.1), (6.1.2), (6.1.4) has to be solved only for the stage values Y_{n+1}^j, $j = 1, \ldots, s$, the system enlarged by (6.1.3) for some

i has to be solved also for the relevant spurious stage values $\tilde{Y}^i_{n+1}$. Indeed, this does not increase the dimension of the system. In fact, by using the K notation

$$K^i_{n+1} = f(t^i_{n+1}, Y^i_{n+1}, \tilde{Y}^i_{n+1}),$$

we get the following system to be solved for K^i_{n+1}, $i = 1, \ldots, s$:

$$\eta(t_n + \theta h_{n+1}) = y_n + h_{n+1} \sum_{i=1}^{s} b_i(\theta) K^i_{n+1}, \quad 0 \le \theta \le 1, \tag{6.1.5}$$

$$K^i_{n+1} = f\big(t^i_{n+1}, y_n + h_{n+1} \sum_{j=1}^{s} a_{ij} K^j_{n+1}, \tilde{Y}^i_{n+1}\big), \quad i = 1, \ldots, s, \tag{6.1.6}$$

where

$$\tilde{Y}^i_{n+1} = y_n + h_{n+1} \sum_{j=1}^{s} b_j \left(c_i - \frac{\tau(t^i_{n+1}, y_n + h_{n+1} \sum_{k=1}^{s} a_{ik} K^k_{n+1})}{h_{n+1}} \right) K^j_{n+1} \tag{6.1.7}$$

if $t^i_{n+1} - \tau(t^i_{n+1}, y_n + h_{n+1} \sum_{k=1}^{s} a_{ik} K^k_{n+1}) > t_n$, and

$$\tilde{Y}^i_{n+1} = \eta \left(t^i_{n+1} - \tau(t^i_{n+1}, y_n + h_{n+1} \sum_{k=1}^{s} a_{ik} K^k_{n+1}) \right) \tag{6.1.8}$$

otherwise.

Most of the methods known in the literature and the relevant existing software for DDEs (and NDDEs) can be framed within this approach.

Despite it not being possible to express all RK methods for DDEs in terms of the stage values Y^i_{n+1} only, there are particular classes, essentially collocation methods (see Section 5.2.1), that allow us to express the spurious stages $\tilde{Y}^i_{n+1}$ in the system (6.1.2) in terms of the Y^i_{n+1}. This is the case for any *natural* CRK method (see Definition 5.1.1) with s distinct abscissae $c_1, \ldots, c_s$ such that $c_i \neq 0$, $i = 1, \ldots, s$, and a continuous extension $\eta(t_n + \theta h_{n+1})$ of degree s. In fact, in this case the polynomial $\eta(t)$ may be written using the Lagrange interpolation formula through the $s+1$ values $y_n (= \eta(t_n))$ and $Y^i_{n+1} (= \eta(t_n + c_i h_{n+1}))$, $i = 1, \ldots, s$, that is

$$\eta(t_n + \theta h_{n+1}) = \ell_0(\theta) y_n + \sum_{i=1}^{s} \ell_i(\theta) Y^i_{n+1}, \tag{6.1.9}$$

where ℓ_j, $j = 0, \ldots, s$ are the Lagrange polynomial coefficients relevant to the nodes $c_0 = 0$ and c_i, $i = 1, \ldots, s$. Therefore $\tilde{Y}^i_{n+1}$, which is equal to $\eta\big(t^i_{n+1} - \tau(t^i_{n+1}, Y^i_{n+1})\big)$, may be written using (6.1.9) for $\theta = \theta^i_{n+1} = c_i - \frac{\tau(t^i_{n+1}, Y^i_{n+1})}{h_{n+1}}$. The Gaussian collocation and Radau IIA methods satisfy the above condition and are natural choices for the construction of DDE methods.

For both the Y and K notations, the method is well-posed for any sufficiently small h_{n+1}, as stated in Theorem 4.3.1.

As for the convergence analysis of the above methods, the following theorem follows from the general theorem, Theorem 4.3.5.

Theorem 6.1.2 *Consider the DDE*

$$\begin{cases} y'(t) = f\Big(t, y(t), y(t-\tau(t, y(t)))\Big), & t_0 \leq t \leq t_f, \\ y(t) = \phi(t), \quad t \leq t_0, \end{cases}$$

where $f(t, y, x)$ is C^p-continuous in $[t_0, t_f] \times \mathbb{R}^d \times \mathbb{R}^d$, the initial function $\phi(t)$ is C^p-continuous and the delay $\tau(t, y)$ is C^p-continuous in $[t_0, t_f] \times \mathbb{R}^d$. Moreover, assume that the mesh $\Delta = \{t_0, t_1, \ldots, t_n, \ldots, t_N = t_f\}$ includes all the discontinuity points of order $\leq p$ lying in $[t_0, t_f]$. If the underlying CRK method has discrete order p and uniform order q, then the DDE method (6.1.1), (6.1.2), (6.1.3), (6.1.4) has discrete global order and uniform global order $q' = \min\{p, q+1\}$; that is

$$\max_{1 \leq n \leq N} \|y(t_n) - y_n\| = O(h^{q'})$$

and

$$\max_{t_0 \leq t \leq t_f} \|y(t) - \eta(t)\| = O(h^{q'}),$$

where $h = \max_{1 < n < N} h_n$.

According to Theorem 6.1.2, if the underlying CRK method has discrete order p and uniform order q, then we can either be satisfied with a DDE method with, possibly lower, uniform global order $q' = \min\{p, q+1\}$, or increase the uniform order of the underlying interpolant up to at least $p - 1$ in order to preserve the uniform global order p.

We can summarize the last option in the following corollary.

Corollary 6.1.3 *Under the hypotheses of Theorem 6.1.2 with $q \geq p - 1$, the continuous numerical solution $\eta(t)$ is such that*

$$\max_{t_0 \leq t \leq t_f} \|y(t) - \eta(t)\| = O(h^p).$$

Theorem 6.1.2 and Corollary 6.1.3 just guarantee that, by using an interpolant of order $p - 1$, the global order p of the discrete method is preserved for any choice of the mesh. A sharper error estimate and convergence analysis of the standard approach reveals that, under some restrictions on the mesh, the condition $q = p - 1$ is no longer necessary for the method to preserve the global order p. In other words, superconvergence is possible. On the other hand, an efficient DDE code ought to be implemented in a variable stepsize mode by performing the error control. In this case, if we try to estimate the local error by a method of higher order $p + 1$, uniform approximation of order $p - 1$ for the deviated arguments $y(t - \tau)$ is not sufficient and must be raised to p. These issues will be examined in the forthcoming sections.

For a deeper insight into the convergence properties it is still worth distinguishing between constant, vanishing and non-vanishing delays. Some convergence results will be stated without proof since they can be obtained as special cases of the general theory developed in Chapter 4.

6.2 DDEs with constant delay. Connections with Bellman's method and superconvergence

When the delay τ is constant, the RK method (6.1.1), (6.1.2), (6.1.4) takes the simpler form

$$\eta(t_n + \theta h_{n+1}) = y_n + h_{n+1} \sum_{i=1}^{s} b_i(\theta) f\left(t_{n+1}^i, Y_{n+1}^i, \eta(t_{n+1}^i - \tau)\right), \quad 0 \leq \theta \leq 1, \tag{6.2.1}$$

$$Y_{n+1}^i = y_n + h_{n+1} \sum_{j=1}^{s} a_{ij} f\left(t_{n+1}^j, Y_{n+1}^j, \eta(t_{n+1}^j - \tau)\right), \quad i = 1, \ldots, s, \tag{6.2.2}$$

where, for $h_{n+1} \leq \tau$, $\eta(t_{n+1}^j - \tau)$ is known for any j. The next theorem follows immediately from the general theorem, Theorem 4.1.2.

Theorem 6.2.1 *Consider the DDE with constant delay*

$$\begin{cases} y'(t) = f\Big(t, y(t), y(t-\tau)\Big), & t_0 \leq t \leq t_f, \\ y(t) = \phi(t), \quad t \leq t_0, \end{cases}$$

where $f(t,y,x)$ is C^p-continuous in $[t_0,t_f] \times \mathbb{R}^d \times \mathbb{R}^d$ and the initial function $\phi(t)$ is C^p-continuous. Moreover, assume that the mesh $\Delta = \{t_0, t_1, \ldots, t_n, \ldots, t_N = t_f\}$ includes the discontinuity points $\xi_i = i\tau$, $i = 1, \ldots, p$, lying in $[t_0, t_f]$ and that the underlying CRK method has discrete order p and uniform order q. Then the DDE method (6.2.1), (6.2.2) has discrete global order and uniform global order $q' = \min\{p, q+1\}$; that is

$$\max_{1 \leq n \leq N} \|y(t_n) - y_n\| = O(h^{q'})$$

and

$$\max_{t_0 \leq t \leq t_f} \|y(t) - \eta(t)\| = O(h^{q'}),$$

where $h = \max_{1 \leq n \leq N} h_n$.

In particular, if the algorithm proceeds with constant stepsize $h = \tau/m$ for some integer $m \geq 1$, then the deviated arguments take the values

$$t_{n+1}^j - \tau = t_{n+1-m}^j = t_{n-m} + c_j h, \quad j = 1, \ldots, s.$$

If the underlying CRK method is natural, then $\eta(t_{n+1}^j) = Y_{n+1}^j$ and also $\eta(t_{n+1}^j - \tau) = Y_{n+1-m}^j$ (see (5.1.8)). Therefore, we get the discrete method

$$y_{n+1} = y_n + h \sum_{i=1}^{s} b_i f\left(t_{n+1}^i, Y_{n+1}^i, Y_{n+1-m}^i\right), \tag{6.2.3}$$

$$Y_{n+1}^i = y_n + h \sum_{j=1}^{s} a_{ij} f\left(t_{n+1}^j, Y_{n+1}^j, Y_{n+1-m}^j\right), \quad i = 1, \ldots, s. \tag{6.2.4}$$

Summarizing, any DDE method (6.2.1), (6.2.2) based on a natural CRK method implemented with constant stepsize $h_{n+1} = h = \tau/m$, $m \geq 1$ integer, takes the form (6.2.3), (6.2.4) which does not make any explicit calls on interpolation. We shall call it the *natural RK method for DDEs.*

As for the convergence and accuracy order of (6.2.3), (6.2.4), we could still use the general results of Theorem 4.1.2. On the basis of that theorem, any RK method for DDEs is convergent with uniform global order $q' = \min\{p, q+1\}$. Therefore, the DDE method preserves the discrete global order p if the underlying interpolant has uniform order $q \geq p-1$. Unfortunately, in general, the interpolant of a natural CRK method has uniform order $q < p-1$. Nevertheless, it can be proved that, for (6.2.3), (6.2.4), the discrete global order p of the underlying natural CRK method is preserved no matter what the uniform order q of the interpolant is.

Theorem 6.2.2 *If the function $f(t, y, x)$ is C^p-continuous in $[t_0, t_f] \times \mathbb{R}^d \times \mathbb{R}^d$, then the natural RK method for DDEs (6.2.3), (6.2.4) of discrete order p is convergent of discrete order p, that is*

$$\max_{1 \leq n \leq N} \|y(t_n) - y_n\| = O(h^p),$$

for any initial function $\phi(t)$ of class C^p in $[t_0 - \tau, t_0]$ and for constant stepsize $h = \tau/m$, $m \geq 1$ integer.

The proof is obtained by simply noting that the nodal approximations given by the scheme (6.2.3), (6.2.4) coincide with those given by the underlying discrete RK scheme, as applied to the ordinary differential system (3.4.2) of Bellman's method of steps with the same stepsize $h = \tau/m$.

Theorem 6.2.2 says, in particular, that a superconvergent CRK method, which is also natural, preserves the superconvergence for DDEs. That is, the attained order at the nodal points is larger than the global uniform order. In view of more general cases involving variable delays, it is worth remarking that this preservation of superconvergence is essentially due to two occurrences which are intrinsic to implementations of natural CRK schemes with constant stepsize $h = \tau/m$, $m \geq 1$ integer:

(i) The deviated argument $t-\tau$ maps any mesh interval $[t_n, t_{n+1}]$ into a previous mesh interval $[t_{n-m}, t_{n+1-m}]$ and, hence, in the local problem

$$\begin{cases} w_{n+1}'(t) = f\big(t, w_{n+1}(t), \eta(t-\tau)\big), & t_n \leq t \leq t_{n+1}, \\ w_{n+1}(t_n) = y_n, \end{cases}$$

the function $\eta(t-\tau)$ is of class C^∞.

(ii) For every abscissa c_i, the deviated argument $t-\tau$ maps the point t_{n+1}^i into the point $t_{n+1}^i - \tau = t_{n+1-m}^i$, so that

$$\eta(t_{n+1}^i - \tau) = Y_{n+1-m}^i.$$

On the other hand, it is easy to check that, for an arbitrary choice of the mesh points $\{\xi_0 = t_0 < t_1 < \cdots < t_m = \xi_1\}$ in the first macro interval $[\xi_0, \xi_1]$, conditions (i) and (ii) still hold in $[t_0, t_f]$ provided that, in the subsequent macro intervals $[\xi_k, \xi_{k+1}]$, $k = 1, 2, \ldots$, the mesh points are defined by

$$t_{km+i} = t_i + k\tau, \quad i = 1, \ldots, m, \quad k = 1, 2, \ldots \tag{6.2.5}$$

A mesh Δ satisfying condition (6.2.5) is called *constrained mesh* (see the forthcoming Definition 6.3.2 relevant to more general delays). It follows that, for a constrained mesh Δ, the method based on natural CRK schemes becomes

$$y_{n+1} = y_n + h_{n+1} \sum_{i=1}^{s} b_i f\left(t_{n+1}^i, Y_{n+1}^i, Y_{n+1-m}^i\right),$$

$$Y_{n+1}^i = y_n + h_{n+1} \sum_{j=1}^{s} a_{ij} f\left(t_{n+1}^j, Y_{n+1}^j, Y_{n+1-m}^j\right), \quad i = 1, \ldots, s,$$

with $h_i = h_{km+i}$, $i = 1, \ldots, m$, $k = 1, 2, \ldots$, and preserves the superconvergence at the mesh points.

Example 6.2.3 Consider equation (1.2.3) and the DDE method (1.2.6), (1.2.7) based on the midpoint rule with constant stepsize $h = 1/(m - \delta)$. Note that, for $\delta = 0$, the corresponding mesh Δ is a constrained mesh, whereas, for $\delta \neq 0$, it is not. Therefore, since the midpoint rule with linear interpolation is a collocation method of order $p = 2$, it is also a natural CRK method, and the resulting DDE method for $\delta = 0$ converges with global order 2. On the other hand, the midpoint rule with linear interpolation is a CRK method of uniform order $q = 1$ and, hence, Theorem 4.1.6 guarantees the global order 2 for any sequence of meshes with constant (or even variable) stepsize. This is the case of $h = 1/m_i$, with $m_i = \frac{5}{3}2^i$, which leads to a non-constrained mesh for any i. The numerical evidence for both constrained and non-constrained meshes is given in Figure 1.10.

The DDE method based on collocation at $\nu > 1$ Gaussian points is different because it is a superconvergent method with global order $p = 2\nu$ and local uniform order $q = \nu$. For $\delta = 0$ the DDE method still preserves the global order 2ν. On the contrary, for $\delta \neq 0$ the mesh is no longer constrained and, hence, by Theorem 4.1.6, the global order decays to the uniform global order $\nu + 1$. For $\nu = 2$, Figure 1.10 illustrates the numerical results with stepsizes that are integer and non-integer submultiples of the delay. ◇

6.3 DDEs with non-vanishing time dependent delay. Constrained mesh method and superconvergence

When the delay depends on t and satisfies hypotheses (H_1) and (H_4), the standard approach consists in computing the discontinuity points $\xi_0 = t_0 < \xi_1 < \cdots < \xi_l = t_f$ by using (2.1.6), and then solving the ODE

$$\begin{cases} w'(t) = f\Big(t, w(t), \eta\big(t - \tau(t)\big)\Big), & \xi_{k-1} \le t \le \xi_k, \\ w(\xi_{k-1}) = \eta(\xi_{k-1}), \end{cases}$$

at each macro interval $[\xi_{k-1}, \xi_k]$.

The RK method (6.1.1), (6.1.2), (6.1.3), (6.1.4) takes the form

$$Y^i_{n+1} = y_n + h_{n+1} \sum_{j=1}^{s} a_{ij} f\Big(t^j_{n+1}, Y^j_{n+1}, \eta\big(t^j_{n+1} - \tau(t^j_{n+1})\big)\Big), \quad i = 1, \ldots, s, \tag{6.3.1}$$

$$\eta(t_n + \theta h_{n+1}) = y_n + h_{n+1} \sum_{i=1}^{s} b_i(\theta) f\Big(t^i_{n+1}, Y^i_{n+1}, \eta\big(t^i_{n+1} - \tau(t^i_{n+1})\big)\Big), \quad 0 \le \theta \le 1, \tag{6.3.2}$$

where hypothesis (H_1) ensures that, for h_{n+1} small enough, the terms $\eta\big(t^i_{n+1} - \tau(t^i_{n+1})\big)$ are known.

By Theorem 4.1.6, we get the following convergence result.

Theorem 6.3.1 *Consider the DDE*

$$\begin{cases} y'(t) = f\Big(t, y(t), y\big(t - \tau(t)\big)\Big), & t_0 \le t \le t_f, \\ y(t) = \phi(t), \quad t \le t_0, \end{cases} \tag{6.3.3}$$

where $f(t,y,x)$ is C^p-continuous in $[t_0, t_f] \times \mathbb{R}^d \times \mathbb{R}^d$, the initial function $\phi(t)$ is C^p-continuous and the delay $\tau(t)$ is C^p-continuous in $[t_0, t_f]$ and fulfills hypothesis (H_1). Moreover, assume that the mesh $\Delta = \{t_0, t_1, \ldots, t_n, \ldots, t_N = t_f\}$ includes all the discontinuity points of order $\le p$ lying in $[t_0, t_f]$. If the underlying CRK method has discrete order p and uniform order q, then the DDE method (6.3.1), (6.3.2) has discrete global order and uniform global order $q' =$ $\min\{p, q+1\}$*; that is*

$$\max_{1 \le n \le N} \|y(t_n) - y_n\| = O(h^{q'})$$

and

$$\max_{t_0 \le t \le t_f} \|y(t) - \eta(t)\| = O(h^{q'}),$$

where $h = \max_{1 \le n \le N} h_n$.

In connection with the constrained mesh selection introduced in the previous section for preserving the superconvergence, consider the following generalization of the definition of a constrained mesh to the case of variable delay.

Definition 6.3.2 *A mesh Δ related to the deviated argument $\alpha(t) = t - \tau(t)$ is called a* constrained mesh *if, for an arbitrary set of mesh points $\{\xi_0 = t_0 < t_1 < \cdots < t_m = \xi_1\}$ in $[\xi_0, \xi_1]$, the mesh is recursively determined in the subsequent intervals $[\xi_k, \xi_{k+1}]$, $k = 1, \ldots, l-1$, by*

$$t_{km+i} = t_{(k-1)m+i} + \tau(t_{km+i}), \qquad i = 1, \ldots, m. \tag{6.3.4}$$

Note that, by starting from $[\xi_0, \xi_1]$, the *forward* construction of the constrained mesh requires the solution of equation (6.3.4) for t_{km+i}, $i = 1, \ldots, m$, $k = 1, \ldots, l-1$. A cheaper construction of a constrained mesh can be done in a *backward* fashion by starting from an arbitrary mesh selection in the last macro interval $[\xi_{l-1}, \xi_l]$ and by explicitly computing the mesh points $t_{(k-1)m+i}$, $k = l-1, l-2, \ldots, 1$, $i = 1, \ldots, m$, in the previous macro intervals.

It is evident that, for a constrained mesh Δ, the characteristic property (i) is still fulfilled. On the contrary, property (ii) is not satisfied even if $b_j(c_i) = a_{ij}$, unless the deviated argument $\alpha(t)$ is a function of the form $\alpha(t) = rt - d$, $0 < r < 1$, $d \geq 0$, $t \geq t_0 > 0$ such as, for example, in the pantograph equation (2.1.1). In fact, in general, we have

$$t_{n+1}^j - \tau(t_{n+1}^j) \neq t_{n-m} + c_j h_{n+1-m} = t_{n+1-m}^j,$$

and hence, even if $\eta(t_{n+1-m}^j) = Y_{n+1-m}^j$, we have $\eta(t_{n+1}^j - \tau(t_{n+1}^j)) \neq Y_{n+1-m}^j$.

Despite the lack of property (ii), which appeared to be crucial in the proof of Theorem 6.2.2 via Bellman's approach, the preservation of superconvergence for CRK methods of uniform order $q < p - 1$ is still possible under the constrained mesh selection strategy.

The preservation of superconvergence for constrained meshes was first proved by Bellen [21] for one-step collocation at Gaussian points (see also Bellen [22], Brunner [52, 53, 55] and the book by Brunner [56] for further generalizations to more general functional differential equations), and later by Vermiglio [257] for a particular one-step subregion method. Finally, Zennaro [279] generalized the result to any RK method interpolated by an NCE. So, the asymptotic orthogonality condition (5.2.3), rather than $b_j(c_i) = a_{ij}$ and, consequently, $\eta(t_{n+1}^i) = Y_{n+1}^i$, is the essential property for interpolation in order to preserve the superconvergence in the constrained mesh implementation of the DDE method.

The following result on superconvergence is a corollary to Theorem 5.2.10 and represents an extension of the original results in [21], valid for one-step collocation, to the whole class of RK methods.

Theorem 6.3.3 *Assume that the DDE method (6.3.1), (6.3.2) is applied to (6.3.3) with a constrained mesh Δ. If the underlying CRK method has discrete order p and the interpolant is an NCE of order q, then the continuous numerical solution $\eta(t)$ is such that*

$$\max_{t_0 \leq t \leq t_f} \|y(t) - \eta(t)\| = O(h^{q'}),$$

where $q' = \min\{q+1, p\}$,

$$\max_{0 \leq n \leq N} \|y(t_n) - \eta(t_n)\| = O(h^p),$$

$$\max_{t_0 \leq t \leq t_f} \|y^{(j)}(t) - \eta^{(j)}(t)\| = O(h^{q-j+1}), \quad j = 1, \ldots, q,$$

$$\max_{0\le n\le N-1}\left\| \int_{t_n}^{t_{n+1}} G(x)[y^{(j)}(x)-\eta^{(j)}(x)]\mathrm{d}x \right\| = O(h^{p+1}), \quad j=0,1,$$

for every sufficiently smooth matrix-valued function $G(x)$.

Proof. The proof is carried out by induction on the macro intervals $[\xi_{k-1},\xi_k]$. The result is clearly true in the first interval $[\xi_0,\xi_1]$ because of the property (5.1.13) of the discrete RK method and of the properties (5.1.14), (5.1.15) and (5.2.5) of the NCE $\eta(t)$.

Then assume that the result is true up to the interval $[\xi_{k-2},\xi_{k-1}]$ and solve

$$\begin{cases} w'(t) = f\Big(t, w(t), \eta\big(t-\tau(t)\big)\Big), & \xi_{k-1}\le t\le \xi_k, \\ w(\xi_{k-1}) = \eta(\xi_{k-1}). \end{cases} \tag{6.3.5}$$

Moreover, consider the auxiliary DDE

$$\begin{cases} z'(t) = f\big(t, z(t), y(t-\tau(t))\big), & \xi_{k-1}\le t\le \xi_k, \\ z(\xi_{k-1}) = \eta(\xi_{k-1}). \end{cases}$$

Since we have assumed $\|y(\xi_{k-1})-\eta(\xi_{k-1})\| = O(h^p)$, it is clear that

$$\max_{\xi_{k-1}\le t\le \xi_k} \|y(t)-z(t)\| = O(h^p). \tag{6.3.6}$$

Because the mesh is constrained, by setting $\alpha(x)=x-\tau(x)$ and by identifying the intervals $[\xi_{k-1},\xi_k]$ with $[\tilde{a},\tilde{b}]$ and $[\xi_{k-2},\xi_{k-1}]$ with $[a,b]$ and the functions $y(t)$ with $u_\Delta(t)$ and $\eta(t)$ with $v_\Delta(t)$, we are in the hypotheses of Theorem 5.2.10. Therefore, formulae (5.2.15), (5.2.16), (5.2.17) and (5.2.18) hold for $w(t)$ and $z(t)$.

Since the RK method and its NCE satisfy (5.1.13), (5.1.14), (5.1.15) and (5.2.5) with respect to the solution $w(t)$ of (6.3.5), the estimate (6.3.6) eventually implies that the result is true also in the interval $[\xi_{k-1},\xi_k]$. ■

Example 6.3.4 Consider the equation

$$\begin{cases} y'(t) = \lambda\frac{t-1}{t}y\big(t-\log(t)-1\big)y(t), & t\ge 1, \\ y(t)=1, \quad 0\le t\le 1. \end{cases} \tag{6.3.7}$$

We want to integrate it in the interval $[1,\xi_2]$, where the breaking points are $\xi_1 = 3.1461932206205825852$ and $\xi_2 = 5.9254498245082464926$ and the solution is

$$y(t) = \begin{cases} \exp\left(\lambda(t-\log(t)-1)\right), & 1\le t\le \xi_1, \\ \exp\left(\lambda + \int_{\xi}^{t} \lambda\frac{s-1}{s}\mathrm{e}^{\lambda(s-\log(s(s-\log(s)-1))-2)}\mathrm{d}s\right), & \xi_1\le t\le \xi_2. \end{cases}$$

To this end, consider the following family of CERK methods:

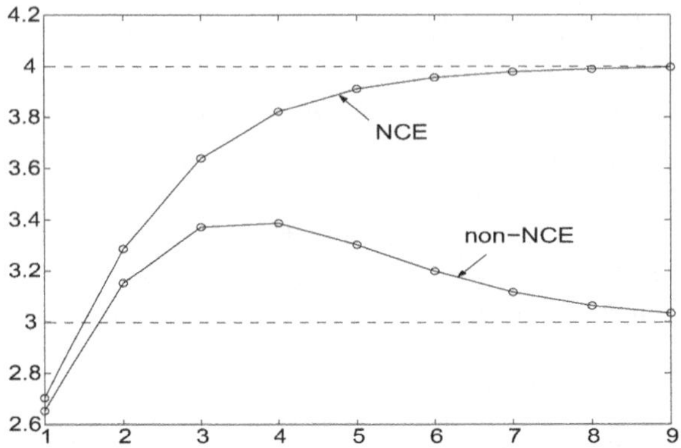

FIG. 6.1. Logarithmic plots of ratios between consecutive errors for the solution of (6.3.7) with $\lambda = 1$, obtained by doubling the number of steps for the explicit RK method of order 4 interpolated to the uniform global order 3 by an NCE and a non-NCE interpolant.

$$\begin{array}{c|cccc} 0 & & & & \\ \frac{1}{2} & \frac{1}{2} & & & \\ \frac{1}{2} & 0 & \frac{1}{2} & & \\ 1 & 0 & 0 & 1 & \\ \hline & \frac{1}{6} & \frac{1}{3} & \frac{1}{3} & \frac{1}{6} \end{array} \qquad \begin{aligned} b_1(\theta) &= -x_1\theta^2 + (x_1 + \tfrac{1}{6})\theta, \\ b_2(\theta) &= -x_2\theta^2 + (x_2 + \tfrac{1}{3})\theta, \\ b_3(\theta) &= -x_3\theta^2 + (x_3 + \tfrac{1}{3})\theta, \\ b_4(\theta) &= -x_4\theta^2 + (x_4 + \tfrac{1}{6})\theta, \end{aligned}$$

of discrete order $p = 4$, uniform order $q = 2$ and uniform global order $q' = 3$, where the parameters x_1, x_2, x_3 and x_4 satisfy the conditions

$$\begin{aligned} x_1 + x_2 + x_3 + x_4 &= 0, \\ x_4 - x_1 &= 1, \\ x_2 + x_3 + 2x_4 &= -1, \end{aligned}$$

derived from Table 5.1.

In particular, for $x_1 = \frac{1}{2}$, $x_4 = -\frac{1}{2}$ and $x_2 = x_3 = 0$, the interpolant is an NCE (see Section 5.2.2) with matrix $B = \big(b_j(c_i)\big) \neq A$. Any other choice of the parameters x_i leads to an interpolant that is not an NCE.

The constrained mesh is built up in a backward fashion with N equidistant mesh points in the last macro interval $[\xi_1, \xi_2]$ and the corresponding images in $[1, \xi_1]$. We have compared the solutions obtained by using the NCE and the interpolant determined by the parameters $x_1 = \frac{1}{2}$, $x_4 = -\frac{1}{2}$ and $x_2 = -x_3 = 1$. In Figure 6.1 the nodal order of the two methods are shown by successively doubling N. As expected, the method performs to order 4 when the interpolant is the NCE, whereas the order is only 3 in the other case. $\diamond$

We conclude this section by remarking that, in Bellman's method of steps for variable delays, system (3.4.3) explicitly depends on $\tau'(t)$ and, hence, formulae

(6.3.1) and (6.3.2) cannot be framed into that scheme. Nevertheless, if $\alpha(t) = rt - d$, $0 < r < 1$, $d \geq 0$ and $t \geq t_0 > 0$, the two approaches lead to the same method for any natural CRK method implemented with a constrained mesh. In fact, given the primary discontinuity points

$$\xi_0 = t_0, \quad \xi_1 = \frac{\xi_0 + d}{r}, \quad \xi_2 = \frac{\xi_1 + d}{r}, \ldots,$$

in the first macro interval $[\xi_0, \xi_1]$ the standard approach on an arbitrary mesh

$$\Delta_1 = \{\xi_0 = t_0, t_1, t_2, \ldots, t_m = \xi_1\}, \qquad h_{n+1} = t_{n+1} - t_n,$$

$n = 0, 1, \ldots, m-1$, equals the first step of the Bellman method (3.4.3) on the same mesh. Now, the second step of the Bellman method on the constrained mesh

$$\Delta_2 = \left\{t_m = \xi_1, \ldots, t_{m+n} = \frac{t_n + d}{r}, \ldots, t_{2m} = \xi_2\right\}, \quad h_{m+n+1} = \frac{h_{n+1}}{r},$$

through to the second macro interval $[\xi_1, \xi_2]$ consists in solving

$$\begin{cases} u'(t) = rf\Big(\alpha(t), u(t), \phi\big(\alpha(\alpha(t))\big)\Big), \\ y'(t) = f\big(t, y(t), u(t)\big), \\ u(\xi_1) = \phi(\xi_0), \\ y(\xi_1) \;\; \text{given by the previous step on } [\xi_0, \xi_1], \end{cases} \tag{6.3.8}$$

where $u(t) = y(rt - d)$.

Owing to the particular form of $\alpha(t) = rt - d$, the integration of (6.3.8) on Δ_2 by a natural RK method provides for $u(t)$ and $y(t)$ the approximate nodal values

$$u_{n+1+m} = u_{n+m} + rh_{n+1+m} \sum_{i=1}^{s} b_i f\big(t_{n+1}^i, U_{n+1+m}^i, \phi(rt_{n+1}^i - d)\big)$$

and

$$y_{n+1+m} = y_{n+m} + h_{n+1+m} \sum_{i=1}^{s} b_i f\big(t_{n+1+m}^i, Y_{n+1+m}^i, U_{n+1+m}^i\big),$$

where

$$U_{n+1+m}^i = u_{n+m} + rh_{n+1+m} \sum_{j=1}^{s} a_{ij} f\big(t_{n+1}^j, U_{n+1+m}^j, \phi(rt_{n+1}^i - d)\big), \quad i = 1, \ldots, s,$$

and

$$Y_{n+1+m}^i = y_{n+m} + h_{n+1+m} \sum_{j=1}^{s} a_{ij} f\big(t_{n+1+m}^j, Y_{n+1+m}^j, U_{n+1+m}^j\big), \quad i = 1, \ldots, s.$$

Since $rh_{n+1+m} = h_{n+1}$ for $n = 0, \ldots, m-1$, we have that u_{n+1+m} and U_{n+1+m}^i coincide with the nodal and stage values y_{n+1} and Y_{n+1}^i of the integration on Δ_1 in the first macro interval $[\xi_0, \xi_1]$. In particular, the initial value $y(\xi_1)$ in (6.3.8) is just y_m, which is equal to u_{2m}.

In conclusion, for DDEs with shifted linear deviated arguments $\alpha(t) = rt - d$, $d \geq 0$ and $t \geq t_0 > 0$, the method of steps with a constrained mesh through the macro intervals $[\xi_{k-1}, \xi_k]$ is equivalent to Bellman's method for any natural CRK method. Therefore, preservation of the discrete global order p of the underlying RK method follows immediately.

6.4 DDEs with vanishing time dependent delay

When the delay depends on t and vanishes at some point $t^* \in [t_0, t_f]$, overlapping necessarily occurs for any mesh interval $[t_n, t_{n+1}]$ including t^*. Therefore, for the method

$$\eta(t_n + \theta h_{n+1}) = y_n + h_{n+1} \sum_{i=1}^{s} b_i(\theta) f(t_{n+1}^i, Y_{n+1}^i, \tilde{Y}_{n+1}^i), \quad 0 \leq \theta \leq 1, \tag{6.4.1}$$

$$Y_{n+1}^i = y_n + h_{n+1} \sum_{j=1}^{s} a_{ij} f(t_{n+1}^j, Y_{n+1}^j, \tilde{Y}_{n+1}^j), \quad i = 1, \ldots, s, \tag{6.4.2}$$

where

$$\tilde{Y}_{n+1}^i = y_n + h_{n+1} \sum_{j=1}^{s} b_j \left(c_i - \frac{\tau(t_{n+1}^i)}{h_{n+1}} \right) f(t_{n+1}^j, Y_{n+1}^j, \tilde{Y}_{n+1}^j) \tag{6.4.3}$$

if $t_{n+1}^i - \tau(t_{n+1}^i) > t_n$, and

$$\tilde{Y}_{n+1}^i = \eta\big(t_{n+1}^i - \tau(t_{n+1}^i)\big) \tag{6.4.4}$$

if $t_{n+1}^i - \tau(t_{n+1}^i) \leq t_n$, the option (6.4.3) is possibly involved and some values $\tilde{Y}_{n+1}^i$ might actually be unknown. This does not cause much trouble and the general results of Theorems 4.2.3 and 4.2.5 extend to the method (6.4.1), (6.4.2), (6.4.3), (6.4.4). In particular, the method is well-posed for any sufficiently small stepsize h_{n+1} and the following convergence result holds.

Theorem 6.4.1 *Consider the DDE*

$$\begin{cases} y'(t) = f\Big(t, y(t), y\big(t - \tau(t)\big)\Big), & t_0 \leq t \leq t_f, \\ y(t) = \phi(t), \quad t \leq t_0, \end{cases}$$

where $f(t, y, x)$ is C^p-continuous in $[t_0, t_f] \times \mathbb{R}^d \times \mathbb{R}^d$, the initial function $\phi(t)$ is C^p-continuous and the delay $\tau(t)$ is C^p-continuous in $[t_0, t_f]$. Moreover, assume that the mesh $\Delta = \{t_0, t_1, \ldots, t_n, \ldots, t_N = t_f\}$ includes all the discontinuity points of order $\leq p$ lying in $[t_0, t_f]$. If the underlying CRK method has discrete order p and uniform order q, then the DDE method (6.4.1), (6.4.2), (6.4.3),

(6.4.4) has discrete global order and uniform global order $q' = \min\{p, q+1\}$; *that is*

$$\max_{1\le n\le N} \|y(t_n) - y_n\| = O(h^{q'})$$

and

$$\max_{t_0\le t\le t_f} \|y(t) - \eta(t)\| = O(h^{q'}),$$

where $h = \max_{1\le n\le N} h_n$.

As for possible superconvergence phenomena, the results of the previous section hold as long as we avoid overlapping. On the contrary, when we pass a vanishing delay point, the previous theory collapses. However, there is a particular case of vanishing delay, namely the *proportional delay* $\tau(t) = (1-r)t$, $0 < r < 1$, typical of the pantograph equation, where the preservation of superconvergence is still possible.

6.4.1 *The pantograph equation*

The pantograph equation

$$\begin{cases} y'(t) = ay(t) + by(rt), & 0 \le t \le t_f, \\ y(0) = y_0, \end{cases} \tag{6.4.5}$$

where $0 < r < 1$, is a very special delay differential equation that arises in quite different fields of pure and applied mathematics such as number theory, dynamical systems, probability, quantum mechanics and electrodynamics. In particular, it was used by Ockendon and Tayler [211] to study how the electric current is collected by the pantograph of an electric locomotive, from where it gets its name.

Well-posedness, stability, oscillation, continuity and asymptotic boundedness of the solution have been studied extensively for equation (6.4.5) and for its neutral counterpart, called the *generalized pantograph equation*,

$$\begin{cases} y'(t) = Ay(t) + By(rt) + Cy'(rt), & 0 \le t \le t_f, \\ y(0) = y_0, \end{cases}$$

where A, B and C are, in general, complex $d \times d$ matrices (see Iserles [145] and the rich bibliography therein).

The very particular form of the deviated argument $\alpha(t) = rt$ entails various peculiarities for equation (6.4.5), which are important for its numerical integration. First of all, no primary discontinuities are present in the solution and, hence, no constraints are imposed on the mesh for its accurate approximation. Secondly, overlapping occurs at the first step $[0, h_1]$ for any value of h_1 and, possibly, at subsequent steps, and therefore the method of steps cannot apply. On the other hand, the deviated argument rt may be viewed as a delayed argument $rt = t - (1-r)t$ with monotonically increasing delay $\tau(t) = (1-r)t$, which vanishes if and only if $t = 0$. Thus, once we have passed the first mesh point $t_1 = h_1$, the delay satisfies hypothesis (H_1) in $[t_1, t_f]$ with $\tau_0 = (1-r)h_1$, i.e.

$$\tau(t) = (1-r)t \geq (1-r)h_1 = \tau_0, \quad t \in [t_1, t_f],$$

and the method of steps possibly applies for continuing the integration.

Then the first integration step is

$$\eta(\theta h_1) = y_0 + h_1 \sum_{i=1}^{s} b_i(\theta) K_1^i, \quad 0 \leq \theta \leq 1,$$

$$K_1^i = a\Big(y_0 + h_1 \sum_{j=1}^{s} a_{ij} K_1^j\Big) + b\Big(y_0 + h_1 \sum_{j=1}^{s} b_j(rc_i) K_1^j\Big), \quad i = 1, \ldots, s.$$

Subsequent integration steps are instead given by

$$\eta(t_n + \theta h_{n+1}) = y_n + h_{n+1} \sum_{i=1}^{s} b_i(\theta) K_{n+1}^i, \quad 0 \leq \theta \leq 1,$$

$$K_{n+1}^i = a\Big(y_n + h_{n+1} \sum_{j=1}^{s} a_{ij} K_{n+1}^j\Big) + b\tilde{Y}_{n+1}^i, \quad i = 1, \ldots, s,$$

where

$$\tilde{Y}_{n+1}^i = y_n + h_{n+1} \sum_{j=1}^{s} b_j \left(\frac{rt_{n+1}^i - t_n}{h_{n+1}} \right) K_{n+1}^j$$

if $rt_{n+1}^i > t_n$, and

$$\tilde{Y}_{n+1}^i = \eta\left(rt_{n+1}^i\right)$$

otherwise.

As for the convergence and accuracy of the method, Theorem 4.2.5 guarantees, on any finite interval $[0, t_f]$, the uniform global error

$$\max_{0 \leq t \leq t_f} |y(t) - \eta(t)| = O(h^{q'}), \quad q' = \min\{p, q+1\},$$

and, on the first interval $[0, h_1]$, the nodal error

$$|y(h_1) - y_1| = O(h_1^{q'+1}) \tag{6.4.6}$$

and the uniform error

$$\max_{0 \leq t \leq h_1} |y(t) - \eta(t)| = O(h_1^{q+1}). \tag{6.4.7}$$

The estimates (6.4.6) and (6.4.7) follow by (4.2.20) and (4.2.19) for $n = 0$ and $e_n = 0$ in the proof of Theorem 4.2.5.

TABLE 6.1. Numerical results for the collocation method at ν Gaussian points applied to equation (6.4.8) with $a = 0$, $b = 1$ and $r = 0.5$.

	$\nu = 2$		$\nu = 3$	
h_1	$\lvert y(h_1) - y_1\rvert$	Ratio with previous error	$\lvert y(h_1) - y_1\rvert$	Ratio with previous error
0.4	$1.91 \cdot 10^{-5}$	—	$3.29 \cdot 10^{-7}$	—
0.2	$1.11 \cdot 10^{-6}$	17.11	$4.91 \cdot 10^{-9}$	67.00
0.1	$6.74 \cdot 10^{-8}$	16.55	$7.50 \cdot 10^{-11}$	65.49
0.05	$4.14 \cdot 10^{-9}$	16.28	$1.16 \cdot 10^{-12}$	64.76
0.025	$2.56 \cdot 10^{-10}$	16.14	—	—

Although the result is optimal for the uniform error in $[0, h_1]$, it is not optimal for the discrete error $|y(h_1) - y_1|$. In fact, some experiments carried out by Brunner [54] on the equation

$$\begin{cases} y'(t) = ay(t) + by(rt), \\ y(0) = 1, \end{cases} \tag{6.4.8}$$

have shown that collocation at ν Gaussian points, for which $p = 2\nu$ and $q = \nu + 1$, exhibits an error $|y(h_1) - y_1| = O(h^{2\nu})$ for any $0 < r < 1$.

In Table 6.1 the errors are displayed, for $\nu = 2$ and $\nu = 3$, for equation (6.4.8) with $a = 0$ with respect to the exact solution

$$y(t) = \sum_{k=0}^{\infty} \frac{r^{\frac{1}{2}(k-1)k}}{k!} (bt)^k.$$

The order 2ν is confirmed by the ratios of consecutive errors in the halving stepsize process.

On the contrary, the estimate (6.4.6) provides an error $O(h^4)$ for $\nu = 2$ and $O(h^5)$ for $\nu = 3$. Whereas the former is optimal and agrees with experiment, the latter is not, and the actual order of the method is underestimated. Indeed, the proof, based on the local order of the discrete method and the uniform order of the interpolant, is designed for a general DDE and is not suitable for analyzing the possible preservation of superconvergence of the method.

In the same paper, Brunner has also proved that, for collocation at $\nu \geq 2$ Gaussian points, the estimate

$$|y(h_1) - y_1| = O(h_1^{2\nu})$$

is the best possible for $0 < r < 1$. The local order of collocation at Gaussian points being $2\nu + 1$, we can say that, when applied to (6.4.8), the method is still superconvergent but one order is missed, essentially due to overlapping. However, optimal nodes for which

$$|y(h_1) - y_1| = O(h_1^{2\nu+1})$$

actually exist, as stated in the following theorem proved by Brunner [54] for $\nu = 2$ and, later, extended by Takama, Muroya and Ishiwata [241] and Ishiwata [151] for $\nu \geq 3$.

Theorem 6.4.2 *The numerical solution y_1 of the DDE (6.4.8) with $a = 0$, provided by collocation at two nodes, c_1 and c_2, satisfies*

$$|y(h_1) - y_1| = O(h_1^5)$$

for every $0 < r \leq 1$ if and only if the nodes $c_1 = c_1(r)$ and $c_2 = c_2(r)$ are the zeros of

$$P_2(2s-1) + \gamma(r)P_1(2s-1),$$

where P_m, $m = 1, 2$, is the Legendre polynomial of degree m and

$$\gamma(r) = \frac{3(1-r)^2}{3-2r}.$$

Further investigations and experiments have been carried out by Brunner [54] and Brunner, Hu and Lin [57] on the preservation of superconvergence for the collocation solution of (6.4.8). Some are devoted to the search for suitable collocation points, others are based on the choice of a particular constrained mesh. For a deeper insight, we refer the interested reader to the book by Brunner [56] and its rich bibliography on the pantograph equation.

Here we describe a hybrid superconvergent method, proposed by Bellen [24]. It is based on the uniform approximation $\phi(t)$ of the solution in some initial interval $[0, t_1]$ furnished, for instance, by truncation of the Taylor expansion

$$y(t) = \sum_{i=0}^{\infty} \frac{t^i}{i!} \prod_{k=0}^{i-1} (a + r^k b),$$

to order p, as well as by any other approximation with uniform error $O(t_1^p)$, and on the method of steps for the problem

$$\begin{cases} z'(t) = az(t) + bz(rt), & t_1 \leq t \leq t_f, \\ z(t) = \phi(t), & 0 \leq t \leq t_1, \end{cases}$$

in the remaining interval $[t_1, t_f]$, where

$$t_1 = r^k t_f$$

for a suitable positive integer k.

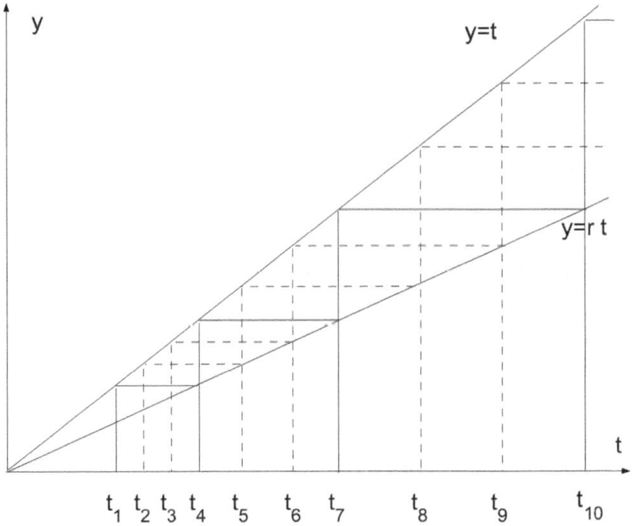

FIG. 6.2. The quasi-geometric mesh Δ_m for $m = 3$.

Note that the solution $z(t)$ is not smooth and, with respect to the solution of (6.4.5), we have

$$\max_{t_1 \le t \le t_f} |y(t) - z(t)| = O(h^p). \tag{6.4.9}$$

On the interval $[t_1, t_f]$, let us consider the sequence of discontinuities

$$\xi_i = r^{k+1-i} t_f, \quad i = 1, \ldots, k+1,$$

and the relevant macro intervals

$$I_i = [r^{k+1-i} t_f, r^{k-i} t_f], \quad i = 1, \ldots, k,$$

of increasing length

$$H_i = t_f r^{k-i}(r-1). \tag{6.4.10}$$

On each interval I_i, let us include $m+1$ equally spaced nodes so that the resulting *quasi-geometric mesh* Δ_m, introduced by Bellen, Guglielmi and Torelli [28], turns out to be a constrained mesh including mk nodes satisfying the property

$$rt_{jm+i} = t_{(j-1)m+i}, \quad j = 1, \ldots, k-1, \quad i = 1, 2, \ldots, m,$$

as illustrated by Figure 6.2.

Therefore, the method based on any natural CRK scheme reads

$$z_{n+1} = z_n + h_{n+1} \sum_{i=1}^{s} b_i(\theta) K_{n+1}^i, \quad 0 \le \theta \le 1,$$

$$K_{n+1}^i = a\left(z_n + h_{n+1} \sum_{j=1}^{s} a_{ij} K_{n+1}^j\right)$$

$$+b\left(z_{n-m}+h_{n+1-m}\sum_{j=1}^{s}a_{ij}K_{n+1-m}^{j}\right), \quad i=1,\dots,s.$$

Note that the maximum stepsize of the mesh Δ_m is attained in the last interval I_k, where the distance between consecutive nodes is

$$h=\frac{t_f(1-r)}{m}. \tag{6.4.11}$$

So, in order to set the first point $t_1 = r^k t_f \approx h$, we take $k = k(m)$ as the minimum integer such that

$$r^k t_f \le \frac{t_f(1-r)}{m},$$

that is

$$k(m)=\left\lceil\frac{\log(1-r)-\log(m)}{\log(r)}\right\rceil,$$

and, hence,

$$k(m)=O(\log m).$$

In conclusion, for any m, we have $k(m)$ macro intervals

$$I_i=[r^{k(m)+1-i}t_f, r^{k(m)-i}t_f], \quad i=1,\dots,k(m),$$

where, for each i, there are m nodes and the stepsize is constant and equal to

$$h_i=r^{k(m)-i}h.$$

Since Δ_m is a constrained mesh, in view of the arguments at the end of Section 6.3, the discrete order p of the underlying natural CRK method is preserved on each I_i, where the discrete global error is essentially $O(h_i^p)$.

We conclude by proving that, despite the number of macro intervals growing unboundedly like $\log(m)$, there is no loss of accuracy order in the overall method. In fact, on each macro interval I_i the maximum error

$$e_i=\max_{(i-1)m\le n\le im}|z(t_n)-z_n|$$

undergoes the bound

$$e_i\le \mathrm{e}^{K_1H_i}e_{i-1}+K_2h_i^p=\mathrm{e}^{K_1H_i}e_{i-1}+K_2(r^p)^{(k(m)-i)}h^p$$

for suitable constants K_1 and $K_2>0$, where H_i is given by (6.4.10) and, hence, by standard arguments,

$$e_{k(m)}\le \mathrm{e}^{K_1(t_f-t_1)}K_2h^p\Big((r^p)^{k(m)}+(r^p)^{k(m)-1}+\dots+1\Big).$$

The last inequality implies that, as $m \to \infty$ and, simultaneously, $h \to 0$ according to (6.4.11), the maximum global error satisfies

$$\max_{1 \le n \le k(m)m} |z(t_n) - z_n| = O(h^p)$$

and thus, by (6.4.9),

$$\max_{1 \le n \le k(m)m} |y(t_n) - z_n| = O(h^p).$$

6.5 RK methods for NDDEs

Consider the DDE of neutral type

$$\begin{cases} y'(t) = f\Big(t, y(t), y(t - \tau(t, y(t))), y'(t - \sigma(t, y(t)))\Big), & t_0 \le t \le t_f, \\ y(t) = \phi(t), \quad t \le t_0. \end{cases} \tag{6.5.1}$$

The standard approach consists in solving step by step the local problems

$$\begin{cases} w'_{n+1}(t) = f\Big(t, w_{n+1}(t), x(t - \tau(t, w_{n+1}(t))), z(t - \sigma(t, w_{n+1}(t)))\Big), \\ \hfill t_n \le t \le t_{n+1}, \\ w_{n+1}(t_n) = \eta(t_n), \end{cases}$$

where

$$x(s) = \begin{cases} \phi(s) & \text{for } s \le t_0, \\ \eta(s) & \text{for } t_0 \le s \le t_n, \\ w_{n+1}(s) & \text{for } t_n \le s \le t_{n+1}, \end{cases}$$

$$z(s) = \begin{cases} \phi'(s) & \text{for } s \le t_0, \\ \lambda(s) & \text{for } t_0 \le s \le t_n, \\ w'_{n+1}(s) & \text{for } t_n \le s \le t_{n+1}, \end{cases}$$

$\eta(t)$ is the continuous approximation of $y(t)$ and $\lambda(t)$ is an approximation of $y'(t)$ given by

$$\lambda(t) = \eta'(t) \tag{6.5.2}$$

or by

$$\lambda(t) = \mathcal{P}\Big(f(\cdot, \eta(\cdot), \eta(\,\cdot - \tau(\cdot, \eta(\cdot))), \lambda(\,\cdot - \sigma(\cdot, \eta(\cdot))))\Big)(t), \tag{6.5.3}$$

where, in each mesh interval $[t_k, t_{k+1}]$, $\mathcal{P}$ is an interpolation operator in a suitable polynomial space of degree possibly other than $\deg(\eta')$ and nodes in $[t_k, t_{k+1}]$.

For the choice (6.5.2), the DDE method (6.0.2), (6.0.3) in the Y notation modifies to the following *RK method for NDDEs*:

$$Y_{n+1}^i = y_n + h_{n+1}\sum_{j=1}^{s} a_{ij} f\left(t_{n+1}^j, Y_{n+1}^j, \tilde{Y}_{n+1}^j, \tilde{Z}_{n+1}^j\right), \quad i = 1, \ldots, s, \tag{6.5.4}$$

$$\eta(t_n + \theta h_{n+1}) = y_n + h_{n+1}\sum_{i=1}^{s} b_i(\theta) f\left(t_{n+1}^i, Y_{n+1}^i, \tilde{Y}_{n+1}^i, \tilde{Z}_{n+1}^i\right),\ 0 \le \theta \le 1, \tag{6.5.5}$$

$$\lambda(t_n + \rho h_{n+1}) = \sum_{i=1}^{s} b_i'(\rho) f\left(t_{n+1}^i, Y_{n+1}^i, \tilde{Y}_{n+1}^i, \tilde{Z}_{n+1}^i\right), \quad 0 \le \rho \le 1, \tag{6.5.6}$$

where

$$\tilde{Y}_{n+1}^j = \eta\left(t_{n+1}^j - \tau(t_{n+1}^j, Y_{n+1}^j)\right),$$
$$\tilde{Z}_{n+1}^j = \lambda\left(t_{n+1}^j - \sigma(t_{n+1}^j, Y_{n+1}^j)\right).$$

Note that, for the arguments $s_j = t_{n+1}^j - \tau(t_{n+1}^j, Y_{n+1}^j)$ and $s_j' = t_{n+1}^j - \sigma(t_{n+1}^j, Y_{n+1}^j)$, the values $\eta(s_j)$ and $\lambda(s_j')$ may or may not be known.

If overlapping occurs, that is if, for some index i, the arguments $s_i > t_n$ and/or $s_i' > t_n$, then the *spurious stages* $\tilde{Y}_{n+1}^i$ and/or $\tilde{Z}_{n+1}^i$ are unknown and are given by (6.5.5) and (6.5.6) for

$$\theta = \theta_{n+1}^i = c_i - \frac{\tau(t_{n+1}^i, Y_{n+1}^i)}{h_{n+1}},$$

and

$$\rho = \rho_{n+1}^i = c_i - \frac{\sigma(t_{n+1}^i, Y_{n+1}^i)}{h_{n+1}},$$

respectively, that is

$$\tilde{Y}_{n+1}^i = y_n + h_{n+1}\sum_{j=1}^{s} b_j(\theta_{n+1}^i) f\left(t_{n+1}^j, Y_{n+1}^j, \tilde{Y}_{n+1}^j, \tilde{Z}_{n+1}^j\right),$$

$$\tilde{Z}_{n+1}^i = \sum_{j=1}^{s} b_j'(\rho_{n+1}^i) f\left(t_{n+1}^j, Y_{n+1}^j, \tilde{Y}_{n+1}^j, \tilde{Z}_{n+1}^j\right).$$

On the contrary, if the arguments of $\eta(s)$ and $\lambda(s)$ lie outside the current interval $[t_n, t_{n+1}]$, then the values $\tilde{Y}_{n+1}^j$ and $\tilde{Z}_{n+1}^j$ are given by the interpolants $\eta(s)$ and $\eta'(s)$ as computed at the past points

$$t_{n+1}^i - \tau(t_{n+1}^i, Y_{n+1}^i) = t_{n+1-m} + \theta h_{n+1-m},$$
$$t_{n+1}^i - \sigma(t_{n+1}^i, Y_{n+1}^i) = t_{n+1-m'} + \rho h_{n+1-m'},$$

for suitable values of m, m', θ and ρ.

As with DDEs with no neutral terms, the spurious stages $\tilde{Y}^i_{n+1}$ and $\tilde{Z}^i_{n+1}$, if any, only apparently increase the dimension of the system to be solved at each step. In fact, by using the K notation

$$K^i_{n+1} = f\left(t^i_{n+1}, Y^i_{n+1}, \tilde{Y}^i_{n+1}, \tilde{Z}^i_{n+1}\right),$$

all the stages Y^i_{n+1}, $\tilde{Y}^i_{n+1}$ and $\tilde{Z}^i_{n+1}$, as well as the arguments θ^i_{n+1} and ρ^i_{n+1}, turn out to depend on K^i_{n+1} only.

Remark 6.5.1 *According to the arguments of Section 6.1, for any natural CRK method with s distinct abscissae $c_1, \ldots, c_s$ such that $c_i \neq 0$, $i = 1, \ldots, s$, and continuous extension $\eta(t)$ of degree s, the system to be solved at each step may be stated in terms of the sole Y^i_{n+1}'s. In fact, the polynomial $\eta(t_n + \theta h_{n+1})$ may be written using the Lagrange interpolation formula through the $s+1$ values $y_n (= \eta(t_n))$ and $Y^i_{n+1} (= \eta(t_n + c_i h_{n+1}))$, $i = 1, \ldots, s$, that is*

$$\eta(t_n + \theta h_{n+1}) = \ell_0(\theta) y_n + \sum_{i=1}^{s} \ell_i(\theta) Y^i_{n+1}, \tag{6.5.7}$$

where ℓ_j, $j = 0, \ldots, s$ are the Lagrange polynomial coefficients on the nodes $c_0 = 0$ and c_i, $i = 1, \ldots, s$. Therefore, $\tilde{Y}^i_{n+1} = \eta\big(t^i_{n+1} - \tau(t^i_{n+1}, Y^i_{n+1})\big)$ may be written by (6.5.7) for $\theta = \theta^i_{n+1} = c_i - \frac{\tau(t^i_{n+1}, Y^i_{n+1})}{h_{n+1}}$. Similarly, $\tilde{Z}^i_{n+1} = \lambda\big(t^i_{n+1} - \sigma(t^i_{n+1}, Y^i_{n+1})\big)$ may be written by the derivative of (6.5.7) for $\theta = \rho^i_{n+1} = c_i - \frac{\sigma(t^i_{n+1}, Y^i_{n+1})}{h_{n+1}}$.

For the choice (6.5.3), the RK method for NDDEs (in the Y notation) is given by (6.5.4), (6.5.5) along with

$$\lambda(t_n + \rho h_{n+1}) = \sum_{i=0}^{s^*} \ell_i(\rho) f\left(\bar{t}^i_{n+1}, U^i_{n+1}, \tilde{U}^i_{n+1}, \tilde{V}^i_{n+1}\right), \quad 0 \le \rho \le 1, \tag{6.5.8}$$

where $\bar{t}^i_{n+1} = t_n + \bar{c}_i h_{n+1}$ and $\ell_i(\theta)$, $i = 0, \ldots, s^*$, are the nodes and the Lagrange polynomial coefficients of the interpolation operator $\mathcal{P}$. Here, besides the values

$$\tilde{Y}^j_{n+1} = \eta\big(t^j_{n+1} - \tau(t^j_{n+1}, Y^j_{n+1})\big),$$
$$\tilde{Z}^j_{n+1} = \lambda\big(t^j_{n+1} - \sigma(t^j_{n+1}, Y^j_{n+1})\big),$$

there are additional values

$$U^j_{n+1} = \eta(\bar{t}^j_{n+1}),$$
$$\tilde{U}^j_{n+1} = \eta\big(\bar{t}^j_{n+1} - \tau(\bar{t}^j_{n+1}, U^j_{n+1}))\big),$$
$$\tilde{V}^j_{n+1} = \lambda\big(\bar{t}^j_{n+1} - \sigma(\bar{t}^j_{n+1}, U^j_{n+1}))\big).$$

Note that, according to the arguments $s_j = t^j_{n+1} - \tau(t^j_{n+1}, Y^j_{n+1})$ and $s'_j = t^j_{n+1} - \sigma(t^j_{n+1}, Y^j_{n+1})$, the values $\eta(s_j)$ and $\lambda(s'_j)$ may or may not be known. If

$s_j > t_n$ and/or $s'_j > t_n$, then $\tilde{Y}^j_{n+1}$ and/or $\tilde{Z}^j_{n+1}$ are unknown and must be computed by (6.5.5) and (6.5.8), respectively. In particular, if $s'_j > t_n$, then for the application of (6.5.8) in the current interval U^j_{n+1}, $\tilde{U}^j_{n+1}$ and $\tilde{V}^j_{n+1}$ need to be known. Here the U^j_{n+1}'s are certainly unknown, whereas knowledge of $\tilde{U}^j_{n+1}$ and $\tilde{V}^j_{n+1}$ depends on the location of the further arguments $\bar{t}^j_{n+1} - \tau(\bar{t}^j_{n+1}, U^j_{n+1})$ and $\bar{t}^j_{n+1} - \sigma(\bar{t}^j_{n+1}, U^j_{n+1})$.

Summarizing, if for some index j some of the arguments are $> t_n$, then the relevant *spurious stages* $\tilde{Y}^j_{n+1}$, $\tilde{Z}^j_{n+1}$, U^j_{n+1}, $\tilde{U}^j_{n+1}$ or $\tilde{V}^j_{n+1}$ are unknown and are given by (6.5.5) and (6.5.8) for suitable values of θ and ρ. More precisely,

$$\tilde{Y}^j_{n+1} = \eta(t_n + \theta^j_{n+1} h_{n+1}) \qquad \text{with } \theta^j_{n+1} = c_j - \frac{\tau(t^j_{n+1}, Y^j_{n+1})}{h_{n+1}},$$

$$\tilde{Z}^j_{n+1} = \lambda(t_n + \rho^j_{n+1} h_{n+1}) \qquad \text{with } \rho^j_{n+1} = c_j - \frac{\sigma(t^j_{n+1}, Y^j_{n+1})}{h_{n+1}},$$

$$U^j_{n+1} = \eta(t_n + \bar{c}^j h_{n+1}),$$

$$\tilde{U}^j_{n+1} = \eta(t_n + \bar{\theta}^j_{n+1} h_{n+1}) \qquad \text{with } \bar{\theta}^j_{n+1} = \bar{c}_j - \frac{\tau(\bar{t}^j_{n+1}, U^j_{n+1})}{h_{n+1}},$$

$$\tilde{V}^j_{n+1} = \lambda(t_n + \bar{\rho}^j_{n+1} h_{n+1}) \qquad \text{with } \bar{\rho}^j_{n+1} = \bar{c}_j - \frac{\sigma(\bar{t}^j_{n+1}, U^j_{n+1})}{h_{n+1}}.$$

The dimension of the system may still be reduced by using the K notation but, unlike the option (6.5.2), besides the K values

$$K^j_{n+1} = f\left(t^j_{n+1}, Y^j_{n+1}, \tilde{Y}^j_{n+1}, \tilde{Z}^j_{n+1}\right), \quad j = 1, \ldots, s,$$

we have the additional values

$$H^j_{n+1} = f\left(\bar{t}^j_{n+1}, U^j_{n+1}, \tilde{U}^j_{n+1}, \tilde{V}^j_{n+1}\right), \quad j = 0, \ldots, s^*.$$

Remark 6.5.2 *Also with the option (6.5.3), the number of unknowns in the system to be solved at each step may be reduced. In fact, if the underlying CRK method is natural and if the interpolation formula (6.5.8) is based on the nodes $\bar{c}_i = c_i$, $i = 1, \ldots, s^* = s$, and on another node $\bar{c}_0 \neq c_i$, then, for $j = 1, \ldots, s$,*

$$Y^j_{n+1} = U^j_{n+1},$$

$$\tilde{Y}^j_{n+1} = \tilde{U}^j_{n+1},$$

$$\tilde{Z}^j_{n+1} = \tilde{V}^j_{n+1}$$

and, therefore, also

$$H^j_{n+1} = K^j_{n+1}.$$

In this case the spurious stages reduce to just Y^j_{n+1}, $\tilde{Y}^j_{n+1}$ and $\tilde{Z}^j_{n+1}$ in the Y notation, and to just

$$K_{n+1}^j = f\left(t_{n+1}^j, Y_{n+1}^j, \tilde{Y}_{n+1}^j, \tilde{Z}_{n+1}^j\right)$$

in the equivalent K notation. Note also that, for the new set of stage values

$$Z_{n+1}^j = \lambda(t_n + c_j h_{n+1}), \quad j = 1, \ldots, s,$$

by (6.5.8) we have

$$Z_{n+1}^j = K_{n+1}^j.$$

On the other hand, independently of the choice of the $\bar{c}_i$'s, if $c_i \neq 0$, $i = 1, \ldots, s$, the arguments of Section 6.1 allow us to express each $\tilde{Y}_{n+1}^j$ in terms of the Y_{n+1}^j's and, hence, the overall method is based on the stage values Y_{n+1}^j and $\tilde{Z}_{n+1}^j$. However, in no case can the RK method reduce to just the Y values.

We give up providing a detailed analysis for the special cases of constant and time dependent delays, as made in the foregoing sections for DDEs with no neutral terms. The analysis is quite similar and the convergence comes from the general results proved in Chapter 4. In particular, Theorem 4.4.3 applies to the procedures described above and may be stated as follows.

Theorem 6.5.3 *Consider the state dependent NDDE (6.5.1), where $f(t, y, x, w)$ is C^p-continuous in $[t_0, t_f] \times \mathbb{R}^d \times \mathbb{R}^d \times \mathbb{R}^d$, the delays $\tau(t, y)$ and $\sigma(t, y)$ are C^p-continuous in $[t_0, t_f] \times \mathbb{R}^d$ and the initial function $\phi(t)$ is C^p-continuous. Moreover, assume that the mesh $\Delta = \{t_0, t_1, \ldots, t_n, \ldots, t_N = t_f\}$ includes all the discontinuity points of order $\leq p$ lying in $[t_0, t_f]$. If the underlying CRK method $(A, b(\theta))$ has discrete order p and uniform order q, and the approximation $\lambda(t)$ has uniform order r, then the resulting RK method for NDDEs has discrete global order and uniform global order $q' = \min\{p, q+1, r+1\}$; that is*

$$\max_{1 \leq n \leq N} \|y(t_n) - y_n\| = O(h^{q'})$$

and

$$\max_{t_0 \leq t \leq t_f} \|y(t) - \eta(t)\| = O(h^{q'}),$$

where $h = \max_{1 \leq n \leq N} h_n$. In particular, if $\lambda(t)$ is given by the option (6.5.2), then $r = q - 1$ and, hence, $q' = q$.

Note that, for the option (6.5.3), $\lambda(t)$ is given by (6.5.8) and the interpolation operator $\mathcal{P}$ has order $r = s^*$. Therefore, on the basis of Theorem 6.5.3, it is useless to take $s^* > q$. On the other hand, the choice $s^* = q$ preserves the optimal order $q' = \min\{p, q+1\}$ and makes the option (6.5.3), along with the conditions in Remark 6.5.2, preferable to (6.5.2).

As for possible superconvergence phenomena, similar results as with the non-neutral equation can be proved. They are essentially based on the general result about the preservation of superconvergence of RK methods stated in Theorem 5.2.10. We leave the interested reader to develop them for constant and time dependent delays, in particular for proportional delays.

6.5.1 *Other formulations of NDDEs*

Now we consider CRK methods applied to non-linear systems of NDDEs in the particular forms

$$\begin{cases} \frac{\mathrm{d}}{\mathrm{d}t}\Big(y(t) - G(t, y(t-\tau(t)))\Big) = H(t, y(t), y(t-\tau(t))), & t_0 \le t \le t_f, \\ y(t) = \phi(t), \quad t \le t_0, \end{cases} \tag{6.5.9}$$

and

$$\begin{cases} y'(t) = F\Big(t, y(t), G\Big(t, y(t-\tau(t)), y'(t-\tau(t))\Big)\Big), & t_0 \le t \le t_f, \\ y(t) = \phi(t), \quad t \le t_0, \end{cases} \tag{6.5.10}$$

where the delay $\tau(t)$ satisfies hypotheses (H_1) and (H_4).

As widely explained in Sections 3.3.2 and 4.5, with

$$\Phi(t) = y(t) - G(t, y(t-\tau(t)))$$

and

$$K(t, y, x) = H(t, y + G(t, x), x),$$

the NDDE (6.5.9) can be rewritten as an ODE

$$\begin{cases} \Phi'(t) = K\left(t, \Phi(t), y(t-\tau(t))\right), & t_0 \le t \le t_f, \\ \Phi(t_0) = \phi(t_0) - G(t_0, \phi(t_0 - \tau(t_0))), \end{cases} \tag{6.5.11}$$

coupled with the algebraic recursion

$$y(t) = \begin{cases} \phi(t), & t \le t_0, \\ \Phi(t) + G(t, y(t-\tau(t))), & t_0 \le t \le t_f. \end{cases} \tag{6.5.12}$$

Analogously, with

$$\Psi(t) = G\Big(t, y(t-\tau(t)), y'(t-\tau(t))\Big)$$

and

$$H(t, y, x) = G\Big(t, y, F(t-\tau(t), y, x)\Big),$$

the NDDE (6.5.10) can be rewritten as an ODE

$$\begin{cases} y'(t) = F\left(t, y(t), \Psi(t)\right), & t_0 \le t \le t_f, \\ y(t_0) = \phi(t_0), \end{cases} \tag{6.5.13}$$

coupled with the algebraic recursion

$$\Psi(t) = \begin{cases} G\Big(t, \phi(t-\tau(t)), \phi'(t-\tau(t))\Big), & t - \tau(t) \le t_0, \\ H\Big(t, y(t-\tau(t)), \Psi(t-\tau(t))\Big), & t - \tau(t) > t_0. \end{cases} \tag{6.5.14}$$

A CRK method applied to the system (6.5.11), (6.5.12) within the method of steps, as explained in Section 4.5 and summarized by Algorithm 4.5.1, aims to provide continuous approximations $\lambda(t)$ and $\eta(t)$ of $\Phi(t)$ and $y(t)$, respectively, with nodal approximations $y_n = \eta(t_n)$ and $\Phi_n = \lambda(t_n)$. It consists in solving step by step for $\Lambda_{n+1}(t)$ the local ODE problem

$$\begin{cases} \Lambda'_{n+1}(t) = K\Big(t, \Lambda_{n+1}(t), \eta\big(t - \tau(t)\big)\Big), & t_n \le t \le t_{n+1}, \\ \Lambda_{n+1}(t_n) = \Phi_n = y_n - G\Big(t_n, \eta\big(t_n - \tau(t_n)\big)\Big), \end{cases} \tag{6.5.15}$$

coupled with the algebraic equation for $w_{n+1}(t)$

$$w_{n+1}(t) = \Lambda_{n+1}(t) + G\Big(t, \eta\big(t - \tau(t)\big)\Big), \quad t_n \le t \le t_{n+1}, \tag{6.5.16}$$

along a mesh $\Delta = \{t_0, t_1, \ldots, t_n, \ldots, t_N = t_f\}$ which includes all the discontinuity points ξ_k.

The adaptation of formulae (5.0.2), (5.0.3), (5.1.6), (5.1.3) to the local problem (6.5.15) yields

$$\Lambda^i_{n+1} = \Phi_n + h_{n+1} \sum_{j=1}^{s} a_{ij} K\Big(t^j_{n+1}, \Lambda^j_{n+1}, \eta(t^j_{n+1} - \tau(t^j_{n+1}))\Big), \quad i = 1, \ldots, s, \tag{6.5.17}$$

$$\lambda(t_n + \theta h_{n+1}) = \Phi_n + h_{n+1} \sum_{i=1}^{s} b_i(\theta) K\Big(t^i_{n+1}, \Lambda^i_{n+1}, \eta(t^i_{n+1} - \tau(t^i_{n+1}))\Big), \quad 0 \le \theta \le 1, \tag{6.5.18}$$

which, for $\theta = 1$, reduces to

$$\Phi_{n+1} = \Phi_n + h_{n+1} \sum_{i=1}^{s} b_i K\Big(t^i_{n+1}, \Lambda^i_{n+1}, \eta(t^i_{n+1} - \tau(t^i_{n+1}))\Big). \tag{6.5.19}$$

The algebraic equation (6.5.16) is approximately solved by

$$\eta(t) = \lambda(t) + G\Big(t, \eta\big(t - \tau(t)\big)\Big), \quad t_n \le t \le t_{n+1}, \tag{6.5.20}$$

or

$$\eta(t) = \mathcal{P}\Big(\lambda(\cdot) + G\Big(\cdot, \eta\big(\cdot - \tau(\cdot)\big)\Big)\Big)(t), \quad t_n \le t \le t_{n+1}, \tag{6.5.21}$$

where $\mathcal{P}$ is an interpolation operator in a suitable polynomial space and nodes in $[t_n, t_{n+1}]$. For $t = t_{n+1}$ both formulae reduce to

$$y_{n+1} = \Phi_{n+1} + G\Big(t_{n+1}, \eta\big(t_{n+1} - \tau(t_{n+1})\big)\Big).$$

As for the NDDEs expressed in the other type of DDAE system (6.5.13), (6.5.14), a CRK method applied within the method of steps, as explained in Section 4.5 and summarized by Algorithm 4.5.3, aims to provide continuous approximations $\lambda(t)$ and $\eta(t)$ of $\Psi(t)$ and $y(t)$, respectively, with nodal approximations $y_n = \eta(t_n)$ and $\Psi_n = \lambda(t_n)$. It consists in solving step by step for $\Lambda_{n+1}(t)$ the algebraic equation

$$\Lambda_{n+1}(t) = G\Big(t, \phi(t-\tau(t)), \phi'(t-\tau(t))\Big), \quad t_n \le t \le t_{n+1}, \tag{6.5.22}$$

if $t - \tau(t) \le t_0$, or

$$\Lambda_{n+1}(t) = H\Big(t, \eta(t-\tau(t)), \lambda(t-\tau(t))\Big), \quad t_n \le t \le t_{n+1}, \tag{6.5.23}$$

if $t - \tau(t) > t_0$, coupled with the local ODE problem for $w_{n+1}(t)$,

$$\begin{cases} w'_{n+1}(t) = F\big(t, w_{n+1}(t), \Lambda_{n+1}(t)\big), & t_n \le t \le t_{n+1}, \\ w_{n+1}(t_n) = y_n, \end{cases} \tag{6.5.24}$$

along a mesh $\Delta = \{t_0, t_1, \ldots, t_n, \ldots, t_N = t_f\}$ which includes all the discontinuity points ξ_k.

The algebraic equation (6.5.22) is computed exactly by setting

$$\lambda(t) = G\Big(t, \phi(t-\tau(t)), \phi'(t-\tau(t))\Big), \quad t_n \le t \le t_{n+1}, \tag{6.5.25}$$

whereas (6.5.23) is again computed exactly by setting

$$\lambda(t) = H\Big(t, \eta(t-\tau(t)), \lambda(t-\tau(t))\Big), \quad t_n \le t \le t_{n+1}, \tag{6.5.26}$$

or approximated by

$$\lambda(t) = H\Big(t, \eta(t-\tau(t)), \mathcal{P}\big(\lambda(\cdot - \tau(\cdot))\big)(t)\Big), \quad t_n \le t \le t_{n+1}, \tag{6.5.27}$$

where $\mathcal{P}$ is an interpolation operator in a suitable polynomial space with nodes in $[t_{k_n}, t_{k_n+1}]$, the mesh interval to which the deviated argument $t_{n+1} - \tau(t_{n+1})$ belongs. For $t = t_{n+1}$ the above formulae reduce to

$$\Psi_{n+1} = G\Big(t_{n+1}, \phi(t_{n+1} - \tau(t_{n+1})), \phi'(t_{n+1} - \tau(t_{n+1}))\Big),$$
$$\Psi_{n+1} = H\Big(t_{n+1}, \eta(t_{n+1} - \tau(t_{n+1})), \lambda(t_{n+1} - \tau(t_{n+1}))\Big),$$
$$\Psi_{n+1} = H\Big(t_{n+1}, \eta(t_{n+1} - \tau(t_{n+1})), \mathcal{P}\big(\lambda(\cdot - \tau(\cdot))\big)(t_{n+1})\Big),$$

respectively.

The adaptation of formulae (5.0.2), (5.0.3), (5.1.6), (5.1.3) to the local problem (6.5.24) yields

$$Y_{n+1}^i = y_n + h_{n+1} \sum_{j=1}^{s} a_{ij} F\Big(t_{n+1}^j, Y_{n+1}^j, \lambda(t_{n+1}^j)\Big), \quad i = 1, \ldots, s, \tag{6.5.28}$$

$$\eta(t_n + \theta h_{n+1}) = y_n + h_{n+1} \sum_{i=1}^{s} b_i(\theta) F\Big(t_{n+1}^i, Y_{n+1}^i, \lambda(t_{n+1}^i)\Big), \quad 0 \le \theta \le 1, \tag{6.5.29}$$

which, for $\theta = 1$, reduces to

$$\Phi_{n+1} = y_n + h_{n+1} \sum_{i=1}^{s} b_i F\Big(t_{n+1}^i, Y_{n+1}^i, \lambda(t_{n+1}^i)\Big).$$

We conclude by remarking that the convergence results for Algorithms 4.5.1 and 4.5.3 based on the CRK methods illustrated above are obtained as particular cases of Theorems 4.5.2 and 4.5.4, respectively. We refer the interested reader to Bellen, Guglielmi and Ruehli [27, 26], Bellen, Guglielmi and Zennaro [29, 30] and Torelli and Vermiglio [250].

6.5.2 *The code RADAR5*

The code RADAR5, developed by Guglielmi and Hairer [112], is a variable stepsize code for the general DDAE

$$\begin{cases} Mu'(t) = F\Big(t, u(t), u\big(t - \tau(t, u(t))\big)\Big), & t_0 \le t \le t_f, \\ u(t) = \phi(t), \quad t \le t_0, \end{cases} \tag{6.5.30}$$

where M is a possibly singular $d' \times d'$ matrix and, according to the simplified form (2.1.7), $u\big(t - \tau(t, u(t))\big)$ stands for the vector

$$\Big[u_1\big(t - \tau_1(t, u(t))\big), \ldots, u_{d'}\big(t - \tau_{d'}(t, u(t))\big)\Big]^T.$$

The code, based on the three-stage Radau IIA method, is a particular implementation of the general procedure we are going to describe.

The Radau IIA methods are s-stage implicit RK methods (A, b), equivalent to collocation on the abscissae c_i, with a non-singular coefficient matrix A and weights satisfying the *stiff accuracy* condition $b_i = a_{si}$, $i = 1, \ldots, s$.

The $(n+1)$st integration step consists in solving, for the U_{n+1}^i,

$$M\big(U_{n+1}^i - u_n\big) = h_{n+1} \sum_{j=1}^{s} a_{ij} F\big(t_{n+1}^j, U_{n+1}^j, \tilde{U}_{n+1}^j\big), \quad i = 1, \ldots, s, \tag{6.5.31}$$

and setting

$$u_{n+1} = U_{n+1}^s.$$

The delayed argument $\tilde{U}_{n+1}^j$, approximating $u\big(t_{n+1}^j - \tau(t_{n+1}^j, U_{n+1}^j)\big)$, is given by

$$\tilde{U}_{n+1}^j = \begin{cases} \phi\big(t_{n+1}^j - \tau(t_{n+1}^j, U_{n+1}^j)\big) & \text{if } t_{n+1}^j - \tau(t_{n+1}^j, U_{n+1}^j) \le t_0, \\ \varphi\big(t_{n+1}^j - \tau(t_{n+1}^j, U_{n+1}^j)\big) & \text{if } t_{n+1}^j - \tau(t_{n+1}^j, U_{n+1}^j) > t_0, \end{cases}$$

where $\varphi(t_m + \theta h_{m+1})$, $0 \le \theta \le 1$, is the collocation polynomial of degree s relevant to the underlying Radau IIA method in the mesh interval $[t_m, t_{m+1}]$.

Since $c_i > 0,\ i = 1, \dots, s$, it may be written by the Lagrange interpolation formula through the $s+1$ values y_m and U^i_{m+1}:

$$\varphi(t_m + \theta h_{m+1}) = \ell_0(\theta) y_m + \sum_{i=1}^{s} \ell_i(\theta) U^i_{m+1}, \tag{6.5.32}$$

where $\ell_i(\theta),\ i = 0, \dots, s$, are the Lagrange polynomial coefficients on the nodes $c_0 = 0$ and $c_j,\ j = 1, \dots, s$.

If some argument $t^j_{n+1} - \tau(t^j_{n+1}, U^j_{n+1})$ is $> t_n$, then the polynomial $\varphi(t)$ to use in order to compute the spurious stage $\tilde{U}^j_{n+1} = \varphi(t^j_{n+1} - \tau(t^j_{n+1}, U^j_{n+1}))$ is not yet known, but is given by the Lagrange formula (6.5.32) for $m = n$ and

$$\theta = \theta^j_{n+1} = c_j - \frac{\tau(t^j_{n+1}, U^j_{n+1})}{h_{n+1}}.$$

Otherwise, if $t^j_{n+1} - \tau(t^j_{n+1}, U^j_{n+1}) \le t_n$, then the value $\tilde{U}^j_{n+1}$ is obtained by evaluating the interpolant φ computed in the previous steps. However, in any case, equation (6.5.31) may be written in terms of the stage values U^j_{n+1} only. This makes the use of Radau IIA, extended by the collocation polynomial, a natural choice for the solution of DDEs.

As a particular case of (6.5.30), we may consider the system

$$\begin{bmatrix} I & 0 \\ 0 & 0 \end{bmatrix} \begin{bmatrix} y'(t) \\ z'(t) \end{bmatrix} = \begin{bmatrix} z(t) \\ z(t) - f\Big(t, y(t), y(t - \tau(t, y(t))), z(t - \sigma(t, y(t)))\Big) \end{bmatrix} \tag{6.5.33}$$

for $t_0 \le t \le t_f$, with the initial conditions

$$\begin{cases} y(t) = \phi(t), \\ z(t) = \phi'(t), \end{cases} \qquad t \le t_0,$$

which is equivalent to the NDDE (6.5.1) transformed by the change of variable $z(t) = y'(t)$.

With the notation

$$\begin{aligned} u(t) &= \left[y(t)^T, z(t)^T\right]^T, \\ U^j_{n+1} &= \left[(Y^j_{n+1})^T, (Z^j_{n+1})^T\right]^T, \\ \tilde{U}^j_{n+1} &= \left[(\tilde{Y}^j_{n+1})^T, (\tilde{Z}^j_{n+1})^T\right]^T, \\ \varphi(t) &= \left[\eta(t)^T, \lambda(t)^T\right]^T, \end{aligned}$$

the method (6.5.31) may be written

$$\begin{bmatrix} Y^i_{n+1} - y_n \\ 0 \end{bmatrix} = h_{n+1} \begin{bmatrix} \sum_{j=1}^{s} a_{ij} Z^j_{n+1} \\ \sum_{j=1}^{s} a_{ij} \big(Z^j_{n+1} - f(t^j_{n+1}, Y^j_{n+1}, \tilde{Y}^j_{n+1}, \tilde{Z}^j_{n+1})\big) \end{bmatrix}.$$

Since the coefficient matrix A of the Radau IIA method is non-singular, the second row yields

$$Z^j_{n+1} = f\big(t^j_{n+1}, Y^j_{n+1}, \tilde{Y}^j_{n+1}, \tilde{Z}^j_{n+1}\big), \tag{6.5.34}$$

whereas the first row remains

$$Y^i_{n+1} = y_n + h_{n+1} \sum_{j=1}^{s} a_{ij} Z^j_{n+1}. \tag{6.5.35}$$

On the other hand, the interpolant (6.5.32) yields

$$\eta(t_m + \theta h_{m+1}) = \ell_0(\theta) y_m + \sum_{i=1}^{s} \ell_i(\theta) Y^i_{m+1}$$

and

$$\lambda(t_m + \rho h_{m+1}) = \ell_0(\rho) z_m + \sum_{i=1}^{s} \ell_i(\rho) Z^i_{m+1}.$$

In particular, in the case of overlapping, for some j we may have

$$\tilde{Y}^j_{n+1} = \eta(t_n + \theta^j_{n+1} h_{n+1}) = \ell_0(\theta^j_{n+1}) y_n + \sum_{i=1}^{s} \ell_i(\theta^j_{n+1}) Y^i_{n+1},$$

and/or

$$\tilde{Z}^j_{n+1} = \lambda(t_n + \rho^j_{n+1} h_{n+1}) = \ell_0(\rho^j_{n+1}) z_n + \sum_{i=1}^{s} \ell_i(\rho^j_{n+1}) Z^i_{n+1},$$

with

$$\theta^j_{n+1} = c_j - \frac{\tau(t^j_{n+1}, Y^j_{n+1})}{h_{n+1}} \quad \text{and} \quad \rho^j_{n+1} = c_j - \frac{\sigma(t^j_{n+1}, Y^j_{n+1})}{h_{n+1}}.$$

In summary, all the approximated delayed arguments $\tilde{Y}^j_{n+1}$ and $\tilde{Z}^j_{n+1}$ may be written in terms of Y^j_{n+1} and Z^j_{n+1} that, because of the uniqueness of the Lagrange interpolation polynomial and in view of Remark 6.5.2, are the same values provided by the standard approach with the option (6.5.3) for the neutral equations in the form (6.5.1). Moreover, by taking equations (6.5.34) and (6.5.35) into account, the overall method may be written in terms of the sole values Z^j_{n+1} that correspond to the K notation of the standard approach.

We can conclude that the two approaches are equivalent and, therefore, that the method converges according to the results stated in Theorem 6.5.3. Since the s-stage Radau IIA method has order $p = 2s - 1$ with interpolants $\eta(t)$ and $\lambda(t)$ of uniform order $q = r = s$, the DDE method converges with global uniform order $q' = s + 1$ for any $s \geq 2$. Note that the order results hold for any stepsize, including the case of stepsizes larger than the delay.

Returning to the particular implementation of the code RADAR5, we mention that it does not track the discontinuities, as explained in Section 4.3.1, but tries to approximate them by means of the stepsize control mechanism. Some more specific details regarding the stepsize control mechanism implemented by the code will be given in Section 7.5, along with some numerical experiments.

7

LOCAL ERROR ESTIMATE AND VARIABLE STEPSIZE

As with ODEs, an efficient DDE code should be implemented in a variable stepsize mode by performing stepsize control. In such a case, even for constant delay, it is not possible to predict a priori where the retarded term $\eta(t-\tau)$ will be computed at each step, and so we must give up exploiting any possible superconvergence property of the underlying continuous RK method. So, it is worthless distinguishing the cases of constant, time dependent and state dependent delay, and we will consider the very general DDE method (6.1.1), (6.1.2), (6.1.3), (6.1.4) or the K notation counterpart (6.1.5), (6.1.6), (6.1.7), (6.1.8), for which the general theorem, Theorem 6.1.2, and Corollary 6.1.3 hold.

As is well known, the variable stepsize strategy aims to ensure, as far as possible, the proportionality of the maximum global error to a given tolerance TOL and to minimize the computational cost. In this context, TOL is meant as a *tolerance per unit step*, and this is what we assume throughout the present chapter. Slightly different stepsize control mechanisms (see, for example, the book by Hairer, Nørsett and Wanner [117]) exist, where TOL is an absolute *tolerance per step*. However, these approaches do not provide an explicit error-tolerance proportionality.

In Section 5.5 we gave some of the most popular ways of estimating the local error for ODEs while using continuous RK methods. In any case, apart from the strategy based on the evaluation of the defect, for which we refer the reader to Enright and Hayashi [90], the stepsize control mechanisms for an *advancing method* of order p are mostly based on the use of a second method, the *error-estimating method*, of different order p', usually $p'=p+1$ or $p'=p-1$.

The extension of such stepsize control mechanisms to DDEs is not straightforward and the mathematical justification is far from obvious. Our analysis, confined to equations without neutral terms, can easily be extended to NDDEs. In particular, we shall see that a correct stepsize selection mechanism for DDEs should control both the discrete and the uniform local error, rather than the sole discrete error. To be more precise, the given tolerance will be asked to bound, as usual, the discrete local error *per unit step* and to bound the uniform local error *tout court*. In doing this one can use a continuous error-estimating method of uniform order other than q, say $q'=q+1$ or $q'=q-1$. However, we shall also see that, under certain hypotheses on the problem and on the method, one may be satisfied with controlling the discrete local error only. This will eventually reflect in a greater proportionality coefficient between the maximum (discrete or uniform) global error and the tolerance TOL.

Here we try, as far as possible, to put the error control mechanism on a firm mathematical basis at the cost of some simplifications. The main simplification is the assumption that the discontinuity points are known a priori and are included as mesh points. This is a reasonable assumption for constant and time dependent delays but not for state dependent delays. On the other hand, in Section 4.3 we saw that, in this case, the actual order of convergence of the methods is still questionable near the discontinuity points and hence the stepsize control mechanism remains very heuristic.

For further analyses on the relationships between the local and global errors of DDE solvers and on the correct order of accuracy required in the interpolation procedure, we refer the reader to Oppelstrupp [212], Oberle and Pesh [210], Arndt [3, 4], Bellen and Zennaro [36], Filippi and Buchacker [98], Higham [128] and Guglielmi and Hairer [112].

7.1 Order of the advancing and error-estimating methods

Assume that the advancing continuous RK method has discrete order p and uniform order $q \geq p-1$, so that the uniform global order is p. Observe that in the last statement we made the important assumption that the discontinuity points have been included as mesh points, as required by the Theorem 6.1.2, otherwise the method would not perform to order p.

The local problem to be solved by the advancing method is

$$\begin{cases} w'_{n+1}(t) = f\big(t, w_{n+1}(t), x(t-\tau(t, w_{n+1}(t)))\big), & t_n \leq t \leq t_{n+1}, \\ w_{n+1}(t_n) = y_n, \end{cases} \tag{7.1.1}$$

$$x(s) = \begin{cases} \phi(s) & \text{for } s \leq t_0, \\ \eta(s) & \text{for } t_0 \leq s \leq t_n, \\ w_{n+1}(s) & \text{for } t_n \leq s \leq t_{n+1}, \end{cases}$$

where, for $s < t_n$, the delayed approximate solution $\eta(s)$ has uniform global order p, that is

$$\max_{s \leq t_n} \|y(s) - \eta(s)\| = O(h^p), \tag{7.1.2}$$

where $h = \max_i h_i$.

Now consider an error-estimating method of order p' (equal to $p-1$ or $p+1$) for the solution of the local problem (7.1.1). It is obvious that the stepsize control mechanism strongly relies on the assumption that both advancing and error-estimating methods actually perform, on the local problem (7.1.1), to their own orders p and p'. Therefore, denoting the relevant approximations by y_{n+1} and $\hat{y}_{n+1}$ respectively, we want to make sure that

$$\|w_{n+1}(t_{n+1}) - y_{n+1}\| = O(h_{n+1}^{p+1}) \tag{7.1.3}$$

and

$$\|w_{n+1}(t_{n+1}) - \hat{y}_{n+1}\| = O(h_{n+1}^{p'+1}) \tag{7.1.4}$$

for fixed $h_1, \ldots, h_n$ and for $h_{n+1} \to 0$. A sufficient condition for this is that the right-hand-side function in (7.1.1) be of class C^{p^*}, where $p^* = \max\{p, p'\}$. In general, the image of the current interval $\alpha([t_n, t_{n+1}])$ in the deviating argument $\alpha(t) = t - \tau(t, w_{n+1}(t))$ may include some previous mesh points and, hence, the interpolant $\eta(t)$ being a piecewise polynomial, the function $\eta(t - \tau(t, w_{n+1}(t)))$ may be just continuous. Note that if the current stepsize h_{n+1} is sufficiently small with respect to the previous steps, then the image $\alpha([t_n, t_{n+1}])$ is entirely contained in some previous mesh interval and, hence, $\eta(t - \tau(t, w_{n+1}(t)))$ is smooth. However, this is an incidental event that unlikely occurs because any step selection mechanism provides meshes satisfying

$$\frac{\max_i h_i}{\min_i h_i} \leq K, \tag{7.1.5}$$

where K is a constant independent of the tolerance TOL. So we must expect to deal with problems (7.1.1) with a non-smooth right-hand side. Nevertheless, conditions (7.1.3) and (7.1.4) may be proved as follows.

Consider the auxiliary problem

$$\begin{cases} \tilde{w}'_{n+1}(t) = f(t, \tilde{w}_{n+1}(t), \tilde{x}(t - \tau(t, \tilde{w}_{n+1}(t)))), & t_n \leq t \leq t_{n+1}, \\ \tilde{w}_{n+1}(t_n) = y_n, \end{cases} \tag{7.1.6}$$

$$\tilde{x}(s) = \begin{cases} \phi(s) & \text{for } s \leq t_0, \\ \tilde{\eta}(s) & \text{for } t_0 \leq s \leq t_n, \\ \tilde{w}_{n+1}(s) & \text{for } t_n \leq s \leq t_{n+1}, \end{cases}$$

where, for $s < t_n$, the non-smooth interpolant $\eta(s)$ is replaced by a function $\tilde{\eta}(s)$ of class C^{p^*}. Moreover, assume that the function $\tilde{\eta}(s)$ approximates $\eta(s)$ to order p^*, that is

$$\max_{s \leq t_n} \|\tilde{\eta}(s) - \eta(s)\| = O(h^{p^*}), \tag{7.1.7}$$

where $h = \max_i h_i$, and observe that, by (7.1.5), (7.1.7) is equivalent to

$$\max_{s \leq t_n} \|\tilde{\eta}(s) - \eta(s)\| = O(h_{n+1}^{p^*}).$$

We denote the numerical solutions of (7.1.6) for the advancing and for the error-estimating methods by z_{n+1} and $\hat{z}_{n+1}$, respectively. Since the problem is sufficiently smooth, we have

$$\|\tilde{w}_{n+1}(t_{n+1}) - z_{n+1}\| = O(h_{n+1}^{p+1}) \tag{7.1.8}$$

and

$$\|\tilde{w}_{n+1}(t_{n+1}) - \hat{z}_{n+1}\| = O(h_{n+1}^{p'+1}). \tag{7.1.9}$$

Furthermore, by (7.1.7) and the Lipschitz continuous dependence of both the exact and numerical solutions of the data, we get

$$\|\tilde{w}_{n+1}(t_{n+1}) - w_{n+1}(t_{n+1})\| = O(h_{n+1}^{p^*+1}), \tag{7.1.10}$$

$$\|y_{n+1} - z_{n+1}\| = O(h_{n+1}^{p^*+1}) \tag{7.1.11}$$

and

$$\|\hat{y}_{n+1} - \hat{z}_{n+1}\| = O(h_{n+1}^{p^*+1}). \tag{7.1.12}$$

Summarizing, we have

$$\begin{aligned}\|w_{n+1}(t_{n+1}) - y_{n+1}\| \le\ & \|w_{n+1}(t_{n+1}) - \tilde{w}_{n+1}(t_{n+1})\| \\ & + \|\tilde{w}_{n+1}(t_{n+1}) - z_{n+1}\| + \|z_{n+1} - y_{n+1}\|\end{aligned}$$

and thus, by (7.1.10), (7.1.8) and (7.1.11), we get (7.1.3).

Analogously,

$$\begin{aligned}\|w_{n+1}(t_{n+1}) - \hat{y}_{n+1}\| \le\ & \|w_{n+1}(t_{n+1}) - \tilde{w}_{n+1}(t_{n+1})\| \\ & + \|\tilde{w}_{n+1}(t_{n+1}) - \hat{z}_{n+1}\| + \|\hat{z}_{n+1} - \hat{y}_{n+1}\|\end{aligned}$$

and thus, by (7.1.10), (7.1.9) and (7.1.12), we get (7.1.4).

Now, the crucial problem is to verify whether or not the smooth function $\tilde{\eta}(s)$ of class C^{p^*} invoked in the auxiliary problem (7.1.6) actually exists that fulfills the accuracy condition (7.1.7). To this end we must handle the two cases $p' = p - 1$ and $p' = p + 1$ separately.

7.1.1 *Error-estimating methods of order $p' = p - 1$*

In this case $p^* = p$ and then, by (7.1.2), we can choose $\tilde{\eta}(s) = y(s)$ which is smooth and complies with (7.1.7). So we can conclude that the uniform order $q = p - 1$ for the interpolant is sufficient.

7.1.2 *Error-estimating methods of order $p' = p + 1$*

In this case $p^* = p + 1$ and thus we obviously assume that the involved functions f, τ and ϕ are of class C^{p+1}. Contrary to the previous case, finding the function $\tilde{\eta}(s)$ satisfying the accuracy condition (7.1.7) is not so easy. In fact, by (7.1.2) it is evident that the choice $\tilde{\eta}(s) = y(s)$ no longer works. Nevertheless, we shall show that, by assuming that the uniform order of the advancing method is raised to $q = p$, the required function $\tilde{\eta}(s)$ exists.

For the sake of simplicity, we restrict ourselves to illustrating the case of time dependent delay $\tau = \tau(t)$ satisfying hypotheses (H_1) and (H_4). This allows us to proceed by induction through the macro intervals $[\xi_{l-1}, \xi_l]$, $l = 1, \ldots, k$, where $[\xi_{k-1}, \xi_k]$ contains the current mesh interval $[t_n, t_{n+1}]$.

Note that assumption (7.1.5) implies that the image in the deviating argument $\alpha(t) = t - \tau(t)$ of any selected step $[t_i, t_{i+1}]$ intersects no more than M previous step intervals, where M is independent of the tolerance TOL and of the mesh Δ.

Therefore, for each $l = k - 1, \ldots, 1$, the image of the current step $[t_n, t_{n+1}]$ iterated $k - l$ times is included between a pair of mesh points t_{n_l} and $t_{n_l + m_l}$ belonging to the macro interval $[\xi_{l-1}, \xi_l]$, with $m_l \le M^{k-l}$.

Now, in the first macro interval $[\xi_0, \xi_1]$, consider the solution $w_{n_1+1}(t)$ to

$$\begin{cases} \tilde{w}'_{n_1+1}(t) = f\big(t, \tilde{w}_{n_1+1}(t), \phi(t-\tau(t))\big), & t_{n_1} \le t \le t_{n_1+m_1}, \\ \tilde{w}_{n_1+1}(t_{n_1}) = y_{n_1}, \end{cases}$$

which is a smooth problem, whose approximate solution is just $\eta(t)$ and has uniform order q. Since $m_1 \le M^{k-1}$, we have

$$\max_{t_{n_1} \le t \le t_{n_1+m_1}} \|\tilde{w}_{n_1+1}(t) - \eta(t)\| = O(h^{q+1})$$

either for $q = p - 1$ or $q = p$.

Then jump into the next macro interval $[\xi_1, \xi_2]$ and consider the problem

$$\begin{cases} w'_{n_2+1}(t) = f\big(t, w_{n_2+1}(t), \eta(t-\tau(t))\big), & t_{n_2} \le t \le t_{n_2+m_2}, \\ w_{n_2+1}(t_{n_2}) = y_{n_2}, \end{cases}$$

whose numerical solution is still $\eta(t)$. The presence of $\eta(t - \tau(t))$ makes the problem non-smooth. So we replace it by the smooth one,

$$\begin{cases} \tilde{w}'_{n_2+1}(t) = f\big(t, \tilde{w}_{n_2+1}(t), \tilde{w}_{n_1+1}(t-\tau(t))\big), & t_{n_2} \le t \le t_{n_2+m_2}, \\ \tilde{z}_{n_2+1}(t_{n_2}) = y_{n_2}, \end{cases} \tag{7.1.13}$$

where $\tilde{w}_{n_1+1}(t - \tau(t))$ is well defined for $t \in [t_{n_2}, t_{n_2+m_2}]$.

Since $m_2 \le M^{k-2}$, we can easily verify that

$$\max_{t_{n_2} \le t \le t_{n_2+m_2}} \|\tilde{w}_{n_2+1}(t) - \eta(t)\| = O(h^{q+1}).$$

This may be done by making use of the numerical solution of (7.1.13) and the triangular inequality by taking into account the Lipschitz continuous dependence of both the exact and numerical solutions of the data.

Now it is clear how the induction works. In particular, the required function $\tilde{\eta}(t)$ of class C^{p+1} satisfying the required condition (7.1.7) is nothing but the solution $\tilde{w}_{n_{k-1}+1}(t)$ of

$$\begin{cases} \tilde{w}'_{n_{k-1}+1}(t) = f\big(t, \tilde{w}_{n_{k-1}+1}(t), \tilde{w}_{n_{k-2}+1}(t-\tau)\big), & t_{n_{k-1}} \le t \le t_{n_{k-1}+m_{k-1}}, \\ \tilde{w}_{n_{k-1}+1}(t_{n_{k-1}}) = y_{n_{k-1}}, \end{cases}$$

provided that $q = p$. So, we can conclude that the uniform order q must be raised to p.

7.1.3 *Continuous error-estimating methods*

As for the estimate of the uniform local error of the advancing method of uniform order q, one can use a continuous error-estimating method of uniform order $q' \neq q$, say $q' = q - 1$ or $q' = q + 1$. The analysis is similar but the proof is simpler. In fact, the choice $\tilde{\eta}(s) = y(s)$ fits for both values of q' and thus the uniform order q of the advancing method is suitable to validate both the higher and lower uniform order of the error-estimating method.

7.2 Error propagation and stepsize control

In this section we analyze how the local discrete and uniform local errors propagate through the integration interval $[t_0, t_f]$ to form the maximum global error. The analysis will suggest which tests the local error estimates must undergo in order to get a global error proportional to the assigned tolerance TOL. For the sake of simplicity, we confine ourselves to the case of time dependent delays $\tau(t)$ and the analysis is carried out in the general case that allows overlapping. We shall use the following elementary version of Gronwall's Lemma (see Hale [121]).

Lemma 7.2.1 *If $\delta(t)$, $\gamma(t)$ and $\beta(t)$ are real-valued continuous functions in $[a,b]$, if $\gamma(t)$ is non-decreasing and if $\beta(t) \geq 0$ with*

$$\delta(t) \leq \gamma(t) + \int_a^t \beta(x)\delta(x)\mathrm{d}x, \quad a \leq t \leq b,$$

then

$$\delta(t) \leq \gamma(t) \exp\left(\int_a^t \beta(x)\mathrm{d}x\right).$$

Let us consider a mesh $\Delta = \{t_0, t_1, \ldots, t_N = t_f\}$ along the integration interval $[t_0, t_f]$ and denote the maximum discrete and uniform global errors by

$$e_n = \max_{i \leq n} \|y(t_i) - y_i\| \quad \text{and} \quad E_n = \max_{t \leq t_n} \|y(t) - \eta(t)\|.$$

Then, for the solution $w_{n+1}(t)$ of the local problem (7.1.1), we have

$$\begin{aligned} w_{n+1}(t) - y(t) = {} & y_n - y(t_n) \\ & + \int_{t_n}^t \Big(f\big(s, w_{n+1}(s), x(\alpha(s))\big) - f\big(s, y(s), y(\alpha(s))\big)\Big)\mathrm{d}s \end{aligned}$$

and, hence,

$$\begin{aligned} \|w_{n+1}(t) - y(t)\| \leq \|y(t_n) - y_n\| & + \int_{t_n}^t L\|w_{n+1}(s) - y(s)\|\mathrm{d}s \\ & + \int_{t_n}^t M\|x(\alpha(s)) - y(\alpha(s))\|\mathrm{d}s, \end{aligned}$$

where L and M are the Lipschitz constants of f. Then Lemma 7.2.1 yields

$$\|w_{n+1}(t) - y(t)\| \leq \Big(e_n + M\int_{t_n}^t \|x(\alpha(s)) - y(\alpha(s))\|\mathrm{d}s\Big)\mathrm{e}^{L(t-t_n)},$$

where

$$\|x(\alpha(s)) - y(\alpha(s))\| \leq \begin{cases} E_n & \text{for } \alpha(s) \leq t_n, \\ \max_{t_n \leq t \leq t_{n+1}} \|w_{n+1}(t) - y(t)\| & \text{for } \alpha(s) > t_n. \end{cases}$$

Therefore, we have

$$\max_{t_n \le t \le t_{n+1}} \|w_{n+1}(t) - y(t)\| \le \frac{e^{Lh_{n+1}}}{1 - Mh_{n+1}e^{Lh}} e_n + \frac{Mh_{n+1}e^{Lh_{n+1}}}{1 - Mhe^{Lh}} E_n \qquad (7.2.1)$$

for $h = \max_i h_i$ sufficiently small.

Note that, whenever possible (or necessary), in what follows we shall substitute the current stepsize h_{n+1} by h in order to get the final formula (7.2.4) for the maximum global error.

Now, let us bound the discrete and uniform global errors in $[t_n, t_{n+1}]$ by

$$\|y(t_{n+1}) - y_{n+1}\| \le h_{n+1}\sigma_{n+1} + \|w_{n+1}(t_{n+1}) - y(t_{n+1})\|$$

and

$$\max_{t_n \le t \le t_{n+1}} \|y(t) - \eta(t)\| \le \Sigma_{n+1} + \max_{t_n \le t \le t_{n+1}} \|w_{n+1}(t) - y(t)\|,$$

where

$$\sigma_{n+1} = \|w_{n+1}(t_{n+1}) - y_{n+1}\|/h_{n+1}$$

and

$$\Sigma_{n+1} = \max_{t_n \le t \le t_{n+1}} \|w_{n+1}(t) - \eta(t)\|$$

are the discrete local error per unit step and the uniform local error, respectively (see Section 5.5).

Then inequality (7.2.1) yields

$$\|y(t_{n+1}) - y_{n+1}\| \le h_{n+1}\sigma_\Delta + \frac{e^{Lh_{n+1}}}{1 - Mh_{n+1}e^{Lh}} e_n + \frac{Mh_{n+1}e^{Lh_{n+1}}}{1 - Mhe^{Lh}} E_n,$$

where $\sigma_\Delta = \max_i \sigma_i$, and, consequently, the following recursion for the maximum discrete global error:

$$e_{n+1} \le h_{n+1}\sigma_\Delta + \frac{e^{Lh_{n+1}}}{1 - Mh_{n+1}e^{Lh}} e_n + \frac{Mh_{n+1}e^{Lh_{n+1}}}{1 - Mhe^{Lh}} E_n. \qquad (7.2.2)$$

Analogously, inequality (7.2.1) yields

$$\max_{t_n \le t \le t_{n+1}} \|y(t) - \eta(t)\| \le \Sigma_\Delta + \frac{e^{Lh}}{1 - Mhe^{Lh}} e_n + \frac{Mhe^{Lh}}{1 - Mhe^{Lh}} E_n,$$

where $\Sigma_\Delta = \max_i \Sigma_i$. Moreover, the monotonicity of e_n and E_n implies

$$E_n \le E_{n+1} \le \Sigma_\Delta + \frac{e^{Lh}}{1 - Mhe^{Lh}} e_n + \frac{Mhe^{Lh}}{1 - Mhe^{Lh}} E_n,$$

and thus, for the maximum uniform global error,

$$E_n \le \frac{1 - Mhe^{Lh}}{1 - 2Mhe^{Lh}} \Sigma_\Delta + \frac{e^{Lh}}{1 - 2Mhe^{Lh}} e_n. \qquad (7.2.3)$$

Finally, by substituting (7.2.2) into the last inequality, after some manipulations we obtain

$$e_{n+1} \leq \mathrm{e}^{Q_\Delta h_{n+1}} e_n + h_{n+1}\Big(\sigma_\Delta + R_\Delta \Sigma_\Delta\Big),$$

where

$$Q_\Delta = L + \frac{3 - 4Mh\mathrm{e}^{Lh}}{1 - 2Mh\mathrm{e}^{Lh}} M\mathrm{e}^{Lh} \qquad \text{and} \qquad R_\Delta = \frac{M\mathrm{e}^{Lh}}{1 - 2Mh\mathrm{e}^{Lh}}.$$

Note that $\lim_{h\to 0} Q_\Delta = L + 3M$ and $\lim_{h\to 0} R_\Delta = M$ so that, for all meshes Δ with $h \leq h_0$ sufficiently small, we have

$$e_{n+1} \leq \mathrm{e}^{Q h_{n+1}} e_n + h_{n+1}\Big(\sigma_\Delta + R \Sigma_\Delta\Big),$$

where Q and R are independent of Δ. Therefore, standard arguments allow us to conclude that there exists a constant $K > 0$ independent of Δ for which the maximum discrete global error satisfies

$$e_n \leq K\Big(\sigma_\Delta + R\Sigma_\Delta\Big) \tag{7.2.4}$$

for all $n = 1, \ldots, N$.

Observe that, by substituting (7.2.4) into (7.2.3), we obtain an analogous bound to the maximum uniform global error too.

The desired proportionality between the global error and the given tolerance TOL is obtained by subjecting the estimates $\tilde{\sigma}_{n+1}$ and $\tilde{\Sigma}_{n+1}$ of the discrete and uniform local errors σ_{n+1} and Σ_{n+1} to the condition

$$\tilde{\sigma}_{n+1} \leq TOL \qquad \text{and} \qquad \tilde{\Sigma}_{n+1} \leq TOL \qquad \text{for all } n. \tag{7.2.5}$$

Remark 7.2.2 *The constants K and R in (7.2.4) have been obtained under the restriction $h \leq h_0$ with $h_0 > 0$ such that $1 - 2Mh_0\mathrm{e}^{Lh_0} > 0$ (see (7.2.3)). This might be a severe restriction in the stepsize, especially for large values of L and M. On the other hand, by replacing L with the one-sided Lipschitz constant (see Section 8.2.3), a sharper analysis can be carried out along with a better estimate of the constants.*

Now we discuss the different effect of the uniform orders $q = p - 1$ and $q = p$ in the advancing RK method with respect to the maximum global error bound (7.2.4). As a consequence, we shall see under which conditions the test (7.2.5) may be limited to the first half regarding the sole discrete local error estimate $\tilde{\sigma}_{n+1}$.

In Section 7.1, we saw that, in both the cases $q = p-1$ and $q = p$, the discrete local error per unit step σ_{n+1} actually satisfies

$$\sigma_{n+1} = O(h_{n+1}^p). \tag{7.2.6}$$

Moreover, by going through the arguments of Section 7.1 again, it is easy to understand that, only in the case $q = p$ does the finer expansion

$$\sigma_{n+1} = \alpha_n h_{n+1}^p + O(h_{n+1}^{p+1}) \tag{7.2.7}$$

hold, where α_n depends continuously on t_n, y_n and $\eta(t)$ for $t \leq t_n$, but is independent of h_{n+1}.

7.2.1 *The uniform order $q = p$*

For the uniform local error Σ_{n+1} it holds that

$$\Sigma_{n+1} = \beta_n h_{n+1}^{p+1} + O(h_{n+1}^{p+2}), \tag{7.2.8}$$

where β_n depends continuously on t_n, y_n and $\eta(t)$ for $t \leq t_n$, but is independent of h_{n+1}. Therefore, by (7.2.7) we have

$$\Sigma_{n+1} = \frac{\beta_n + O(h_{n+1})}{\alpha_n + O(h_{n+1})} h_{n+1}\sigma_{n+1}$$

and thus, under the mild assumption that there exists $\kappa > 0$ such that

$$\frac{\beta_n}{\alpha_n} \leq \kappa$$

for $h = \max_i h_i$ sufficiently small (i.e. for y_n and $\eta(t)$ sufficiently close to $y(t_n)$ and $y(t)$) and uniformly with respect to t_n in $[t_0, t_f]$, *local extrapolation* works to yield

$$\Sigma_\Delta = hO(\sigma_\Delta). \tag{7.2.9}$$

In conclusion, as the tolerance TOL is reduced, h becomes smaller and smaller and the influence of the uniform local error Σ_Δ in (7.2.4) decreases. Therefore the maximum global error is mostly determined by the discrete local error per unit step σ_Δ. In such a situation, it is reasonable to impose the sole condition

$$\tilde{\sigma}_{n+1} \leq TOL \qquad \text{for all } n \tag{7.2.10}$$

instead of (7.2.5). On the contrary, when the uniform local error Σ_Δ is not negligible with respect to σ_Δ, the complete test (7.2.5) would be more effective. This is the case in problems for which the tolerance imposed by the user in the local error test (7.2.10) leads to large stepsizes.

These occurrences will be widely illustrated in the forthcoming Examples 7.3.1 and 7.4.1.

7.2.2 *The uniform order $q = p - 1$*

For the uniform local error Σ_{n+1} it holds that

$$\Sigma_{n+1} = \beta_n h_{n+1}^p + O(h_{n+1}^{p+1}),$$

where β_n depends continuously on t_n, y_n and $\eta(t)$ for $t \leq t_n$, but is independent of h_{n+1}.

TABLE 7.1. Uniform order conditions and test of tolerance for an advancing method of order p and error-estimating method of order p'.

p'	q	Test	Expected behavior of the global errors
$p+1$	p	$\sigma_\Delta \leq TOL$	$e_N = O(TOL)$ $E_N = O(TOL)$
$p-1$	$p-1$	$\sigma_\Delta \leq TOL$ $\Sigma_\Delta \leq TOL$ (*) (*) Not necessary but advisable	$e_N = O(TOL)$ $E_N = O(TOL)$

This time the order of Σ_{n+1} equals the order of σ_{n+1} and thus, in place of (7.2.9), the result is just

$$\Sigma_\Delta = O(\sigma_\Delta). \tag{7.2.11}$$

On the other hand, in general (7.2.7) does not hold and only (7.2.6) is available. This fact makes even the validity of (7.2.11) not so evident as in the previous case $q = p$.

In conclusion, the influence of the uniform local error Σ_Δ and of the discrete local error per unit step σ_Δ in (7.2.4) remains comparable as the tolerance TOL becomes smaller and smaller. Therefore, it is strongly recommended to subject both the discrete and uniform local error estimates to the complete test (7.2.5).

7.2.3 *Conclusions*

Although with slightly different general lines, the foregoing analysis is consistent with the relevant results available in the literature. In particular, we quote Oppelstrupp [212] and, later, Higham [128]. The former proved that an asymptotic expansion of the global error with at least one term (i.e. at least up to order $p+1$) exists only if the uniform order is $q = p$, so as to allow its correct estimate by means of Richardson extrapolation. The latter proved that, by controlling only the discrete local error, it is necessary and sufficient that the uniform order q of the interpolation be equal to p in order to ensure error-tolerance proportionality.

We also remark that, in order to illustrate the influence of the interpolation procedure on the global error of a DDE solver, Arndt [3] constructed a particular example with uniform order $q = p - 1$, where the size of the global error is determined only by the uniform local errors, whereas the discrete local errors are always equal to zero. Even though the choice of the equation is very special, his example shows that, in the case $q = p - 1$, control of the sole discrete local error may be very unreliable.

In Table 7.1 we have summarized the minimum conditions in order to ensure the asymptotic (i.e. for sufficiently small maximum stepsize h) error-tolerance proportionality in the two cases $p' = p+1$ and $p' = p-1$.

7.3 Implementation without overlapping

In this section we illustrate the foregoing theory by means of a sample method and related numerical experiments. We are not going to present a completely detailed code, so that some parameters are not fixed a priori. Their value has to be chosen by the reader when he/she wants to actually implement the methods we propose.

We illustrate the strategy based on an advancing method of uniform order $q = p$ endowed with a discrete error-estimating method of order $p' = p+1$, obtained by means of a quadrature formula of the type (5.5.2).

In this setting, we apply the procedure described in Section 7.2 to explicit RK methods of order $p = 4$, namely the classical four-stage RK method:

$$\begin{array}{c|cccc} 0 & & & & \\ \frac{1}{2} & \frac{1}{2} & & & \\ \frac{1}{2} & 0 & \frac{1}{2} & & \\ 1 & 0 & 0 & 1 & \\ \hline & \frac{1}{6} & \frac{1}{3} & \frac{1}{3} & \frac{1}{6} \end{array}$$

We consider DDEs with time dependent delays $\tau = \tau(t)$ satisfying hypothesis (H_1), so that overlapping may be avoided simply by choosing the stepsize sufficiently small. Moreover, we assume that the discontinuity points are computed a priori and are included in the mesh.

The local problem to solve in the current step $[t_n, t_{n+1}]$ is

$$\begin{cases} w'_{n+1}(t) = f\big(t, w_{n+1}(t), \eta(t-\tau(t))\big), & t_n \le t \le t_{n+1}, \\ w_{n+1}(t_n) = y_n. \end{cases} \tag{7.3.1}$$

With the K notation, the considered RK method reads

$$\begin{aligned} K^1_{n+1} &= f\big(t_n, y_n, \eta(t_n - \tau(t_n))\big), \\ K^2_{n+1} &= f\big(t_n + \tfrac{1}{2}h_{n+1}, y_n + \tfrac{1}{2}h_{n+1}K^1_{n+1}, \eta(t_n + \tfrac{1}{2}h_{n+1} - \tau(t_n + \tfrac{1}{2}h_{n+1}))\big), \\ K^3_{n+1} &= f\big(t_n + \tfrac{1}{2}h_{n+1}, y_n + \tfrac{1}{2}h_{n+1}K^2_{n+1}, \eta(t_n + \tfrac{1}{2}h_{n+1} - \tau(t_n + \tfrac{1}{2}h_{n+1}))\big), \\ K^4_{n+1} &= f\big(t_n + h_{n+1}, y_n + h_{n+1}K^3_{n+1}, \eta(t_n + h_{n+1} - \tau(t_n + h_{n+1}))\big), \end{aligned}$$

$$y_{n+1} = y_n + h_{n+1}\big(\tfrac{1}{6}K^1_{n+1} + \tfrac{1}{3}K^2_{n+1} + \tfrac{1}{3}K^3_{n+1} + \tfrac{1}{6}K^4_{n+1}\big). \tag{7.3.2}$$

As explained in Example 5.2.7, the method possesses an NCE of uniform order $q = 3$. Nevertheless, since we want to upgrade the uniform order of interpolation up to $q = p = 4$ by using the procedure of Section 5.3.1, we start from the cubic

Hermite interpolant $\eta_{(0)}(t)$ at the end-points t_n and t_{n+1}, which also makes use of the extra stage

$$K_{n+1}^5 = K_{n+2}^1 = f\big(t_{n+1}, y_{n+1}, \eta(t_{n+1} - \tau(t_{n+1}))\big),$$

available for free from the next step $[t_{n+1}, t_{n+2}]$. Therefore, we have

$$\eta_{(0)}(t_n + \theta h_{n+1}) = d_1(\theta) y_n + d_2(\theta) y_{n+1} + d_3(\theta) h_{n+1} K_{n+1}^1 + d_4(\theta) h_{n+1} K_{n+1}^5$$

or, by substituting y_{n+1} from (7.3.2), the equivalent form of the type (5.1.4)

$$\begin{aligned}\eta_{(0)}(t_n + \theta h_{n+1}) = y_n + h_{n+1}\Big(&\big(\tfrac{1}{6} d_2(\theta) + d_3(\theta)\big) K_{n+1}^1 + \tfrac{1}{3} d_2(\theta) K_{n+1}^2 \\ &+ \tfrac{1}{3} d_2(\theta) K_{n+1}^3 + \tfrac{1}{6} d_2(\theta) K_{n+1}^4 + d_4(\theta) K_{n+1}^5\Big),\end{aligned}$$

where

$$\begin{aligned} d_1(\theta) &= (\theta - 1)^2 (2\theta + 1), \\ d_2(\theta) &= \theta^2 (3 - 2\theta), \\ d_3(\theta) &= \theta(\theta - 1)^2, \\ d_4(\theta) &= \theta^2(\theta - 1). \end{aligned}$$

In order to construct the uniform correction $\eta_{(1)}(t)$ of order $q = 4$, to be assumed as the advancing continuous approximation $\eta(t)$ in the current step $[t_n, t_{n+1}]$, we consider the extra abscissa $\theta_1 \in (0, 1)$, $\theta_1 \neq \frac{1}{2}$, and the extra stage

$$K_{n+1}^6 = f\big(t_n + \theta_1 h_{n+1}, \eta_{(0)}(t_n + \theta_1 h_{n+1}), \eta(t_n + \theta_1 h_{n+1} - \tau(t_n + \theta_1 h_{n+1}))\big).$$

Thus, with the notation of Section 5.3.1, $\eta_{(1)}(t)$ is defined as the fourth-degree Hermite–Birkhoff interpolant relevant to the conditions on $P(0)$, $P(1)$, $P'(0)$, $P'(1)$ and $P'(\theta_1)$. Note that $\theta_1 \neq \frac{1}{2}$ because the interpolation condition on $P'(\frac{1}{2})$ is linearly dependent on the other four conditions. We obtain

$$\begin{aligned}\eta_{(1)}(t_n + \theta h_{n+1}) = d_1(\theta) y_n &+ d_2(\theta) y_{n+1} + d_3(\theta) h_{n+1} K_{n+1}^1 \\ &+ d_4(\theta) h_{n+1} K_{n+1}^5 + d_5(\theta) h_{n+1} K_{n+1}^6\end{aligned}$$

or, by substituting y_{n+1} from (7.3.2), the equivalent form of the type (5.1.4)

$$\begin{aligned}\eta_{(1)}(t_n + \theta h_{n+1}) = y_n + h_{n+1}\big(&\big(\tfrac{1}{6} d_2(\theta) + d_3(\theta)\big) K_{n+1}^1 + \tfrac{1}{3} d_2(\theta) K_{n+1}^2 \\ &+ \tfrac{1}{3} d_2(\theta) K_{n+1}^3 + \tfrac{1}{6} d_2(\theta) K_{n+1}^4 + d_4(\theta) K_{n+1}^5 + d_5(\theta) K_{n+1}^6\big),\end{aligned}$$

where

$$\begin{aligned} d_1(\theta) &= \tfrac{1}{2\theta_1 - 1} (\theta - 1)^2 \big(-3\theta^2 + 2(2\theta_1 - 1)\theta + 2\theta_1 - 1 \big), \\ d_2(\theta) &= \tfrac{1}{2\theta_1 - 1} \theta^2 \big(3\theta^2 - 4(\theta_1 + 1)\theta + 6\theta_1 \big), \\ d_3(\theta) &= \tfrac{1}{2\theta_1(2\theta_1 - 1)} \theta(\theta - 1)^2 \big((1 - 3\theta_1)\theta + 2\theta_1(2\theta_1 - 1) \big), \\ d_4(\theta) &= \tfrac{1}{2(\theta_1 - 1)(2\theta_1 - 1)} \theta^2 (\theta - 1) \big((2 - 3\theta_1)\theta + \theta_1(4\theta_1 - 3) \big), \\ d_5(\theta) &= \tfrac{1}{2\theta_1(2\theta_1 - 1)(\theta_1 - 1)} \theta^2 (\theta - 1)^2. \end{aligned}$$

In order to estimate the discrete local error by means of an error-estimating method of higher order $p' = p + 1 = 5$, we construct an approximation $\tilde{y}_{n+1}$ by using formula (5.5.2). More precisely, we consider the Lobatto abscissae $\pi_1 = \frac{5-\sqrt{5}}{10}$ and $\pi_2 = \frac{5+\sqrt{5}}{10}$ and the additional stages

$$K_{n+1}^7 = f\big(t_n + \pi_1 h_{n+1}, \eta_{(1)}(t_n + \pi_1 h_{n+1}), \eta(t_n + \pi_1 h_{n+1} - \tau(t_n + \pi_1 h_{n+1}))\big)$$

and

$$K_{n+1}^8 = f\big(t_n + \pi_2 h_{n+1}, \eta_{(1)}(t_n + \pi_2 h_{n+1}), \eta(t_n + \pi_2 h_{n+1} - \tau(t_n + \pi_2 h_{n+1}))\big).$$

Then we define

$$\tilde{y}_{n+1} = y_n + h_{n+1}\big(\tfrac{1}{12}K_{n+1}^1 + \tfrac{5}{12}K_{n+1}^7 + \tfrac{5}{12}K_{n+1}^8 + \tfrac{1}{12}K_{n+1}^5\big) \tag{7.3.3}$$

and, as is usually done (see (5.5.1)), we estimate the discrete local error per unit step by

$$\tilde{\sigma}_{n+1} = \|\tilde{y}_{n+1} - y_{n+1}\|/h_{n+1}.$$

The current step is accepted if and only if

$$\tilde{\sigma}_{n+1} \leq TOL, \tag{7.3.4}$$

where TOL is the user-supplied tolerance.

The new stepsize h_{new}, to be used either as h_{n+2} in the next step $[t_{n+1}, t_{n+2}]$ if (7.3.4) is satisfied or, again, as h_{n+1} in the current step $[t_n, t_{n+1}]$ if (7.3.4) is violated, is computed by using the common formula

$$h_{\text{new}} = \max\left\{\omega_{\min}, \min\left\{\omega_{\max}, \rho\Big(\frac{TOL}{\tilde{\sigma}_{n+1}}\Big)^{1/4}\right\}\right\} \cdot h_{n+1}, \tag{7.3.5}$$

where $\omega_{\min} \in (0,1)$, $\omega_{\max} > 1$ and $\rho \in (0,1)$ are suitable *safety factors.*

A maximum number of rejections per step may be fixed by the user as well.

Note that, when (7.3.4) is satisfied and $t_{n+1} + h_{\text{new}} > \xi_k$ for a discontinuity point ξ_k which must be included in the mesh, we set $t_{n+2} = \xi_k$ and $h_{n+2} = \xi_k - t_{n+1}$.

Since the uniform order is $q = p$, controlling the discrete local error is sufficient for the stepsize selection. Nevertheless, in the light of the conclusion in Section 7.2.1, we also propose an additional optional control of the uniform local error Σ_{n+1} by a continuous error-estimating method of lower order $q' = q-1 = 3$, by employing local extrapolation.

More precisely, we consider the computable quantity

$$\hat{\Sigma}_{n+1} = \max_{0\leq\theta\leq1} \|\eta_{(0)}(t_n + \theta h_{n+1}) - \eta_{(1)}(t_n + \theta h_{n+1})\|. \tag{7.3.6}$$

It holds that

$$\hat{\Sigma}_{n+1} = \gamma_n h_{n+1}^4 + O(h_{n+1}^5),$$

where γ_n depends continuously on t_n, y_n and $\eta(t)$ for $t \leq t_n$, but is independent of h_{n+1}. On the other hand, the uniform local error Σ_{n+1} satisfies (7.2.8), that is

$$\Sigma_{n+1} = \beta_n h_{n+1}^5 + O(h_{n+1}^6),$$

where β_n depends continuously on t_n, y_n and $\eta(t)$ for $t \le t_n$, but is independent of h_{n+1}. Therefore, we have

$$\Sigma_{n+1} = \frac{\beta_n + O(h_{n+1})}{\gamma_n + O(h_{n+1})}\, h_{n+1}\hat{\Sigma}_{n+1}.$$

Since for $h = \max_i h_i$ sufficiently small (i.e. for y_n and $\eta(t)$ sufficiently close to $y(t_n)$ and $y(t)$) there exists $\kappa > 0$ such that

$$\frac{\beta_n}{\gamma_n} \le \kappa$$

uniformly with respect to t_n in $[t_0, t_f]$, local extrapolation works and we can conclude that

$$\Sigma_\Delta = O(TOL) \tag{7.3.7}$$

is assured if we impose

$$h_{n+1}\hat{\Sigma}_{n+1} \le TOL. \tag{7.3.8}$$

Note that, by (7.2.4), relation (7.3.7) is necessary to ensure error-tolerance proportionality.

The test (7.3.8) is made whenever (7.3.4) holds. If (7.3.8) is not satisfied, then we reduce the stepsize by using the formula

$$h_{\text{new2}} = \max\left\{\omega_{\min}, \rho\Big(\frac{TOL}{h_{n+1}\hat{\Sigma}_{n+1}}\Big)^{1/5}\right\}\cdot h_{n+1} \tag{7.3.9}$$

and repeat the computations in the current step $[t_n, t_{n+1}]$ with $h_{n+1} = h_{\text{new2}}$.

In any case, even if (7.3.4) is satisfied, the stepsize h_{new2} suggested by (7.3.9) is used to bound the next step h_{n+2}, so that (7.3.5) is substituted by

$$h_{n+2} = \max\left\{\omega_{\min}, \min\left\{\omega_{\max}, \rho\Big(\frac{TOL}{\tilde{\sigma}_{n+1}}\Big)^{1/4}, \rho\Big(\frac{TOL}{h_{n+1}\hat{\Sigma}_{n+1}}\Big)^{1/5}\right\}\right\}\cdot h_{n+1}.$$

In order to compute $\hat{\Sigma}_{n+1}$ from (7.3.6), we set

$$\delta(\theta) = \eta_{(0)}(t_n + \theta h_{n+1}) - \eta_{(1)}(t_n + \theta h_{n+1})$$

and observe that

$$\begin{aligned}\delta(\theta) = h_{n+1}\frac{\theta^2(\theta-1)^2}{2(2\theta_1-1)}\Big(&\frac{2\theta_1-1}{\theta_1}K_{n+1}^1 - (2K_{n+1}^2 + 2K_{n+1}^3 + K_{n+1}^4)\\ &+\frac{3\theta_1-2}{\theta_1-1}K_{n+1}^5 + \frac{1}{\theta_1(\theta_1-1)}K_{n+1}^6\Big),\end{aligned}$$

so that

$$\hat{\Sigma}_{n+1} = \|\delta(\tfrac{1}{2})\|$$

$$= \frac{h_{n+1}}{32|2\theta_1 - 1|} \left\| \frac{2\theta_1 - 1}{\theta_1} K^1_{n+1} - (2K^2_{n+1} + 2K^3_{n+1} + K^4_{n+1}) + \frac{3\theta_1 - 2}{\theta_1 - 1} K^5_{n+1} + \frac{1}{\theta_1(\theta_1 - 1)} K^6_{n+1} \right\|.$$

Example 7.3.1 In order to illustrate and test the effectiveness of the foregoing procedure, let us consider the constructed equation

$$\begin{cases} y'(t) = -y(t) + y(t-20) + \frac{1}{20}\cos\left(\frac{t}{20}\right) + \sin\left(\frac{t}{20}\right) - \sin\left(\frac{t-20}{20}\right), & 0 \le t \le 1000, \\ y(t) = \sin\left(\frac{t}{20}\right), \quad -20 \le t \le 0, \end{cases} \tag{7.3.10}$$

with smooth solution $y(t) = \sin\left(\frac{t}{20}\right)$. The solution being of class C^∞, no discontinuities have to be included as mesh points and the sole restriction on the stepsize is given by $h_{n+1} \le 20$, so as to avoid overlapping. However, for small tolerances this restriction disappears as well. In such a way, the procedure provides a mesh selection which is completely determined by the estimate of the local errors and by the test of tolerance (7.2.5), or (7.2.10), with no external disturbances.

We have chosen $\theta_1 = \frac{1}{3}$ for the computation of K^6_{n+1}, and $\omega_{\min} = 0.5$, $\omega_{\max} = 1.5$, $\rho = 0.9$ in (7.3.5) and (7.3.9).

Numerical experiments confirm that, for sufficiently small stepsizes, the maximum uniform local error Σ_Δ is negligible with respect to σ_Δ. Consequently, whenever the condition $\tilde{\sigma}_{n+1} \le TOL$ produces a small stepsize h_{n+1}, the second condition $h_{n+1}\hat{\Sigma}_{n+1} \le TOL$ is automatically satisfied. On the contrary, whenever the condition $\tilde{\sigma}_{n+1} \le TOL$ produces a large stepsize h_{n+1}, the second test on $h_{n+1}\ddot{\Sigma}_{n+1}$ might actually fail.

The number of rejections due to the test either on $\tilde{\sigma}_{n+1}$ or on $h_{n+1}\hat{\Sigma}_{n+1}$ are displayed in the columns under σ-rejections and Σ-rejections in Table 7.2, respectively.

However, the ratio between the maximum discrete global error e_N and the given tolerance TOL tends to a constant value.

The results for the code performing the test of tolerance (7.2.10) on the sole discrete error estimate σ_Δ are displayed in Table 7.3.

By comparing Tables 7.2 and 7.3 it is noteworthy that, for large values of the tolerance TOL, the control based on the sole discrete error is much less reliable. In fact, it performs many more steps (including the rejections) to compute the solution with an error one hundred times as large. Figure 7.1 displays the solution of equation (7.3.10) for both the double and single tests of tolerance with $TOL = 10^{-2}$ relevant to the second row of Tables 7.2 and 7.3. ◇

7.4 The general delay case

Now we show how to adapt the procedure illustrated in Section 7.3 to the case of DDEs of more general type, even with state dependent delay, for which overlapping may occur. We still consider the classical four-stage RK method of order $p = 4$.

TABLE 7.2. Numerical results for equation (7.3.10) with the double test of tolerance (7.2.5).

TOL	σ-rejections	Σ-rejections	Steps	$\frac{e_N}{TOL}$
10^{-1}	0	144	385	0.85
10^{-2}	0	101	427	0.66
10^{-3}	0	64	569	0.93
10^{-4}	1	39	794	1.78
10^{-5}	30	17	1264	2.29
10^{-6}	35	0	2213	2.14
10^{-7}	28	0	3993	2.06

TABLE 7.3. Numerical results for equation (7.3.10) with the single test of tolerance (7.2.10).

TOL	σ-rejections	Steps	$\frac{e_N}{TOL}$
10^{-1}	209	439	44.38
10^{-2}	203	469	55.45
10^{-3}	187	487	65.01
10^{-4}	67	721	45.18
10^{-5}	37	1261	2.32
10^{-6}	35	2213	2.14
10^{-7}	28	3993	2.06

Then assume that we are solving the local problem

$$\begin{cases} w'_{n+1}(t) = f\Big(t, w_{n+1}(t), x\big(t-\tau(t, w_{n+1}(t))\big)\Big), & t_n \le t \le t_{n+1}, \\ w_{n+1}(t_n) = y_n, \end{cases} \tag{7.4.1}$$

where

$$x(s) = \begin{cases} \phi(s) & \text{for } s \le t_0, \\ \eta(s) & \text{for } t_0 \le s \le t_n, \\ w_{n+1}(s) & \text{for } t_n \le s \le t_{n+1}. \end{cases}$$

We also assume to have computed an interpolant $\eta(t)$ of order $q = p = 4$ up to the current left end-point t_n.

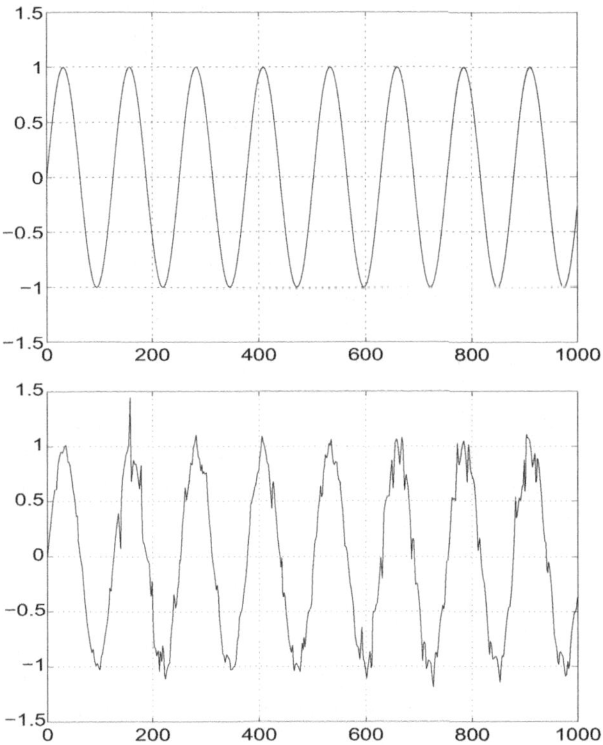

FIG. 7.1. Numerical solution of (7.3.10) for the double (top) and single (bottom) test of tolerance (7.2.5) and (7.2.10) with $TOL = 10^{-2}$.

If the delay is not state dependent and overlapping does not occur in the current step $[t_n, t_{n+1}]$ (i.e. the local problem (7.4.1) equals (7.3.1)), the procedure is obviously the same as before.

On the other hand, if the delay is state dependent but overlapping does not occur in the current step $[t_n, t_{n+1}]$, the procedure is still the same as before, except that the stages become

$$
\begin{aligned}
K_{n+1}^1 &= f\big(t_n, y_n, \eta(t_n - \tau(t_n, y_n))\big),\\
\tilde{Y}_{n+1}^2 &= \eta\big(t_n + \tfrac{1}{2}h_{n+1} - \tau(t_n + \tfrac{1}{2}h_{n+1}, y_n + \tfrac{1}{2}h_{n+1}K_{n+1}^1)\big),\\
K_{n+1}^2 &= f\big(t_n + \tfrac{1}{2}h_{n+1}, y_n + \tfrac{1}{2}h_{n+1}K_{n+1}^1, \tilde{Y}_{n+1}^2\big),\\
\tilde{Y}_{n+1}^3 &= \eta\big(t_n + \tfrac{1}{2}h_{n+1} - \tau(t_n + \tfrac{1}{2}h_{n+1}, y_n + \tfrac{1}{2}h_{n+1}K_{n+1}^2)\big),\\
K_{n+1}^3 &= f\big(t_n + \tfrac{1}{2}h_{n+1}, y_n + \tfrac{1}{2}h_{n+1}K_{n+1}^2, \tilde{Y}_{n+1}^3\big),\\
\tilde{Y}_{n+1}^4 &= \eta\big(t_n + h_{n+1} - \tau(t_n + h_{n+1}, y_n + h_{n+1}K_{n+1}^3)\big),\\
K_{n+1}^4 &= f\big(t_n + h_{n+1}, y_n + h_{n+1}K_{n+1}^3, \tilde{Y}_{n+1}^4\big),\\
K_{n+1}^5 &= K_{n+2}^1 = f\big(t_{n+1}, y_{n+1}, \eta(t_{n+1} - \tau(t_{n+1}, y_{n+1}))\big),
\end{aligned}
$$

$$\begin{aligned}
\tilde{Y}^6_{n+1} &= \eta(t_n + \theta_1 h_{n+1} - \tau(t_n + \theta_1 h_{n+1}, \eta_{(0)}(t_n + \theta_1 h_{n+1}))), \\
K^6_{n+1} &= f(t_n + \theta_1 h_{n+1}, \eta_{(0)}(t_n + \theta_1 h_{n+1}), \tilde{Y}^6_{n+1}), \\
\tilde{Y}^7_{n+1} &= \eta(t_n + \pi_1 h_{n+1} - \tau(t_n + \pi_1 h_{n+1}, \eta_{(1)}(t_n + \pi_1 h_{n+1}))), \\
K^7_{n+1} &= f(t_n + \pi_1 h_{n+1}, \eta_{(1)}(t_n + \pi_1 h_{n+1}), \tilde{Y}^7_{n+1}), \\
\tilde{Y}^8_{n+1} &= \eta(t_n + \pi_2 h_{n+1} - \tau(t_n + \pi_2 h_{n+1}, \eta_{(1)}(t_n + \pi_2 h_{n+1}))), \\
K^8_{n+1} &= f(t_n + \pi_2 h_{n+1}, \eta_{(1)}(t_n + \pi_2 h_{n+1}), \tilde{Y}^8_{n+1}).
\end{aligned}$$

Moreover, the procedure described in Section 4.3.1 may be used in order to compute and include the discontinuity points in the mesh.

Finally, if overlapping occurs in the current step $[t_n, t_{n+1}]$, then the procedure necessarily changes, even if the delay is not state dependent. We illustrate it for the most general case of state dependent delay.

In view of formula (4.2.20) for $n = 0$ in the proof of Theorem 4.2.5, it turns out that, in order to maintain the order $p = 4$ of the discrete RK method, when implementing formulae (6.1.5), (6.1.6), (6.1.7), (6.1.8) it is sufficient to use an interpolant of order $p - 1 = 3$ in (6.1.5) and (6.1.7) (i.e. in the current step $[t_n, t_{n+1}]$).

Although the underlying discrete RK method is explicit, computation of the stage values leads to a fully implicit system (in general non-linear), whose unknowns are those K^i_{n+1} values for which the relevant $\tilde{Y}^i_{n+1}$ is given by formula (6.1.7). For this reason, as interpolant of uniform order 3 in the current step $[t_n, t_{n+1}]$, it is more convenient to consider the NCE of degree 3 presented in Example 5.2.7 in place of the cubic Hermite interpolant which makes use also of the extra stage K^5_{n+1}. Therefore, the first part of the procedure modifies to

$$\hat{\eta}_{(0)}(t_n + \theta h_{n+1}) = y_n + h_{n+1} \sum_{i=1}^{4} b_i(\theta) K^i_{n+1}, \quad 0 \le \theta \le 1, \tag{7.4.2}$$

where

$$\begin{aligned}
b_1(\theta) &= (\tfrac{2}{3}\theta^2 - \tfrac{3}{2}\theta + 1)\theta, \\
b_2(\theta) &= (-\tfrac{2}{3}\theta + 1)\theta^2, \\
b_3(\theta) &= (-\tfrac{2}{3}\theta + 1)\theta^2, \\
b_4(\theta) &= (\tfrac{2}{3}\theta - \tfrac{1}{2})\theta^2,
\end{aligned}$$

and the stage values are

$$K^1_{n+1} = f(t_n, y_n, \eta(t_n - \tau(t_n, y_n)));$$

$$K^2_{n+1} = f(t_n + \tfrac{1}{2}h_{n+1}, y_n + \tfrac{1}{2}h_{n+1}K^1_{n+1}, \tilde{Y}^2_{n+1}),$$

with

$$\tilde{Y}^2_{n+1} = \hat{\eta}_{(0)}(t_n + \tfrac{1}{2}h_{n+1} - \tau(t_n + \tfrac{1}{2}h_{n+1}, y_n + \tfrac{1}{2}h_{n+1}K^1_{n+1}))$$

$$= y_n + h_{n+1} \sum_{i=1}^{4} b_i \Big(\frac{1}{2} - \frac{\tau(t_n + \frac{1}{2}h_{n+1}, y_n + \frac{1}{2}h_{n+1}K^1_{n+1})}{h_{n+1}}\Big) K^i_{n+1}$$

if $t_n + \frac{1}{2}h_{n+1} - \tau(t_n + \frac{1}{2}h_{n+1}, y_n + \frac{1}{2}h_{n+1}K^1_{n+1}) > t_n$, and

$$\tilde{Y}^2_{n+1} = \eta\big(t_n + \tfrac{1}{2}h_{n+1} - \tau(t_n + \tfrac{1}{2}h_{n+1}, y_n + \tfrac{1}{2}h_{n+1}K^1_{n+1})\big)$$

otherwise;

$$K^3_{n+1} = f\big(t_n + \tfrac{1}{2}h_{n+1}, y_n + \tfrac{1}{2}h_{n+1}K^2_{n+1}, \tilde{Y}^3_{n+1}\big),$$

with

$$\begin{aligned}\tilde{Y}^3_{n+1} &= \hat{\eta}_{(0)}\big(t_n + \tfrac{1}{2}h_{n+1} - \tau(t_n + \tfrac{1}{2}h_{n+1}, y_n + \tfrac{1}{2}h_{n+1}K^2_{n+1})\big) \\ &= y_n + h_{n+1} \sum_{i=1}^{4} b_i \Big(\frac{1}{2} - \frac{\tau(t_n + \frac{1}{2}h_{n+1}, y_n + \frac{1}{2}h_{n+1}K^2_{n+1})}{h_{n+1}}\Big) K^i_{n+1}\end{aligned}$$

if $t_n + \frac{1}{2}h_{n+1} - \tau(t_n + \frac{1}{2}h_{n+1}, y_n + \frac{1}{2}h_{n+1}K^2_{n+1}) > t_n$, and

$$\tilde{Y}^3_{n+1} = \eta\big(t_n + \tfrac{1}{2}h_{n+1} - \tau(t_n + \tfrac{1}{2}h_{n+1}, y_n + \tfrac{1}{2}h_{n+1}K^2_{n+1})\big)$$

otherwise;

$$K^4_{n+1} = f\big(t_n + h_{n+1}, y_n + h_{n+1}K^3_{n+1}, \tilde{Y}^4_{n+1}\big),$$

with

$$\begin{aligned}\tilde{Y}^4_{n+1} &= \hat{\eta}_{(0)}\big(t_n + h_{n+1} - \tau(t_n + h_{n+1}, y_n + h_{n+1}K^3_{n+1})\big) \\ &= y_n + h_{n+1} \sum_{i=1}^{4} b_i \Big(1 - \frac{\tau(t_n + h_{n+1}, y_n + h_{n+1}K^3_{n+1})}{h_{n+1}}\Big) K^i_{n+1}\end{aligned}$$

if $t_n + h_{n+1} - \tau(t_n + h_{n+1}, y_n + h_{n+1}K^3_{n+1}) > t_n$, and

$$\tilde{Y}^4_{n+1} = \eta\big(t_n + h_{n+1} - \tau(t_n + h_{n+1}, y_n + h_{n+1}K^3_{n+1})\big)$$

otherwise.

For $\theta = 1$, (7.4.2) gives the same value y_{n+1} as (7.3.2).

Once we have computed the discrete approximation y_{n+1}, we do not consider the interpolant $\hat{\eta}_{(0)}(t)$ any more, but we switch again to the cubic Hermite interpolant $\eta_{(0)}(t)$ by using the extra stage

$$K^5_{n+1} = f\big(t_{n+1}, y_{n+1}, \tilde{Y}^5_{n+1}\big),$$

where

$$\begin{aligned}\tilde{Y}^5_{n+1} &= \hat{\eta}_{(0)}\big(t_{n+1} - \tau(t_{n+1}, y_{n+1})\big) \\ &= y_n + h_{n+1} \sum_{i=1}^{4} b_i \Big(1 - \frac{\tau(t_{n+1}, y_{n+1})}{h_{n+1}}\Big) K^i_{n+1}\end{aligned}$$

if $t_{n+1} - \tau(t_{n+1}, y_{n+1}) > t_n$, and

$$\tilde{Y}^5_{n+1} = \eta(t_{n+1} - \tau(t_{n+1}, y_{n+1}))$$

otherwise.

Note that, since overlapping occurs in the current step, in most cases $t_{n+1} - \tau(t_{n+1}, y_{n+1}) > t_n$ and, thus, K^5_{n+1} differs from $K^1_{n+2} = f\big(t_{n+1}, y_{n+1}, \eta_{(1)}(t_{n+1} - \tau(t_{n+1}, y_{n+1}))\big)$.

From now on the procedure is again the same as in the case without overlapping, except that the remaining stage values are defined as follows:

$$K^6_{n+1} = f\big(t_n + \theta_1 h_{n+1}, \eta_{(0)}(t_n + \theta_1 h_{n+1}), \tilde{Y}^6_{n+1}\big),$$

where

$$\tilde{Y}^6_{n+1} = \eta_{(0)}\big(t_n + \theta_1 h_{n+1} - \tau(t_n + \theta_1 h_{n+1}, \eta_{(0)}(t_n + \theta_1 h_{n+1}))\big)$$

if $t_n + \theta_1 h_{n+1} - \tau(t_n + \theta_1 h_{n+1}, \eta_{(0)}(t_n + \theta_1 h_{n+1})) > t_n$, and

$$\tilde{Y}^6_{n+1} = \eta\big(t_n + \theta_1 h_{n+1} - \tau(t_n + \theta_1 h_{n+1}, \eta_{(0)}(t_n + \theta_1 h_{n+1}))\big)$$

otherwise;

$$K^7_{n+1} = f\big(t_n + \pi_1 h_{n+1}, \eta_{(1)}(t_n + \pi_1 h_{n+1}), \tilde{Y}^7_{n+1}\big),$$

where

$$\tilde{Y}^7_{n+1} = \eta_{(1)}\big(t_n + \pi_1 h_{n+1} - \tau(t_n + \pi_1 h_{n+1}, \eta_{(1)}(t_n + \pi_1 h_{n+1}))\big)$$

if $t_n + \pi_1 h_{n+1} - \tau(t_n + \pi_1 h_{n+1}, \eta_{(1)}(t_n + \pi_1 h_{n+1})) > t_n$, and

$$\tilde{Y}^7_{n+1} = \eta\big(t_n + \pi_1 h_{n+1} - \tau(t_n + \pi_1 h_{n+1}, \eta_{(1)}(t_n + \pi_1 h_{n+1}))\big)$$

otherwise;

$$K^8_{n+1} = f\big(t_n + \pi_2 h_{n+1}, \eta_{(1)}(t_n + \pi_2 h_{n+1}), \tilde{Y}^8_{n+1}\big),$$

where

$$\tilde{Y}^8_{n+1} = \eta_{(1)}\big(t_n + \pi_2 h_{n+1} - \tau(t_n + \pi_2 h_{n+1}, \eta_{(1)}(t_n + \pi_2 h_{n+1}))\big)$$

if $t_n + \pi_2 h_{n+1} - \tau(t_n + \pi_2 h_{n+1}, \eta_{(1)}(t_n + \pi_2 h_{n+1})) > t_n$, and

$$\tilde{Y}^8_{n+1} = \eta\big(t_n + \pi_2 h_{n+1} - \tau(t_n + \pi_2 h_{n+1}, \eta_{(1)}(t_n + \pi_2 h_{n+1}))\big)$$

otherwise.

Moreover, the quadrature formula (7.3.3) must be changed to

$$\tilde{y}_{n+1} = y_n + h_{n+1}\big(\tfrac{1}{12}K^1_{n+1} + \tfrac{5}{12}K^7_{n+1} + \tfrac{5}{12}K^8_{n+1} + \tfrac{1}{12}\bar{K}^5_{n+1}\big),$$

where, as we have already observed,

$$\bar{K}^5_{n+1} = K^1_{n+2} = f\big(t_{n+1}, y_{n+1}, \eta_{(1)}(t_{n+1} - \tau(t_{n+1}, y_{n+1}))\big)$$

may or may not equal the already available stage K^5_{n+1}.

TABLE 7.4. Numerical results for equation (7.4.3) with the double test of tolerance (7.2.5).

TOL	σ-rejections	Σ-rejections	Steps	Overlappings	$\frac{e_N}{TOL}$
10^{-1}	2	158	276	276	1.47
10^{-2}	1	238	415	415	2.86
10^{-3}	16	238	573	462	3.61
10^{-4}	128	112	780	446	10.38
10^{-5}	209	0	1196	509	16.40
10^{-6}	235	0	2121	650	15.64
10^{-7}	253	0	3740	771	17.19

Example 7.4.1 Let us consider the equation

$$\begin{cases} y'(t) = -y(t) + y(\alpha(t)) + \frac{t}{20}\cos\left(\frac{t}{20}\right) + \sin\left(\frac{t}{20}\right) - \sin\left(\frac{\alpha(t)}{20}\right), & 0 \le t \le 1000, \\ y(t) = \sin\left(\frac{t}{20}\right), \quad -20 \le t \le 0, \end{cases} \tag{7.4.3}$$

with the variable deviated argument $\alpha(t) = t-1+\sin(t)$ corresponding to a delay $\tau(t) = 1 - \sin(t)$ which periodically vanishes at the points $\frac{\pi}{2} + 2k\pi$, k positive integer. The exact solution $y(t) = \sin\left(\frac{t}{20}\right)$ is smooth and hence no discontinuity points have to be included as mesh points for preserving the accuracy of the methods. Contrary to the previous example, overlapping occurs in the proximity of every point where the delay vanishes, and possibly somewhere else.

As with the previous example, we have chosen $\theta_1 = \frac{1}{3}$ for the computation of K_{n+1}^6, and $\omega_{\min} = 0.5$, $\omega_{\max} = 1.5$, $\rho = 0.9$ in (7.3.5) and (7.3.9).

Numerical results for the method described above are displayed in Tables 7.4 and 7.5 for the double and single tests of tolerance (7.2.5) and (7.2.10), respectively.

Comparing Tables 7.4 and 7.5, as well as the graphs in Figure 7.2, brings us to the same conclusion as in Example 7.3.1. So we can say that, for implementations with and without overlapping, the performance of the error control and of the varying stepsize mechanism described in this section is essentially the same. $\diamond$

Example 7.4.2 The following equation, taken from the test set by Paul [217], was analyzed by May [199] as a model for the dynamics of baleen whale population:

$$\begin{cases} y'(t) = -y(t) + y(t-2)\left(2.5 - 1.5\left(\frac{y(t-2)}{1000}\right)^{2.5}\right), & 0 \le t \le 40, \\ y(t) = 999, \quad -2 \le t \le 0. \end{cases} \tag{7.4.4}$$

TABLE 7.5. Numerical results for equation (7.4.3) with the single test of tolerance (7.2.10).

TOL	σ-rejections	Steps	Overlappings	$\frac{e_N}{TOL}$
10^{-1}	5	17	17	32.17
10^{-2}	108	200	200	25.34
10^{-3}	210	429	429	26.72
10^{-4}	208	687	430	11.03
10^{-5}	253	1211	523	14.50
10^{-6}	280	2120	649	20.28
10^{-7}	265	3741	770	20.12

Since $y'^{+}(0) \simeq 3.74$, the solution is not smooth at any even integer $\xi_i = 2i$, $i = 1, 2, \ldots$, where it has a discontinuity of order i. Let us integrate (7.4.4) without including the discontinuities as mesh points and by setting $\theta_1 = \frac{1}{3}$ for the computation of K^6_{n+1}, and $\omega_{\min} = 0.5$, $\omega_{\max} = 1.5$, $\rho = 0.9$ in (7.3.5) and (7.3.9). The graphic of the solution obtained by the double test of tolerance (7.2.5) with the mild tolerance $TOL = 10^{-2}$ is displayed in Figure 7.3. It includes 162 nodal points and coincides, at the plot resolution, with the reference solution obtained by $TOL = 10^{-11}$. ◇

Example 7.4.3 The following equation, again taken from the test set by Paul [217], was analyzed by Smith and Kuang [234] as a state dependent delay model with a slowly oscillating periodic solution:

$$\begin{cases} y'(t) = -ky(t-1-|y(t)|)(1-y(t)^2), & 0 \le t \le 20, \\ y(t) = \frac{1}{2}, & t \le 0. \end{cases} \tag{7.4.5}$$

Even though the previous theory for error-estimating methods was rigorously developed only for delays that are not state dependent, the same procedure as in the Example 7.4.2 performs satisfactorily also on (7.4.5) (see Figure 7.4). ◇

7.5 Stepsize control in the code RADAR5

Now we briefly outline the stepsize control mechanism implemented by the code RADAR5, developed by Guglielmi and Hairer [112], which was introduced in Section 6.5.2, whose aim is the efficient numerical solution of the very general class of stiff systems of delay equations (6.5.30).

Indeed, it is not completely consistent with the theory and the approach described in the previous sections. In fact, the error-estimating method has order $p' = p - 2 = 3$, $p = 5$ being the discrete order of the advancing method, whereas

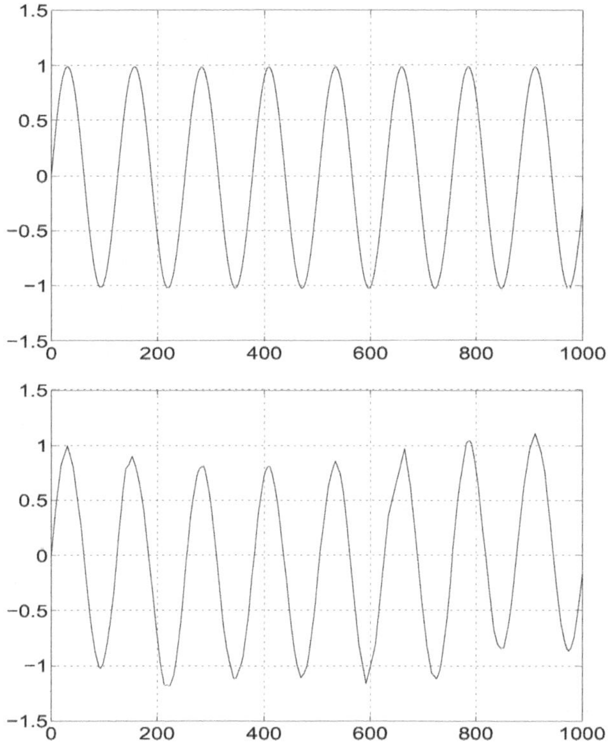

FIG. 7.2. Numerical solution of (7.4.3) for the double (top) and single (bottom) test of tolerance (7.2.5) and (7.2.10) with $TOL = 10^{-2}$.

a *tolerance per step* is used to control the local error of the lower order method. Moreover, the uniform order of the advancing method is only $q = p - 2 = 3$. Therefore, on the basis of the theory developed in Section 7.1, it turns out that the discrete local error of the advancing method may sometimes fail to actually be $O(h_{n+1}^{p+1}) = O(h_{n+1}^{6})$. On the contrary, this would always be assured if the advancing method had uniform order ≥ 4. The continuous error-estimating method has order $q' = q - 1 = 2$ and the continuous error estimate is used to correct the discrete error estimate by means of a linear combination, whose coefficients are decided by the user in connection with the particular features and the degree of stiffness of the underlying problem. As already observed in Section 6.5.2, the resulting method for DDEs has global order $\min\{p, q+1\} = 4$.

In any case, since the code is designed mainly for solving stiff problems, the previous considerations, which are based on the analysis of the classical order, may not be so relevant and do not compromise its efficiency.

The local problem to solve by means of formulae (6.5.31) and (6.5.32) in the step $[t_n, t_{n+1}]$ is

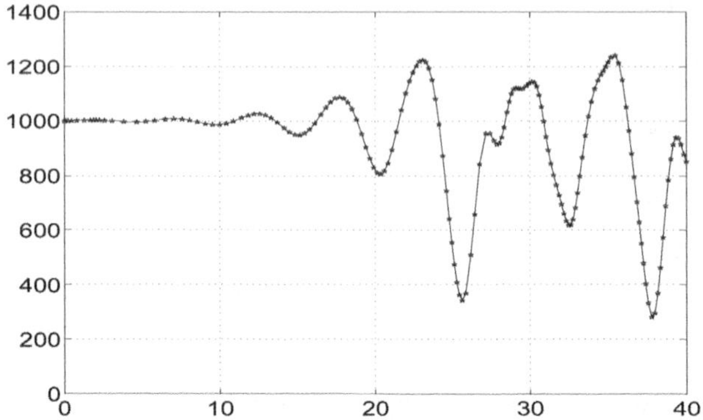

FIG. 7.3. Solution of (7.4.4).

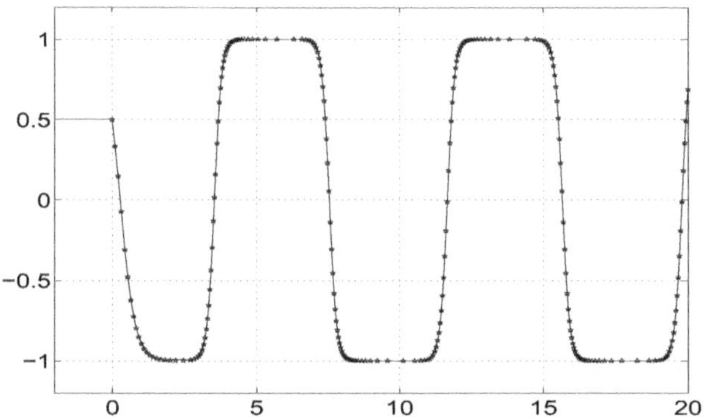

FIG. 7.4. Solution of (7.4.5) with $k = 4$.

$$\begin{cases} Mw'_{n+1}(t) = F\Big(t, w_{n+1}(t), x\big(t - \tau(t, w_{n+1}(t))\big)\Big), & t_n \le t \le t_{n+1}, \\ w_{n+1}(t_n) = u_n, \end{cases}$$

where

$$x(s) = \begin{cases} \phi(s) & \text{for } s \le t_0, \\ \varphi(s) & \text{for } t_0 \le s \le t_n, \\ w_{n+1}(s) & \text{for } t_n \le s \le t_{n+1}. \end{cases}$$

The discrete advancing method (three-stage Radau IIA) has order $p = 5$ and is given by

$$\begin{array}{c|ccc} \frac{4-\sqrt{6}}{10} & \frac{88-7\sqrt{6}}{360} & \frac{296-169\sqrt{6}}{1800} & \frac{-2+3\sqrt{6}}{225} \\ \frac{4+\sqrt{6}}{10} & \frac{296+169\sqrt{6}}{1800} & \frac{88+7\sqrt{6}}{360} & \frac{-2-3\sqrt{6}}{225} \\ 1 & \frac{16-\sqrt{6}}{36} & \frac{16+\sqrt{6}}{36} & \frac{1}{9} \\ \hline & \frac{16-\sqrt{6}}{36} & \frac{16+\sqrt{6}}{36} & \frac{1}{9} \end{array}$$

The error-estimating method has order $p' = 3$ and is constructed by using the same stages

$$K_{n+1}^i = F\big(t_{n+1}^i, U_{n+1}^i, \tilde{U}_{n+1}^i\big), \quad i = 1, 2, 3,$$

plus the additional explicit stage

$$K_{n+1}^4 = F(t_n, u_n, \varphi(t_n - \tau(t_n, u_n))).$$

Therefore,

$$M\bar{u}_{n+1} = Mu_n + h_{n+1} \sum_{i=1}^{4} \bar{b}_i K_{n+1}^i,$$

where

$$\begin{aligned}
\bar{b}_1 &= \tfrac{16-\sqrt{6}}{36} - \tfrac{2+3\sqrt{6}}{6}\gamma_0, \\
\bar{b}_2 &= \tfrac{16+\sqrt{6}}{36} - \tfrac{2-3\sqrt{6}}{6}\gamma_0, \\
\bar{b}_3 &= \tfrac{1}{9} - \tfrac{1}{3}\gamma_0, \\
\bar{b}_4 &= \gamma_0,
\end{aligned}$$

and

$$\gamma_0 = \frac{6 - \sqrt[3]{9} + 3\sqrt[3]{3}}{30}$$

is the real eigenvalue of the coefficient matrix A of the method (see also Hairer and Wanner [118, Section IV.8]).

Unfortunately, as the stiffness of the problem and/or the stepsize h_{n+1} grow, the classical error estimate $\sigma_{n+1} = \|\bar{u}_{n+1} - u_{n+1}\|$ is unbounded and, consequently, it dramatically overestimates the true local error. Therefore, in order to overcome this inconvenience, the code utilizes the alternative quantity

$$\bar{\sigma}_{n+1} = \|P_{n+1}(\bar{u}_{n+1} - u_{n+1})\| = O(h_{n+1}^4),$$

or even $\bar{\sigma}_{n+1} = \|P_{n+1}^2(\bar{u}_{n+1} - u_{n+1})\|$, where the projection matrix

$$P_{n+1} = (M - h_{n+1}\gamma_0 J_{n+1})^{-1} M$$

is constructed by using the Jacobian

$$\begin{aligned}
J_{n+1} = {} & \frac{\partial F}{\partial u}(t_n, u_n, \varphi(t_n - \tau(t_n, u_n))) \\
& - \frac{\partial F}{\partial v}(t_n, u_n, \varphi(t_n - \tau(t_n, u_n)))\varphi'(t_n - \tau(t_n, u_n))\frac{\partial \tau}{\partial u}(t_n, u_n) \\
& + H_{n+1},
\end{aligned}$$

where H_{n+1} denotes additional terms that are present only if overlapping occurs. The LU-decomposition of $M - h_{n+1}\gamma_0 J_{n+1}$ is already available from the simplified Newton iterations needed for the solution of the non-linear equations (recall that the method is implicit).

This alternative estimate of the local error has the advantage that, while asymptotically equivalent to the classical estimate for vanishing stiffness and stepsize h_{n+1}, it remains bounded (or even vanishes with the second choice) as the stiffness of the problem and/or the stepsize h_{n+1} grow.

However, control of the sole discrete local error may sometimes turn out to be unsatisfactory. In such cases, the uniform local error of the collocation polynomial (6.5.32), whose order is $q = 3$, is controlled by comparing it with the following interpolant of lower order $q' = 2$:

$$\bar{\varphi}(t_n + \theta h_{n+1}) = \sum_{i=1}^{3} \bar{\ell}_i(\theta) U_{n+1}^i,$$

where $\bar{\ell}_i(\theta)$, $i = 1, 2, 3$, are the Lagrange polynomial coefficients on the nodes c_j, $j = 1, 2, 3$. It turns out that

$$\begin{aligned} \bar{\Sigma}_{n+1} &= \max_{0 \le \theta \le 1} \|\bar{\varphi}(t_n + \theta h_{n+1}) - \varphi(t_n + \theta h_{n+1})\| \\ &= \|\bar{\varphi}(t_n) - \varphi(t_n)\| = O(h_{n+1}^3), \end{aligned}$$

which is an easily computable quantity.

In conclusion, the local error estimate used for the stepsize control mechanism is

$$est_{n+1} = \alpha_1 \bar{\sigma}_{n+1} + \alpha_2 h_{n+1} \bar{\Sigma}_{n+1} = O(h_{n+1}^4), \tag{7.5.1}$$

where the parameters $\alpha_1 \ge 0$ and $\alpha_2 \ge 0$ may be tuned by the user.

In general, the test for the acceptance of the current step involves two parameters for each component of the local errors: $ATOL_i$, the absolute tolerance for the ith component, and $RTOL_i$, the relative tolerance for the ith component. However, by choosing all the parameters $ATOL_i = TOL$ and $RTOL_i$ to be very small if compared with TOL (they are not allowed to be zero), the test takes the form

$$est_{n+1} \le TOL.$$

In order to predict the new stepsize h_{new}, either to repeat the current step $[t_n, t_{n+1}]$ or to advance to the next step $[t_{n+1}, t_{n+2}]$, the user may choose the classical formula

$$h_{\text{new}} = \max\left\{\omega_{\min}, \min\left\{\omega_{\max}, \rho\left(\frac{TOL}{est_{n+1}}\right)^{1/4}\right\}\right\} \cdot h_{n+1} \tag{7.5.2}$$

for suitable values of the safety factors $\omega_{\min}$, $\omega_{\max}$ and ρ, or, alternatively, a somewhat more complicated version of it inspired by the work of Gustafsson, Lundh and Söderlind [115] and Gustafsson [114] (see also [118, Section IV.2]). In any case, the safety factor ρ also takes into account the number $Newt$ of

Newton iterations in the current step and the maximum number $k_{\max}$ of Newton iterations and is given by

$$\rho = (2k_{\max} + 1)\rho'/(2k_{\max} + Newt) \tag{7.5.3}$$

for a suitable parameter $\rho' < 1$. Finally, in order to minimize the number of Jacobian computations, the stepsize is kept unchanged whenever formula (7.5.2) leads to

$$h_{\text{new}} \in [\beta_1 h_{n+1}, \beta_2 h_{n+1}], \tag{7.5.4}$$

where $\beta_1 \leq 1$ and $\beta_2 \geq 1$ are chosen by the user.

We conclude the chapter with some numerical experiments on two real-life problems.

Example 7.5.1 We consider the following very stiff state dependent 6×6 system of DDEs, which models the antibody response to antigen challenge (Waltman [263]):

$$\begin{cases} y_1'(t) = -5\cdot 10^4 y_1(t)y_2(t) - 10^5 y_1(t)y_4(t), \\ y_2'(t) = -5\cdot 10^4 y_1(t)y_2(t) + 9\cdot 10^4 y_1(y_5(t))y_2(y_5(t))H(t-35), \\ y_3'(t) = 5\cdot 10^4 y_1(t)y_2(t), \\ y_4'(t) = -10^5 y_1(t)y_4(t) - 2\cdot 10^{-3} y_4(t) + 10^6 y_1(y_6(t))y_2(y_6(t))H(t-197), \\ y_5'(t) = H(t-35)\dfrac{y_1(t)y_2(t) + y_3(t)}{y_1(y_5(t))y_2(y_5(t)) + y_3(y_5(t))}, \\ y_6'(t) = H(t-197)\dfrac{10^{-12} + y_2(t) + y_3(t)}{10^{-12} + y_2(y_6(t)) + y_3(y_6(t))}, \end{cases} \tag{7.5.5}$$

for $0 \leq t \leq 300$, with initial conditions

$$\begin{cases} y_1(t) \equiv 5\cdot 10^{-6}, \\ y_2(t) \equiv 10^{-15}, \\ y_3(t) \equiv 0, \\ y_4(t) \equiv 0, \\ y_5(t) \equiv 0, \\ y_6(t) \equiv 0, \end{cases} \qquad t \leq 0,$$

$H(s)$ being the Heavyside function ($H(s) = 0$ if $s < 0$ and $H(s) = 1$ if $s \geq 0$).

This problem has the form (6.5.30) with $M = I$, the identity matrix. It has some peculiarities, which make its numerical solution a difficult task. The delays $t - y_5(t)$ and $t - y_6(t)$ are state dependent and vanish asymptotically. Moreover, owing to the presence of the Heavyside functions $H(t-37)$ and $H(t-197)$, the component $y_2(t)$ is very steep next to the secondary discontinuity point $t = 37$, whereas the components $y_4(t)$ and $y_6(t)$ are very steep next to the secondary discontinuity point $t = 197$.

We solve this problem by using the code RADAR5 with the following choices: $\alpha_1 = \alpha_2 = 0.5$ in (7.5.1), $\omega_{\min} = 0.2$ and $\omega_{\max} = 8$ in (7.5.2), $\rho' = 0.9$ and

TABLE 7.6. Numerical results for the system (7.5.5) by the code RADAR5.

TOL	Rejections	Steps	$\frac{e_N}{TOL}$
10^{-19}	7	696	$1.22 \cdot 10^{13}$
10^{-20}	7	1080	$1.06 \cdot 10^{13}$
10^{-21}	28	1528	$2.10 \cdot 10^{13}$
10^{-22}	37	2046	$1.69 \cdot 10^{13}$
10^{-23}	55	2614	$1.44 \cdot 10^{13}$
10^{-24}	64	3222	$1.86 \cdot 10^{13}$

$k_{\max} = 7$ in (7.5.3), $\beta_1 = 1$ and $\beta_2 = 1.2$ in (7.5.4). Of course, the secondary discontinuity points $t = 37$ and $t = 197$ are included in the mesh.

Since the scale of the first four components is much smaller than the scale of the last two components (the deviating arguments) of an average factor of about 10^{-15}, we measure the local errors in the weighted Euclidean norm

$$\|y\| = \sqrt{|y_1|^2 + |y_2|^2 + |y_3|^2 + |y_4|^2 + 10^{-30}(|y_5|^2 + |y_6|^2)}. \tag{7.5.6}$$

The analytical form of the exact solution is not available. Therefore, in order to evaluate the maximum global error e_N (measured in the weighted Euclidean norm (7.5.6)), we computed the reference values at ten equidistant points $t_k^* = 30k$, $k = 1, \ldots, 10$, ($t_{30}^* = t_f = 300$) by using very small tolerances. Such reference points were included in the mesh.

The numerical results are displayed in Table 7.6. They show a very good performance of the code with an almost perfect proportional response of the global error to the tolerance TOL, even if we did not exploit all its capabilities, such as different absolute and relative tolerances for the various components and the more sophisticated stepsize prediction based on the results in [115] and [114]. Furthermore, we remark that, the system being of neutral type, there are many primary and secondary discontinuity points spread forward by the delays at which the exact solution is just continuous. Although these discontinuities are not included in the mesh (recall that the code does not try to track them, but it handle them simply by suitably reducing the stepsize in their proximity), the approximate solution does not seem to suffer from a lack of smoothness. ◇

Example 7.5.2 We consider the following 2×2 system of NDDEs with constant delay, taken from the test set by Paul [217], which is a particular instance of a predator–prey model due to Kuang [182]:

$$\begin{cases} y_1'(t) = y_1(t)\Big(1 - y_1(t-\frac{21}{50}) - \frac{29}{10}y_1'(t-\frac{21}{50})\Big) - \dfrac{y_1(t)^2 y_2(t)}{y_1(t)^2+1}, \\ y_2'(t) = \Big(\dfrac{y_1(t)^2}{y_1(t)^2+1} - \dfrac{1}{10}\Big)y_2(t), \end{cases} \qquad 0 \le t \le 6, \tag{7.5.7}$$

with initial conditions

$$\begin{cases} y_1(t) = \frac{33}{100} - \frac{1}{10}t, \\ y_2(t) = \frac{111}{50} + \frac{1}{10}t, \end{cases} \qquad t \le 0.$$

In order to solve the problem numerically by means of the code RADAR5, we have to transform it according to (6.5.33). However, since the delayed derivative of the second component $y_2'(t-\frac{21}{50})$ is not involved in the right-hand side, it is useless to make the complete substitution $z(t) = y'(t)$, but it is sufficient to add the sole variable

$$z_1(t) = y_1'(t).$$

So we obtain the 3×3 system of DDAEs

$$\begin{cases} y_1'(t) = z_1(t), \\ y_2'(t) = \Big(\dfrac{y_1(t)^2}{y_1(t)^2+1} - \dfrac{1}{10}\Big)y_2(t), \\ 0 = y_1(t)\Big(1 - y_1(t-\frac{21}{50}) - \frac{29}{10}z_1(t-\frac{21}{50})\Big) - \dfrac{y_1(t)^2 y_2(t)}{y_1(t)^2+1}, \end{cases} \qquad 0 \le t \le 6, \tag{7.5.8}$$

with initial conditions

$$\begin{cases} y_1(t) = \frac{33}{100} - \frac{1}{10}t, \\ y_2(t) = \frac{111}{50} + \frac{1}{10}t, \\ z_1(t) = -\frac{1}{10}, \end{cases} \qquad t \le 0.$$

With $u(t) = [y_1(t), y_2(t), z_1(t)]^T$, this system has the form (6.5.30) with

$$M = \begin{bmatrix} 1 & 0 & 0 \\ 0 & 1 & 0 \\ 0 & 0 & 0 \end{bmatrix}.$$

In using the code RADAR5, we made the following choices: $\alpha_1 = \alpha_2 = 0.5$ in (7.5.1), $\omega_{\min} = 0.2$ and $\omega_{\max} = 8$ in (7.5.2), $\rho' = 0.9$ and $k_{\max} = 7$ in (7.5.3), $\beta_1 = 1$ and $\beta_2 = 1.2$ in (7.5.4).

Because the delay is constant, the discontinuity points $\xi_k = \frac{21}{50}k$, $k = 1, \dots, 14$, are known a priori and thus were included in the mesh.

The analytical form of the exact solution is not available. Therefore, in order to evaluate the maximum global error (measured in the standard Euclidean norm), we computed the reference values at the 14 middle points $t_k^* = \xi_k - \frac{21}{100}k$ of the macro intervals $[\xi_{k-1}, \xi_k]$, $k = 1, \dots, 14$, ($\xi_0 = 0$) and at the final point $t_f = 6$. Such reference points were included in the mesh.

TABLE 7.7. Numerical results for the system (7.5.8) by the code RADAR5.

TOL	Rejections	Steps	$\frac{e_N}{TOL}$	$\frac{e'_N}{TOL}$
10^{-9}	100	791	$4.41 \cdot 10^7$	$3.12 \cdot 10^7$
10^{-10}	131	1344	$9.08 \cdot 10^7$	$6.45 \cdot 10^7$
10^{-11}	159	2045	$1.01 \cdot 10^8$	$7.18 \cdot 10^7$
10^{-12}	189	3410	$5.07 \cdot 10^7$	$3.59 \cdot 10^7$
10^{-13}	216	5778	$1.94 \cdot 10^8$	$9.50 \cdot 10^7$
10^{-14}	247	8800	$3.55 \cdot 10^8$	$1.20 \cdot 10^8$

The numerical results are displayed in Table 7.7, where we have reported both the global error e_N relevant to the whole three-dimensional solution $u(t)$ of (7.5.8) and the global error e'_N relevant to the sole first two components $y_1(t)$ and $y_2(t)$, which represent the solution $y(t)$ of the original problem (7.5.7).

We note a good performance of the code, with a satisfactory proportionality of the global error to the tolerance TOL. ◇

8

STABILITY ANALYSIS OF RUNGE–KUTTA METHODS FOR ODES

In this chapter we review a few of the most important classical stability concepts and results for ODE methods, assuming that the reader is already familiar with this topic. Such a review includes the notions of *contractivity*, or *dissipativity*, related to the linear growth of the error, and *asymptotic stability* for various kinds of test equations. In particular, some connections between the two concepts will be illustrated by following the line of Hairer and Zennaro [120], who have investigated more deeply the so-called *error growth functions* and the asymptotic stability properties of Runge–Kutta (RK) methods. Moreover, we recall the concept of stability introduced by Bellen and Zennaro [37] for the interpolants and, in view of the application to the numerical methods for DDEs, we also review the concept of *stability with respect to forcing terms*, which have been introduced and investigated by Torelli [247, 248], Bellen and Zennaro [38] and Zennaro [282].

8.1 Linear error growth

In the efficient construction of step-by-step numerical methods for the accurate integration of IVPs for ODEs such as

$$\begin{cases} y'(t) = g(t, y(t)), & t_0 \le t \le t_f, \\ y(t_0) = y_0, \end{cases} \tag{8.1.1}$$

one essentially tries to control the propagation of the error in terms of the length of the integration interval $[t_0, t_f]$ along a mesh $\Delta = \{t_0, t_1, \ldots, t_n, \ldots, t_N = t_f\}$ with as large as possible stepsizes.

It is known that, in the mesh interval $[t_n, t_{n+1}]$, a one-step method provides the approximation y_{n+1}, which is the numerical solution of the local problem

$$\begin{cases} w'_{n+1}(t) = g(t, w_{n+1}(t)), & t_n \le t \le t_{n+1}, \\ w_{n+1}(t_n) = y_n. \end{cases}$$

Consider also the fictitious local problem

$$\begin{cases} z'_{n+1}(t) = g(t, z_{n+1}(t)), & t_n \le t \le t_{n+1}, \\ z_{n+1}(t_n) = y(t_n), \end{cases} \tag{8.1.2}$$

whose solution is just $z_{n+1}(t) \equiv y(t)$, the solution of (8.1.1), and let z_{n+1} be its numerical solution (which is not actually computed).

In analogy to the notation introduced in Chapter 7, let

$$\sigma_{n+1}^* = \|y(t_{n+1}) - z_{n+1}\|/h_{n+1}$$

be the local error per unit step (with respect to the fictitious local problem (8.1.2)) and let

$$\sigma_\Delta^* = \max_{1 \le n \le N} \sigma_n^*.$$

With this notation, the triangular inequality

$$\|y(t_{n+1}) - y_{n+1}\| \le \|y(t_{n+1}) - z_{n+1}\| + \|z_{n+1} - y_{n+1}\|$$

implies

$$\|y(t_{n+1}) - y_{n+1}\| \le \|z_{n+1} - y_{n+1}\| + h_{n+1}\sigma_\Delta^*.$$

Therefore, if the propagated error $\|z_{n+1} - y_{n+1}\|$ satisfies the *stability condition*

$$\|z_{n+1} - y_{n+1}\| \le \|y(t_n) - y_n\| \tag{8.1.3}$$

for all $n \le N - 1$, since $e_0 = 0$, then a simple recursion immediately yields

$$e_N \le (t_f - t_0)\sigma_\Delta^*,$$

where, for all $n = 0, 1, \ldots, N$,

$$e_n = \max_{0 \le i \le n} \|y(t_i) - y_i\|$$

is the maximum global error (up to the nodal point t_n).

In other words, the maximum global error e_N cannot grow more than linearly with respect to $t_f - t_0$. More precisely, the growth of the global error is essentially given by the sum of the local errors across the elapsed steps. This is a desirable property as long as we are integrating on intervals where σ_Δ^* remains small even for large values of the stepsize.

Verifying whether or not the stability condition (8.1.3) holds for stable equations is, in general, a difficult task. It depends on the equation we are dealing with, on the particular method we are considering and on the stepsize we are using. When the property is satisfied independently of the stepsize whenever the underlying test equation is stable, the method is said to be *stable* as well. More precisely, according to the particular choice of the test equation, the property will be referred to as *A-stability*, *AN-stability* and *BN-stability* for constant and variable coefficient linear scalar equations and for non-linear systems, respectively.

In this context, the stability of both the equation and numerical method is viewed as *contractivity* or *dissipativity* and is related to the concept of *stiffness*. Avoiding trying to give a rigorous definition of stiffness, we say that a *stiff* situation occurs whenever the use of non-stable methods leads, in the current step, to a global error e_{n+1} which is largely dominated by the propagated error $\|z_{n+1} - y_{n+1}\|$ rather than by the local error $\|y(t_{n+1}) - z_{n+1}\|$. As a consequence, any adapting stepsize mechanism should be forced to dominate the propagated error by a dramatic reduction in the stepsize. This is a serious drawback and any efficient code should just make use, in this instance, of stable methods.

8.2 Some classical stability results for ODE methods

Many results reported in this section, without a proof, are well known and may be found, for example, in Butcher [62], Dekker and Verwer [74] and Hairer and Wanner [118].

8.2.1 *A-stability*

The simplest concept of stability for ODE methods is related to the linear scalar autonomous test equation

$$\begin{cases} y'(t) = \lambda y(t), & t \geq t_0, \\ y(t_0) = y_0, \end{cases} \tag{8.2.1}$$

where $\lambda \in \mathbb{C}$.

It is known that the solution is $y(t) = \mathrm{e}^{\lambda(t-t_0)}y_0$ and, hence,

$$|y(t)| = \mathrm{e}^{\Re(\lambda)(t-t_0)}|y_0|.$$

Therefore, whenever $y_0 \neq 0$, the condition

$$\Re(\lambda) \leq \lambda_0 \tag{8.2.2}$$

is equivalent to

$$|y(t)| \leq \mathrm{e}^{\lambda_0(t-t_0)}|y_0|, \quad t \geq t_0. \tag{8.2.3}$$

In particular, the condition $\Re(\lambda) < 0$ is equivalent to the asymptotic stability property

$$\lim_{t\to+\infty} y(t) = 0, \tag{8.2.4}$$

whereas the slightly weaker condition $\Re(\lambda) \leq 0$ is equivalent to the contractivity, or dissipativity, property

$$|y(t)| \leq |y_0|, \quad t \geq t_0. \tag{8.2.5}$$

Definition 8.2.1 *The* A-stability region *of a numerical method for ODEs is the set S_A of complex numbers $\alpha = h\lambda$ such that the numerical solution $\{y_n\}_{n\geq 0}$ of (8.2.1) obtained with the constant stepsize h satisfies* $\lim_{n\to\infty} y_n = 0$.

Definition 8.2.2 *A numerical method for ODEs is* A-stable *if*

$$S_A \supseteq \{\alpha \in \mathbb{C} \mid \Re(\alpha) < 0\}.$$

In other words, a numerical method for ODEs is A-stable if, when implemented with constant stepsize h, it preserves the asymptotic stability properties of the solution $y(t)$ of (8.2.1).

We recall that the numerical solution of (8.2.1) produced by an RK method (5.0.2), (5.0.3) is defined recursively by

$$y_{n+1} = R(h_{n+1}\lambda)y_n. \tag{8.2.6}$$

Here, the rational function

$$R(\alpha) = 1 + \alpha b^T(I - \alpha A)^{-1}e, \tag{8.2.7}$$

where $b = [b_1, \ldots, b_\nu]^T$, $A = [a_{ij}]_{i,j=1}^{\nu}$ and $e = [1, \ldots, 1]^T$ is the unit ν-vector and I is the $\nu \times \nu$ identity matrix, is the so-called *A-stability function*, and may be rewritten in the equivalent form

$$R(\alpha) = \frac{\det[I - \alpha A + \alpha e b^T]}{\det[I - \alpha A]}. \tag{8.2.8}$$

Therefore, in the case of an RK method (5.0.2), (5.0.3), the A-stability region is

$$S_A = \{\alpha \in \mathbb{C} \mid |R(\alpha)| < 1\}. \tag{8.2.9}$$

It is worth remarking that an explicit RK method cannot be A-stable. In other words, only implicit RK methods can have the best stability properties with respect to the test equation (8.2.1).

The following result by Scherer and Müller [226] provides a useful characterization of A-stable RK methods.

Theorem 8.2.3 *Assume that* $\det[A] \neq 0$ *and that the A-stability function* $R(\alpha) = p(\alpha)/q(\alpha)$ *is such that* $p(\alpha)$ *and* $q(\alpha)$ *have no common roots. Then the RK method (5.0.2), (5.0.3) is A-stable if and only if a symmetric positive semi-definite matrix* $\mathcal{R}$ *exists such that* $\mathcal{R}e = b$ *and*

$$\mathcal{R}A + A^T\mathcal{R} - bb^T \tag{8.2.10}$$

is positive semi-definite.

The concept of A-stability is also useful for analyzing the asymptotic behavior of RK methods when applied to autonomous linear systems of the form

$$\begin{cases} y'(t) = Ly(t), & t \geq t_0, \\ y(t_0) = y_0, \end{cases} \tag{8.2.11}$$

where L is a complex $d \times d$ matrix.

It is known that the solution is $y(t) = \mathrm{e}^{(t-t_0)L}y_0$. Therefore,

$$\lim_{t \to +\infty} y(t) = 0$$

for all initial values y_0 if and only if $\Re(\lambda) < 0$ for all eigenvalues λ of L.

When we apply the RK method (5.0.2), (5.0.3) to the system (8.2.11) in the mesh interval $[t_n, t_{n+1}]$, we get

$$y_{n+1} = R(h_{n+1}L)y_n,$$

where we formally substitute the matrix $h_{n+1}L$ into the argument of the rational function $R(\alpha)$. Therefore, the following result is easily obtained.

Theorem 8.2.4 *The numerical solution $\{y_n\}_{n\geq 0}$ of (8.2.11) furnished by an RK method (5.0.2), (5.0.3) with constant stepsize h satisfies $\lim_{n\to\infty} y_n = 0$ for all initial values y_0 if and only if $h\lambda \in S_A$ for all eigenvalues λ of L.*

In view of future considerations, given an inner product $\langle\cdot,\cdot\rangle$ in $\mathbb{C}^d$ and the corresponding norm $\|\cdot\|$, let λ_0 be the smallest real number such that

$$\Re(\langle Ly, y\rangle) \leq \lambda_0\|y\|^2 \quad \text{for all } y \in \mathbb{C}^d. \tag{8.2.12}$$

We recall that λ_0 equals the so-called *logarithmic norm* (induced by the chosen inner product norm) of the matrix L, which is usually denoted by $\mu[L]$ and is defined by

$$\mu[L] = \lim_{\epsilon\to 0^+} \frac{\|I+\epsilon L\| - 1}{\epsilon}. \tag{8.2.13}$$

Condition (8.2.12) is equivalent to

$$\|y(t)\| \leq \mathrm{e}^{\lambda_0(t-t_0)}\|y_0\|, \quad t \geq t_0, \tag{8.2.14}$$

for all initial values y_0. In particular, the condition $\lambda_0 \leq 0$ is equivalent to the contractivity property

$$\|y(t)\| \leq \|y_0\|, \quad t \geq t_0,$$

for all initial values y_0.

8.2.2 *AN-stability*

The second stability concept in importance for ODE methods is related to the linear scalar non-autonomous test equation

$$\begin{cases} y'(t) = \lambda(t)y(t), & t \geq t_0, \\ y(t_0) = y_0, \end{cases} \tag{8.2.15}$$

where $\lambda(t)$ is a continuous complex valued function.

It is known that the solution is $y(t) = \exp\left(\int_{t_0}^t \lambda(x)\mathrm{d}x\right) y_0$. Therefore, if

$$\Re(\lambda(t)) \leq \Lambda_0, \quad t \geq t_0, \tag{8.2.16}$$

then

$$|y(t)| \leq \mathrm{e}^{\Lambda_0(t-t_0)}|y_0|, \quad t \geq t_0. \tag{8.2.17}$$

In particular, if

$$\Re(\lambda(t)) \leq 0, \quad t \geq t_0, \tag{8.2.18}$$

then (8.2.5) holds. Moreover, if

$$\Lambda_0 < 0, \tag{8.2.19}$$

then the solution $y(t)$ asymptotically vanishes, that is (8.2.4) holds.

As far as the necessity of conditions (8.2.16), (8.2.18) and (8.2.19) is concerned, it is easy to see that the situation is different from the autonomous case. More precisely, (8.2.16) and (8.2.18) are necessary (and also sufficient) for

$$|y(t_2)| \leq e^{\Lambda_0(t_2-t_1)}|y(t_1)|, \quad \text{for all } t_2 \geq t_1 \geq t_0,$$

and

$$|y(t_2)| \leq |y(t_1)|, \quad \text{for all } t_2 \geq t_1 \geq t_0,$$

respectively, which are clearly stronger properties than (8.2.17) and (8.2.5). Unlike the autonomous case, condition (8.2.19) is not necessary for the asymptotic stability property (8.2.4).

Although there exists the concept of *circle contractivity*, concerning the regions of stability of numerical methods with respect to the test equation (8.2.15), we focus directly on those methods that completely preserve the contractivity properties of the exact solution $y(t)$.

Definition 8.2.5 *A numerical method for ODEs is* AN-stable *if the numerical solution $\{y_n\}_{n\geq 0}$ of (8.2.15) satisfying (8.2.18) is such that $|y_{n+1}| \leq |y_n|$ for any mesh Δ.*

It is clear that the concept of AN-stability is stronger than the concept of A-stability.

We recall that the numerical solution of (8.2.15) produced by an RK method (5.0.2), (5.0.3) is defined recursively by

$$y_{n+1} = K(h_{n+1}\lambda(t^1_{n+1}), \ldots, h_{n+1}\lambda(t^\nu_{n+1}))y_n. \tag{8.2.20}$$

Here, the rational function

$$K(\alpha_1, \ldots, \alpha_\nu) = 1 + b^T(I - ZA)^{-1}Ze, \tag{8.2.21}$$

where $Z = \text{diag}(\alpha_1, \ldots, \alpha_\nu)$, is the so-called *AN-stability function.*

Therefore, in the case of an RK method (5.0.2), (5.0.3), AN-stability is equivalent to

$$|K(\alpha_1, \ldots, \alpha_\nu)| \leq 1 \quad \forall \Re(\alpha_i) \leq 0, \tag{8.2.22}$$

under the restriction that $\alpha_h = \alpha_k$ if $c_h = c_k$.

8.2.3 *BN-stability*

The most general concept of stability for ODE methods is related to the general system

$$\begin{cases} y'(t) = g(t, y(t)), & t \geq t_0, \\ y(t_0) = y_0, \end{cases} \tag{8.2.23}$$

where $g \in C^0\left([t_0, +\infty) \times \mathbb{C}^d, \mathbb{C}^d\right)$. First of all, we observe that, for the linear test equations (8.2.1), (8.2.11) and (8.2.15), both the true and the numerical solutions are equal to the *zero function* whenever $y_0 = 0$. This fact, along with the linearity of the problems, allowed us to study, without loss of generality, the contractivity and asymptotic stability properties of both the true and the numerical solutions simply by comparing them with the zero function.

On the contrary, in the case of the general non-linear system (8.2.23), it is not straightforward to do the same simplification, especially as far as the numerical solution is concerned. Therefore, we consider another system

$$\begin{cases} \tilde{y}'(t) = g(t, \tilde{y}(t)), & t \geq t_0, \\ \tilde{y}(t_0) = \tilde{y}_0, \end{cases} \tag{8.2.24}$$

with the same right-hand side function $g(t, y)$ but different initial value $\tilde{y}_0$.

Given an inner product $\langle \cdot, \cdot \rangle$ in $\mathbb{C}^d$ and the corresponding norm $\| \cdot \|$, we assume that a continuous function $Y(t)$ exists such that

$$Y(t) \geq \sup_{y_1 \neq y_2} \frac{\Re(\langle g(t, y_1) - g(t, y_2), y_1 - y_2 \rangle)}{\|y_1 - y_2\|^2}. \tag{8.2.25}$$

Note that, if the function $g(t, y)$ is differentiable with respect to y, then the smallest number $Y(t)$ satisfying the inequality (8.2.25) equals the logarithmic norm (see (8.2.13)) of the Jacobian matrix $\frac{\partial g}{\partial y}(t, y)$.

It is known that if

$$Y(t) \leq Y_0, \quad t \geq t_0, \tag{8.2.26}$$

then, for the solutions $y(t)$ and $\tilde{y}(t)$ of the ODEs (8.2.23) and (8.2.24), it holds that

$$\|\delta y(t)\| \leq \mathrm{e}^{Y_0(t-t_0)} \|\delta y_0\|, \quad t \geq t_0, \tag{8.2.27}$$

where

$$\delta y(t) = \tilde{y}(t) - y(t),$$
$$\delta y_0 = \tilde{y}_0 - y_0.$$

The number Y_0 is called the *one-sided Lipschitz constant* of the function $g(t, y)$ with respect to the inner product $\langle \cdot, \cdot \rangle$.

In particular, if

$$Y(t) \leq 0, \quad t \geq t_0, \tag{8.2.28}$$

then

$$\|\delta y(t)\| \leq \|\delta y_0\|, \quad t \geq t_0. \tag{8.2.29}$$

Moreover, if

$$Y_0 < 0, \tag{8.2.30}$$

then the difference between the two solutions $\delta y(t)$ vanishes asymptotically, that is

$$\lim_{t \to +\infty} \delta y(t) = 0. \tag{8.2.31}$$

Conditions (8.2.26) and (8.2.28) are necessary (and also sufficient) for

$$\|\delta y(t_2)\| \leq \mathrm{e}^{Y_0(t_2-t_1)} \|\delta y(t_1)\| \quad \text{for all } t_2 \geq t_1 \geq t_0,$$

and

$$\|\delta y(t_2)\| \le \|\delta y(t_1)\| \quad \text{for all } t_2 \ge t_1 \ge t_0,$$

respectively, which are clearly stronger properties than (8.2.27) and (8.2.29). The condition (8.2.30) is not necessary for the asymptotic stability property (8.2.31).

Now let us consider a numerical method for ODEs and let $\{y_n\}_{n\ge 0}$ and $\{\tilde{y}_n\}_{n\ge 0}$ be the numerical solutions of (8.2.23) and (8.2.24), respectively. Then we define

$$\delta y_n = \tilde{y}_n - y_n$$

and give the following definition of contractivity.

Definition 8.2.6 *A numerical method for ODEs is* BN-stable *if the difference* $\{\delta y_n\}_{n\ge 0}$ *between the numerical solutions of the systems (8.2.24) and (8.2.23) is such that* $\|\delta y_{n+1}\| \le \|\delta y_n\|$ *for any mesh* Δ *whenever (8.2.28) holds.*

In other words, a numerical method for ODEs is BN-stable if it preserves the contractivity properties of the solutions of the ODE (8.2.23) with respect to inner product norms. It is clear that the concept of BN-stability is stronger than the concept of AN-stability.

8.2.4 *Algebraic stability*

For an RK method (5.0.2), (5.0.3), AN-stability and BN-stability can be characterized by means of the following further stability concept.

Definition 8.2.7 *An RK method (5.0.2), (5.0.3) is* algebraically stable *if the weights* $b_i \ge 0$, $i = 1, \ldots, \nu$, *and the matrix*

$$M = \mathrm{diag}(b_1, \ldots, b_\nu)A + A^T \mathrm{diag}(b_1, \ldots, b_\nu) - bb^T \tag{8.2.32}$$

is positive semi-definite.

Observe that the matrix (8.2.32) is a particular instance of (8.2.10) for $\mathcal{R} = \mathrm{diag}(b_1, \ldots, b_\nu)$.

Theorem 8.2.8 *If an RK method (5.0.2), (5.0.3) is algebraically stable, then it is also BN-stable.*

In connection with the characterization (8.2.22) of AN-stability, we recall the following definition.

Definition 8.2.9 *An RK method (5.0.2), (5.0.3) is* non-confluent *if* $c_i \ne c_j$ *whenever* $i \ne j$. *Otherwise it is* confluent.

In other words, an RK method is confluent if it has at least two equal abscissae.

Theorem 8.2.10 *Assume that the RK method (5.0.2), (5.0.3) is non-confluent. Then, if it is AN-stable, it is also algebraically stable.*

Since a BN-stable method is also AN-stable, Theorems 8.2.8 and 8.2.10 immediately yield the following equivalence result.

Corollary 8.2.11 *For a non-confluent RK method (5.0.2), (5.0.3) the following statements are equivalent:*

- *the method is AN-stable;*
- *the method is algebraically stable;*
- *the method is BN-stable.*

The above equivalence partially holds also for most confluent methods. In order to illustrate such a generalization, we need to recall the concept of *reducibility* for RK methods. Indeed, there are two concepts of reducibility, namely *DJ-reducibility* and *S-reducibility*, that do not imply each other. However, both of them imply the following simple property, which we assume as the definition of reducibility.

Definition 8.2.12 *An RK method (5.0.2), (5.0.3) is* reducible *if it is equivalent to another method with fewer stages, in the sense that they provide the same numerical solution for all systems of the type (8.2.23). Otherwise it is* irreducible.

Theorem 8.2.13 *If an irreducible RK method (5.0.2), (5.0.3) is BN-stable, then it is also algebraically stable.*

Since this is the converse of Theorem 8.2.8, we have the following equivalence result.

Corollary 8.2.14 *For an irreducible RK method (5.0.2), (5.0.3) the following statements are equivalent:*

- *the method is algebraically stable;*
- *the method is BN-stable.*

In what follows, without any practical restriction, we shall consider only irreducible methods. This is not restrictive because we could replace the coefficients of the reducible method by the coefficients of the equivalent irreducible method, without changing the numerical solution.

In view of formulae (8.3.9) and (8.3.12) in Section 8.3, we recall that the weights of an algebraically stable irreducible RK method satisfy

$$b_i > 0, \quad i = 1, \ldots, \nu.$$

8.3 Asymptotic stability and contractivity of RK methods

We have seen that A-stability is a demand for asymptotic stability of the numerical solution for constant stepsize, whereas AN-stability and BN-stability sound more as a demand for contractivity with respect to the initial data.

Indeed, it is well known that, in the case of RK methods, A-stability can also be defined as a contractivity property. In fact, we have that

$$\overline{S_A} = \{\alpha \in \mathbb{C} \mid |R(\alpha)| \leq 1\}$$

and, in particular, the following result holds.

Proposition 8.3.1 *An RK method (5.0.2), (5.0.3) is A-stable if the numerical solution* $\{y_n\}_{n\geq 0}$ *of (8.2.1) satisfies* $|y_{n+1}| \leq |y_n|$ *whenever* $\Re(\lambda) \leq 0$ *for any mesh* Δ.

However, following Hairer and Zennaro [120], in this section we shall see that, for the case of RK methods, under some mild hypotheses, AN-stability and BN-stability actually also imply the asymptotic stability of the numerical solution for constant stepsize.

In what follows we do not consider the regions of stability, but only methods that are A-stable, AN-stable, or BN-stable.

8.3.1 *Constant and variable stepsize*

First of all we observe that, even if an RK method is algebraically stable, it is not guaranteed that the numerical solution $\{y_n\}_{n\geq 0}$ of the test equation (8.2.1) with $\Re(\lambda) < 0$ asymptotically vanishes if the mesh Δ is not uniform and uses stepsizes that are not uniformly bounded. This may happen when the rational function (8.2.7) is such that $\lim_{|\alpha|\to+\infty} |R(\alpha)| = 1$. This is the case, for example, of all the collocation methods at Gaussian points.

In fact, in such a case we can easily choose a diverging sequence of stepsizes h_n such that, even if $|R(h_n\lambda)| < 1$ for all n, the product $\prod_{k=1}^{n} |R(h_k\lambda)|$ tends to some limit $r \neq 0$ as $n \to \infty$. Since $|y_{n+1}| = |R(h_{n+1}\lambda)| \cdot |y_n|$, it turns out that $\lim_{n\to\infty} |y_n| = r|y_0| \neq 0$.

What we just observed motivates the following definitions.

Definition 8.3.2 *An A-stable numerical method for ODEs is* asymptotically A-stable *if the numerical solution* $\{y_n\}_{n\geq 0}$ *of (8.2.1) satisfies* $\lim_{n\to\infty} y_n = 0$ *for any mesh* Δ *whenever* $\Re(\lambda) < 0$.

Definition 8.3.3 *An AN-stable numerical method for ODEs is* asymptotically AN-stable *if the numerical solution* $\{y_n\}_{n\geq 0}$ *of (8.2.15) satisfies* $\lim_{n\to\infty} y_n = 0$ *for any mesh* Δ *whenever (8.2.19) holds.*

Definition 8.3.4 *A BN-stable numerical method for ODEs is* asymptotically BN-stable *if the difference* $\{\delta y_n\}_{n\geq 0}$ *of the numerical solutions of the systems (8.2.23) and (8.2.24) is such that* $\lim_{n\to\infty} \delta y_n = 0$ *for any mesh* Δ *whenever (8.2.30) holds.*

We can conclude that a necessary condition for an RK method (5.0.2), (5.0.3) to be asymptotically A-stable (and, *a fortiori*, asymptotically AN- or BN-stable) is

$$|R(\infty)| = \lim_{|\alpha|\to+\infty} |R(\alpha)| < 1. \tag{8.3.1}$$

8.3.2 *Error growth functions*

It is interesting to see how the exponential bounds to the initial errors displayed in (8.2.3), (8.2.14), (8.2.17) and (8.2.27) modify under discretization by means of an RK method. The analogs of such exponential bounds are the so-called *error growth functions*.

Definition 8.3.5 *The* A-error growth function *of an A-stable RK method (5.0.2), (5.0.3) is the smallest value* $\Phi_A(\lambda_0, h_{n+1})$ *such that the numerical solution* $\{y_n\}_{n\geq 0}$ *of (8.2.1) satisfies*

$$|y_{n+1}| \leq \Phi_A(\lambda_0, h_{n+1})|y_n|,$$

whenever (8.2.2) holds.

Definition 8.3.6 *The* AN-error growth function *of an AN-stable RK method (5.0.2), (5.0.3) is the smallest value* $\Phi_{AN}(\Lambda_0, h_{n+1})$ *such that the numerical solution* $\{y_n\}_{n\geq 0}$ *of (8.2.15) satisfies*

$$|y_{n+1}| \leq \Phi_{AN}(\Lambda_0, h_{n+1})|y_n|$$

whenever (8.2.16) holds.

Definition 8.3.7 *The* BN-error growth function *of a BN-stable RK method (5.0.2), (5.0.3) is the smallest value* $\Phi_{BN}(Y_0, h_{n+1})$ *such that the difference* $\{\delta y_n\}_{n\geq 0}$ *of the numerical solutions of the systems (8.2.23) and (8.2.24) satisfies*

$$\|\delta y_{n+1}\| \leq \Phi_{BN}(Y_0, h_{n+1})\|\delta y_n\|$$

whenever (8.2.26) holds.

It is immediately seen that the following ordering holds among the various error growth functions of a BN-stable RK method:

$$\Phi_A(x,h) \leq \Phi_{AN}(x,h) \leq \Phi_{BN}(x,h) \leq 1 \quad \text{for all } x \leq 0 \text{ and } h > 0. \tag{8.3.2}$$

As for the A- and AN-error growth functions, the following results are obvious in view of (8.2.6) and (8.2.20), respectively.

Proposition 8.3.8 *Let (5.0.2), (5.0.3) be an A-stable RK method. Then*

$$\Phi_A(\lambda_0, h) = \sup_{\Re(\alpha)\leq h\lambda_0} |R(\alpha)|,$$

where $R(\alpha)$ *is the A-stability function (8.2.7).*

Proposition 8.3.9 *Let (5.0.2), (5.0.3) be an AN-stable RK method. Then*

$$\Phi_{AN}(\Lambda_0, h) \leq \Phi_K(\Lambda_0, h), \tag{8.3.3}$$

where

$$\Phi_K(\Lambda_0, h) = \sup_{\Re(\alpha_1)\leq h\Lambda_0,\ldots,\Re(\alpha_\nu)\leq h\Lambda_0} |K(\alpha_1,\ldots,\alpha_\nu)| \tag{8.3.4}$$

and $K(\alpha_1,\ldots,\alpha_\nu)$ *is the AN-stability function (8.2.21).*

Moreover, for non-confluent RK methods, equality holds in (8.3.3).

From a theorem of Von Neumann it is possible to derive the following result for autonomous linear systems (8.2.11).

Theorem 8.3.10 *The numerical solution $\{y_n\}_{n\geq 0}$ of (8.2.11) furnished by an A-stable RK method (5.0.2), (5.0.3) satisfies*

$$\|y_{n+1}\| \leq \Phi_A(\lambda_0, h_{n+1})\|y_n\|,$$

whenever (8.2.12) holds.

Hairer and Zennaro [120] have proved the following result concerning the BN-error growth function.

Proposition 8.3.11 *Let (5.0.2), (5.0.3) be a BN-stable RK method. Then*

$$\Phi_{BN}(Y_0, h) = \sup_{\mu[A_1]\leq hY_0,\ldots,\mu[A_\nu]\leq hY_0} \|K(A_1, \ldots, A_\nu)\|,$$

where $A_1, \ldots, A_\nu$ are $d\times d$ matrices, $\mu[\cdot]$ stands for the logarithmic norm (8.2.13) and $K(\alpha_1, \ldots, \alpha_\nu)$ is the AN-stability function (8.2.21).

Moreover,

$$\Phi_{BN}(Y_0, h) \geq \Phi_K(Y_0, h), \tag{8.3.5}$$

where $\Phi_K(Y_0, h)$ is given by (8.3.4).

Hairer and Wanner [119] have shown that, for one-stage and (irreducible) two-stage RK methods, equality holds in (8.3.5).

Propositions 8.3.8, 8.3.9 and 8.3.11 immediately yield the following corollary.

Corollary 8.3.12 *The following statements hold:*

- *Let (5.0.2), (5.0.3) be an A-stable RK method. Then the A-error growth function $\Phi_A(\lambda_0, h)$ is non-increasing with respect to h for any $\lambda_0 \leq 0$.*
- *Let (5.0.2), (5.0.3) be an AN-stable RK method. Then the function $\Phi_K(\Lambda_0, h)$ is non-increasing with respect to h for any $\Lambda_0 \leq 0$.*
- *Let (5.0.2), (5.0.3) be a BN-stable RK method. Then the BN-error growth function $\Phi_{BN}(Y_0, h)$ is non-increasing with respect to h for any $Y_0 \leq 0$.*

Propositions 8.3.8, 8.3.9 and 8.3.11 give a theoretical quantification of the error growth functions that, in the general case, are unlikely to be useful for an actual computation. Now, we give instead some elementary upper bounds that are easily expressed in terms of the coefficients of the RK method.

To this aim, we consider also the case when the classical Lipschitz constant L exists such that

$$\|g(t, y_1) - g(t, y_2)\| \leq L\|y_1 - y_2\| \tag{8.3.6}$$

for all $t \geq t_0$ and $y_1, y_2 \in \mathbb{C}^d$. It is clear that, under this condition, the problem (8.2.23) has limited stiffness and that

$$L \geq |Y_0|. \tag{8.3.7}$$

The following proposition, whose proof may be found in [120], gives an upper bound to the error growth of an algebraically stable RK method, which depends also on the classical Lipschitz constant L.

Proposition 8.3.13 *Let the RK method (5.0.2), (5.0.3) be algebraically stable. Then the difference* $\{\delta y_n\}_{n\geq 0}$ *between the numerical solutions of the systems (8.2.23) and (8.2.24) satisfying the conditions* $Y_0 \leq 0$ *and (8.3.6) is such that*

$$\|\delta y_{n+1}\| \leq \Psi_1(L, Y_0, h_{n+1})\|\delta y_n\|,$$

where

$$\Psi_1(L, Y_0, h) = \sqrt{1 + \frac{2hY_0}{(1+hL\gamma)^2}} \tag{8.3.8}$$

and

$$\gamma = \sqrt{\sum_{j=1}^{\nu} b_j^{-1}\left(\sum_{i=1}^{\nu} b_i a_{ij}\right)^2}. \tag{8.3.9}$$

The next proposition gives an upper bound to the BN-error growth function which, contrary to the previous bound (8.3.8), is independent of the classical Lipschitz constant L.

Lemma 8.3.14 *Let the RK method (5.0.2), (5.0.3) be algebraically stable. Then there exists*

$$\lim_{\epsilon\to 0} b^T(A+\epsilon I)^{-1} = v^T \in \mathbb{R}^\nu.$$

Clearly, if the coefficient matrix A is non-singular, then $v^T = b^T A^{-1}$. Moreover, in any case, it holds that

$$R(\infty) = \lim_{|\alpha|\to+\infty} R(\alpha) = 1 - v^T e. \tag{8.3.10}$$

Incidentally, it is interesting to remark that there actually exist algebraically stable irreducible RK methods that are characterized by a singular coefficient matrix A. An example is given by the following three-stage RK method of order 2 (completely useless for practical purposes), for which $M = 0$ in (8.2.32):

$$\begin{array}{c|ccc}
\frac14 - \frac{\sqrt3}{12} & \frac18 & \frac18 - \frac{\sqrt3}{12} & 0 \\
\frac14 + \frac{\sqrt3}{4} & \frac18 + \frac{\sqrt3}{12} & \frac18 & \frac{\sqrt3}{6} \\
\frac34 - \frac{\sqrt3}{12} & \frac14 & \frac14 - \frac{\sqrt3}{12} & \frac14 \\
\hline
 & \frac14 & \frac14 & \frac12
\end{array}$$

Proposition 8.3.15 *Let the RK method (5.0.2), (5.0.3) be algebraically stable. Then, for all* $Y_0 \leq 0$ *the BN-error growth function satisfies*

$$\Phi_{BN}(Y_0, h) \leq \Psi_2(Y_0, h),$$

where

$$\Psi_2(Y_0, h) = \frac{-2hY_0\Gamma|R(\infty)| + \sqrt{1 - 2(1 - |R(\infty)|^2)hY_0\Gamma}}{1 - 2hY_0\Gamma} \tag{8.3.11}$$

and

$$\Gamma = \left(\sum_{j=1}^{\nu} b_j^{-1} v_j^2 \right)^{-1}. \tag{8.3.12}$$

8.3.3 *Superexponential functions*

In order to establish the asymptotic stability of algebraically stable RK methods, we need the following concepts, introduced by Hairer and Zennaro [120] and Zennaro [284].

Definition 8.3.16 *A continuous function* $\Phi : [0, +\infty) \longrightarrow \mathbb{R}$ *is* superexponential *if it satisfies the following properties:*

(se1) $\Phi(0) = 1$;

(se2) $\Phi(h_1)\Phi(h_2) \le \Phi(h_1 + h_2)$ *for all* $h_1, h_2 \ge 0$.

The next proposition justifies the name given to such functions, as they turn out to approximate from above an exponential function.

Proposition 8.3.17 *Let the continuous function* $\Phi : [0, +\infty) \longrightarrow \mathbb{R}$ *be superexponential. Then*

$$\Phi(h) \ge e^{\alpha h} \quad \textit{for all } h \ge 0, \tag{8.3.13}$$

where $\alpha = \limsup_{h \to 0^+} \frac{\Phi(h) - \Phi(0)}{h}$.

Observe that, in particular, a superexponential function always assumes positive values.

Definition 8.3.18 *A continuous function* $\Phi : [0, +\infty) \longrightarrow \mathbb{R}$ *is* negative superexponential *if it is superexponential and satisfies the additional property*

(nse) $\Phi(h) < 1$ *for all* $h > 0$.

Observe that, for a negative superexponential function, it necessarily holds that $\alpha < 0$ in (8.3.13). This justifies the name given to such functions.

Definition 8.3.19 *A continuous function* $\Phi : [0, +\infty) \longrightarrow \mathbb{R}$ *is* asymptotically negative superexponential *if it is negative superexponential and satisfies the additional property*

(anse) $\limsup_{h \to +\infty} \Phi(h) < 1$.

For the elementary bounds (8.3.8) and (8.3.11) we have the following results (see [120]).

Proposition 8.3.20 *For given* $L > 0$ *and* $Y_0 < 0$ *satisfying (8.3.7), the function* $\Psi_1(L, Y_0, h)$ *defined by (8.3.8) is a negative superexponential function of* h, *provided that* $\gamma \ge \frac{8}{15}$.

Moreover, it takes its minimum $\min_{\Psi_1} = \sqrt{1 + Y_0/(2L\gamma)}$ *for* $h = L^{-1}\gamma^{-1}$, *is strictly decreasing for* $h < L^{-1}\gamma^{-1}$ *and strictly increasing for* $h > L^{-1}\gamma^{-1}$ *and satisfies*

$$\lim_{h \to +\infty} \Psi_1(L, Y_0, h) = 1.$$

Proposition 8.3.21 *Assume that (8.3.1) holds and let* $Y_0 < 0$ *be given. Then* $\Psi_2(Y_0, h)$ *defined by (8.3.11) is an asymptotically negative superexponential function of* h.

Moreover, it is strictly decreasing for all $h \geq 0$ *and satisfies*

$$\lim_{h \to +\infty} \Psi_2(Y_0, h) = |R(\infty)| < 1.$$

The next theorem states that the error growth functions are superexponential (see again [120]).

Theorem 8.3.22 *The following statements hold:*

- *The A-error growth function* $\Phi_A(\lambda_0, h)$ *of an A-stable RK method (5.0.2), (5.0.3) is a superexponential function of* h *for all* $\lambda_0 \leq 0$. *Moreover, if (8.3.1) holds and* $\lambda_0 < 0$, *then it is asymptotically negative superexponential.*
- *The function* $\Phi_K(\Lambda_0, h)$ *of an AN-stable RK method (5.0.2), (5.0.3) is a superexponential function of* h *for all* $\Lambda_0 \leq 0$. *Moreover, if the method is algebraically stable, if (8.3.1) holds and* $\Lambda_0 < 0$, *then it is asymptotically negative superexponential.*
- *The BN-error growth function* $\Phi_{BN}(Y_0, h)$ *of a BN-stable RK method (5.0.2), (5.0.3) is a superexponential function of* h *for all* $Y_0 \leq 0$. *Moreover, if (8.3.1) holds and* $Y_0 < 0$, *then it is asymptotically negative superexponential.*

8.3.4 *Asymptotic stability*

The following theorem illustrates the role of the negative superexponential functions in the asymptotic stability analysis of the numerical ODE methods.

Theorem 8.3.23 *Given the interval* $[t_0, +\infty)$, *let* F *be a functional defined on the set of all meshes* Δ *that associates a real number* $F_n(\Delta) \geq 0$ *to each point* $t_n \in \Delta$. *Moreover, let the functional* F *be such that*

$$F_{n+1}(\Delta) \leq \Phi(h_{n+1}) F_n(\Delta) \quad \textit{for all } n \geq 0, \tag{8.3.14}$$

where $\Phi(h)$ *is a negative superexponential function. Then*

$$\lim_{n \to \infty} F_n(\Delta) = 0 \tag{8.3.15}$$

for all meshes Δ *such that* $h = \sup_{n \geq 1} h_n < +\infty$, *and the convergence rate is at least of exponential type, i.e. at least like* $\mathrm{e}^{-\alpha(t_n - t_0)}$ *for some* $\alpha > 0$.

Moreover, if $\Phi(h)$ is asymptotically negative superexponential, then (8.3.15) holds for any mesh Δ (even with unbounded stepsizes h_n), but this time the rate of convergence is not guaranteed to be of exponential type.

Proof. Since the mesh Δ is such that $h = \sup_{n\geq 1} h_n < +\infty$, possibly by suppressing some suitable mesh points, we can find a submesh $\Delta' = \{t'_0 = t_0, t'_1, \ldots, t'_k, t'_{k+1}, \ldots\} \subseteq \Delta$ for which all the stepsizes h'_k are such that $h'_{\min} \leq h'_k \leq h'_{\max}$ for suitable $h'_{\min}, h'_{\max} > 0$. By condition (nse) and by the uniform continuity of Φ in the interval $[h'_{\min}, h'_{\max}]$, we have that

$$\Phi(h'_k) \leq \rho \quad \text{for some } \rho < 1 \tag{8.3.16}$$

for all $k \geq 1$ and thus, for the sequence of real numbers $\delta'_k = \prod_{s=1}^{k} \Phi(h'_s)$, we get

$$\delta'_k \leq \rho^k \tag{8.3.17}$$

and, consequently,

$$\lim_{k\to\infty} \delta'_k = 0. \tag{8.3.18}$$

If for each $k \geq 0$ we denote by n_k the index such that $t_{n_k} = t'_k$, then by condition (se2) and by (8.3.14) we easily get

$$F_{n_k}(\Delta) \leq \delta'_k F_0(\Delta) \quad \text{for all } k \geq 1. \tag{8.3.19}$$

Finally, because of (8.3.14), the sequence $\{F_n(\Delta)\}$ is monotonically decreasing and, therefore, (8.3.18) and (8.3.19) imply (8.3.15).

In order to prove the exponential convergence rate, observe that, if $t_n \in [t_{n_k}, t_{n_{k+1}})$, then $k > \frac{t_n - t_0}{h'_{\max}} - 1$ and $F_n(\Delta) \leq F_{n_k}(\Delta)$. Therefore, by (8.3.17) and (8.3.19) we get

$$F_n(\Delta) < \rho^{-1} \rho^{\frac{t_n - t_0}{h'_{\max}}} F_0(\Delta)$$

and, hence, the exponential rate with exponent $\alpha = \frac{-\log(\rho)}{h'_{\max}} > 0$.

Now assume that Φ satisfies the further condition (anse). In this case, in order to fulfill (8.3.16) and to prove (8.3.15), it is sufficient to find a submesh $\Delta' \subseteq \Delta$ for which the stepsizes h'_k are such that $h'_k \geq h'_{\min}$ for a suitable $h'_{\min} > 0$. Therefore, we are allowed to consider also meshes Δ with unbounded stepsizes h_n. However, in this case we cannot give a lower bound to k when we assume that $t_n \in [t_{n_k}, t_{n_{k+1}})$ and, thus, we cannot guarantee an exponential rate of convergence. ∎

Theorem 8.3.23 together with Propositions 8.3.13 and 8.3.20 easily establishes the asymptotic stability of algebraically stable RK methods for problems with limited stiffness (i.e. for which (8.3.6) holds) by using meshes Δ with stepsizes that are uniformly bounded from above (in particular, constant stepsize).

Theorem 8.3.24 *Let the RK method (5.0.2), (5.0.3) be BN-stable. Then the difference $\{\delta y_n\}_{n\geq 0}$ between the numerical solutions of the systems (8.2.23) and*

(8.2.24) satisfying (8.2.30) and (8.3.6) is such that $\lim_{n\to\infty} \delta y_n = 0$ *for any mesh* Δ *with* $h = \sup_{n\geq 1} h_n < +\infty$*, and the rate of convergence of such a limit is at least of exponential type, i.e. at least like* $\mathrm{e}^{-\alpha(t_n - t_0)}$ *for some* $\alpha > 0$.

Theorem 8.3.25 *Let the RK method (5.0.2), (5.0.3) be algebraically stable. Then the numerical solution* $\{y_n\}_{n\geq 0}$ *of (8.2.15) satisfying*

$$-L \leq \Re(\lambda(t)) \leq \Lambda_0 < 0, \quad t \geq t_0,$$

for some $L > 0$ *and* $\Lambda_0 < 0$ *is such that* $\lim_{n\to\infty} y_n = 0$ *for any mesh* Δ *with* $h = \sup_{n\geq 1} h_n < +\infty$*, and the rate of convergence of such a limit is at least of exponential type, i.e. at least like* $\mathrm{e}^{-\alpha(t_n - t_0)}$ *for some* $\alpha > 0$.

Observe that Theorems 8.3.24 and 8.3.25 state that, in the particular case of constant stepsize, algebraically stable RK methods satisfy for more general ODEs with limited stiffness such as (8.2.23) and (8.2.15), respectively, the same stability property required in Definition 8.2.1 in A-stable RK methods for the simpler test equation (8.2.1).

In order to get asymptotic stability results for problems with unlimited stiffness (i.e. for which (8.3.6) does not hold) or, in any case, in using meshes Δ with possibly unbounded stepsize, it is sufficient to impose the additional property (8.3.1) to the A-stability function. In fact, in view of Theorems 8.3.22 and 8.3.23, we immediately get the following results.

Theorem 8.3.26 *Let the RK method (5.0.2), (5.0.3) be BN-stable and satisfy (8.3.1). Then it is asymptotically BN-stable.*

Theorem 8.3.27 *Let the RK method (5.0.2), (5.0.3) be algebraically stable and satisfy (8.3.1). Then it is asymptotically AN-stable.*

Theorem 8.3.28 *Let the RK method (5.0.2), (5.0.3) be A-stable and satisfy (8.3.1). Then it is asymptotically A-stable.*

By also using Theorem 8.3.10, we get the analogous result for autonomous linear systems (8.2.11).

Theorem 8.3.29 *Let the RK method (5.0.2), (5.0.3) be A-stable and satisfy (8.3.1). Then the numerical solution* $\{y_n\}_{n\geq 0}$ *of (8.2.11) satisfies* $\lim_{n\to\infty} y_n = 0$ *for any mesh* Δ *whenever* $\lambda_0 < 0$ *in (8.2.12).*

8.4 Stability of interpolants of RK methods

In this section we recall what, to our knowledge, was the first concept of stability for interpolants of RK methods. It was introduced by Bellen and Zennaro [37] by considering the linear test equation (8.2.1), that is

$$\begin{cases} y'(t) = \lambda y(t), & t \geq t_0, \\ y(t_0) = y_0, \end{cases} \tag{8.4.1}$$

where $\lambda \in \mathbb{C}$.

Definition 8.4.1 *The interpolant (5.1.3) is* stable *with respect to the (discrete) RK method (5.0.2), (5.0.3) if there exists a constant $M > 0$ such that the numerical solution $\eta(t)$ of (8.4.1) satisfies*

$$\max_{0\leq\theta\leq 1} |\eta(t_n + \theta h_{n+1})| \leq M \max\{|y_n|, |y_{n+1}|\},$$

for any mesh Δ, whenever $\Re(\lambda) \leq 0$.

This kind of analysis aims to determine those interpolants that do not qualitatively deteriorate the A-stability properties of the discrete RK method they extend.

Observe that, if the (discrete) RK method (5.0.2), (5.0.3) is A-stable, then a stable interpolant of it is such that

$$\max_{t\geq t_0} |\eta(t)| \leq M|y_0|$$

for any mesh Δ.

8.4.1 *Characterization and existence of stable interpolants*

It is possible to give a precise characterization of stable interpolants, as was done in Bellen and Zennaro [37]. Although the idea behind it is simple, in [37] a somewhat complicated formalism has been introduced in order to be able to present the theory fluently and to carry out the proofs of the results. Anyway, it goes beyond the scope of this book and, therefore, we avoid the formalism in [37].

We begin by observing that, as the nodal value y_{n+1} is given by

$$y_{n+1} = R(h_{n+1}\lambda)y_n,$$

where

$$R(\alpha) = 1 + \alpha b^T(I - \alpha A)^{-1}e$$

is the A-stability function (8.2.7), and the interpolant $\eta(t_n + \theta h_{n+1})$ is given by

$$\eta(t_n + \theta h_{n+1}) = R^{(\theta)}(h_{n+1}\lambda)y_n,$$

where

$$R^{(\theta)}(\alpha) = 1 + \alpha b(\theta)^T(I - \alpha A)^{-1}e \tag{8.4.2}$$

is a rational function of α for each value of $\theta \in [0, 1]$. Observe that

$$R^{(1)}(\alpha) = R(\alpha).$$

As is usually done, we say that, for a rational function R, the *degree* $\deg(R)$ is equal to the difference in the degrees of its numerator and denominator.

Proposition 8.4.2 *The interpolant (5.1.3) is stable with respect to the (discrete) RK method (5.0.2), (5.0.3) if and only if the following two conditions hold:*

(i) $\max_{0\le\theta\le 1} \deg\left(R^{(\theta)} - 1\right) \le \max\{0, \deg(R-1)\}$;

(ii) *for any $\theta \in (0,1)$, each pole z of $R^{(\theta)}$ with multiplicity μ such that $\Re(z) \le 0$, if any, is also a pole of R with multiplicity $\mu' \ge \mu$.*

For all RK methods, the existence of stable interpolants is guaranteed. It is easily seen that the linear interpolant is always stable. Of course, for methods of order $p \ge 3$, the linear interpolant is not sufficient for accuracy reasons. In [37] the following result is proved.

Theorem 8.4.3 *The NCE defined by formulae (5.2.8) and (5.2.9) is stable with respect to the (discrete) RK method (5.0.2), (5.0.3) of order $p \ge 3$.*

Finally, we remark that the procedure for implicit uniform corrections illustrated in Section 5.3.3 can be suitably used in order to produce higher order stable NCEs of an RK method (5.0.2), (5.0.3) starting, for example, from the stable NCE defined by formulae (5.2.8) and (5.2.9). However, the actual procedure is rather technical and, therefore, we refer the interested reader to [37].

We conclude this section by an illustrative example, taken from [37], showing that conditions (i) and (ii) in Proposition 8.4.2 are both necessary for the stability of the interpolant.

Example 8.4.4 Let us consider the system of linear equations

$$\begin{cases} y_1'(t) = -3y_1(t) - 2y_2(t) + f_1(t), \\ y_2'(t) = \frac{3601}{1800}y_1(t) + y_2(t) + f_2(t), \\ y_3'(t) = -y_1(t) - y_2(t) - 2y_3(t) + \frac{101}{20}y_4(t) + f_3(t), \\ y_4'(t) = -y_1(t) + y_2(t) - \frac{1}{5}y_3(t) - 4y_4(t) + f_4(t), \end{cases} \tag{8.4.3}$$

with initial condition

$$[y_1(0), y_2(0), y_3(0), y_4(0)]^T = [1, 0, 1, 0]^T,$$

where

$$\begin{aligned} f_1(t) &= \tfrac{9}{10}e^{-\frac{t}{10}} + 2e^{-\frac{t}{20}}, \\ f_2(t) &= -\tfrac{1621}{1800}e^{-\frac{t}{10}} - \tfrac{21}{20}e^{-\frac{t}{20}}, \\ f_3(t) &= e^{-\frac{t}{20}} + 2, \\ f_4(t) &= 2e^{-\frac{t}{10}} - e^{-\frac{t}{20}} + \tfrac{1}{5}, \end{aligned}$$

the solution of which is

$$\begin{aligned} y_1(t) &= e^{-\frac{t}{10}}, \\ y_2(t) &= -e^{-\frac{t}{10}} + e^{-\frac{t}{20}}, \\ y_3(t) &= 1, \\ y_4(t) &= 0. \end{aligned}$$

The coefficients of the system (8.4.3) are constant and form a matrix whose eigenvalues are $\lambda_{1,2} = -1 \pm \frac{1}{30}i$ and $\lambda_{3,4} = -3 \pm \frac{1}{10}i$. Therefore, the solution is asymptotically stable and any A-stable RK method should provide reasonable approximations even for large stepsizes. Let us consider, in particular, the trapezoidal rule

$$\begin{array}{c|cc} 0 & 0 & 0 \\ 1 & \frac{1}{2} & \frac{1}{2} \\ \hline & \frac{1}{2} & \frac{1}{2} \end{array} \tag{8.4.4}$$

and the two-stage RK method

$$\begin{array}{c|cc} \frac{1}{3} & \frac{1}{12} & \frac{1}{4} \\ \frac{7}{9} & \frac{25}{36} & \frac{1}{12} \\ \hline & \frac{5}{8} & \frac{3}{8} \end{array} \tag{8.4.5}$$

which are both of order 2, are A-stable and have the same A-stability function

$$R(\alpha) = \frac{1+\alpha/2}{1-\alpha/2}.$$

Each method possesses an NCE $\eta(t)$ of order 2, which is given by the variable weights

$$\begin{aligned} b_1(\theta) &= -\tfrac{1}{2}\theta(\theta-2), \\ b_2(\theta) &= \tfrac{1}{2}\theta^2, \end{aligned} \tag{8.4.6}$$

for (8.4.4) and by

$$\begin{aligned} b_1(\theta) &= -\tfrac{1}{8}\theta(9\theta-14), \\ b_2(\theta) &= \tfrac{3}{8}\theta(3\theta-2), \end{aligned} \tag{8.4.7}$$

for (8.4.5).

The resulting rational functions (8.4.2) are

$$R^{(\theta)}(\alpha) = \frac{1+(\theta-\frac{1}{2})\alpha+\frac{1}{2}(\theta^2-\theta)\alpha^2}{1-\frac{1}{2}\alpha} \tag{8.4.8}$$

for (8.4.4) and

$$R^{(\theta)}(\alpha) = \frac{1+(\theta-\frac{1}{6})\alpha+\frac{1}{2}(\theta^2-\frac{1}{3}\theta-\frac{1}{3})\alpha^2}{(1-\frac{1}{2}\alpha)(1+\frac{1}{3}\alpha)} \tag{8.4.9}$$

for (8.4.5).

Now observe that, in both cases, $R^{(\theta)}$ does not satisfy at least one of the conditions (i) and (ii) of Proposition 8.4.2. In fact, R has only a simple pole at $\alpha = 2$ and is such that $\deg(R-1) = 0$. For (8.4.8) we have $\max_{0\le\theta\le1} \deg\big(R^{(\theta)} - 1\big) = 1$, which violates (i). On the contrary, (8.4.9) fulfills condition (i), but has a pole with negative real part at $\alpha = -3$, and this violates (ii).

Tables 8.1 and 8.2 give the discrete error $e = \|y(h) - \eta(h)\|$, the maximum uniform error $E = \max_{0\le\theta\le1} \|y(\theta h) - \eta(\theta h)\|$ and the ratio E/e for the numerical

TABLE 8.1. Numerical results for equation (8.4.3) using the CRK method (8.4.4), (8.4.6).

h	$e = \|y(h) - \eta(h)\|$	$E = \max_{0\le\theta\le 1} \|y(\theta h) - \eta(\theta h)\|$	E/e
0.05	$1.3 \cdot 10^{-8}$	$1.3 \cdot 10^{-8}$	$1.0 \cdot 10^{0}$
0.1	$1.0 \cdot 10^{-7}$	$1.0 \cdot 10^{-7}$	$1.0 \cdot 10^{0}$
0.2	$17.8 \cdot 10^{-7}$	$7.8 \cdot 10^{-7}$	$1.0 \cdot 10^{0}$
$\vdots$	$\vdots$	$\vdots$	$\vdots$
1	$6.8 \cdot 10^{-5}$	$7.2 \cdot 10^{-5}$	$1.1 \cdot 10^{0}$
10	$1.0 \cdot 10^{-2}$	$2.6 \cdot 10^{-2}$	$2.6 \cdot 10^{0}$
100	$2.9 \cdot 10^{-2}$	$2.1 \cdot 10^{0}$	$7.2 \cdot 10^{1}$
1000	$4.7 \cdot 10^{-2}$	$2.7 \cdot 10^{1}$	$5.7 \cdot 10^{2}$
10000	$5.1 \cdot 10^{-2}$	$2.8 \cdot 10^{2}$	$5.5 \cdot 10^{3}$

approximation of (8.4.3) using the two considered CRK methods with increasing values of h. In Table 8.1, relevant to the CRK method (8.4.4), (8.4.6), violation of (i) results in a fast growth of E/e at infinity. In Table 8.2, relevant to the CRK method (8.4.5), (8.4.7), the ratio E/e is asymptotically stationary and violation of (ii) results in two abnormal peaks of the ratio around the values $h = 1$ and $h = 3$. Since the matrix has eigenvalues $\lambda_{1,2} = -1 \pm \frac{1}{30}i$ and $\lambda_{3,4} = -3 \pm \frac{1}{10}i$, some of the products $\alpha = h\lambda_i$ become very close to -3 for $h = 1$ and $h = 3$. This explains the instability around these values of the stepsize h. $\diamond$

8.5 Stability concepts with respect to forcing terms

In this section we analyze the influence of the *forcing term* $u(t)$ on the exact and numerical solutions of an ODE such as

$$\begin{cases} y'(t) = f(t, y(t), u(t)), & t \ge t_0, \\ y(t_0) = y_0, \end{cases} \tag{8.5.1}$$

where $f \in C^0\left([t_0, +\infty) \times \mathbb{C}^d \times \mathbb{C}^d, \mathbb{C}^d\right)$. This analysis will allow us to determine some sufficient conditions for the contractivity and asymptotic stability of the solutions of DDEs and their numerical approximation as well. To this end, similarly as we did in Section 8.2.3, we consider another ODE

$$\begin{cases} \tilde{y}'(t) = f(t, \tilde{y}(t), \tilde{u}(t)), & t \ge t_0, \\ \tilde{y}(t_0) = \tilde{y}_0, \end{cases} \tag{8.5.2}$$

and define

$$\delta y(t) = \tilde{y}(t) - y(t),$$

TABLE 8.2. Numerical results for equation (8.4.3) using the CRK method (8.4.5), (8.4.7).

h	$e = \|y(h) - \eta(h)\|$	$E = \max_{0\le\theta\le 1} \|y(\theta h) - \eta(\theta h)\|$	E/e
0.05	$3.1 \cdot 10^{-7}$	$3.6 \cdot 10^{-7}$	$1.2 \cdot 10^{0}$
0.1	$2.3 \cdot 10^{-6}$	$2.9 \cdot 10^{-6}$	$1.3 \cdot 10^{0}$
0.2	$1.7 \cdot 10^{-5}$	$2.4 \cdot 10^{-5}$	$1.4 \cdot 10^{0}$
$\vdots$	$\vdots$	$\vdots$	$\vdots$
0.9	$9.5 \cdot 10^{-4}$	$2.0 \cdot 10^{-1}$	$2.1 \cdot 10^{2}$
1.0	$1.2 \cdot 10^{-3}$	$3.3 \cdot 10^{0}$	$2.8 \cdot 10^{3}$
1.1	$1.5 \cdot 10^{-3}$	$5.0 \cdot 10^{-1}$	$3.3 \cdot 10^{2}$
$\vdots$	$\vdots$	$\vdots$	$\vdots$
2.0	$6.2 \cdot 10^{-3}$	$1.9 \cdot 10^{-1}$	$3.1 \cdot 10^{1}$
$\vdots$	$\vdots$	$\vdots$	$\vdots$
2.9	$1.4 \cdot 10^{-2}$	$1.1 \cdot 10^{1}$	$7.9 \cdot 10^{2}$
3.0	$1.5 \cdot 10^{-2}$	$2.1 \cdot 10^{1}$	$1.4 \cdot 10^{3}$
3.1	$1.6 \cdot 10^{-2}$	$1.0 \cdot 10^{1}$	$6.3 \cdot 10^{2}$
$\vdots$	$\vdots$	$\vdots$	$\vdots$
4	$2.7 \cdot 10^{-2}$	$2.3 \cdot 10^{-1}$	$8.6 \cdot 10^{0}$
5	$4.2 \cdot 10^{-2}$	$4.9 \cdot 10^{-2}$	$1.2 \cdot 10^{0}$
10	$1.4 \cdot 10^{-1}$	$1.4 \cdot 10^{-1}$	$1.0 \cdot 10^{0}$
100	$9.9 \cdot 10^{-1}$	$9.9 \cdot 10^{-1}$	$1.0 \cdot 10^{0}$
1000	$1.0 \cdot 10^{0}$	$1.1 \cdot 10^{0}$	$1.1 \cdot 10^{0}$
10000	$1.0 \cdot 10^{0}$	$1.1 \cdot 10^{0}$	$1.1 \cdot 10^{0}$

$$\delta u(t) = \tilde{u}(t) - u(t),$$
$$\delta y_0 = \tilde{y}_0 - y_0.$$

8.5.1 *The behavior of the test equations*

The following result has been given by Zennaro [284], but the proof was omitted. So, we give it here.

Proposition 8.5.1 *Consider the systems of ODEs (8.5.1) and (8.5.2), where $u(t)$ and $\tilde{u}(t)$ are continuous functions. Given an inner product $\langle \cdot, \cdot \rangle$ in $\mathbb{C}^d$ and the corresponding norm $\|\cdot\|$, let $Y(t)$ and $X(t)$ be continuous functions such that*

$$Y(t) \geq \sup_{x, y_1 \neq y_2} \frac{\Re(\langle f(t, y_1, x) - f(t, y_2, x), y_1 - y_2 \rangle)}{\|y_1 - y_2\|^2} \tag{8.5.3}$$

and

$$X(t) \geq \sup_{y, x_1 \neq x_2} \frac{\|(f(t, y, x_1) - f(t, y, x_2)\|}{\|x_1 - x_2\|}. \tag{8.5.4}$$

If

$$Y(t) \leq 0, \quad t \geq t_0, \tag{8.5.5}$$

and if there exists a bounded function $r(t) < 0$, integrable in any bounded interval, such that

$$X(t) = r(t)Y(t), \quad t \geq t_0, \tag{8.5.6}$$

then the difference $\delta y(t)$ in the solutions satisfies

$$\|\delta y(t)\| \leq E(t_0, t)\|\delta y_0\| + (1 - E(t_0, t)) \sup_{t_0 \leq x \leq t} (|r(x)| \cdot \|\delta u(x)\|), \quad t \geq t_0, \tag{8.5.7}$$

where $E(t_1, t_2) = \exp\left(\int_{t_1}^{t_2} Y(x)\mathrm{d}x\right)$.

For the proof, we need the following lemma.

Lemma 8.5.2 *Let $\delta(t) \geq 0$ be a C^1-continuous real function such that*

$$\begin{cases} \delta'(t) \leq -a(t)\big(\delta(t) - \gamma(t)\big), & t^* \leq t \leq t^* + H, \\ \delta(t^*) \leq \delta^*, \end{cases}$$

where $\gamma(t) \geq 0$ is integrable and bounded and $a(t) \geq 0$ is continuous. Then it holds that

$$\delta(t) \leq \mathrm{e}^{-A(t)}\delta^* + (1 - \mathrm{e}^{-A(t)}) \sup_{t^* \leq x \leq t} \gamma(x),$$

for all $t \in [t^*, t^* + H]$, where $A(t) = \int_{t^*}^{t} a(x)\mathrm{d}x$.

Proof. It is clear that there exists an integrable and bounded function $\rho(t) \geq 0$ such that $\delta'(t) = -a(t)\big(\delta(t) - \gamma(t)\big) - \rho(t)$ for $t^* \leq t \leq t^* + H$. Therefore, we get

$$\begin{aligned} \delta(t) &= \mathrm{e}^{-A(t)}\left(\delta(t^*) + \int_{t^*}^{t} \mathrm{e}^{A(x)}(a(x)\gamma(x) - \rho(x))\mathrm{d}x\right) \\ &\leq \mathrm{e}^{-A(t)}\left(\delta^* + \int_{t^*}^{t} \mathrm{e}^{A(x)}a(x)\gamma(x)\mathrm{d}x\right) \\ &\leq \mathrm{e}^{-A(t)}\delta^* + (1 - \mathrm{e}^{-A(t)}) \sup_{t^* \leq x \leq t} \gamma(x). \quad \blacksquare \end{aligned}$$

Proof of Proposition 8.5.1. First of all, observe that (8.5.5) implies

$$0 < E(t_1, t_2) \leq 1 \quad \text{whenever } t_2 \geq t_1 \geq t_0. \tag{8.5.8}$$

Define $\delta(t) = \|\delta y(t)\|$, choose a positive real number $K > \|\delta y_0\|$ and consider the function

$$M(t) = E(t_0, t)K + (1 - E(t_0, t)) \sup_{t_0 \leq x \leq t} (|r(x)| \cdot \|u(x)\|). \tag{8.5.9}$$

Now assume, by contradiction, that there exists $\bar{t} > t_0$ such that $\delta(\bar{t}) > M(\bar{t})$. Then there exists $\tilde{t} = \inf\{t \geq t_0 \mid \delta(t) > M(t)\}$. The function $M(t)$ may not be continuous at $\tilde{t}$. Thus we can only say that $\delta(t) \leq M(t)$ for all $t \in [t_0, \tilde{t})$, whereas nothing can be said about the sign of the difference $\delta(\tilde{t}) - M(\tilde{t})$. On the other hand, (8.5.8) implies $M(t) \geq E(t_0, t)K$ for all $t \geq t_0$ and, thus, since $\delta(t)$ and $E(t_0, t)$ are continuous functions, it follows that $\delta(\tilde{t}) \geq E(t_0, \tilde{t})K > 0$. Moreover, $\delta(t_0) = \|\delta y_0\| < K = M(t_0)$ and, therefore, again by the continuity of $\delta(t)$, there exists an interval $[t^*, t^* + H]$, with $H > 0$, $t_0 \leq t^* \leq \tilde{t} < t^* + H$ and $\delta(t^*) \leq M(t^*)$, where $\delta(t) > 0$.

Now find the difference between (8.5.2) and (8.5.1), subtract and add $f(t, y(t), \tilde{u}(t))$ to the right-hand side and take the real part of the scalar product by $\delta y(t)$, so as to get

$$\begin{aligned}\Re(\langle \delta y'(t), \delta y(t)\rangle) &= \Re(\langle f(t, \tilde{y}(t), \tilde{u}(t)) - f(t, y(t), \tilde{u}(t)), \delta y(t)\rangle) \\ &\quad + \Re(\langle f(t, y(t), \tilde{u}(t)) - f(t, y(t), u(t)), \delta y(t)\rangle).\end{aligned}$$

Since $\delta'(t) = \frac{\Re(\langle \delta y'(t), \delta y(t)\rangle)}{\delta(t)}$ whenever $\delta(t) > 0$, by (8.5.3), (8.5.4) and (8.5.6) we get

$$\begin{cases} \delta'(t) \leq Y(t)\big(\delta(t) + r(t)\|\delta u(t)\|\big), & t^* \leq t \leq t^* + H, \\ \delta(t^*) \leq M(t^*). \end{cases}$$

Since $Y(t) \leq 0$ and $r(t) \leq 0$, with $a(t) = -Y(t)$, $\gamma(t) = -r(t)\|\delta u(t)\|$ and $\delta^* = M(t^*)$, Lemma 8.5.2 yields

$$\delta(t) \leq E(t^*, t)M(t^*) + (1 - E(t^*, t)) \sup_{t^* \leq x \leq t} (|r(x)| \cdot \|\delta u(x)\|) \tag{8.5.10}$$

for all $t \in [t^*, t^* + H]$. On the other hand, by substituting the actual value of $M(t^*)$ from (8.5.9), the inequality (8.5.8) and the equality $E(t_0, t^*)E(t^*, t) = E(t_0, t)$ immediately imply that the right-hand side of (8.5.10) is bounded by $M(t)$, yielding $\delta(t) \leq M(t)$ for all $t \leq t^* + H$, which contradicts the definition of $\tilde{t}$. Therefore, we can conclude that $\delta(t) \leq M(t)$ for all $t \geq t_0$.

Finally, since the result holds for any $K > \|\delta y_0\|$, the proof is complete. ■

The result of Proposition 8.5.1 can be generalized to any norm as follows.

Proposition 8.5.3 *Proposition 8.5.1 holds for an arbitrary vector norm $\|\cdot\|$ in $\mathbb{C}^d$ where, with respect to the corresponding logarithmic norm, condition (8.5.3) on the continuous function $Y(t)$ is substituted by*

$$Y(t) \geq \sup_{y, u \in \mathbb{C}^d} \mu[f_y(t, y, u)]. \tag{8.5.11}$$

The proof is based on Theorem 3.1 in Bellen [23], and Lemma 8.5.2. The next corollary to Propositions 8.5.1 and 8.5.3 has been also proved directly by Torelli [247] for norms derived by inner products.

Corollary 8.5.4 *Under the hypotheses of Proposition 8.5.1 or 8.5.3, the difference $\delta y(t)$ in the solutions of (8.5.2) and (8.5.1) satisfies*

$$\|\delta y(t)\| \leq \max\left\{\|\delta y_0\|, \sup_{t_0 \leq x \leq t} (|r(x)| \cdot \|\delta u(x)\|)\right\}, \quad t \geq t_0, \tag{8.5.12}$$

in a suitable norm.

It is useful to rephrase Proposition 8.5.1 and Corollary 8.5.4 in the particular cases of linear equations.

Proposition 8.5.5 *Consider the scalar ODE with the forcing term*

$$\begin{cases} y'(t) = \lambda(t)y(t) + \Re(\lambda(t))g(t), & t \geq t_0, \\ y(t_0) = y_0, \end{cases} \tag{8.5.13}$$

where $\lambda(t)$ and $g(t)$ are continuous complex valued functions. If $\Re(\lambda(t)) \leq 0$, $t \geq t_0$, then

$$|y(t)| \leq \Lambda(t_0, t)|y_0| + (1 - \Lambda(t_0, t)) \max_{t_0 \leq x \leq t} |g(x)|, \quad t \geq t_0, \tag{8.5.14}$$

where $\Lambda(t_1, t_2) = \exp\left(\int_{t_1}^{t_2} \Re(\lambda(x))\mathrm{d}x\right)$.

Corollary 8.5.6 *Under the hypotheses of Proposition 8.5.5, the solution $y(t)$ of (8.5.13) satisfies*

$$|y(t)| \leq \max\left\{|y_0|, \max_{t_0 \leq x \leq t} |g(x)|\right\}, \quad t \geq t_0. \tag{8.5.15}$$

Proposition 8.5.7 *Consider the scalar ODE with the forcing term*

$$\begin{cases} y'(t) = \lambda y(t) + \Re(\lambda)g(t), & t \geq t_0, \\ y(t_0) = y_0, \end{cases} \tag{8.5.16}$$

where $\lambda \in \mathbb{C}$ and $g(t)$ is a continuous complex valued function. If $\Re(\lambda) \leq 0$, then

$$|y(t)| \leq \mathrm{e}^{\Re(\lambda)(t-t_0)}|y_0| + \left(1 - \mathrm{e}^{\Re(\lambda)(t-t_0)}\right) \max_{t_0 \leq x \leq t} |g(x)|, \quad t \geq t_0. \tag{8.5.17}$$

Corollary 8.5.8 *Under the hypotheses of Proposition 8.5.7, the solution $y(t)$ of (8.5.16) satisfies (8.5.15).*

8.5.2 *Stability definitions for RK methods*

Stronger concepts of stability for RK methods, based on the test equations (8.5.16), (8.5.13) and (8.5.1) with forcing terms, has been introduced by Torelli [247, 248], Bellen and Zennaro [38] and Zennaro [282].

Definition 8.5.9 *The RK method (5.0.2), (5.0.3) is* $\mathrm{A_f}$-stable *if the numerical solution $\{y_n\}_{n\geq 0}$ of (8.5.16) satisfies*

$$|y_{n+1}| \leq \max\left\{|y_n|, \max_{1 \leq i \leq \nu} |g(t_{n+1}^i)|\right\} \tag{8.5.18}$$

for any mesh Δ, whenever $\Re(\lambda) \leq 0$.

Definition 8.5.10 *The CRK method (5.0.2), (5.1.6), (5.1.3) is* semi-A$_\mathrm{f}$-stable *if the numerical solution $\eta(t)$ of (8.5.16) satisfies*

$$|\eta(t_n + \theta h_{n+1})| \leq \max\left\{|y_n|, \max_{1\leq i\leq s} |g(t^i_{n+1})|\right\} \tag{8.5.19}$$

for $\theta = c_1, \ldots, c_s, 1$ and for any mesh Δ, whenever $\Re(\lambda) \leq 0$.

The method is A$_\mathrm{f}$-stable *if (8.5.19) holds for all $\theta \in [0, 1]$.*

Definition 8.5.11 *The RK method (5.0.2), (5.0.3) is* AN$_\mathrm{f}$-stable *if the numerical solution $\{y_n\}_{n\geq 0}$ of (8.5.13) satisfies (8.5.18) for any mesh Δ, whenever $\Re(\lambda(t)) \leq 0$, $t \geq t_0$.*

Definition 8.5.12 *The CRK method (5.0.2), (5.1.6), (5.1.3) is* semi-AN$_\mathrm{f}$-stable *if the numerical solution $\eta(t)$ of (8.5.13) satisfies (8.5.19) for $\theta = c_1, \ldots, c_s, 1$ and for any mesh Δ, whenever $\Re(\lambda(t)) \leq 0$, $t \geq t_0$.*

The method is AN$_\mathrm{f}$-stable *if (8.5.19) holds for all $\theta \in [0, 1]$.*

Remark 8.5.13 *Conditions (8.5.18) and (8.5.19) for the numerical method are apparently stronger than (8.5.15) for the exact solutions of (8.5.16) and (8.5.13). In fact, for the most common RK methods the abscissae $c_i \in [0, 1], i = 1, \ldots, s$, and therefore*

$$\max_{1\leq i\leq s} |g(t^i_{n+1})| \leq \max_{t_n\leq x\leq t_{n+1}} |g(x)|. \tag{8.5.20}$$

On the other hand, the numerical solution $\eta(t_n + \theta h_{n+1})$ depends only on the values of the forcing term $g(t)$ at the points $t^i_{n+1}, i = 1, \ldots, s$. Since $g(t)$ is arbitrary, it might be modified outside the set of such points without changing the numerical results, so that it is not restrictive to assume that (8.5.20) is, indeed, an equality.

In conclusion, the property required for the numerical solution is not more restrictive than the one satisfied by the exact solution.

It is clear that requiring A$_\mathrm{f}$-stability or AN$_\mathrm{f}$-stability is more than requiring A-stability or AN-stability, respectively. In fact, A-stability and AN-stability are obtained as particular cases when the forcing term $g(t)$ is identically zero in the test equations (8.5.16) and (8.5.13), respectively.

Definition 8.5.14 *The RK method (5.0.2), (5.0.3) is* BN$_\mathrm{f}$-stable *if, under conditions (8.5.5) and (8.5.6), the difference $\{\delta y_n\}_{n\geq 0}$ in the numerical solutions of (8.5.2) and (8.5.1) satisfies*

$$\|\delta y_{n+1}\| \leq \max\left\{\|\delta y_n\|, \max_{1\leq i\leq \nu} (|r(t^i_{n+1})| \cdot \|\delta u(t^i_{n+1})\|)\right\}$$

for any mesh Δ.

Definition 8.5.15 *The CRK method (5.0.2), (5.1.6), (5.1.3) is* semi-BN$_f$-stable *if, under conditions (8.5.5) and (8.5.6), the difference $\delta\eta(t)$ in the numerical solutions of (8.5.2) and (8.5.1) satisfies*

$$\|\delta\eta(t_n + \theta h_{n+1})\| \le \max\left\{\|\delta y_n\|, \max_{1\le i\le s}(|r(t^i_{n+1})| \cdot \|\delta u(t^i_{n+1})\|)\right\} \qquad (8.5.21)$$

for $\theta = c_1, \ldots, c_s, 1$, for any mesh Δ.

The method is BN$_f$-stable *if (8.5.21) holds for all $\theta \in [0,1]$.*

Remark 8.5.16 *Similar considerations may be assumed as in Remark 8.5.13, so that, without loss of generality, we can assume that*

$$\max_{1\le i\le s}(|r(t^i_{n+1})| \cdot \|\delta u(t^i_{n+1})\|) = \sup_{t_n\le x\le t_{n+1}}(|r(x)| \cdot \|\delta u(x)\|).$$

8.5.3 *Analysis of RK methods*

We now give the necessary and sufficient conditions on the coefficients of the underlying CRK method for its A$_f$-stability and AN$_f$-stability. The application of formulae (5.0.2), (5.1.6), (5.1.3) to the test equation with forcing term (8.5.16) yields

$$Y^i_{n+1} = y_n + h_{n+1}\sum_{j=1}^{s} a_{ij}\big(\lambda Y^j_{n+1} + \Re(\lambda)g(t^j_{n+1})\big), \quad i = 1, \ldots, s,$$

$$\eta(t_n + \theta h_{n+1}) = y_n + h_{n+1}\sum_{i=1}^{s} b_i(\theta)\big(\lambda Y^i_{n+1} + \Re(\lambda)g(t^i_{n+1})\big).$$

Therefore, we get

$$\eta(t_n + \theta h_{n+1}) = R^{(\theta)}(\alpha)y_n + \sum_{i=1}^{s}\Re(\alpha)\big(b(\theta)^T(I - \alpha A)^{-1}\big)_i g(t^i_{n+1}), \qquad (8.5.22)$$

where $R^{(\theta)}(\alpha)$ is given by (8.4.2), $\alpha = h_{n+1}\lambda$ and $(x)_i$ stands for the ith component of the s vector x.

Observe that, for $\theta = 1$, (8.5.22) reduces to

$$y_{n+1} = R(\alpha)y_n + \sum_{i=1}^{\nu}\Re(\alpha)(b^T(I - \alpha A)^{-1})_i g(t^i_{n+1}),$$

where $R(\alpha)$ is the A-stability function (8.2.7).

In order to treat properly the case of confluent RK methods, we give the following definition.

Definition 8.5.17 *The* confluency partition *of the index set of an RK method (5.0.2), (5.0.3) is the partition $\{I_1, \ldots, I_\rho\}$, $\rho \le \nu$, of the index set $\{1, \ldots, \nu\}$ such that $\{i \in I_h$ and $j \in I_k$* for *$h \ne k\} \Longrightarrow c_i \ne c_j$ and $i, j \in I_k \Longrightarrow c_i = c_j$.*

Note that, whenever the method is non-confluent, the confluency partition is just $\{\{1\},\ldots,\{\nu\}\}$.

Now it is easy to prove the following result.

Proposition 8.5.18 *The discrete RK method (5.0.2), (5.0.3) is A_f-stable if and only if*

$$|R(\alpha)| + \sum_{k=1}^{\rho} \left| \sum_{i_k \in I_k} \Re(\alpha)\big(b^T(I-\alpha A)^{-1}\big)_{i_k} \right| \leq 1 \quad \forall \Re(\alpha) \leq 0, \tag{8.5.23}$$

where $\{I_1,\ldots,I_\rho\}$ is the confluency partition of the index set $\{1,\ldots,\nu\}$.

In particular, if the RK method is non-confluent, then condition (8.5.23) reduces to

$$|R(\alpha)| + \|\Re(\alpha)b^T(I-\alpha A)^{-1}\|_1 \leq 1 \quad \forall \Re(\alpha) \leq 0, \tag{8.5.24}$$

where $\|x\|_1 = \sum_{i=1}^{\nu} |x_i|$ for any ν-dimensional row vector x.

Proposition 8.5.19 *Consider a CRK method (5.0.2), (5.1.6), (5.1.3). Then we have:*

- *The method is semi-A_f-stable if and only if*

$$|R^{(\theta)}(\alpha)| + \sum_{k=1}^{r} \left| \sum_{i_k \in I_k} \Re(\alpha)\big(b(\theta)^T(I-\alpha A)^{-1}\big)_{i_k} \right| \leq 1 \quad \forall \Re(\alpha) \leq 0 \tag{8.5.25}$$

  holds for $\theta = c_1,\ldots,c_s,1$, where $\{I_1,\ldots,I_r\}$ is the confluency partition of the index set $\{1,\ldots,s\}$.
- *The method is A_f-stable if and only if (8.5.25) holds for all $\theta \in [0,1]$.*

In particular, if the CRK method is non-confluent, then condition (8.5.25) reduces to

$$|R^{(\theta)}(\alpha)| + \|\Re(\alpha)b(\theta)^T(I-\alpha A)^{-1}\|_1 \leq 1 \quad \forall \Re(\alpha) \leq 0. \tag{8.5.26}$$

The application of formulae (5.0.2), (5.1.6), (5.1.3) to the test equation with forcing term (8.5.13) yields instead

$$Y_{n+1}^i = y_n + h_{n+1}\sum_{j=1}^{s} a_{ij}\big(\lambda(t_{n+1}^j)Y_{n+1}^j + \Re(\lambda(t_{n+1}^j))g(t_{n+1}^j)\big), \quad i=1,\ldots,s,$$

$$\eta(t_n+\theta h_{n+1}) = y_n + h_{n+1}\sum_{i=1}^{s} b_i(\theta)(\lambda(t_{n+1}^i)Y_{n+1}^i + \Re(\lambda(t_{n+1}^i))g(t_{n+1}^i)).$$

Therefore, we get

$$\eta(t_n+\theta h_{n+1}) = (1+b(\theta)^T(I-ZA)^{-1}Ze)y_n + \sum_{i=1}^{s}\big(b(\theta)^T(I-ZA)^{-1}\Re(Z)\big)_i g(t_{n+1}^i), \tag{8.5.27}$$

where $Z = (\alpha_1,\ldots,\alpha_s)$, $\alpha_i = h_{n+1}\lambda(t_{n+1}^i)$, $i=1,\ldots,s$.

Observe that, for $\theta = 1$, (8.5.27) reduces to

$$y_{n+1} = K(\alpha_1, \ldots, \alpha_\nu) y_n + \sum_{i=1}^{\nu} \left(b^T (I - ZA)^{-1} \Re(Z)\right)_i g(t^i_{n+1}),$$

where $K(\alpha_1, \ldots, \alpha_\nu)$ is the AN-stability function (8.2.21).

Now it is easy to prove the following result.

Proposition 8.5.20 *The discrete RK method (5.0.2), (5.0.3) is AN_f-stable if and only if*

$$|1 + b^T (I - ZA)^{-1} Ze| + \sum_{k=1}^{\rho} \left| \sum_{i_k \in I_k} \left(b^T (I - ZA)^{-1} \Re(Z)\right)_{i_k} \right| \leq 1 \quad \forall \Re(\alpha_i) \leq 0 \tag{8.5.28}$$

under the restriction that $i, j \in I_k \Longrightarrow \alpha_i = \alpha_j$, where $\{I_1, \ldots, I_\rho\}$ is the confluency partition of the index set $\{1, \ldots, \nu\}$.

In particular, if the RK method is non-confluent, then condition (8.5.28) reduces to

$$|1 + b^T (I - ZA)^{-1} Ze| + \|b^T (I - ZA)^{-1} \Re(Z)\|_1 \leq 1 \quad \forall \Re(\alpha_i) \leq 0. \tag{8.5.29}$$

Proposition 8.5.21 *Consider a CRK method (5.0.2), (5.1.6), (5.1.3). Then we have:*

- *The method is semi-AN_f-stable if and only if*

$$|1 + b(\theta)^T (I - ZA)^{-1} Ze| + \sum_{k=1}^{r} \left| \sum_{i_k \in I_k} \left(b(\theta)^T (I - ZA)^{-1} \Re(Z)\right)_{i_k} \right| \leq 1$$
$$\forall \Re(\alpha_i) \leq 0 \tag{8.5.30}$$

  holds for $\theta = c_1, \ldots, c_s, 1$, under the restriction that $i, j \in I_k \Longrightarrow \alpha_i = \alpha_j$, where $\{I_1, \ldots, I_r\}$ is the confluency partition of the index set $\{1, \ldots, s\}$.
- *The method is AN_f-stable if and only if (8.5.30) holds for all $\theta \in [0, 1]$.*

In particular, if the CRK method is non-confluent, then condition (8.5.30) reduces to

$$|1 + b(\theta)^T (I - ZA)^{-1} Ze| + \|b(\theta)^T (I - ZA)^{-1} \Re(Z)\|_1 \leq 1 \quad \forall \Re(\alpha_i) \leq 0. \tag{8.5.31}$$

Sometimes it is more convenient to consider conditions (8.5.23) and (8.5.28) in the equivalent forms

$$|1 + b^T (I - \alpha A)^{-1} (\alpha I + W) e| \leq 1 \quad \forall \Re(\alpha) \leq -|\beta_i| \tag{8.5.32}$$

under the restriction that $i, j \in I_k \Longrightarrow \beta_i = \beta_j$, where $\{I_1, \ldots, I_\rho\}$ is the confluency partition of the index set $\{1, \ldots, \nu\}$ and $W = \mathrm{diag}(\beta_1, \ldots, \beta_\nu)$ denotes a diagonal matrix with complex coefficients, and

$$|1 + b^T (I - ZA)^{-1} (Z + W) e| \leq 1 \quad \forall \Re(\alpha_i) \leq -|\beta_i| \tag{8.5.33}$$

under the restriction that $i, j \in I_k \Longrightarrow \{\alpha_i = \alpha_j \text{ and } \beta_i = \beta_j\}$, respectively.

Checking conditions (8.5.24) and (8.5.29) or, equivalently, (8.5.32) and (8.5.33), is not an easy task. Unfortunately, so far no equivalent easier conditions have been found to perform this task. This happens in spite of the fact that (8.5.29) and (8.5.33) are modifications of the condition for AN-stability (to see this, put $W = 0$ in (8.5.33)). Recall from Section 8.2.4 that algebraic stability is an equivalent condition to AN-stability; it is very easy to verify whenever the method is non-confluent.

However, we observe that condition (8.3.1) is necessary for $\mathrm{A_f}$-, $\mathrm{AN_f}$- and $\mathrm{BN_f}$-stable RK methods.

Proposition 8.5.22 *Let the RK method (5.0.2), (5.0.3) be of order $p \geq 1$ and A_f-stable. Then it satisfies condition (8.3.1).*

In particular, if $c_i \in [-1, 1]$, $i = 1, \ldots, \nu$, then $R(\infty) = 0$.

Proof. By Proposition 8.5.18, an $\mathrm{AN_f}$-stable RK method satisfies the inequality (8.5.23), where $\{I_1, \ldots, I_\rho\}$ is the confluency partition of the index set $\{1, \ldots, \nu\}$.

No matter whether the matrix A is singular or not, there exists $\epsilon_0 > 0$ such that the perturbed matrix $A + \epsilon I$ is non-singular for all $|\epsilon| < \epsilon_0$, $\epsilon \neq 0$. Therefore, by setting $\alpha = -\epsilon^{-1}$, $\epsilon > 0$, and letting $\epsilon \to 0$ in (8.5.23), we have that there exist

$$\lim_{\epsilon \to 0} \sum_{i_k \in I_k} (b^T (A + \epsilon I)^{-1})_{i_k} = v_k^* \in \mathbb{R}, \quad k = 1, \ldots, \rho, \tag{8.5.34}$$

and that

$$\left| 1 - \sum_{k=1}^{\rho} v_k^* \right| + \sum_{k=1}^{\rho} |v_k^*| \leq 1. \tag{8.5.35}$$

Since $Ae = c$, where $c = [c_1, \ldots, c_\nu]^T$, we have that $(A + \epsilon I)e = c + \epsilon e$. Moreover, because the method is of order $p \geq 1$, we get

$$1 = b^T e = b^T (A + \epsilon I)^{-1} c + \epsilon b^T (A + \epsilon I)^{-1} e.$$

For each $k = 1, \ldots, \rho$, let $i_k \in I_k$. Then, letting $\epsilon \to 0$, the existence of the limits (8.5.34) implies

$$\sum_{k=1}^{\rho} v_k^* c_{i_k} = 1, \tag{8.5.36}$$

which, in turn, yields $\sum_{k=1}^{\rho} |v_k^*| \neq 0$. Therefore, since $R(\infty) = 1 - \sum_{k=1}^{\rho} v_k^*$, (8.5.35) yields (8.3.1).

Moreover, if $c_i \in [-1, 1]$, $i = 1, \ldots, \nu$, (8.5.35) and (8.5.36) yield

$$1 = \sum_{k=1}^{\rho} v_k^* c_{i_k} \leq \sum_{k=1}^{\rho} |v_k^*| \leq 1 - |R(\infty)|,$$

and thus $R(\infty) = 0$. ■

An immediate consequence of Proposition 8.5.22 is that Gaussian collocation is not $\mathrm{A_f}$-stable, since $|R(\infty)| = 1$.

In fact, the set of $\mathrm{A_f}$-stable RK methods is very meager.

In the class of one-stage CRK methods of order 1, the only one that is $\mathrm{AN_f}$-stable (and $\mathrm{A_f}$-stable) is the *backward Euler* method with linear interpolation.

In the class of two-stage CRK methods of order 2, we have the following class of $\mathrm{A_f}$-stable methods (see Bellen and Zennaro [38]):

$$\begin{array}{c|cc} c_1 & b_1 & c_1 - b_1 \\ 1 & b_1 & 1 - b_1 \\ \hline & b_1 & 1 - b_1 \end{array}$$

where $0 \leq c_1 \leq 1/3$ and $b_1 = \frac{1}{2(1-c_1)}$, with linear interpolation. Within this class, the only one that is $\mathrm{AN_f}$-stable is the *Lobatto IIIC* method, obtained for $c_1 = 0$. Although no proof exists on the equivalence between $\mathrm{AN_f}$-stability and $\mathrm{BN_f}$-stability, the authors also provided a direct proof of $\mathrm{BN_f}$-stability both for backward Euler and two-stage Lobatto IIIC.

So far, no $\mathrm{AN_f}$-stable discrete RK method of order $p = 3$ is known and In 't Hout [136] proved that no $\mathrm{AN_f}$-stable method can exceed order $p = 4$. In particular, Zennaro [282] proved that no three-stage discrete RK method of order $p \geq 3$ with abscissae in $[0, 1]$ exists which is $\mathrm{AN_f}$-stable. This suggests the existence of a possibly more restrictive order barrier.

Being satisfied with scalar real equations, the previously hyphenated definitions of stability are restricted to the case of real parameters and functions and are denoted by $A_f(0)$-, $AN_f(0)$- and $BN_f(0)$-*stability*. The conditions for stability are obviously relaxed and a possibly larger class of stable methods are expected. The theory was developed first in [38] and later in [282], where the following three-stage $\mathrm{AN_f}(0)$-stable CRK method of discrete order 3 and uniform order 2 is presented:

$$\begin{array}{c|ccc} 0 & \frac{5}{2} & -2 & -\frac{1}{2} \\ \frac{1}{2} & -1 & 2 & -\frac{1}{2} \\ 1 & \frac{1}{6} & \frac{2}{3} & \frac{1}{6} \\ \hline & \frac{1}{6} & \frac{2}{3} & \frac{1}{6} \end{array} \qquad \begin{aligned} b_1(\theta) &= -\tfrac{5}{6}\theta^2 + \theta, \\ b_2(\theta) &= -\tfrac{2}{3}\theta^2, \\ b_3(\theta) &= \tfrac{1}{6}\theta^2. \end{aligned}$$

8.5.4 *Further properties of stable RK methods*

The conditions required in the definitions of $\mathrm{A_f}$-, $\mathrm{AN_f}$- and $\mathrm{BN_f}$-stability are the discrete analogs of conditions (8.5.15) and (8.5.12) satisfied by the exact solutions and are related to the concept of contractivity. However, we know that the exact solutions satisfy sharper conditions, namely (8.5.17), (8.5.14) and (8.5.7). Now we shall see that also the discrete analogs of these last conditions hold for $\mathrm{A_f}$-, $\mathrm{AN_f}$- and $\mathrm{BN_f}$-stable methods. All these results will be exploited in the asymptotic stability analysis of DDEs and related numerical methods.

As an immediate consequence of Proposition 8.5.18, we have the following discrete analog of (8.5.17) (valid also for confluent RK methods).

Theorem 8.5.23 *Let the RK method (5.0.2), (5.0.3) be A_f-stable. Then, for any mesh Δ, the numerical solution $\{y_n\}_{n\geq 0}$ of (8.5.16) with $\Re(\lambda) \leq 0$ satisfies*

$$|y_{n+1}| \leq \Psi^A_{n+1}|y_n| + (1 - \Psi^A_{n+1}) \max_{1\leq i\leq \nu} |g(t^i_{n+1})|,$$

where $\Psi^A_{n+1} = |R(h_{n+1}\lambda)|$.

Observe that, by Definition 8.3.5, if (8.2.2) holds, then

$$0 \leq \Psi^A_{n+1} \leq \Phi_A(\lambda_0, h_{n+1}). \tag{8.5.37}$$

Analogously, Proposition 8.5.20 yields the discrete analog of (8.5.14).

Theorem 8.5.24 *Let the RK method (5.0.2), (5.0.3) be AN_f-stable. Then, for any mesh Δ, the numerical solution $\{y_n\}_{n\geq 0}$ of (8.5.13) with $\Re(\lambda(t)) \leq 0$ satisfies*

$$|y_{n+1}| \leq \Psi^{AN}_{n+1}|y_n| + (1 - \Psi^{AN}_{n+1}) \max_{1\leq i\leq \nu} |g(t^i_{n+1})|,$$

where $\Psi^{AN}_{n+1} = |K(h_{n+1}\lambda(t^1_{n+1}), \ldots, h_{n+1}\lambda(t^\nu_{n+1}))|$.

Observe that, by Definition 8.3.6, if (8.2.16) holds, then

$$0 \leq \Psi^{AN}_{n+1} \leq \Phi_{AN}(\Lambda_0, h_{n+1}). \tag{8.5.38}$$

The discrete analog of (8.5.7) is given by the following theorem. The proof can be found in Zennaro [284] and is omitted. On the other hand, it is quite similar to the proof of the forthcoming Theorem 8.5.27.

Theorem 8.5.25 *Let the RK method (5.0.2), (5.0.3) be BN_f-stable. Then, for any mesh Δ, under conditions (8.2.26) with $Y_0 \leq 0$ and (8.5.6), the difference $\{\delta y_n\}_{n\geq 0}$ in the numerical solutions of (8.5.2) and (8.5.1) satisfies*

$$\|\delta y_{n+1}\| \leq \Psi^{BN}_{n+1}\|\delta y_n\| + (1 - \Psi^{BN}_{n+1}) \max_{1\leq i\leq \nu} (|r(t^i_{n+1})| \cdot \|\delta u(t^i_{n+1})\|), \tag{8.5.39}$$

where

$$0 \leq \Psi^{BN}_{n+1} \leq \Phi_{BN}(Y_0, h_{n+1}). \tag{8.5.40}$$

We now consider an alternative form of (8.5.39), extended to continuous RK methods, which stresses the asymptotic proportionality to the stepsize h_{n+1} (as $h_{n+1} \to 0$) in the contribution due to the forcing term. It will be used at the beginning of Chapter 10.

To this end, we assume that the function $f(t, y, x)$ in (8.5.1) satisfies the classical Lipschitz condition

$$\|f(t, y_1, x) - f(t, y_2, x)\| \leq L\|y_1 - y_2\| \tag{8.5.41}$$

for all $t \geq t_0$ and $y_1, y_2, x \in \mathbb{C}^d$, for some constant $L > 0$.

Lemma 8.5.26 *Consider a CRK method (5.0.2), (5.1.6), (5.1.3) for the numerical solution of the local problems*

$$\begin{cases} w'_{n+1}(t) = f(t, w_{n+1}(t), u(t)), & t_n \le t \le t_{n+1}, \\ w_{n+1}(t_n) = y_n, \end{cases} \tag{8.5.42}$$

and

$$\begin{cases} z'_{n+1}(t) = f(t, z_{n+1}(t), \tilde{u}(t)), & t_n \le t \le t_{n+1}, \\ z_{n+1}(t_n) = y_n, \end{cases} \tag{8.5.43}$$

perturbed from each other only in the forcing term, and assume that the function $f(t, y, x)$ *satisfies (8.5.3), (8.5.4), (8.5.6) and (8.5.41). Then there exist a constant* $\gamma^* > 0$ *and a stepsize* $h^* > 0$*, independent of the mesh* Δ*, of the nodal approximation* y_n *and of the forcing terms* $u(t)$ *and* $\tilde{u}(t)$*, such that the continuous numerical solutions* $\eta(t)$ *and* $\zeta(t)$ *satisfy*

$$\max_{0\le\theta\le1} \|\zeta(t_n + \theta h_{n+1}) - \eta(t_n + \theta h_{n+1})\| = h_{n+1}\gamma^*_{n+1} \max_{1\le i\le s} (|r(t^i_{n+1})| \cdot \|\delta u(t^i_{n+1})\|) \tag{8.5.44}$$

with

$$0 \le \gamma^*_{n+1} \le \gamma^* \tag{8.5.45}$$

for all stepsizes $h_{n+1} \le h^*$.

Moreover, if the CRK method is BN_f*-stable and condition (8.2.26) holds with* $Y_0 \le 0$*, then*

$$0 \le \gamma^*_{n+1} \le \gamma \quad \textit{and} \quad h_{n+1}\gamma^*_{n+1} \le 1 \tag{8.5.46}$$

for any stepsize h_{n+1}*, where* $\gamma = \max\{\gamma^*, 1/h^*\}$.

Proof. By setting $\tilde{g}(t, y) = f(t, y, \tilde{u}(t))$ and $g(t, y) = f(t, y, u(t))$ and by using the same notation as in the proof of Proposition 5.1.2 (in Section 5.1), the inequality (5.1.11) yields

$$\begin{aligned} &\|\zeta(t_n + \theta h_{n+1}) - \eta(t_n + \theta h_{n+1})\| \\ &\quad = h_{n+1}\|\Psi(y_n; \theta, \tilde{g}, t_n, t_{n+1}) - \Psi(y_n; \theta, g, t_n, t_{n+1})\| \\ &\quad \le h_{n+1}\delta_g \max_{1\le i\le s} \|\tilde{g}(t^i_{n+1}, \tilde{Y}^i_{n+1}) - g(t^i_{n+1}, Y^i_{n+1})\| \\ &\quad \le h_{n+1}\delta_g \max_{1\le i\le s} \sup_{t_n\le t\le t_{n+1},\ y\in\mathbb{C}^d} \|f(t^i_{n+1}, y, \tilde{u}(t^i_{n+1})) - f(t^i_{n+1}, y, u(t^i_{n+1}))\| \end{aligned}$$

whenever $h_{n+1} \le h_g$, where δ_g and h_g are determined by the Lipschitz constant L in (8.5.41).

Therefore, since (8.5.3) and (8.5.41) easily imply $|Y(t)| \le L$ for all $t \ge t_0$, (8.5.4) and (8.5.6) yield (8.5.44), (8.5.45) with $\gamma^* = \delta_g L$ and $h^* = h_g$.

If, in addition, the CRK method (5.0.2), (5.1.6), (5.1.3) is BN$_{\mathfrak{f}}$-stable and condition (8.2.26) holds with $Y_0 \leq 0$, then the difference in the continuous numerical solutions of (8.5.43) and (8.5.42) is such that

$$\max_{0 \leq \theta \leq 1} \|\zeta(t_n + \theta h_{n+1}) - \eta(t_n + \theta h_{n+1})\| \leq \max_{1 \leq i \leq s} (|r(t^i_{n+1})| \cdot \|\delta u(t^i_{n+1})\|)$$

for any stepsize h_{n+1}. Therefore, the number γ^*_{n+1} satisfies (8.5.46). ■

Theorem 8.5.27 *Let the CRK method (5.0.2), (5.1.6), (5.1.3) be BN$_f$-stable. Then, under conditions (8.2.26) with $Y_0 \leq 0$, (8.5.6) and (8.5.41), there exists a constant $\gamma > 0$, independent of the mesh Δ, such that the difference $\delta\eta(t)$ in the continuous numerical solutions of (8.5.2) and (8.5.1) satisfies*

$$\|\delta\eta(t_n + \theta h_{n+1})\| \leq (1 - h_{n+1}\gamma_{n+1})\|\delta y_n\| + h_{n+1}\gamma_{n+1} \max_{1 \leq i \leq s} (|r(t^i_{n+1})| \cdot \|\delta u(t^i_{n+1})\|), \tag{8.5.47}$$

where

$$0 \leq \gamma_{n+1} \leq \gamma \quad \textit{and} \quad h_{n+1}\gamma_{n+1} \leq 1. \tag{8.5.48}$$

Proof. Consider the local problem

$$\begin{cases} \tilde{w}'_{n+1}(t) = f(t, \tilde{w}_{n+1}(t), \tilde{u}(t)), & t_n \leq t \leq t_{n+1}, \\ \tilde{w}_{n+1}(t_n) = \tilde{y}_n, \end{cases} \tag{8.5.49}$$

whose continuous numerical solution is $\tilde{\eta}(t)$, the continuous numerical solution of (8.5.2). Since the CRK method is BN$_{\mathfrak{f}}$-stable and condition (8.2.26) holds with $Y_0 \leq 0$, the difference $\delta\eta(t) = \tilde{\eta}(t) - \eta(t)$ between the continuous numerical solutions of (8.5.49) and (8.5.42) satisfies

$$\max_{0 \leq \theta \leq 1} \|\delta\eta(t_n + \theta h_{n+1})\| \leq \max\left\{ \|\delta y_n\|, \max_{1 \leq i \leq s} (|r(t^i_{n+1})| \cdot \|\delta u(t^i_{n+1})\|) \right\}. \tag{8.5.50}$$

If $\|\delta y_n\| \geq \max_{1 \leq i \leq s}(|r(t^i_{n+1})| \cdot \|\delta u(t^i_{n+1})\|)$, then (8.5.50) clearly implies (8.5.47) with $\gamma_{n+1} = 0$.

Otherwise, if $\|\delta y_n\| < \max_{1 \leq i \leq s}(|r(t^i_{n+1})| \cdot \|\delta u(t^i_{n+1})\|)$, we define

$$\rho = \frac{\|\delta y_n\|}{\max_{1 \leq i \leq s}(|r(t^i_{n+1})| \cdot \|\delta u(t^i_{n+1})\|)} < 1 \tag{8.5.51}$$

and consider the further local problem

$$\begin{cases} x'_{n+1}(t) = f(t, x_{n+1}(t), u(t) + \rho\delta u(t)), & t_n \leq t \leq t_{n+1}, \\ x_{n+1}(t_n) = \tilde{y}_n, \end{cases} \tag{8.5.52}$$

and denote by $\lambda(t)$ its continuous numerical solution.

By Lemma 8.5.26, for the difference between the continuous numerical solutions of (8.5.49) and (8.5.52) we obtain

$$\begin{aligned}\max_{0\le\theta\le1} &\|\tilde{\eta}(t_n+\theta h_{n+1})-\lambda(t_n+\theta h_{n+1})\| \\ &= h_{n+1}\gamma^*_{n+1}(1-\rho)\max_{1\le i\le s}(|r(t^i_{n+1})|\cdot\|\delta u(t^i_{n+1})\|),\end{aligned}$$

where γ^*_{n+1} satisfies condition (8.5.46).

On the other hand, (8.5.51) implies

$$\|\delta y_n\| = \rho\max_{1\le i\le s}(|r(t^i_{n+1})|\cdot\|\delta u(t^i_{n+1})\|)$$

and, hence, the difference between the continuous numerical solutions of (8.5.52) and (8.5.42) is such that

$$\begin{aligned}\max_{0\le\theta\le1} &\|\lambda(t_n+\theta h_{n+1})-\eta(t_n+\theta h_{n+1})\| \\ &\le \max\left\{\|\delta y_n\|,\rho\max_{1\le i\le s}(|r(t^i_{n+1})|\cdot\|\delta u(t^i_{n+1})\|)\right\} \\ &= (1-h_{n+1}\gamma^*_{n+1})\|\delta y_n\| + h_{n+1}\gamma^*_{n+1}\rho\max_{1\le i\le s}(|r(t^i_{n+1})|\cdot\|\delta u(t^i_{n+1})\|).\end{aligned}$$

Thus we can conclude again with (8.5.47) for $\gamma_{n+1}=\gamma^*_{n+1}$, which satisfies (8.5.48) because of (8.5.46). ■

9

STABILITY ANALYSIS OF DDES

The stability analysis of DDEs have been investigated extensively, although not completely developed for the more complicated cases yet, and we refer the interested reader to the books cited in Chapter 1. Here we resume with some definitions and results on *contractivity*, or *dissipativity*, *boundedness* and *asymptotic stability* of DDEs and NDDEs that are useful for deriving classes of numerical methods that preserve the same property. In the stability analysis of DDEs, there are two possible approaches that have been considered in the literature in view of the construction of correspondingly stable numerical methods. One approach consists in finding conditions on the right-hand-side function f such that the problem is stable for all or for some classes of delays, typically for all constant delays. The second approach consists in finding weaker conditions on f such that the desired stability property is guaranteed for the specific given (in general constant) delay. The two concepts of stability are actually different and we shall refer to the former as *stability for all delays*, or *delay independent stability*, and to the latter as *stability for fixed delay*, or *delay dependent stability*.

For the sake of completeness, recall that in the early Russian literature the asymptotic stability for fixed delay was simply called *asymptotic stability*, whereas the asymptotic stability for all delays was called *absolute asymptotic stability* for constant delays and *strongly absolute asymptotic stability* for continuous variable delays.

It is evident that the class of problems that are stable for all delays is smaller than that for fixed delay. Moreover, characterizing the class of problems that are stable for all delays and finding numerical methods that are stable on that class is easier than for fixed delay. On the contrary, the analysis for fixed delay is sharper and the resulting class of stable problems is larger. Characterizing this class and finding stable numerical methods turns out to be much harder.

The results holding for all delays are mainly derived by those of Section 8.5, where we have developed an *ad hoc* stability analysis for equations such as $y'(t) = f(t, y(t), u(t))$ in view of substituting the forcing term $u(t)$ with the delayed term $y(t-\tau)$. Some of the results we obtain by following this line may be derived by other approaches, such as Razumikhin-like theorems, Lyapunov techniques, spectral analysis, etc.; however, some others are new. However, they are obtained in a constructive way by the method of steps, which straightforwardly applies to the stability analysis of numerical methods developed in Chapter 10. A second advantage of this approach is that it often allows us to treat equations with variable delays $\tau = \tau(t)$ satisfying some of the hypotheses stated in Chapter 2, which are worth recalling here:

(H_1) There exists a constant $\tau_0 > 0$ such that $\tau(t) \geq \tau_0$ for all $t \geq t_0$.

(H_2) $\lim_{t \to +\infty} t - \tau(t) = +\infty$.

(H_3) There exists a constant $\tau_1 > 0$ such that $\tau(t) \leq \tau_1$ for all $t \geq t_0$.

(H_4) The deviating argument $t - \tau(t)$ is a strictly increasing function for all $t \geq t_0$.

For the class of linear systems with constant coefficients and constant delay, the asymptotic analysis can also be carried out by analyzing the characteristic equation. In general, the results are sharper but the resulting conditions on the coefficients are difficult to handle and are often insufficient for describing the stability region and useless for determining a class of stable numerical methods. Also, by the spectral analysis, general results are often confined to the stronger request of stability for all delays. However, in the special case of the scalar DDE $y'(t) = \lambda y(t) + \mu y(t - \tau)$, a complete and satisfactory description of the stability region is instead obtained for a fixed delay, both for real and complex coefficients λ and μ.

More complicated is the stability analysis for the class of so-called *uncertain delay systems*, that arise in control theory, which are systems where the delay has an unknown constant value ranging in some interval. Despite analysis for all delays being suitable for this case, the resulting conditions are too restrictive and delay dependent results are often expected in practical situations when the delay has an upper bound. This analysis goes beyond the scope of this book and we refer the interested reader to Niculescu [208].

9.1 General non-linear DDEs

Consider the system of DDEs

$$\begin{cases} y'(t) = f(t, y(t), y(t - \tau(t))), & t \geq t_0, \\ y(t) = \phi(t), & t \leq t_0, \end{cases} \tag{9.1.1}$$

where $f \in C^0\left([t_0, t_f] \times \mathbb{C}^d \times \mathbb{C}^d, \mathbb{C}^d\right)$ and the time dependent delay $\tau(t)$ is assumed to be continuous and to satisfy hypothesis (H_1).

Then consider another system

$$\begin{cases} \tilde{y}'(t) = f(t, \tilde{y}(t), \tilde{y}(t - \tau(t))), & t \geq t_0, \\ \tilde{y}(t) = \tilde{\phi}(t), & t \leq t_0, \end{cases} \tag{9.1.2}$$

with the same right-hand side $f(t, y, x)$ but with different initial function $\tilde{\phi}(t)$, and define

$$\delta y(t) = \tilde{y}(t) - y(t),$$

$$\delta\phi(t) = \tilde{\phi}(t) - \phi(t).$$

Definition 9.1.1 *We say that the system of DDEs (9.1.1) is* contractive *or* dissipative *if, for any other initial function $\tilde{\phi}(t)$, the difference $\delta y(t)$ in the solutions of (9.1.2) and (9.1.1) fulfils*

$$\|\delta y(t)\| \leq \max_{x \leq t_0} \|\delta\phi(x)\|, \quad t \geq t_0. \tag{9.1.3}$$

Definition 9.1.2 *We say that the system of DDEs (9.1.1) or, equivalently, that its solution $y(t)$ is* asymptotically stable *if, for any other initial function $\tilde{\phi}(t)$, the difference $\delta y(t)$ in the solutions of (9.1.2) and (9.1.1) fulfils*

$$\lim_{t \to +\infty} \delta y(t) = 0. \tag{9.1.4}$$

We begin with the following result, essentially due to Torelli [247]. However, our formulation is slightly different and more general. It could be proved in an indirect way by using the Razumikhin type theorems presented in the book by Hale [121], but we think it interesting to consider a direct proof, since the same idea is applied when analyzing the stability of the numerical methods. Such a direct proof makes use of the stability theory with respect to forcing terms developed in Section 8.5 and, in particular, of Corollary 8.5.4 applied inductively to the macro intervals $[\bar{\xi}_{k-1}, \bar{\xi}_k]$, where the $\bar{\xi}_k$'s are the principal discontinuity points (see Definition 2.1.4).

Theorem 9.1.3 *Given an inner product $\langle \cdot, \cdot \rangle$ in $\mathbb{C}^d$ and the corresponding norm $\| \cdot \|$, consider the systems of DDEs (9.1.1) and (9.1.2). If*

$$Y(t) + X(t) \leq 0, \quad t \geq t_0, \tag{9.1.5}$$

where $Y(t)$ and $X(t)$ are the continuous functions (8.5.3) and (8.5.4), then the system (9.1.1) is contractive (i.e. (9.1.3) holds) for all initial functions $\phi(t)$ and for all delays $\tau(t)$ satisfying (H_1).

Proof. It can be easily seen that condition (9.1.5) is entirely equivalent to conditions (8.5.5) and (8.5.6) with $-1 \leq r(t) \leq 0$.

It is clear that $\|\delta y(t)\| \leq \max_{x \leq t_0} \|\delta\phi(x)\|$ holds for $t \leq t_0$. Now assume by induction that (9.1.3) holds for $t \leq \bar{\xi}_{k-1}$ and consider the solution $y(t)$ for $t \in [\bar{\xi}_{k-1}, \bar{\xi}_k]$. Since $|r(x)| \leq 1$, by Corollary 8.5.4 we get

$$\begin{aligned}
\|\delta y(t)\| &\leq \max\left\{ \|\delta y(\bar{\xi}_{k-1})\|, \sup_{\bar{\xi}_{k-1} \leq x \leq \bar{\xi}_k} \left(|r(x)| \cdot \|\delta y(x - \tau(x))\| \right) \right\} \\
&\leq \max\left\{ \|\delta y(\bar{\xi}_{k-1})\|, \max_{x \leq \bar{\xi}_{k-1}} \|\delta y(x)\| \right\}.
\end{aligned}$$

Hence, by the inductive hypothesis, (9.1.3) also holds for $t \in [\bar{\xi}_{k-1}, \bar{\xi}_k]$ and, thus, the proof is complete. ■

We remark that (8.5.5) weakens the stronger hypothesis $Y(t) < 0$, $t \geq t_0$, in [247].

Theorem 9.1.3 gives sufficient conditions in order that the system of DDEs (9.1.1) be dissipative, but it does not say when it is asymptotically stable. This will be done by the next theorem, proved by Zennaro [284].

Theorem 9.1.4 *Consider the system of DDEs (9.1.1). If the functions $Y(t)$ and $X(t)$ defined by (8.5.3) and (8.5.4) are such that*

$$Y(t) \leq Y_0 < 0, \quad t \geq t_0, \tag{9.1.6}$$

and, for some non-negative real number $R < 1$,

$$R \cdot Y(t) + X(t) \leq 0, \quad t \geq t_0, \tag{9.1.7}$$

then problem (9.1.1), besides being contractive, is also asymptotically stable (i.e. (9.1.4) holds) for all initial functions $\phi(t)$ and for all delays $\tau(t)$ satisfying hypotheses (H_1) and (H_2). Moreover, if also hypothesis (H_3) is satisfied, then the convergence rate in (9.1.4) is at least of exponential type, i.e. at least like $e^{-\alpha(t-t_0)}$ for some $\alpha > 0$.

Proof. The proof proceeds inductively on the macro intervals between the principal discontinuity points.

Set $\bar{\xi}_0 = t_0$ and, for $t \leq \bar{\xi}_0$, define the constant function

$$\delta_0(t) = \max_{x \leq \bar{\xi}_0} \|\delta\phi(x)\|$$

and the numbers

$$\Delta_{-1} = \Delta_0 = \delta_0(\bar{\xi}_0).$$

Observe that the interesting case is $\Delta_{-1} = \Delta_0 > 0$, otherwise (9.1.4) is obvious.

Then, for $k \geq 1$, on each interval $[\bar{\xi}_{k-1}, \bar{\xi}_k]$ inductively define the functions

$$\delta_k(t) = R\Delta_{h(k-1)} + e^{Y_0(t-\bar{\xi}_{k-1})}(\Delta_{k-1} - R\Delta_{h(k-1)}) \tag{9.1.8}$$

and the numbers

$$\Delta_k = \delta_k(\bar{\xi}_k), \tag{9.1.9}$$

where

$$h(k-1) = \max\{h \mid \bar{\xi}_h \leq t - \tau(t) \ \forall t \in [\bar{\xi}_{k-1}, \bar{\xi}_k]\}.$$

Observe that, since (H_1) implies

$$\bar{\xi}_k - \bar{\xi}_{k-1} \geq \tau_0, \tag{9.1.10}$$

by (H_2) we have

$$\bar{\xi}_k \to +\infty \quad \text{and} \quad h(k) \to \infty \quad \text{as } k \to \infty. \tag{9.1.11}$$

Of course, we have $\delta_k(\bar{\xi}_{k-1}) = \Delta_{k-1}$, so that the piecewise exponential function $\delta(t)$, defined on $[\xi_{-1}, +\infty)$ by setting $\delta(t) = \delta_k(t)$ if $t \in [\bar{\xi}_{k-1}, \bar{\xi}_k]$, $k \geq 0$, is continuous.

Now, by the inductive hypothesis, assume that $\Delta_{k-1} - R\Delta_{h(k-1)} > 0$ and $\Delta_{j-1} \geq \Delta_j, \forall j \leq k-1$ (these inequalities clearly hold for $k = 1$). Then, by (9.1.8) and (9.1.9), we have

$$\begin{aligned}\Delta_k - R\Delta_{h(k)} &= R\Delta_{h(k-1)} + \mathrm{e}^{Y_0(\bar{\xi}_k - \bar{\xi}_{k-1})}(\Delta_{k-1} - R\Delta_{h(k-1)}) - R\Delta_{h(k)} \\ &= R(\Delta_{h(k-1)} - \Delta_{h(k)}) + \mathrm{e}^{Y_0(\bar{\xi}_k - \bar{\xi}_{k-1})}(\Delta_{k-1} - R\Delta_{h(k-1)}) > 0\end{aligned}$$

and

$$\begin{aligned}\Delta_{k-1} - \Delta_k &= \Delta_{k-1} - R\Delta_{h(k-1)} - \mathrm{e}^{Y_0(\bar{\xi}_k - \bar{\xi}_{k-1})}(\Delta_{k-1} - R\Delta_{h(k-1)}) \\ &= (\Delta_{k-1} - R\Delta_{h(k-1)})(1 - \mathrm{e}^{Y_0(\bar{\xi}_k - \bar{\xi}_{k-1})}) > 0.\end{aligned}$$

Therefore, we have proved that the sequence $\{\Delta_k\}$ and, consequently, the function $\delta(t)$ are monotonically decreasing and that

$$\Delta_k - R\Delta_{h(k)} > 0, \quad k \geq 0. \tag{9.1.12}$$

Now we want to prove that

$$\|\delta y(t)\| \leq \delta(t), \quad t \geq \bar{\xi}_{-1}. \tag{9.1.13}$$

The above inequality clearly holds for $t \in [\bar{\xi}_{-1}, \bar{\xi}_0]$. Now assume by induction that it holds for $t \in [\bar{\xi}_{-1}, \bar{\xi}_{k-1}]$ and consider the function $\delta y(t)$ for $t \in [\bar{\xi}_{k-1}, \bar{\xi}_k]$. It can easily be seen that condition (9.1.7) implies condition (8.5.6) with $-R \leq r(t) \leq 0$. Therefore, by Proposition 8.5.1 and by the inductive hypothesis, we get

$$\begin{aligned}\|\delta y(t)\| &\leq E(\bar{\xi}_{k-1}, t)\|\delta y(\bar{\xi}_{k-1})\| \\ &\quad + \left(1 - E(\bar{\xi}_{k-1}, t)) \sup_{\bar{\xi}_{k-1} \leq x \leq \bar{\xi}_k} (|r(x)| \cdot \|\delta y(x - \tau(x))\|\right) \\ &\leq E(\bar{\xi}_{k-1}, t)\|y(\bar{\xi}_{k-1})\| \\ &\quad +(1 - E(\bar{\xi}_{k-1}, t))R \sup_{\bar{\xi}_{h(k-1)} \leq x \leq \bar{\xi}_{k-1}} \|\delta y(x)\| \\ &\leq E(\bar{\xi}_{k-1}, t)\Delta_{k-1} + (1 - E(\bar{\xi}_{k-1}, t))R \sup_{\bar{\xi}_{h(k-1)} \leq x \leq \bar{\xi}_{k-1}} \delta(x) \\ &= E(\bar{\xi}_{k-1}, t)\Delta_{k-1} + (1 - E(\bar{\xi}_{k-1}, t))R\Delta_{h(k-1)} \\ &= R\Delta_{h(k-1)} + E(\bar{\xi}_{k-1}, t)(\Delta_{k-1} - R\Delta_{h(k-1)}),\end{aligned}$$

where $E(t_1, t_2) = \exp\left(\int_{t_1}^{t_2} Y(x)\mathrm{d}x\right)$. Finally, by (9.1.6), (9.1.8) and (9.1.12), we obtain

$$\|\delta y(t)\| \leq \delta_k(t), \quad \bar{\xi}_{k-1} \leq t \leq \bar{\xi}_k,$$

and, hence, (9.1.13) holds also for $t \in [\bar{\xi}_{k-1}, \bar{\xi}_k]$.

To conclude the proof of (9.1.4), we have to show that the function $\delta(t)$ vanishes as $t \to +\infty$. Because of (9.1.11), it is sufficient to prove that $\Delta_k \to 0$. To this end, observe that, by (9.1.8), (9.1.9), (9.1.12) and (9.1.10), the sequence $\{\Delta_k\}_{k\geq -1}$ is such that

$$\begin{cases} \Delta_k \leq S\Delta_{k-1} + R(1-S)\Delta_{h(k-1)}, & k \geq 1, \\ \Delta_{-1} = \Delta_0 > 0, \end{cases} \tag{9.1.14}$$

where $S = \mathrm{e}^{Y_0\tau_0} < 1$. This is a variable order linear difference inequality with two positive constant coefficients, the sum of which is < 1. Therefore, since the sequence $\{\Delta_k\}$ is convergent (because it is decreasing and positive), its limit can not be strictly positive. Thus $\Delta_k \to 0$ as $k \to \infty$.

Now, assume hypothesis (H_3) is satisfied. In this case, by (9.1.10) it turns out that $h(k-1) \geq k - m - 1$, where $m = \lceil \frac{\tau_1}{\tau_0} \rceil$. So, since the sequence $\{\Delta_k\}$ is decreasing, the difference inequality (9.1.14) yields

$$\Delta_k \leq S\Delta_{k-1} + R(1-S)\Delta_{k-m-1}, \quad k \geq m,$$

which is a linear difference inequality of order $m+1$ with constant coefficients. Since the sum of the (positive) coefficients in the right-hand side is < 1, all the characteristic roots are < 1 in modulus. Thus, if β is the greatest one, then we can conclude that $\Delta_k \to 0$ at least as $|\beta|^k$.

Finally, for a given $t \geq t_0$, we have that $t \in [\bar{\xi}_{k-1}, \bar{\xi}_k]$ for some $k \geq \frac{t-t_0}{\tau_1}$. Therefore, by (9.1.8) and (9.1.9), it follows that $\delta(t) \to 0$ at least as $\mathrm{e}^{-\alpha(t-t_0)}$, where $\alpha = -(1/\tau_1)\log|\beta|$. ■

Example 9.1.5 Let us apply the foregoing results to the equation

$$y'(t) = -ay(t) + \frac{by(t-\tau)}{1 + [y(t-\tau)]^n},$$

where $a > 0$ and b are real parameters and n is an even positive integer. This equation has already been considered in Example 1.1.7 where, for certain values of the parameters and of the delay, the solution was shown to oscillate chaotically. On the other hand, it can be seen that, for $n = 2, 4$, it always holds that $|\frac{\partial f(t,y,x)}{\partial x}| \leq |b|$. Therefore, using the foregoing Theorems 9.1.3 and 9.1.4, we can conclude that for $n = 2, 4$ the model is dissipative whenever $a \geq |b|$, and even asymptotically stable whenever $a > |b|$, independently of the particular value of the delay τ. ◇

Remark 9.1.6 *Theorems 9.1.3 and 9.1.4, as well as Proposition 8.5.1, which is used in their proofs, deal with norms derived by inner products, but they may be proved in the same way for arbitrary norms just by using Proposition 8.5.3 instead of Proposition 8.5.1. Although this is a useful improvement for the stability analysis of delay equations, it seems to be of little help for the analysis of numerical methods. In fact, as with ODEs, results on numerical stability for arbitrary norms are much more restrictive than for inner product norms (see, for example, Spijker [236]).*

Now we need to stress the fact that the bound (9.1.13) is sharp with respect to the whole class of DDEs satisfying the hypotheses (9.1.6) and (9.1.7). In order to see this, consider the piecewise linear delay

$$\tau^*(t) = \begin{cases} t-(k-2) & \text{for } k-1 \le t < k, \\ t-(k-1) & \text{for } t = k, \end{cases} \quad k = 1, 2, \ldots, \tag{9.1.15}$$

so that the deviating argument is the piecewise constant function

$$t - \tau^*(t) = \begin{cases} k-2 & \text{for } k-1 \le t < k, \\ k-1 & \text{for } t = k, \end{cases} \quad k = 1, 2, \ldots.$$

Now it is easy to see that, for the linear equation $y'(t) = \lambda y(t) + \mu y(t - \tau^*(t))$ with real constant coefficients $\lambda = Y_0 < 0$ and $\mu = -R \cdot Y_0 > 0$ and initial function $\phi(t) \equiv 1$, the solution $y(t)$ is given just by the corresponding function $\delta(t)$ introduced in the proof of Theorem 9.1.4.

The delay $\tau^*(t)$ fulfils hypothesis (H_1), but is not continuous. Nevertheless, it is clear that it can be viewed as the pointwise limit of a sequence of continuous delays. In other words, there exist continuous delays that fulfill (H_1) and are arbitrarily close (pointwise) to (9.1.15). As a consequence, there exist linear DDEs whose solutions $y(t)$ are arbitrarily close to the corresponding function $\delta(t)$.

It is also worth remarking that hypothesis (H_2) is necessary for the asymptotic stability with respect to the whole class of DDEs satisfying the hypotheses (9.1.6) and (9.1.7). In fact, if it does not hold, there exist only a finite number of discontinuity points $\bar{\xi}_k$. So, if $\bar{\xi}_K$ is the last one, on the last interval $[\bar{\xi}_K, +\infty)$ the function $\delta(t) = \delta_{K+1}(t)$ is bounded from below by the quantity $R\Delta_{h(K)} > 0$ (see (9.1.8)).

Since conditions (9.1.6) and (9.1.7) are more restrictive than (9.1.5), they guarantee asymptotic stability along with contractivity. On the other hand, contractivity is not necessary for asymptotic stability even in the linear constant coefficient case, as shown by the equation

$$\begin{cases} y'(t) = \frac{1}{2} y(t) - y(t-1), & t \ge 0, \\ y(t) = \phi(t) = t + 1, & -1 \le t \le 0. \end{cases} \tag{9.1.16}$$

According to the theory developed in the forthcoming sections, the solution $y(t)$ is asymptotically stable but, in a right neighborhood of 0, the solution is larger than $\max_{x \le 0} |\phi(x)| = 1$. In fact, $y(0) = 1$ and $y'(0)^+ = \frac{1}{2} > 0$.

It is also worth remarking that the conditions assumed in Theorems 9.1.3 and 9.1.4 are sufficient to guarantee the contractivity and asymptotic stability independently of the delay $\tau(t)$. In other words, under those conditions, the properties hold *for all delays*. Are those conditions also necessary to guarantee the properties *for all delays*? In the next section it will be proved that the answer is *yes* even for linear equations. On the contrary, if we want the stability properties to hold for a given constant delay, the mentioned conditions are still necessary for contractivity but not for asymptotic stability.

9.2 Linear scalar test equations

We distinguish the class of scalar linear DDEs with variable coefficients

$$\begin{cases} y'(t) = \lambda(t)y(t) + \mu(t)y(t-\tau(t)), & t \geq t_0, \\ y(t) = \phi(t), & t \leq t_0, \end{cases} \tag{9.2.1}$$

and with constant coefficients

$$\begin{cases} y'(t) = \lambda y(t) + \mu y(t-\tau(t)), & t \geq t_0, \\ y(t) = \phi(t), & t \leq t_0, \end{cases} \tag{9.2.2}$$

respectively. Observe that, if the delay τ is constant, then equation (9.2.2) is a linear autonomous equation.

In such a context, condition (9.1.5) becomes

$$\Re(\lambda(t)) + |\mu(t)| \leq 0, \quad t \geq t_0,$$

for equation (9.2.1) and

$$\Re(\lambda) + |\mu| \leq 0$$

for equation (9.2.2), whereas conditions (9.1.6) and (9.1.7) become

$$\Re(\lambda(t)) \leq \Lambda_0 < 0, \quad t \geq t_0,$$

and

$$R \cdot \Re(\lambda(t)) + |\mu(t)| \leq 0, \quad t \geq t_0,$$

respectively, for equation (9.2.1) and the sole condition

$$\Re(\lambda) + |\mu| < 0$$

for equation (9.2.2).

Owing to linearity, without any restriction, equations (9.2.1) and (9.2.2) may be compared just with

$$\begin{cases} \tilde{y}'(t) = \lambda(t)\tilde{y}(t) + \mu(t)\tilde{y}(t-\tau(t)), & t \geq t_0, \\ \tilde{y}(t) = 0, & t \leq t_0, \end{cases}$$

and

$$\begin{cases} \tilde{y}'(t) = \lambda\tilde{y}(t) + \mu\tilde{y}(t-\tau(t)), & t \geq t_0, \\ \tilde{y}(t) = 0, & t \leq t_0, \end{cases}$$

respectively, whose solution is the zero function $\tilde{y}(t) \equiv 0$. Therefore, for both equations, (9.1.3) reduces to

$$|y(t)| \leq \max_{x \leq t_0} |\phi(x)|, \quad t \geq t_0, \tag{9.2.3}$$

and (9.1.4) reduces to

$$\lim_{t \to +\infty} y(t) = 0. \tag{9.2.4}$$

So, for the linear equations, we can restate Theorems 9.1.3 and 9.1.4 as follows.

Theorem 9.2.1 *If the coefficients of equations (9.2.1) and (9.2.2) are such that*

$$\Re(\lambda(t)) + |\mu(t)| \leq 0, \quad t \geq t_0,$$

and

$$\Re(\lambda) + |\mu| \leq 0, \tag{9.2.5}$$

respectively, then the equations are contractive (i.e. (9.2.3) holds) for any initial function $\phi(t)$ and for all delays $\tau(t)$ satisfying hypothesis (H_1).

Theorem 9.2.2 *If the coefficients of equation (9.2.1) are such that*

$$\Re(\lambda(t)) \leq \Lambda_0 < 0, \quad t \geq t_0,$$

and

$$R \cdot \Re(\lambda(t)) + |\mu(t)| \leq 0, \quad t \geq t_0,$$

for some non-negative real number $R < 1$, and the coefficients of equation (9.2.2) are such that

$$\Re(\lambda) + |\mu| < 0, \tag{9.2.6}$$

then the solution $y(t)$, besides being contractive, is asymptotically stable (i.e. (9.2.4) holds) for any initial function $\phi(t)$ and for all delays $\tau(t)$ satisfying hypotheses (H_1) and (H_2). Moreover, if also hypothesis (H_3) is satisfied, then the convergence rate is at least of exponential type, i.e. at least like $e^{-\alpha(t-t_0)}$ for some $\alpha > 0$.

It is useful to point out that, for the constant delay instance of (9.2.2)

$$\begin{cases} y'(t) = \lambda y(t) + \mu y(t-\tau), & t \geq t_0, \\ y(t) = \phi(t), & t \leq t_0, \end{cases} \tag{9.2.7}$$

the stability analysis may be done directly by studying the roots of the *characteristic equation*

$$\zeta - \lambda - \mu e^{-\tau\zeta} = 0. \tag{9.2.8}$$

It is known (see El'sgol'ts and Norkin [81]) that such an equation has infinitely many roots ζ_i, each of which has a certain multiplicity m_i. They lie in the complex half-plane $\Re(\lambda) < \alpha$ for some real α and their real parts accumulate at $-\infty$. Therefore, in any vertical strip of the complex plane there are only a finite number of roots.

It is also known that the solution to (9.2.2) has an expansion of the form

$$y(t) = \sum_{i=1}^{\infty} \sum_{n_i=0}^{m_i-1} \alpha_{i,n_i} t^{n_i} e^{\zeta_i t}, \tag{9.2.9}$$

where the coefficients α_{i,n_i} are determined by the initial function $\phi(t)$. In view of the representation (9.2.9), it is easy to understand that a necessary and sufficient condition for the asymptotic stability of (9.2.2) is that all the roots ζ_i of (9.2.8) be such that $\Re(\zeta_i) < 0$.

Now, we shall see that such a condition is guaranteed if (9.2.6) holds. In fact, if we assume by contradiction that there exists a root ζ^* of (9.2.8) such that $\Re(\zeta^*) \geq 0$, we have

$$
\begin{aligned}
0 \leq \Re(\zeta^*) &= \Re(\lambda) + \Re(\mu \mathrm{e}^{-\tau\zeta^*}) \\
&\leq \Re(\lambda) + |\mu| \cdot |\mathrm{e}^{-\tau\zeta^*}| \\
&\leq \Re(\lambda) + |\mu|,
\end{aligned}
$$

which gives a contradiction.

So, we can say that results obtained by Theorem 9.2.2 on asymptotic stability for all delays of (9.2.2) are not worse than those obtained by the characteristic equation, at least for constant delays.

We have seen that Theorems 9.2.1 and 9.2.2 give conditions that are sufficient for contractivity and asymptotic stability independently of the particular delay $\tau(t)$. From a theoretical and practical point of view, it is interesting to investigate whether or not all or some of the above conditions are also necessary.

Again we must distinguish two cases: either when a particular delay $\tau(t)$ is considered or when the contractivity and/or asymptotic stability properties are requested for all delays of a certain class (for example, constant delays).

We confine ourselves to considering the test equation (9.2.2), where the delay $\tau(t)$ satisfies the additional hypothesis (H_4).

Proposition 9.2.3 *Assume that the contractivity property (9.2.3) holds for equation (9.2.2) with a fixed delay $\tau(t)$ satisfying hypotheses (H_1) and (H_4). Then the coefficients λ and μ satisfy condition (9.2.5).*

Proof. Since hypotheses (H_1) and (H_4) are satisfied, the deviating argument $\alpha(t) = t - \tau(t)$ is a one-to-one map from $[t_0, \xi_1]$ to $[t_0 - \tau(t_0), t_0]$. Thus, we can consider the inverse function $\beta(t)$ from $[t_0 - \tau(t_0), t_0]$ onto $[t_0, \xi_1]$ and define the initial function

$$
\phi^*(t) = \begin{cases} \mathrm{e}^{i\Im(\lambda)(\beta(t)-t_0)} & \text{for } t_0 - \tau(t_0) \leq t < t_0, \\ \mu/|\mu| & \text{for } t = t_0, \end{cases}
$$

which may be discontinuous at t_0. The solution in the first interval $[t_0, \xi_1]$ is given by

$$
\begin{aligned}
y^*(t) &= \mathrm{e}^{\lambda(t-t_0)} \left(\phi^*(t_0) + \mu \int_{t_0}^{t} \mathrm{e}^{-\lambda(x-t_0)} \phi^*(x - \tau(x)) \mathrm{d}x \right) \\
&= \mathrm{e}^{\lambda(t-t_0)} \left(\mu/|\mu| + \mu \int_{t_0}^{t} \mathrm{e}^{-\Re(\lambda)(x-t_0)} \mathrm{d}x \right)
\end{aligned}
$$

and, therefore, we get

$$
|y^*(\xi_1)| = \begin{cases} 1 + |\mu|(\xi_1 - t_0) & \text{if } \Re(\lambda) = 0, \\ \left| \dfrac{\Re(\lambda)\mathrm{e}^{\Re(\lambda)(\xi_1 - t_0)} + |\mu|(\mathrm{e}^{\Re(\lambda)(\xi_1 - t_0)} - 1)}{\Re(\lambda)} \right| & \text{if } \Re(\lambda) \neq 0. \end{cases}
$$

TABLE 9.1. Contractivity scheme for $y'(t) = \lambda y(t) + \mu y(t - \tau(t))$ with $\lambda, \mu \in \mathbb{C}$.

$$|y(t)| \leq \max_{x \leq t_0} |\phi(x)| \text{ for all delays } \tau(t) \text{ satisfying } (H_1)$$
$$\Updownarrow$$
$$\Re(\lambda) + |\mu| \leq 0$$
$$\Updownarrow$$
$$|y(t)| \leq \max_{x \leq t_0} |\phi(x)| \text{ for fixed delay } \tau(t) \text{ satisfying } (H_1) \text{ and } (H_4)$$

Now it can be easily seen that, if (9.2.5) does not hold, i.e. $\Re(\lambda) + |\mu| > 0$, then $|y^*(\xi_1)| > 1$. As we observed, the initial function $\phi^*(t)$ may not be continuous at t_0. However, it is clear that, for any $\epsilon > 0$, we can find a continuous initial function $\phi_\epsilon(t)$ such that the corresponding solution $y_\epsilon(t)$ satisfies the inequality $\max_{t_0 \leq t \leq \xi_1} |y_\epsilon(t) - y^*(t)| < \epsilon$. This completes the proof. ■

The foregoing proposition states that condition (9.2.5) is necessary for contractivity for a particular fixed delay satisfying (H_1) and (H_4) and, *a fortiori*, for all delays satisfying (H_1). We can summarize the results on contractivity for equation (9.2.2) in Table 9.1.

A different case is that of asymptotic stability where, in order to analyze the necessity of condition (9.2.6), we confine our analysis to the case of constant delays. To do this, we need a complete description of the stability region $\mathcal{S}_\tau$ defined as follows.

Definition 9.2.4 *For any fixed τ, the* asymptotic stability region $\mathcal{S}_\tau$ *of equation (9.2.7) is the set of pairs (λ, μ) such that the corresponding solution is asymptotically stable for any initial function $\phi(t)$.*

An efficient technique for finding the asymptotic stability region $\mathcal{S}_\tau$ is based on the method of *D-partitions* introduced by Naimark [203] and widely used in the Russian literature since then (see, for example, El'sgol'ts and Norkin [81] and Kolmanovskii and Nosov [173]). The method consists in determining, in the (λ, μ)-space of parameters, the *root locus curve*, where the roots of the characteristic equation $\zeta - \lambda - \mu e^{-\tau\zeta} = 0$ are purely imaginary, and in analyzing the sign of $\Re(\zeta)$ inside the regions determined by the curve branches. The technique extends to more general equations and possesses a numerical analog that is usually termed the *boundary locus* method. It has been applied by Baker and Ford [10] and Baker and Paul [13] to various test equations and related numerical methods. In particular, the analysis of the asymptotic stability region $\mathcal{S}_\tau$ has been thoroughly investigated and completely described by Guglielmi [106, 108], Guglielmi and Hairer [111] and Maset [195] for real and complex λ and μ.

9.2.1 *Description of the asymptotic stability region $\mathcal{S}_\tau$ for real coefficients*

Let us first consider the case of real λ and μ. A deeper analysis of the roots of the characteristic equation (9.2.8) reveals that, for a fixed value of the delay τ, the region of stability is larger than the cone $\lambda + |\mu| < 0$ derived by the inequality (9.2.6). In fact, the region of asymptotic stability $\mathcal{S}_\tau$ is given by the set of pairs (λ, μ) such that (see Hayes [127])

$$\lambda < -\mu \quad \text{and} \quad \sqrt{\mu^2 - \lambda^2} < \frac{1}{\tau} \arccos(-\lambda/\mu).$$

In Figure 9.1 we show, for real λ and μ, the form of the asymptotic stability region in the (λ, μ)-plane.

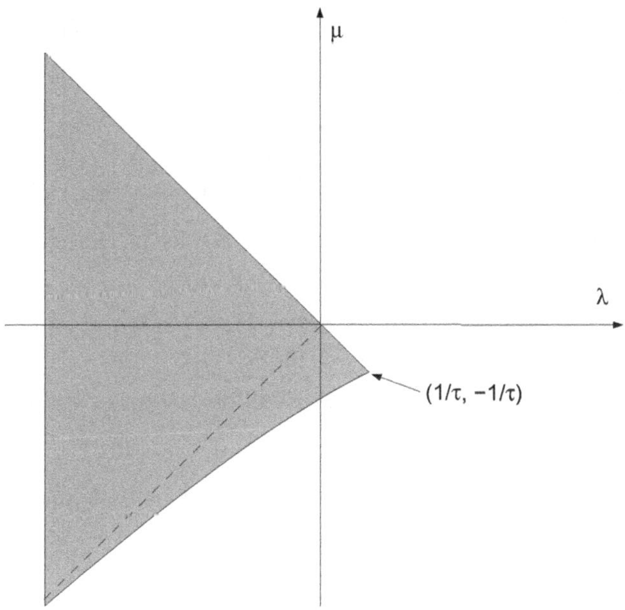

FIG. 9.1. Asymptotic stability region $\mathcal{S}_\tau$ for equation (9.2.2) with constant delay τ in the real (λ, μ)-plane.

Equation (9.1.16), where $\lambda = \frac{1}{2}$, $\mu = -1$ and $\tau = 1$, provides an example where the solution goes to zero despite the coefficients not fulfilling condition (9.2.6). Thus, for asymptotic stability condition (9.2.6) is not necessary for a fixed constant delay τ. However, when we let the delay τ go to $+\infty$, we see that the region of asymptotic stability tends to the region described by

$$\lambda \leq \mu < -\lambda. \tag{9.2.10}$$

Therefore, condition (9.2.10), which is slightly weaker than condition (9.2.6), is necessary for the asymptotic stability of equation (9.2.2) for all constant delays τ. The results are summarized in Table 9.2.

9.2.2 *Description of the asymptotic stability region $\mathcal{S}_\tau$ for complex coefficients*

Now let us describe the set $\mathcal{S}_\tau$ for complex λ and μ. Remember that, by using the characteristic equation (9.2.8), it is

TABLE 9.2. Asymptotic stability scheme for $y'(t) = \lambda y(t) + \mu y(t-\tau)$ with $\lambda, \mu \in \mathbb{R}$.

$$\lim_{t\to+\infty} y(t) = 0 \text{ for all constant delays } \tau$$
$$\Updownarrow$$
$$\lambda \leq \mu < -\lambda$$
$$\Downarrow$$
$$(\lambda, \mu) \in \mathcal{S}_\tau$$
$$\Updownarrow$$
$$\lim_{t\to+\infty} y(t) = 0 \text{ for fixed constant delay } \tau$$

$$\mathcal{S}_\tau = \left\{(\lambda, \mu) \in \mathbb{C}^2 \mid \zeta - \lambda - \mu e^{-\tau\zeta} = 0 \;\Rightarrow\; \Re(\zeta) < 0\right\}.$$

In order to simplify the notation for describing $\mathcal{S}_\tau$, let us change t into τt so that equation (9.2.2) becomes

$$\begin{cases} y'(t) = ay(t) + by(t-1), & t \geq t_0/\tau, \\ y(t) = \phi(t), & t \leq t_0/\tau, \end{cases}$$

where $a = \lambda\tau$ and $b = \mu\tau$. Therefore, we shall describe the stability set

$$S = \left\{(a, b) \in \mathbb{C}^2 \mid \zeta - a - be^{\zeta} = 0 \;\Rightarrow\; \Re(\zeta) < 0\right\}$$

by following the line of Maset [195]. To this end, let us fix b and consider the set

$$S_b = \left\{\zeta - be^{-\zeta} \mid \zeta \in \mathbb{C},\; \Re(\zeta) \geq 0\right\}.$$

It is evident that the pair (a, b) belongs to S if and only if $a \notin S_b$. So, let $b \neq 0$ with $b = Re^{i\theta}$, and describe the set S_b for $R \geq 0$.

To this purpose, if Q is a set of complex numbers and $x \in \mathbb{C}$, we denote by $Q + x$ the shifted set of complex numbers $q + x$ with $q \in Q$. It is immediate to verify the following results.

Proposition 9.2.5 *We have*

$$S_b = S_R + i\theta, \tag{9.2.11}$$

where

$$S_R = \left\{\zeta - Re^{-\zeta} \mid \Re(\zeta) \geq 0\right\}.$$

Proposition 9.2.6 *We have*

$$S_R = \bigcup_{k \in Z} \left(S_R^0 + 2\pi k i\right), \tag{9.2.12}$$

where

$$S_R^0 = \left\{\zeta - Re^{-\zeta} \mid \Re(\zeta) \geq 0,\; \Im(\zeta) \in [-\pi, \pi]\right\}.$$

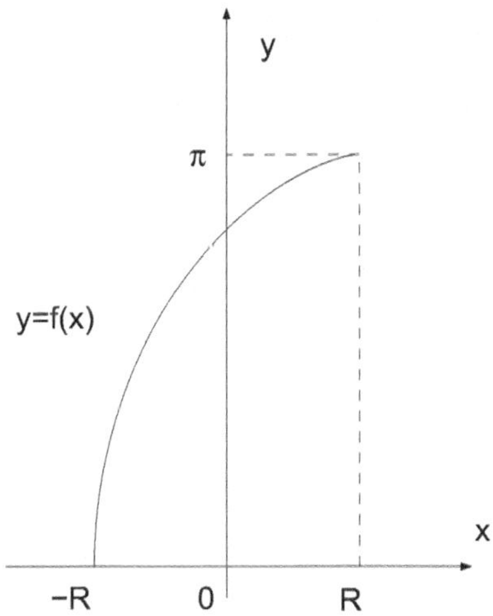

FIG. 9.2. Graph of the function f for $R \leq 1$.

Proposition 9.2.7 *We have*

$$S_R^0 = S_R^{0+} \cup \overline{S_R^{0+}}, \tag{9.2.13}$$

where

$$S_R^{0+} = \{\zeta - R\mathrm{e}^{-\zeta} \mid \Re(\zeta) \geq 0, \quad \Im(\zeta) \in [0, \pi]\}$$

and $\overline{S_R^{0+}}$ *is the set of the conjugates of the elements of* S_R^{0+}.

So, it is sufficient to describe the set S_R^{0+}, which is the image of the function $g(\zeta) = \zeta - R\mathrm{e}^{-\zeta}$ on the unbounded strip

$$\Sigma = \{\zeta \in \mathbb{C} \mid \Re(\zeta) \geq 0, \quad \Im(\zeta) \in [0, \pi]\}.$$

The image of the half-line $\{x + i\pi \mid x \geq 0\}$ is

$$A = \left\{x + i\pi - R\mathrm{e}^{-(x+i\pi)} \mid x \geq 0\right\} = \{x + R\mathrm{e}^{-x} + i\pi \mid x \geq 0\}.$$

If $R \leq 1$, then the function $x \mapsto x + R\mathrm{e}^{-x}$ is increasing in $[0, +\infty)$ and, thus,

$$A = \{x + i\pi \mid x \geq R\}.$$

If $R > 1$, then the function is decreasing in $[0, \log R]$ and increasing in $[\log R, +\infty)$, so that

$$A = \{x + i\pi \mid x \geq \log R + 1\}.$$

The image of the segment $\{iy \mid 0 \leq y \leq \pi\}$ is

$$B = \left\{iy - R\mathrm{e}^{-iy} \mid 0 \leq y \leq \pi\right\}$$

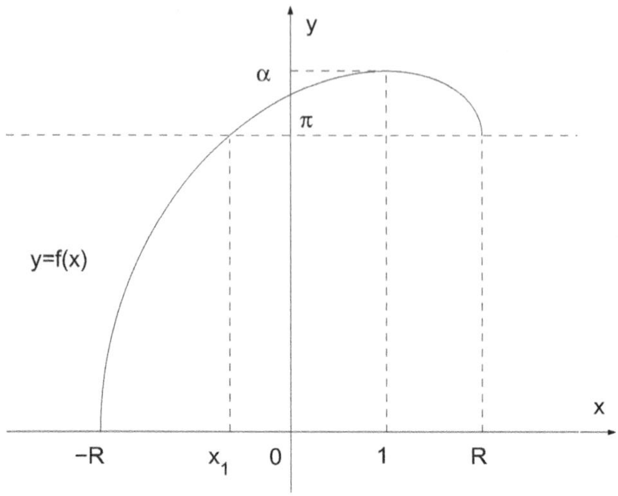

FIG. 9.3. Graph of the function f for $R > 1$.

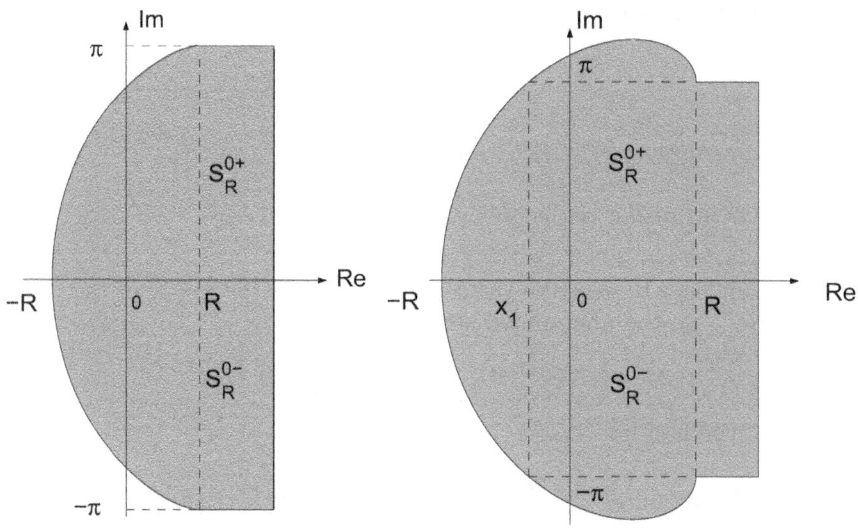

FIG. 9.4. The set S_R^0 for $R \leq 1$ (left) and $R > 1$ (right).

$$= \{R\cos\varphi + i\,(\pi - \varphi + R\sin\varphi) \mid 0 \leq \varphi \leq \pi\}.$$

By setting $x = R\cos\varphi$, we see that B is the graph of the function

$$f\,(x) = \pi - \arccos\left(\frac{x}{R}\right) + \sqrt{R^2 - x^2}, \quad x \in [-R, R].$$

If $R \leq 1$, then f is increasing in $[-R, R]$ with image $[0, \pi]$ (see Figure 9.2). If $R > 1$, then f is increasing in $[-R, 1]$ with image $[0, \alpha]$, where

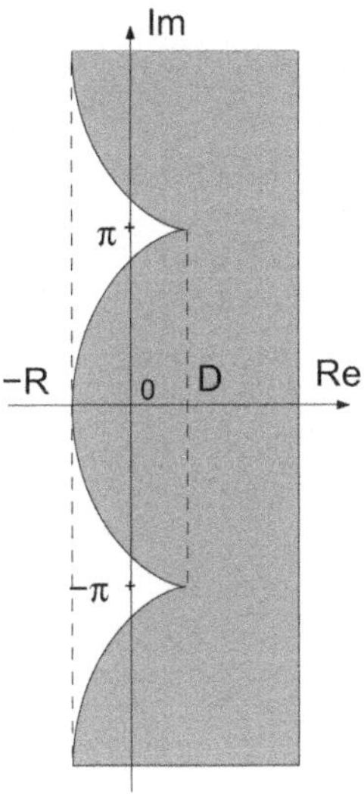

FIG. 9.5. The set S_R for $R \leq 1$.

$$\alpha = \pi - \arccos\left(\frac{1}{R}\right) + \sqrt{R^2 - 1},$$

and decreasing in $[1, R]$ with image $[\pi, \alpha]$ (see Figure 9.3). Note that, in the last case, B intersects the line $\{\zeta \in \mathbb{C} \mid \Im(\zeta) = \pi\}$ at the point of real abscissa $x_1 = R\cos\varphi_1$, where φ_1 is the solution of

$$\varphi = R\sin\varphi, \quad \varphi \in (0, \pi).$$

The image of the last piece of border, i.e. the half-line $\{x \,|\, x \geq 0\}$, is

$$C = \left\{x - R\mathrm{e}^{-x} \mid x \geq 0\right\}.$$

Since the function $x \mapsto x - R\mathrm{e}^{-x}$ is increasing in $[0, +\infty)$, we obtain

$$C = [-R, +\infty).$$

Since the set S_R^{0+} has lower bounded real parts, by (9.2.13) we can say that S_R^0 is the set shown in Figure 9.4.

By (9.2.12), we obtain that S_R is the set shown in Figure 9.5. The real abscissa $D = D(R)$ is given by

$$D = \begin{cases} R & \text{if } 0 \leq R \leq 1, \\ x_1(R) & \text{if } R > 1. \end{cases}$$

Finally, by (9.2.11), the set S_b is obtained by shifting S_R by θ along the imaginary axis. The set of complex numbers a for which $(a, b) \in S$ is the complementary set of S_b. Note that this set is the union of the cone $\Re(a) < -|b|$ and of a 2π-periodic "crest".

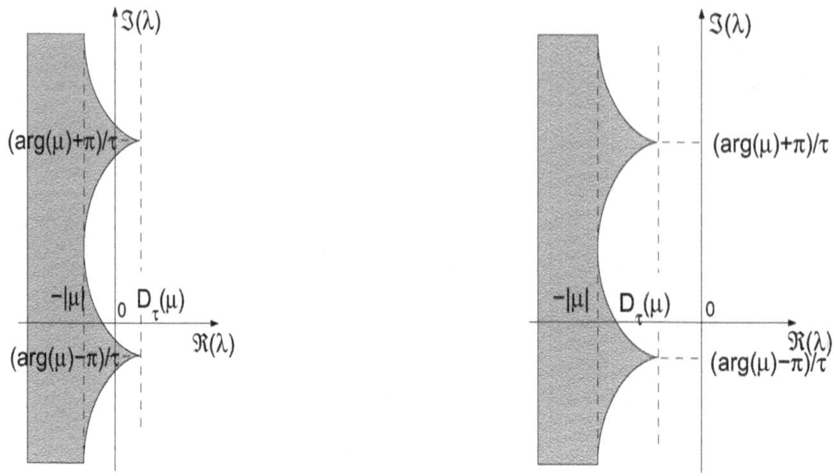

FIG. 9.6. The restriction to the complex λ-plane of the asymptotic stability region $\mathcal{S}_\tau$ for a fixed value of μ with $1/\tau < |\mu| < \pi/(2\tau)$ (left) and $\pi/(2\tau) < |\mu|$ (right).

Since $S_{-R} = S_R + i\pi$, the graph of the function $D = D(R)$ is the border of the asymptotic stability region depicted in Figure 9.1 for $\tau = 1$, $a = \lambda$, $b = \mu$ and $b < 0$.

Returning to equation (9.2.2) and to the relevant asymptotic stability set $\mathcal{S}_\tau$, it turns out that, for a given μ, the set of possible λ is sufficiently well described and is given in Figure 9.6. The peak points of the crest have a real part $\Re(\lambda) = D_\tau(\mu)$, where the pair $(D_\tau(\mu), -|\mu|)$ belongs to the border of the real stability region depicted in Figure 9.1.

As for the region of asymptotic stability for all delays, observe that the value $D_\tau(|\mu|)$ tends to $-|\mu|$ as $\tau \to +\infty$. So, the border of the region moves to the line $\Re(\lambda) = -|\mu|$ and then, for any pair (λ, μ) such that $\Re(\lambda) < -|\mu|$, the solution asymptotically vanishes for all delays. The complete region is obtained by adding the following subset of the border (see Zennaro [280, Proposition 14]):

$$\{(\lambda, \mu) \in \mathbb{C}^2 \mid \lambda \in \mathbb{R},\ |\mu| = -\lambda,\ \lambda + \mu \neq 0\}. \tag{9.2.14}$$

The results are summarized in Table 9.3.

An alternative approach for describing the set $\mathcal{S}_\tau$ consists in fixing a value of λ and looking for the possible values of μ (see Guglielmi [108]). The analysis, based on the boundary locus technique, is equivalent to the previous approach but it turns out to be more convenient to describe the case of *pure* DDEs, i.e. $\lambda = 0$, which will be considered in Section 9.7.3.

The parametric equations of the boundary of the stability regions are

TABLE 9.3. Asymptotic stability scheme for $y'(t) = \lambda y(t) + \mu y(t-\tau)$ with $\lambda, \mu \in \mathbb{C}$.

$\lim_{t\to+\infty} y(t) = 0$ for all constant delays τ
$\Updownarrow$
$\Re(\lambda) < -\lvert\mu\rvert$ or $\lambda \in \mathbb{R},\ \lvert\mu\rvert = -\lambda,\ \lambda + \mu \neq 0$
$\Downarrow$
$(\lambda, \mu) \in \mathcal{S}_\tau$
$\Updownarrow$
$\lim_{t\to+\infty} y(t) = 0$ for fixed constant delay τ

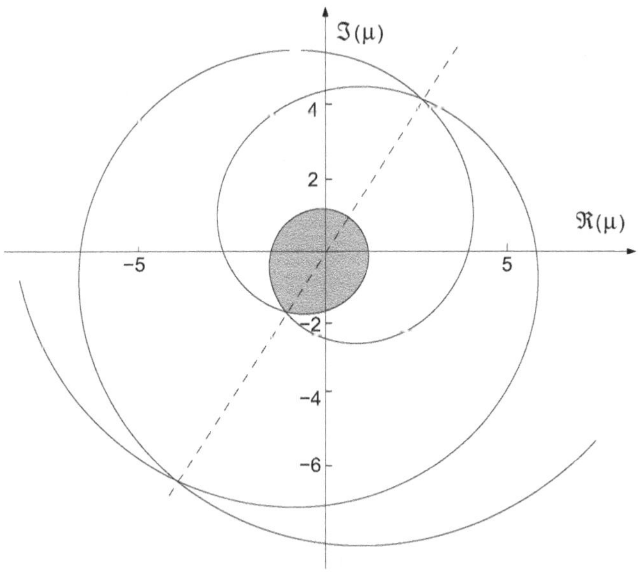

FIG. 9.7. Asymptotic stability region for equation (9.2.2) with constant delay $\tau = 1$ in the complex μ-plane for $\lambda = -1 + i$.

$$\begin{cases} \Re(\mu) = -\Re(\lambda)\cos\theta + \Im(\lambda)\sin\theta - \frac{\theta\sin\theta}{\tau}, \\ \Im(\mu) = -\Im(\lambda)\cos\theta - \Re(\lambda)\sin\theta + \frac{\theta\cos\theta}{\tau}, \end{cases}$$

where the parameter θ varies in $[0, +\infty)$. Observe that the boundary is the "sum" of a circle centered in the origin with radius $|\lambda|$ and of the transcendental curve $\left(-\frac{\theta\sin\theta}{\tau}, \frac{\theta\cos\theta}{\tau}\right)$, which does not depend on λ.

The form of the stability region in the complex μ-plane is reported in Figure 9.7 for a given complex value of λ.

9.3 Linear systems of DDEs

Consider the linear system

$$\begin{cases} y'(t) = L(t)y(t) + M(t)y(t-\tau(t)), & t \geq t_0, \\ y(t) = \phi(t), & t \leq t_0, \end{cases} \tag{9.3.1}$$

where $L(t)$ and $M(t)$ are complex valued $d \times d$ matrices.

A first result on contractivity and asymptotic stability for equation (9.3.1) can be obtained by Theorems 9.1.3 and 9.1.4. Now the quantities $Y(t)$ and $X(t)$, as defined by (8.5.3) and (8.5.4), take the form

$$Y(t) = \mu[L(t)] \quad \text{and} \quad X(t) = \|M(t)\|,$$

where $\mu[L(t)]$ is the logarithmic norm of $L(t)$ previously defined in (8.2.13) with respect to the norm $\|\cdot\|$ derived by the inner product $\langle\cdot,\cdot\rangle$ in $\mathbb{C}^d$. In this setting Theorems 9.1.3 and 9.1.4 read as follows.

Theorem 9.3.1 *If the system (9.3.1) satisfies the condition*

$$\mu[L(t)] + \|M(t)\| \leq 0, \quad t \geq t_0,$$

then it is contractive, that is

$$\|y(t)\| \leq \max_{x \leq t_0} \|\phi(x)\|, \quad t \geq t_0,$$

for any initial function $\phi(t)$ and for all delays $\tau(t)$ satisfying hypothesis (H_1).

Theorem 9.3.2 *If the system (9.3.1) satisfies the conditions*

$$\mu[L(t)] \leq \mu_0 < 0, \quad t \geq t_0,$$

and

$$R \cdot \mu[L(t)] + \|M(t)\| \leq 0, \quad t \geq t_0,$$

for some non-negative real number $R < 1$, then the solution $y(t)$, besides being contractive, is asymptotically stable, that is

$$\lim_{t \to +\infty} y(t) = 0,$$

for any initial function $\phi(t)$ and for all delays $\tau(t)$ satisfying hypotheses (H_1) and (H_2). Moreover, if also hypothesis (H_3) is satisfied, then the convergence rate is at least of exponential type, i.e. at least like $\mathrm{e}^{-\alpha(t-t_0)}$ for some $\alpha > 0$.

For the autonomous equation with constant delay

$$\begin{cases} y'(t) = Ly(t) + My(t-\tau), & t \geq t_0, \\ y(t) = \phi(t), & t \leq t_0, \end{cases} \tag{9.3.2}$$

sharper results can be obtained by using the characteristic equation. It is known (see El'sgol'ts and Norkin [81]) that the solution to (9.3.2) has an expansion of the form

$$y(t) = \sum_{i=1}^{\infty} \sum_{n_i=0}^{m_i-1} \alpha_{i,n_i} t^{n_i} \mathrm{e}^{\zeta_i t},$$

where the vectors α_{i,n_i} are determined by the initial function $\phi(t)$, and the characteristic equation has the form

$$\det[\zeta I - L - \mathrm{e}^{-\tau\zeta} M] = 0. \tag{9.3.3}$$

As with the scalar equation (9.2.7), the roots of (9.3.3) lie in the complex half-plane $\Re(\lambda) < \alpha$ for some real α and their real parts accumulate at $-\infty$. Therefore, in any vertical strip of the complex plane there are only a finite number of roots. Thus, it is evident that for the asymptotic stability of (9.3.1) it is necessary and sufficient that all the roots ζ_i of (9.3.3) be such that $\Re(\zeta_i) < 0$.

In general, this is not easy to verify. It can be seen that, if the matrices L and M are simultaneously diagonalizable, then the usual change of variable involving the common eigenvectors leads to a decoupled system of d independent scalar equations of the form (9.2.7). Therefore, the asymptotic stability of the system (9.3.2) is, in this case, completely described by the eigenvalues of the two matrices. A deeper stability analysis of the matrices L and M which are not simultaneously diagonalizable can be found in Hu and Mitsui [141].

More complicated is the case when the eigenspaces of L and M are different. It has been proved (see In 't Hout [135]) that a sufficient condition for the asymptotic stability of (9.3.2) for all constant delays τ is

$$\begin{cases} \lambda \in \sigma[L] \Rightarrow \Re(\lambda) < 0, \\ \sup\limits_{\Re(\xi)=0} \rho[(\xi I - L)^{-1} M] < 1, \end{cases} \tag{9.3.4}$$

where $\sigma[\cdot]$ and $\rho[\cdot]$ denote the spectrum and the spectral radius of a matrix, respectively. The proof is given in Theorem 9.5.2 for the more general equation with a neutral term. Similarly to the scalar case, it was also proved that condition (9.3.4) is not far from being necessary for the asymptotic stability of (9.3.2) for all delays. In fact, it is proved in [135] that the following slightly weaker condition on L and M is necessary and sufficient for the asymptotic stability for all delays:

$$\begin{cases} \lambda \in \sigma[L] \Rightarrow \Re(\lambda) < 0, \\ \rho[(\xi I - L)^{-1} M] < 1 \quad \text{whenever } \Re(\xi) = 0, \ \xi \neq 0, \\ -1 \notin \sigma[L^{-1} M]. \end{cases}$$

Observe that the resulting asymptotic stability region consists in adding some part of the border to the region defined by (9.3.4). As expected, when L and M are scalar we get the region described in Table 9.3.

As for the asymptotic stability for fixed delay, to date no explicit conditions are known that are suitable to describe the stability region of (9.3.2) and to

TABLE 9.4. Asymptotic stability scheme for $y'(t) = Ly(t) + My(t-\tau)$.

$$\lim_{t\to+\infty} y(t) = 0 \text{ for all constant delays } \tau$$

$$\Updownarrow$$

$$\begin{cases} \lambda \in \sigma[L] \Rightarrow \Re(\lambda) < 0, \\ \rho[(\xi I - L)^{-1}M] < 1 \quad \text{whenever } \Re(\xi) = 0,\ \xi \neq 0, \\ -1 \notin \sigma[L^{-1}M] \end{cases}$$

$$\Downarrow$$

$$\lim_{t\to+\infty} y(t) = 0 \text{ for fixed constant delay } \tau$$

analyze numerical methods. The special case $L = 0$ and constant M is instead successfully analyzed in Section 9.7.3. The results are summarized in Table 9.4.

9.4 General non-linear NDDEs

Consider the system of NDDEs

$$\begin{cases} y'(t) = f\Big(t, y(t), y(t-\tau(t)), y'(t-\tau(t))\Big), & t \geq t_0, \\ y(t) = \phi(t), \quad t \leq t_0, \end{cases} \tag{9.4.1}$$

where $f \in C^0\left([t_0, t_f] \times \mathbb{C}^d \times \mathbb{C}^d \times \mathbb{C}^d, \mathbb{C}^d\right)$.

For such a problem no contractivity and stability results similar to those available for DDEs of the form (9.1.1) and reported in Section 9.1 are known, which involve in a simple way the Lipschitz constants of the right-hand-side function $f(t, y, u, v)$ and which, at the same time, could be reproduced in an analogous way for the numerical methods.

On the other hand, in Chapter 3 we have seen that, besides the standard form (9.4.1), equations of neutral type may exist that, by a suitable change of variable, reduce to an ODE coupled to an algebraic recursion. The absence of the neutral term $y'(t-\tau(t))$ in the resulting system of DDAEs allows us to view them as ODEs with a forcing term and to exploit the theory developed in Section 8.5.

We first consider the neutral equation in the form

$$\begin{cases} \frac{\mathrm{d}}{\mathrm{d}t}\Big(y(t) - G(t, y(t-\tau(t)))\Big) = H(t, y(t), y(t-\tau(t))), & t \geq t_0, \\ y(t) = \phi(t), \quad t \leq t_0. \end{cases} \tag{9.4.2}$$

Setting

$$\Phi(t) = y(t) - G(t, y(t-\tau(t)))$$

(see Sections 3.3.2, 4.5 and 6.5.1) allows us to express (9.4.2) as the ODE

$$\begin{cases} \Phi'(t) = K\left(t, \Phi(t), y(t-\tau(t))\right), & t \geq t_0, \\ \Phi(t_0) = \phi(t_0) - G(t_0, \phi(t_0 - \tau(t_0))), \end{cases} \tag{9.4.3}$$

where

$$K(t, y, x) = H(t, y + G(t, x), x),$$

coupled to the algebraic recursion

$$y(t) = \begin{cases} \phi(t), & t \leq t_0, \\ \Phi(t) + G(t, y(t - \tau(t))), & t \geq t_0. \end{cases} \tag{9.4.4}$$

By looking at (9.4.3) as an ODE with a forcing term, the stability of the global system of DDAEs (9.4.3), (9.4.4) can be analyzed, for delays $\tau(t)$ satisfying hypotheses (H_1) and (H_4), by proceeding with the method of steps across the macro intervals $[\xi_{k-1}, \xi_k]$ determined by the primary discontinuities.

To this end, we consider another system of DDAEs

$$\begin{cases} \tilde{\Phi}'(t) = K\left(t, \tilde{\Phi}(t), \tilde{y}(t - \tau(t))\right), & t \geq t_0, \\ \tilde{\Phi}(t_0) = \tilde{\phi}(t_0) - G(t_0, \tilde{\phi}(t_0 - \tau(t_0))), \end{cases} \tag{9.4.5}$$

$$\tilde{y}(t) = \begin{cases} \tilde{\phi}(t), & t \leq t_0, \\ \tilde{\Phi}(t) + G(t, \tilde{y}(t - \tau(t))), & t \geq t_0, \end{cases} \tag{9.4.6}$$

with the same right-hand sides $K(t, y, x)$ and $G(t, x)$, but different initial function $\tilde{\phi}(t)$, and define

$$\begin{aligned} \delta y(t) &= \tilde{y}(t) - y(t), \\ \delta\Phi(t) &= \tilde{\Phi}(t) - \Phi(t), \\ \delta\phi(t) &= \tilde{\phi}(t) - \phi(t). \end{aligned}$$

The following theorem was proved by Torelli and Vermiglio [250] and is essentially based on Proposition 8.5.1.

Theorem 9.4.1 *Consider the systems of DDAEs (9.4.3), (9.4.4) and (9.4.5), (9.4.6) and assume that, for an inner product $\langle \cdot, \cdot \rangle$ and induced norm $\| \cdot \|$ in $\mathbb{C}^d$, there exist continuous functions $Y(t)$, $X(t)$ and $Z(t)$ such that*

$$Y(t) \geq \sup_{y_1 \neq y_2, x} \frac{\Re(\langle K(t, y_1, x) - K(t, y_2, x), y_1 - y_2 \rangle)}{\|y_1 - y_2\|^2},$$

$$X(t) \geq \sup_{y, x_1 \neq x_2} \frac{\|K(t, y, x_1) - K(t, y, x_2)\|}{\|x_1 - x_2\|},$$

$$Z(t) \geq \sup_{x_1 \neq x_2} \frac{\|G(t, x_1) - G(t, x_2)\|}{\|x_1 - x_2\|}.$$

If there exist a bounded function $r(t) \leq 0$, $r(t_0) < 0$, integrable in any bounded interval and such that

$$X(t) = r(t)Y(t), \quad t \geq t_0, \tag{9.4.7}$$

and a non-negative constant $\rho \leq 1$ such that

$$\sup_{t_0 \le x \le t} |r(x)| + Z(t) \le \rho, \quad t \ge t_0, \tag{9.4.8}$$

then, for all delays $\tau(t)$ satisfying hypotheses (H_1) and (H_4) and for all initial functions $\phi(t)$ and $\tilde{\phi}(t)$, the bound

$$\|\delta y(t)\| \le \max\left\{ \sup_{x \le t_0} \|\delta\phi(x)\|, \frac{\|\delta\Phi(t_0)\|}{-r(t_0)} \right\}, \quad t \ge t_0, \tag{9.4.9}$$

holds.

Moreover, if $\rho < 1$ and

$$Y(t) \le Y_0 < 0, \quad t \ge t_0, \tag{9.4.10}$$

then the system (9.4.3), (9.4.4), and thus (9.4.2), is asymptotically stable, that is

$$\lim_{t \to +\infty} \delta y(t) = 0$$

for any initial functions $\phi(t)$ and $\tilde{\phi}(t)$ and for all delays $\tau(t)$ also satisfying hypothesis (H_2). Finally, if also hypothesis (H_3) holds, then the convergence rate is at least of exponential type, i.e. at least as $\mathrm{e}^{-\alpha(t-t_0)}$ for some $\alpha > 0$.

Proof. By Proposition 8.5.1, for any $t \in I_k = [\xi_{k-1}, \xi_k]$ we have

$$\|\delta\Phi(t)\| \le E(\xi_{k-1}, t)\|\delta\Phi(\xi_{k-1})\| + (1 - E(\xi_{k-1}, t)) \sup_{\xi_{k-1} \le x \le t} (|r(x)| \cdot \|\delta y(x - \tau(x))\|), \tag{9.4.11}$$

where $E(t_1, t_2) = \exp\left(\int_{t_1}^{t_2} Y(x)\mathrm{d}x\right) \le 1$. In particular, for $t = \xi_k$,

$$\|\delta\Phi(\xi_k)\| \le E(\xi_{k-1}, \xi_k)\|\delta\Phi(\xi_{k-1})\| + (1 - E(\xi_{k-1}, \xi_k)) \sup_{x \in I_k} (|r(x)| \cdot \|\delta y(x - \tau(x))\|), \tag{9.4.12}$$

where $E(\xi_{k-1}, \xi_k) \le 1$. Now, for any vector function $v(x)$, set

$$|||v|||_0 = \max_{x \le t_0} \|v(x)\|$$

and, for any integer $l \ge 1$,

$$|||v|||_l = \max_{x \in I_l} \|v(x)\|.$$

Moreover, for any scalar function $u(x)$, set

$$u^*(t) = \sup_{t_0 \le x \le t} |u(x)|.$$

Then, by (9.4.4) and (9.4.6) we have

$$\|\delta y(t)\| \le \|\delta\Phi(t)\| + Z(t)\|\delta y(t - \tau(t))\|, \quad t \ge t_0,$$

and hence, by (9.4.11) and (9.4.8), taking hypothesis (H_4) into account,

$$\begin{aligned}\|\delta y(t)\| &\leq E(\xi_{k-1},t)\|\delta\Phi(\xi_{k-1})\| + \Big((1-E(\xi_{k-1},t))r^*(t)+Z(t)\Big)|||\delta y|||_{k-1}\\ &\leq E(\xi_{k-1},t)r^*(t)\tfrac{\|\delta\Phi(\xi_{k-1})\|}{r^*(\xi_{k-1})} + \Big(\rho - E(\xi_{k-1},t)r^*(t)\Big)|||\delta y|||_{k-1}\end{aligned} \tag{9.4.13}$$

for all $t \in I_k$.

For $\rho = 1$, the last term is a convex combination of $\frac{\|\delta\Phi(\xi_{k-1})\|}{r^*(\xi_{k-1})}$ and $|||\delta y|||_{k-1}$. Therefore,

$$|||\delta y|||_k \leq \max\left\{\frac{\|\delta\Phi(\xi_{k-1})\|}{r^*(\xi_{k-1})}, |||\delta y|||_{k-1}\right\}. \tag{9.4.14}$$

On the other hand, by (9.4.12),

$$\|\delta\Phi(\xi_k)\| \leq E(\xi_{k-1},\xi_k)\|\delta\Phi(\xi_{k-1})\| + (1-E(\xi_{k-1},\xi_k))r^*(\xi_k)|||\delta y|||_{k-1}.$$

Thus, by the monotonicity of $r^*(t)$,

$$\frac{\|\delta\Phi(\xi_k)\|}{r^*(\xi_k)} \leq E(\xi_{k-1},\xi_k)\frac{\|\delta\Phi(\xi_{k-1})\|}{r^*(\xi_{k-1})} + (1-E(\xi_{k-1},\xi_k))|||\delta y|||_{k-1} \tag{9.4.15}$$

and, hence,

$$\frac{\|\delta\Phi(\xi_k)\|}{r^*(\xi_k)} \leq \max\left\{\frac{\|\delta\Phi(\xi_{k-1})\|}{r^*(\xi_{k-1})}, |||\delta y|||_{k-1}\right\}. \tag{9.4.16}$$

Finally, (9.4.14) and (9.4.16) imply that the sequence of numbers

$$\max\left\{\frac{\|\delta\Phi(\xi_k)\|}{r^*(\xi_k)}, |||\delta y|||_k\right\}, \qquad k = 0, 1, \ldots,$$

is non-increasing and, hence, (9.4.9) is proved.

In order to prove asymptotic stability, assume $\rho < 1$ and observe that the pair of inequalities (9.4.15) and (9.4.13), evaluated at a suitable point $t = s_k \in I_k$ which maximizes the left-hand side $\|\delta y(t)\|$, can be rewritten componentwise as

$$Q_k \leq W_k Q_{k-1},$$

where $Q_k = \left(\frac{\|\delta\Phi(\xi_k)\|}{r^*(\xi_k)}, |||\delta y|||_k\right)^T$ and

$$W_k = \begin{bmatrix} E(\xi_{k-1},\xi_k) & 1-E(\xi_{k-1},\xi_k) \\ E(\xi_{k-1},s_k)r^*(s_k) & \rho - E(\xi_{k-1},s_k)r^*(s_k) \end{bmatrix}.$$

By hypothesis (H_1) and (9.4.10), it holds that

$$E(\xi_{k-1},\xi_k) \leq e^{Y_0\tau_0} < 1 \tag{9.4.17}$$

for all $k \geq 1$, and then we can find a 2×2 matrix norm $\|\cdot\|$ such that $\|W_k\| \leq \nu_* < 1$ for all $k \geq 1$. To this end, define

$$\|W_k\| = \|P^{-1}W_kP\|_\infty \quad \text{with } P = \begin{bmatrix} 1 & 0 \\ 0 & \epsilon \end{bmatrix} \text{ and } \epsilon < 1.$$

Direct computation of $\|W_k\|$ leads to

$$\|W_k\| = \max\left\{\epsilon + (1-\epsilon)E(\xi_{k-1},\xi_k)\ ,\ (\epsilon^{-1}-1)E(\xi_{k-1},s_k)r^*(s_k) + \rho\right\}$$

and, by (9.4.8) and (9.4.17), to

$$\|W_k\| \le \max\left\{\epsilon + (1-\epsilon)\mathrm{e}^{Y_0\tau_0}\ ,\ \epsilon^{-1}\rho\right\} \quad \forall k \ge 1.$$

With the choice $\epsilon^* = \frac{1+\rho}{2}$, we have

$$\|W_k\| \le \nu_* = \max\left\{\frac{1+\rho}{2} + \frac{1-\rho}{2}\mathrm{e}^{Y_0\tau_0}\ ,\ \frac{2\rho}{1+\rho}\right\} < 1 \quad \forall k \ge 1,$$

and then $\|Q_k\| \to 0$ as $k \to +\infty$ at least like ν_*^k. Since the sequence of ξ_k's diverges because of hypotheses (H_1) and (H_2), the asymptotic stability is proved.

Finally, assume also that hypothesis (H_3) holds. In this case, for a given $t \ge t_0$, we have that $t \in I_k$ for some $k \ge (t-t_0)/\tau_1$. Therefore, it follows that $\|\delta y(t)\| \to 0$ at least as $\mathrm{e}^{-\alpha(t-t_0)}$, where $\alpha = -\log(\nu_*)/\tau_1$. ■

Besides the particular form (9.4.2), the NDDE (9.4.1) may sometimes have the form (see again Sections 3.3.2, 4.5 and 6.5.1)

$$\begin{cases} y'(t) = F\Big(t, y(t), G\big(t, y(t-\tau(t)), y'(t-\tau(t))\big)\Big), & t \ge t_0, \\ y(t) = \phi(t), \quad t \le t_0. \end{cases} \tag{9.4.18}$$

Setting

$$\Psi(t) = G\Big(t, y(t-\tau(t)), y'(t-\tau(t))\Big)$$

allows us to express (9.4.18) as the ODE

$$\begin{cases} y'(t) = F\left(t, y(t), \Psi(t)\right), & t \ge t_0, \\ y(t_0) = \phi(t_0), \end{cases} \tag{9.4.19}$$

coupled with the algebraic recursion

$$\Psi(t) = \begin{cases} G\Big(t, \phi(t-\tau(t)), \phi'(t-\tau(t))\Big), & t-\tau(t) \le t_0, \\ H\Big(t, y(t-\tau(t)), \Psi(t-\tau(t))\Big), & t-\tau(t) > t_0, \end{cases} \tag{9.4.20}$$

where

$$H(t,y,x) = G\Big(t, y, F(t-\tau(t), y, x)\Big).$$

Observe that $\Psi(t)$, and hence $y'(t)$, is not continuous from the left at the discontinuity points ξ_k.

As in the case of the formulation (9.4.2), by looking at (9.4.19) as an ODE with a forcing term, the stability of the global delay DAE system (9.4.19), (9.4.20) can be analyzed, for delays $\tau(t)$ satisfying hypotheses (H_1) and (H_4), by proceeding with the method of steps across the macro intervals $[\xi_{k-1}, \xi_k]$ determined by the primary discontinuity points. To this end, we consider again another system

$$\begin{cases} \tilde{y}'(t) = F\left(t, \tilde{y}(t), \tilde{\Psi}(t)\right), & t \geq t_0, \\ \tilde{y}(t_0) = \tilde{\phi}(t_0), \end{cases} \tag{9.4.21}$$

coupled with the algebraic recursion

$$\tilde{\Psi}(t) = \begin{cases} G\Big(t, \tilde{\phi}(t-\tau(t)), \tilde{\phi}'(t-\tau(t))\Big), & t-\tau(t) \leq t_0, \\ H\Big(t, \tilde{y}(t-\tau(t)), \tilde{\Psi}(t-\tau(t))\Big), & t-\tau(t) > t_0, \end{cases} \tag{9.4.22}$$

with the same right-hand sides $F(t,y,x)$, $H(t,y,x)$ and $G(t,x)$, but different initial function $\tilde{\phi}(t)$, and define

$$\begin{aligned} \delta y(t) &= \tilde{y}(t) - y(t), \\ \delta\Psi(t) &= \tilde{\Psi}(t) - \Psi(t), \\ \delta\phi(t) &= \tilde{\phi}(t) - \phi(t). \end{aligned}$$

The following theorem can be found in Bellen, Guglielmi and Zennaro [30] and its proof is similar to the proof of Theorem 9.4.1.

Theorem 9.4.2 *Consider the systems of DDAEs (9.4.19), (9.4.20) and (9.4.21), (9.4.22) and assume that, for an inner product $\langle \cdot, \cdot \rangle$ and induced norm $\| \cdot \|$ in $\mathbb{C}^d$, there exist continuous functions $Y(t)$, $X(t)$, $W(t)$ and $Z(t)$ such that*

$$Y(t) \geq \sup_{y_1 \neq y_2, x} \frac{\Re(\langle F(t,y_1,x) - F(t,y_2,x), y_1 - y_2 \rangle)}{\|y_1 - y_2\|^2},$$

$$X(t) \geq \sup_{y, x_1 \neq x_2} \frac{\|F(t,y,x_1) - F(t,y,x_2)\|}{\|x_1 - x_2\|},$$

$$W(t) \geq \sup_{y_1 \neq y_2, x} \frac{\|H(t,y_1,x) - H(t,y_2,x)\|}{\|y_1 - y_2\|},$$

$$Z(t) \geq \sup_{y, x_1 \neq x_2} \frac{\|H(t,y,x_1) - H(t,y,x_2)\|}{\|x_1 - x_2\|}.$$

If there exists a bounded function $r(t) < 0$, integrable in any bounded interval and such that

$$X(t) = r(t)Y(t), \quad t \geq t_0, \tag{9.4.23}$$

and

$$-r(t)\Big(W(t) + \frac{Z(t)}{-r(t-\tau(t))}\Big) \leq 1, \quad t \geq t_0, \tag{9.4.24}$$

then, for all delays $\tau(t)$ satisfying hypotheses (H_1) and (H_4) and for all initial functions $\phi(t)$ and $\tilde{\phi}(t)$, the bound

$$\|\delta y(t)\| \le \max\left\{ \|\delta\phi(t_0)\|, \sup_{t_0 \le x \le \xi_1} (|r(x)| \cdot \|\delta\Psi(x)\|) \right\}, \quad t \ge t_0, \tag{9.4.25}$$

holds, where ξ_1 is the first primary discontinuity point (i.e. $\xi_1 - \tau(\xi_1) = t_0$).

Moreover, if

$$Y(t) \le Y_0 < 0, \quad t \ge t_0, \tag{9.4.26}$$

and if there exist two non-negative constants $\nu < 1$ and $k < 1$ such that

$$\frac{r(t)}{r(t-\tau(t))} Z(t) \le \nu, \quad t \ge t_0, \tag{9.4.27}$$

and

$$-r(t)W(t) \le k\Big(1 - \frac{r(t)}{r(t-\tau(t))} Z(t)\Big), \quad t \ge t_0, \tag{9.4.28}$$

then the system (9.4.19), (9.4.20), and thus (9.4.18), is asymptotically stable, that is

$$\lim_{t\to+\infty} \delta y(t) = 0$$

for any initial functions $\phi(t)$ and $\tilde{\phi}(t)$ and for all delays $\tau(t)$ also satisfying hypothesis (H_2). Finally, if also hypothesis (H_3) holds, then the convergence rate is at least of exponential type, i.e. at least as $\mathrm{e}^{-\alpha(t-t_0)}$ for some $\alpha > 0$.

9.5 Linear systems of NDDEs

The linear version of (9.4.1) is

$$\begin{cases} y'(t) = L(t)y(t) + M(t)y(t-\tau(t)) + N(t)y'(t-\tau(t)), & t \ge t_0, \\ y(t) = \phi(t), \quad t \le t_0, \end{cases} \tag{9.5.1}$$

where $L(t)$, $M(t)$ and $N(t)$ are complex valued $d \times d$ matrices.

For the stability analysis of (9.5.1), there are various equivalent formulations that avoid explicitly using the delay term $y'(t-\tau)$. Integration of the new problem by the method of steps leads to different conditions on the coefficients for the boundedness and/or asymptotic stability of the solution.

Whenever the linear NDDE is given in or may be reduced to the form

$$\begin{cases} \frac{\mathrm{d}}{\mathrm{d}t}\Big(y(t) - N(t)y(t-\tau(t))\Big) = L(t)y(t) + M(t)y(t-\tau(t)), & t \ge t_0, \\ y(t) = \phi(t), \quad t \le t_0, \end{cases} \tag{9.5.2}$$

the first approach proposed in Section 9.4 consists in the substitution

$$\Phi(t) = y(t) - N(t)y(t-\tau(t)),$$

for which the coupled systems (9.4.3), (9.4.4) take the form

$$\begin{cases} \Phi'(t) = L(t)\Phi(t) + \Big(L(t)N(t) + M(t)\Big)y(t - \tau(t)), & t_0 \le t, \\ \Phi(t_0) = \phi(t_0) - N(t_0)\phi(t_0 - \tau(t_0)), \end{cases}$$

and

$$y(t) = \begin{cases} \phi(t), & t \le t_0, \\ \Phi(t) + N(t)y(t - \tau(t)), & t \ge t_0. \end{cases}$$

The corresponding boundedness conditions (9.4.7) and (9.4.8) become

$$\mu[L(t)] < 0$$

and

$$\max_{t_0 \le x \le t} \frac{\|L(x)N(x) + M(x)\|}{-\mu[L(x)]} + \|N(t)\| \le 1, \quad t \ge t_0, \tag{9.5.3}$$

whereas the bound (9.4.9) becomes

$$\|y(t)\| \le \max\left\{ \sup_{x \le t_0} \|\phi(x)\|, \frac{\|M(t_0)\| \cdot \|\phi(t_0) - N(t_0)\phi(t_0 - \tau(t_0))\|}{-\mu[L(t_0)]} \right\}. \tag{9.5.4}$$

The stronger conditions for asymptotic stability become

$$\mu[L(t)] \le Y_0 < 0$$

and

$$\max_{t_0 \le x \le t} \frac{\|L(x)N(x) + M(x)\|}{-\mu[L(x)]} + \|N(t)\| \le \rho < 1, \quad t \ge t_0. \tag{9.5.5}$$

Observe that the coefficient $N(t)$ has different roles in the two formulations (9.5.2) and (9.5.1). In fact, in (9.5.2) it appears under differentiation.

In the case when the NDDE is written in the standard form (9.5.1), we can rewrite it as

$$\begin{cases} y'(t) = L(t)y(t) + \Big(M(t) + N(t)L(t - \tau(t))\Big)y(t - \tau(t)) \\ \qquad + N(t)\Big(y'(t - \tau(t)) - L(t - \tau(t))y(t - \tau(t))\Big), \quad t \ge t_0, \\ y(t) = \phi(t), \quad t \le t_0, \end{cases}$$

which is the linear counterpart of (9.4.18). Therefore, by setting

$$\Psi(t) = y'(t) - L(t)y(t)$$

(see (9.4.20)), the system (9.5.1) is equivalent to

$$\begin{cases} y'(t) = L(t)y(t) + \Psi(t), & t \ge t_0, \\ y(t_0) = \phi(t_0), \end{cases}$$

along with the algebraic recursion

$$\Psi(t) = \begin{cases} M(t)\phi(t - \tau(t)) + N(t)\phi'(t - \tau(t)), & t - \tau(t) \le t_0, \\ \Big(M(t) + N(t)L(t - \tau(t))\Big)y(t - \tau(t)) + N(t)\Psi(t - \tau(t)), \\ \hfill t - \tau(t) > t_0. \end{cases}$$

The corresponding boundedness conditions (9.4.23) and (9.4.24) become

$$\mu[L(t)] < 0 \tag{9.5.6}$$

and

$$\frac{\|N(t)L(t-\tau(t)) + M(t)\| - \mu[L(t-\tau(t))]\|N(t)\|}{-\mu[L(t)]} \le 1, \quad t \ge t_0, \tag{9.5.7}$$

whereas the bound (9.4.25) becomes

$$\|y(t)\| \le \max\left\{\|\phi(t_0)\|, \max_{t_0 \le x \le \xi_1} \frac{\|M(x)\phi(x-\tau(x)) + N(x)\phi'(x-\tau(x))\|}{-\mu[L(x)]}\right\}. \tag{9.5.8}$$

The stronger conditions (9.4.26), (9.4.27) and (9.4.28) become

$$\mu[L(t)] \le Y_0 < 0,$$

$$\frac{\mu[L(t-\tau(t))]}{\mu[L(t)]}\|N(t)\| \le \nu < 1, \quad t \ge t_0, \tag{9.5.9}$$

and

$$\frac{\|N(t)L(t-\tau(t)) + M(t)\|}{-\mu[L(t)]} \le k\left(1 - \frac{\mu[L(t-\tau(t))]}{\mu[L(t)]}\|N(t)\|\right), \quad t \ge t_0, \tag{9.5.10}$$

for a constant $k < 1$, and imply asymptotic stability.

For the constant coefficients and constant delay equation

$$y'(t) = Ly(t) + My(t-\tau) + Ny'(t-\tau) \tag{9.5.11}$$

satisfying

$$\mu[L] < 0,$$

the boundedness conditions (9.5.3) and (9.5.7) reduce to

$$\frac{\|LN + M\|}{-\mu[L]} + \|N\| \le 1$$

and

$$\frac{\|NL + M\|}{-\mu[L]} + \|N\| \le 1, \tag{9.5.12}$$

respectively, whereas the bounds (9.5.4) and (9.5.8) become

$$\|y(t)\| \le \max\left\{\sup_{x \le t_0}\|\phi(x)\|, \frac{\|M\| \cdot \|\phi(t_0) - N\phi(t_0 - \tau(t_0))\|}{-\mu[L]}\right\}$$

and

$$\|y(t)\| \le \max\left\{\|\phi(t_0)\|, \max_{t_0 \le x \le \xi_1} \frac{\|M\phi(x-\tau(x)) + N\phi'(x-\tau(x))\|}{-\mu[L]}\right\},$$

respectively. Moreover, the asymptotic stability condition (9.5.5) reduces to

$$\frac{\|LN + M\|}{-\mu[L]} + \|N\| < 1,$$

whereas conditions (9.5.9) and (9.5.10) become

$$\frac{\|NL + M\|}{-\mu[L]} + \|N\| < 1. \tag{9.5.13}$$

Similar criteria for the asymptotic stability of the solutions of (9.5.11) have been found by Hu and Mitsui [140], namely

$$\mu[L] + \|M\| + \frac{\|NL\| + \|NM\|}{1 - \|N\|} < 0,$$

and by Slemrod and Infante [233], namely

(i) $\rho[N] < 1$;

(ii) $L + L^T \pm M^T N \pm N^T M$ is negative definite;

(iii) $H = \begin{bmatrix} L + L^T - M^T N - N^T M & 2\,(M - L^T N) \\ 2\,(M^T - N^T L) & L + L^T - M^T N - N^T M \end{bmatrix}$
is negative definite.

The four conditions for asymptotic stability displayed above are not comparable with one another. There are cases in which each of them is less restrictive than the others.

Sharper results can be obtained for the asymptotic stability analysis by using the characteristic equation

$$\det[\zeta I - L - (M + N\zeta)\mathrm{e}^{-\tau\zeta}] = 0.$$

This equation has infinitely many solutions ζ_i that, for $N \neq 0$, lie in the vertical strip $\alpha < \Re(\zeta) < \beta$ for some real α and β, and $|\lambda_i| \to \infty$ as $i \to \infty$. Therefore, even if all the roots of the characteristic equation satisfy $\Re(\zeta_i) < 0$, they could stay arbitrarily close to the imaginary axis and may prevent asymptotic stability. Indeed, Snow [235] has shown that, in this case, it is possible for the solution $y(t)$ to be unbounded as $t \to +\infty$. It was proved by Miranker [202] that equation (9.5.11) is stable if the characteristic roots are bounded away from the imaginary axis, that is

$$\det[\zeta I - L - (M + N\zeta)\mathrm{e}^{-\tau\zeta}] = 0 \quad \Longrightarrow \quad \Re(\zeta) \leq \delta < 0.$$

If the characteristic roots ζ_i are such that $\Re(\zeta_i) < 0$, then in order to keep them bounded away from the imaginary axis, it is sufficient that $\rho[N] < 1$. In fact, the following theorem holds.

Theorem 9.5.1 *If $\rho[N] < 1$ in equation (9.5.11) and if all the characteristic roots ζ_i are such that $\Re(\zeta_i) < 0$, then there exists $\delta < 0$ such that $\Re(\zeta_i) \leq \delta$.*

Proof. Assume, by contradiction, that $|\zeta_i| \to \infty$ and $\Re(\zeta_i) \to 0$. Then we have

$$\det\left[\zeta_i I - L - (M + N\zeta_i)e^{-\tau\zeta_i}\right] = \det\left[\zeta_i e^{-\tau\zeta_i} I\right] \cdot \det\left[e^{\tau\zeta_i} I - \frac{e^{\tau\zeta_i}}{\zeta_i} L - \frac{1}{\zeta_i} M - N\right] = 0.$$

Hence,

$$\det[\rho_i\, I - (B_i + N)] = 0,$$

where

$$|\rho_i| = |e^{\tau\zeta_i}| \quad \text{and} \quad B_i = \frac{e^{\tau\zeta_i}}{\zeta_i} L + \frac{1}{\zeta_i} M$$

are such that

$$\lim_{i\to\infty} |\rho_i| = 1 \quad \text{and} \quad \lim_{i\to\infty} B_i = 0. \tag{9.5.14}$$

Therefore, given an arbitrary norm $\|\cdot\|$ in $\mathbb{C}^d$, there exists a sequence of eigenvectors x_i with $\|x_i\| = 1$ such that

$$(B_i + N)x_i = \rho_i x_i$$

and, consequently,

$$\|B_i + N\| \geq |\rho_i|.$$

Thus, (9.5.14) yields

$$\|N\| \geq 1$$

which, by the arbitrariness of the norm, contradicts $\rho[N] < 1$. ■

A sufficient condition for the asymptotic stability of (9.5.11) for all constant delays is given by the following theorem that generalizes condition (9.3.4) relevant to the non-neutral case.

Theorem 9.5.2 *The system (9.5.11) is asymptotically stable for all constant delays if*

$$\begin{cases} \lambda \in \sigma[L] \Rightarrow \Re(\lambda) < 0, \\ \sup\limits_{\Re(\xi)=0} \rho[(\xi I - L)^{-1}(M + \xi N)] < 1. \end{cases} \tag{9.5.15}$$

Proof. Since $\xi I - L$ is non-singular for all ξ such that $\Re(\xi) \geq 0$, all the eigenvalues $\lambda(\xi)$ of $(\xi I - L)^{-1}(M + \xi N)$ are analytic functions of ξ. Therefore, we can apply the maximum modulus principle to obtain

$$\sup_{\Re(\xi)=0} \rho\left[(\xi I - L)^{-1}(M + \xi N)\right] = \sup_{\Re(\xi)\geq 0} \rho\left[(\xi I - L)^{-1}(M + \xi N)\right].$$

We are now in a position to prove that

$$\det\left[\zeta I - L - (M + N\zeta)e^{-\tau\zeta}\right] = 0 \quad \Longrightarrow \quad \Re(\zeta) < 0.$$

By contradiction, assume to the contrary that $\Re(\zeta) \geq 0$. Consequently, the matrix $\zeta I - L$ is non-singular and the equality

$$\det\left[\zeta I - L - (M + N\zeta)e^{-\tau\zeta}\right]$$

$$= \det[\zeta I - L] \cdot \det\left[I - (\zeta I - L)^{-1}(M + \zeta N)e^{-\tau\zeta}\right] = 0$$

implies that $e^{\tau\zeta}$ is an eigenvalue of $(\zeta I - L)^{-1}(M + \zeta N)$. On the other hand, it holds that $|e^{\tau\zeta}| \geq 1$, against the hypothesis (9.5.15).

In order to apply Theorem 9.5.1, we are left to prove that $\rho[N] < 1$. To this end, let us consider the imaginary numbers ix, $x \neq 0$, and observe that, by hypothesis (9.5.15),

$$\sup_{x\neq 0} \rho\left[(ixI - L)^{-1}(M + ixN)\right] = \sup_{x\neq 0} \rho\left[\left(I - \frac{L}{ix}\right)^{-1}\left(\frac{M}{ix} + N\right)\right] < 1.$$

Passing to the limit for $|x| \to \infty$ yields the desired inequality $\rho[N] < 1$. ■

Condition (9.5.15), which coincides with condition (9.3.4) for non-neutral cases, is not necessary for the asymptotic stability of (9.5.11) for all delays. Actually, characterizing the asymptotic stability region of the neutral system (9.5.11) is more complicated than the analogous case for $N = 0$. However, since it is satisfied with equations where $\rho[N] < 1$, by a slight modification of the proof given in Kuang, Xiang and Tian [181] based on Theorem 9.5.1, one can prove that the following weaker conditions on L, M and N are necessary and sufficient for asymptotic stability for all delays:

$$\begin{cases} \lambda \in \sigma[L] \Rightarrow \Re(\lambda) < 0, \\ \rho\left[(\xi I - L)^{-1}(M + \xi N)\right] < 1 \quad \text{whenever } \Re(\xi) = 0,\ \xi \neq 0, \\ -1 \notin \sigma[L^{-1}M]. \end{cases} \tag{9.5.16}$$

By still confining our interest to a subclass of problems (9.5.11), the following theorem holds.

Theorem 9.5.3 *The system (9.5.11) is asymptotically stable for all constant delays if*

$$\rho[N] < 1 \tag{9.5.17}$$

and

$$\zeta \in \sigma\left[(I - \xi N)^{-1}(L + \xi M)\right] \quad \Longrightarrow \quad \Re(\zeta) < 0 \tag{9.5.18}$$

holds for all $\xi \in \mathbb{C}$ such that $|\xi| \leq 1$.

Proof. In view of Theorem 9.5.1, it is sufficient to prove

$$\det\left[\zeta I - L - (M + N\zeta)e^{-\tau\zeta}\right] = 0 \quad \Longrightarrow \quad \Re(\zeta) < 0.$$

Assume, by contradiction, that $\Re(\zeta) \geq 0$ and, hence, that $|e^{-\tau\zeta}| \leq 1$. Therefore, $\det\left[I - e^{-\tau\zeta}N\right] \neq 0$ and

$$\begin{aligned} &\det\left[\zeta I - L - (M + N\zeta)e^{-\tau\zeta}\right] \\ &\quad = \det\left[\zeta(I - e^{-\tau\zeta}N) - (L + e^{-\tau\zeta}M)\right] \end{aligned}$$

TABLE 9.5. Asymptotic stability scheme for $y'(t) = Ly(t)+My(t-\tau)+Ny'(t-\tau)$ with $\rho[N] < 1$.

$$\lim_{t\to+\infty} y(t) = 0 \text{ for all constant delays } \tau$$

$$\Updownarrow$$

$$\begin{cases} \lambda \in \sigma[L] \Rightarrow \Re(\lambda) < 0, \\ \rho\big[(\xi I - L)^{-1}(M + \xi N)\big] < 1 \quad \text{whenever } \Re(\xi) = 0,\ \xi \neq 0, \\ -1 \notin \sigma[L^{-1}M] \end{cases}$$

$$\Updownarrow$$

$$\begin{cases} \lambda \in \sigma[L] \Rightarrow \Re(\lambda) < 0, \\ \zeta \in \sigma\big[(I - \xi N)^{-1}(L + \xi M)\big],\ \zeta \neq 0 \Longrightarrow \Re(\zeta) < 0 \quad \text{for all } \xi \in \mathbb{C},\ |\xi| = 1, \\ -1 \notin \sigma[L^{-1}M] \end{cases}$$

$$= \det\left[I - e^{-\tau\zeta}N\right] \cdot \det\left[\zeta I - (I - e^{-\tau\zeta}N)^{-1}(L + e^{-\tau\zeta}M)\right] = 0,$$

which implies

$$\det\left[\zeta I - (I - e^{-\tau\zeta}N)^{-1}(L + e^{-\tau\zeta}M)\right] = 0.$$

This means that ζ is an eigenvalue of $(I - e^{-\tau\zeta}N)^{-1}(L + e^{-\tau\zeta}M)$ with a non-negative real part, against the hypothesis (9.5.18). ■

The foregoing result is what Hu and Mitsui essentially proved in [140].

In the class of problems (9.5.11) where $\rho[N] < 1$, Qiu, Yang and Kuang [221] proved that the following condition, slightly weaker than (9.5.18), is necessary and sufficient for asymptotic stability for all delays and, hence, equivalent to (9.5.16):

$$\begin{cases} \lambda \in \sigma[L] \Rightarrow \Re(\lambda) < 0, \\ \zeta \in \sigma\big[(I - \xi N)^{-1}(L + \xi M)\big],\ \zeta \neq 0 \Longrightarrow \Re(\zeta) < 0 \quad \text{for all } \xi \in \mathbb{C},\ |\xi| = 1, \\ -1 \notin \sigma[L^{-1}M]. \end{cases}$$

The results for equation (9.5.11) are summarized in Table 9.5.

9.6 Linear scalar NDDEs

Conditions for the boundedness and asymptotic stability of linear scalar equations with a neutral term and with delay $\tau(t)$ satisfying (H_1) and (H_4) come directly from the results of Section 9.5, according to the various formulations of the problem.

In particular, for the constant coefficient and constant delay case

$$\begin{cases} y'(t) = \lambda y(t) + \mu y(t-\tau) + \nu y'(t-\tau), & t \geq t_0, \\ y(t) = \phi(t), & t \leq t_0, \end{cases} \tag{9.6.1}$$

with $\lambda, \mu, \nu \in \mathbb{R}$ and $|\nu| < 1$, the conditions in Table 9.5 for asymptotic stability read

$$\lambda \le \mu < -\lambda \quad \Longleftrightarrow \quad \lim_{t\to+\infty} y(t) = 0 \quad \text{for all } \tau,$$

as essentially proved by Brayton and Willoughby [49].

For complex coefficients with $|\nu| < 1$, the asymptotic stability condition can be stated in various equivalent forms deduced by the conditions displayed in Table 9.5 for systems. For example, straight computations on the first condition lead to

$$\begin{cases} \Re(\lambda) < 0, \\ \left(\Im(\mu\nu + \lambda)\right)^2 + (1 - |\nu|^2)(|\mu|^2 - |\lambda|^2) \le 0, \\ \mu \ne -\lambda, \end{cases}$$

whereas rather long computations on the second condition yield

$$\begin{cases} \Re(\lambda) < 0, \\ |\lambda\nu + \mu| < \left|2\Re(\lambda) + |\lambda\bar{\nu} - \bar{\mu}|\right|, \\ \mu \ne -\lambda. \end{cases} \tag{9.6.2}$$

Despite appearing different, the two expressions are equivalent.

Bellen, Jackiewicz and Zennaro [31] proved the following sufficient condition for asymptotic stability:

$$|\lambda\bar{\nu} - \bar{\mu}| + |\lambda\nu + \mu| < -2\Re(\lambda), \tag{9.6.3}$$

which can easily be obtained by choosing the option

$$2\Re(\lambda) + |\lambda\bar{\nu} - \bar{\mu}| \le 0$$

in (9.6.2). On the other hand, it is not difficult to verify that

$$\lambda \in \mathbb{R} \quad \Longrightarrow \quad 2\Re(\lambda) + |\lambda\bar{\nu} - \bar{\mu}| \le 0,$$

so that (9.6.3) characterizes the asymptotic stability for $\lambda \in \mathbb{R}$ and $\mu, \nu \in \mathbb{C}$. Observe that (9.6.3) turns out to be equivalent to $|\mu| < -\lambda$ and $|\nu| < 1$ for real coefficients, and to $|\mu| < -\Re(\lambda)$ if $\nu = 0$, both of which are also necessary for asymptotic stability apart from some points on the border.

As for the asymptotic stability analysis of (9.6.1) for fixed delay, we have the following definition.

Definition 9.6.1 *For any fixed τ, the* asymptotic stability region $\mathcal{NS}_\tau$ *of equation (9.6.1) is the set of triplets (λ, μ, ν) such that the corresponding solution is asymptotically stable for any initial function $\phi(t)$.*

For the real case, Kuang [183] described the region $\mathcal{NS}_\tau$ as the subset of $\mathbb{R}^3$ bounded by the planes of equations $\nu = 1$, $\nu = -1$ and $\lambda + \mu = 0$, and by the transcendental surface

$$\Gamma_\tau = \left\{ (\lambda, \mu, \nu) \;\middle|\; \nu x^2 + \mu^2 = \lambda^2 + x^2, \; x = \frac{1}{\tau} \arctan \frac{x(\mu + \lambda\nu)}{\lambda\mu - \nu x^2}, \; x \in \mathbb{R} \right\}.$$

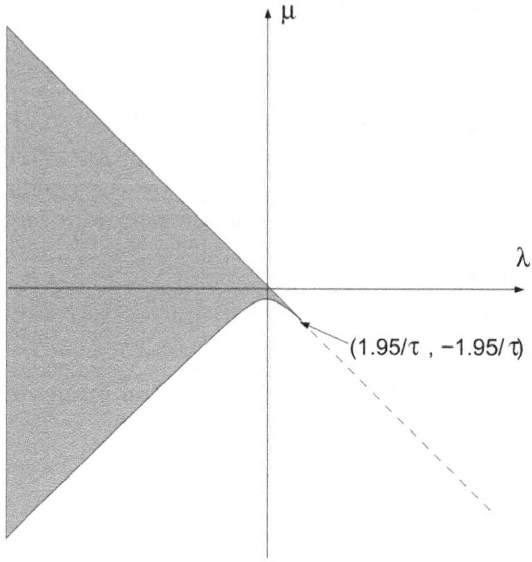

FIG. 9.8. Section of the asymptotic stability region $\mathcal{NS}_\tau$ of equation (9.6.1) for $\nu = -0.95$ in the real (λ, μ)-plane.

For any fixed value of ν with $|\nu| < 1$, the corresponding section of Γ_τ is

$$\Gamma_\tau(\nu) = \Big\{ \big(\lambda(\nu, \theta), \mu(\nu, \theta)\big) \ \mid \ \theta \in (0, \pi) \Big\},$$

where

$$\begin{cases} \lambda(\nu, \theta) = \dfrac{\theta}{\tau \sin \theta} (\cos \theta - \nu), \\ \\ \mu(\nu, \theta) = \dfrac{\theta}{\tau \sin \theta} (\nu \cos \theta - 1). \end{cases}$$

Note that the curve $\Gamma_\tau(\nu)$ intersects the line $\lambda + \mu = 0$ at the point

$$\Big(\frac{1-\nu}{\tau}, -\frac{1-\nu}{\tau} \Big),$$

where the border of the stability region has a cusp (see Figure 9.8).

9.7 Further classes of test equations

In this section we investigate the contractivity, boundedness and asymptotic stability of a larger class of delay equations, namely DDEs and NDDEs with vanishing delay and *pure* DDEs.

9.7.1 *DDEs with vanishing delay. The pantograph equation*

Let us consider the class of DDEs

$$\begin{cases} y'(t) = f(t, y(t), y(\alpha(t))), & t \geq t_0, \\ y(t) = \phi(t), & t \leq t_0, \end{cases} \tag{9.7.1}$$

where $\alpha(t) = t - \tau(t)$ with

$$\tau(t_0) \geq 0 \quad \text{and} \quad \tau(t) > 0 \text{ for all } t > t_0. \tag{9.7.2}$$

In other words, assume that t_0 is the sole point where the delay may vanish.

If $\tau(t_0) = 0$ and the deviating argument is such that $t_0 \leq \alpha(t) < t$ for all $t > t_0$, then the initial condition reduces to $y(t_0) = y_0$.

The stability analysis for this class of deviated arguments, which may not satisfy (H_1), can be done by suitably exploiting the results obtained for delays that do fulfill such a hypothesis (namely, Theorem 9.1.3 and subsequent Remark 9.1.6), and is illustrated by the next theorem, Theorem 9.7.1. To this end, we consider also another system

$$\begin{cases} \tilde{y}'(t) = f(t, \tilde{y}(t), \tilde{y}(\alpha(t))), & t \geq t_0, \\ \tilde{y}(t) = \tilde{\phi}(t), & t \leq t_0, \end{cases} \tag{9.7.3}$$

with the same right-hand side $f(t, y, x)$ but different initial function $\tilde{\phi}(t)$, and define

$$\begin{aligned} \delta y(t) &= \tilde{y}(t) - y(t), \\ \delta\phi(t) &= \tilde{\phi}(t) - \phi(t), \\ \delta y_0 &= \tilde{y}_0 - y_0. \end{aligned}$$

Theorem 9.7.1 *Given a norm $\|\cdot\|$ in $\mathbb{C}^d$, consider the systems of DDEs (9.7.1) and (9.7.3). If*

$$Y(t) + X(t) \leq 0, \quad t \geq t_0, \tag{9.7.4}$$

where $Y(t)$ and $X(t)$ are the continuous functions (8.5.11) and (8.5.4), then the system (9.7.1) is contractive (i.e. (9.1.3) holds) for all initial functions $\phi(t)$ and for all delays $\tau(t)$ satisfying the condition (9.7.2).

On the other hand, if

$$Y(t) \leq Y_0 < 0, \quad t \geq t_0, \tag{9.7.5}$$

and, for some non-negative real number $R < 1$,

$$R \cdot Y(t) + X(t) \leq 0, \quad t \geq t_0, \tag{9.7.6}$$

then problem (9.7.1), besides being contractive, is also asymptotically stable (i.e. (9.1.4) holds) for all initial functions $\phi(t)$ and for all delays $\tau(t) = t - \alpha(t)$ satisfying hypothesis (9.7.2) and hypothesis (H_1) in $[t_0+H, +\infty)$ for some $H > 0$. Moreover, if also hypothesis (H_3) is satisfied, then the stability is of exponential type, i.e. the convergence rate in (9.1.4) is at least like $\mathrm{e}^{-\alpha(t-t_0)}$ for some $\alpha > 0$.

Proof. Condition (9.7.2) implies hypothesis (H_1) on any finite interval $[t_0+H, t_f]$ with $H > 0$ and arbitrary $t_f > t_0 + H$. On the contrary, (H_1) is not satisfied on $[t_0, t_0 + H]$ if $\tau(t_0) = 0$.

Therefore, we fix $\epsilon > 0$ and, by the continuity of $\delta y(t)$, in any case we can choose $H > 0$ in such a way that

$$\|\delta y(t)\| \leq \|\delta y_0\| + \epsilon \tag{9.7.7}$$

for all $t \in [t_0, t_0 + H]$.

By Corollary 8.5.4 with $|r(x)| \leq 1$, we have

$$\|\delta y(t)\| \leq \max\left\{\|\delta y_0\|, \max_{x\leq t_0}\|\delta\phi(x)\|, \max_{t_0\leq x\leq t}\|\delta y(\alpha(x))\|\right\}, \quad t_0 \leq t \leq t_0 + H,$$

and hence, by (9.7.7),

$$\max_{t_0\leq t\leq t_0+H}\|\delta y(t)\| \leq \max_{x\leq t_0}\|\delta\phi(x)\| + \epsilon.$$

Now, Theorem 9.1.3 (and subsequent Remark 9.1.6) can be applied on any finite interval $[t_0 + H, t_f]$ in order to obtain

$$\max_{t_0+H\leq t\leq t_f}\|\delta y(t)\| \leq \max_{x\leq t_0}\|\delta\phi(x)\| + \epsilon$$

and, consequently,

$$\max_{t_0\leq t\leq t_f}\|\delta y(t)\| \leq \max_{x\leq t_0}\|\delta\phi(x)\| + \epsilon.$$

In conclusion, by the arbitrariness of ϵ and t_f, the contractivity is proved.

As for asymptotic stability, the results of Theorem 9.1.4 still hold for the system (9.7.1) with delays $\tau(t) = t - \alpha(t)$. In fact, the possible complications due to the vanishing delay are confined to a right neighborhood of the initial point t_0, that have no influence at all on the asymptotic behavior of the solution. So the proof is complete. ■

An important instance of equation (9.7.1) is given by the pantograph equation

$$\begin{cases} y'(t) = L(t)y(t) + M(t)y(qt), & t \geq 0, \\ y(0) = y_0, & t \leq 0, \end{cases} \tag{9.7.8}$$

where $0 < q < 1$, $t_0 = 0$ and the deviated argument $\alpha(t) \equiv qt$ may be viewed as $t - \tau(t)$ with time dependent delay $\tau(t) = (1 - q)t$. This is a very special instance of time dependent delays, namely a proportional delay, which makes the pantograph equation peculiar in the wide panorama of DDEs. Many results on the dynamics and stability analysis of the exact and numerical solutions of (9.7.8), of the forthcoming equation (9.7.13) and of more general DDEs with vanishing delays have been presented in a series of papers by Buhmann and Iserles [58, 59, 60], Iserles [146, 147] and others.

In particular, for equation (9.7.8), the inequality (9.7.4) reads

$$\mu[L(t)] + \|M(t)\| \leq 0, \quad t \geq 0, \tag{9.7.9}$$

and implies the contractivity condition

$$\|y(t)\| \leq \|y_0\| \quad \text{for all } t \geq 0 \tag{9.7.10}$$

for any $q \in (0,1)$.

The asymptotic stability conditions (9.7.5) and (9.7.6) read

$$\mu[L(t)] \leq Y_0 < 0, \quad t \geq t_0,$$

and

$$R \cdot \mu[L(t)] + \|M(t)\| \leq 0, \quad t \geq t_0, \quad \text{for some} \quad R < 1,$$

and imply

$$\lim_{t \to +\infty} y(t) = 0$$

for any $q \in (0,1)$ and for all y_0. To be more precise, $\|y(t)\| \to 0$ monotonically but the decay is not necessarily exponential, since hypothesis (H_3) is violated.

Many results on the asymptotic stability of the pantograph equations have been given by various authors since the original paper by Ockendon and Tayler [211]. In particular, Kato and McLeod [168] proved that the scalar equation

$$\begin{cases} y'(t) = \lambda y(t) + \mu y(qt), & t \geq 0, \\ y(0) = y_0, & t \leq 0, \end{cases}$$

with $\lambda \in \mathbb{R}$ and $\mu \in \mathbb{C}$, is asymptotically stable for all $q \in (0,1)$ if and only if

$$\lambda < 0 \quad \text{and} \quad |\mu| < -\lambda.$$

For the vector equation

$$\begin{cases} y'(t) = Ly(t) + My(qt), & t \geq 0, \\ y(0) = y_0, & t \leq 0, \end{cases}$$

Iserles [145, 149] proved that the zero solution is asymptotically stable if and only if

$$\begin{cases} \lambda \in \sigma[L] \Rightarrow \Re(\lambda) < 0, \\ \rho[L^{-1}M] < 1. \end{cases} \tag{9.7.11}$$

9.7.2 *NDDEs with vanishing delay. The generalized pantograph equation*

By following the arguments of Section 9.7.1, namely by splitting the integration interval $[t_0, +\infty)$ into a first subinterval $[t_0, t_0+H]$ plus the remaining subinterval

$[t_0 + H, +\infty)$, contractivity, boundedness and asymptotic stability results for NDDEs with vanishing delay such as

$$\begin{cases} y'(t) = f(t, y(t), y(\alpha(t)), y'(\alpha(t))), & t \geq t_0, \\ y(t) = \phi(t), \quad t \leq t_0, \end{cases} \tag{9.7.12}$$

can be obtained, where $\alpha(t) = t - \tau(t)$ with

$$\tau(t_0) \geq 0 \quad \text{and} \quad \tau(t) > 0 \text{ for all } t > t_0.$$

The technique works whenever the NDDE (9.7.12) can be rewritten in the form (9.4.2) and/or (9.4.18), so that the results of Section 9.4 can be applied in the subinterval $[t_0 + H, +\infty)$.

The generalized pantograph equation deserves special attention:

$$\begin{cases} y'(t) = L(t)y(t) + M(t)y(qt) + N(t)y'(qt), & t \geq 0, \\ y(0) = y_0. \end{cases} \tag{9.7.13}$$

For such an equation, by splitting the interval $[0, +\infty)$ into $[0, H] \cup [H, +\infty)$ for some $H > 0$ and by assuming (9.5.6) and (9.5.7), i.e.

$$\mu[L(t)] < 0$$

and

$$\frac{\|N(t)L(qt) + M(t)\| - \mu[L(qt)]\|N(t)\|}{-\mu[L(t)]} \leq 1, \quad t \geq 0, \tag{9.7.14}$$

we get the bound

$$\|y(t)\| \leq \max \left\{ \|y_0\|, \max_{0 \leq x \leq H} \frac{\|M(x)y(qx) + N(x)y'(qx)\|}{-\mu[L(x)]} \right\} \tag{9.7.15}$$

for all $t \geq H$ (see (9.5.8)). Now, since the existence and uniqueness assumptions for the solution of (9.7.13) imply that

$$y'(0) = \big(I - N(0)\big)^{-1}\big(M(0) + L(0)\big)y_0,$$

by evaluating (9.7.14) for $t = 0$, we can easily see that, for $H \to 0^+$, the bound (9.7.15) becomes

$$\|y(t)\| \leq \|y_0\| \quad \text{for all } t \geq 0 \tag{9.7.16}$$

for any $q \in (0, 1)$.

In addition, if we assume the stronger conditions

$$\mu[L(t)] \leq Y_0 < 0,$$

$$\frac{\mu[L(qt)]}{\mu[L(t)]}\|N(t)\| \leq \nu < 1, \quad t \geq 0,$$

and

$$\frac{\|N(t)L(qt) + M(t)\|}{-\mu[L(t)]} \leq k\left(1 - \frac{\mu[L(qt)]}{\mu[L(t)]}\|N(t)\|\right), \quad t \geq 0,$$

for a constant $k < 1$, we have that

$$\lim_{t \to +\infty} y(t) = 0$$

for any $q \in (0, 1)$.

By using different techniques, results on the NDDE (9.7.12) have been obtained by Iserles [145, 148], Iserles and Terjeki [150] and Liu [191, 189]. In particular, by assuming that

$$\mu[L(t)] \leq 0$$

and

$$\max_{t_0 \leq x \leq t} \|N(x)\| \leq \nu^* < 1, \quad t \geq 0,$$

and by setting

$$\omega^*(t) = \max_{t_0 \leq x \leq t} \|M(x) + N(x)L(qx)\|,$$

Iserles and Terjeki [150] proved that the condition

$$\omega^*(t) + \mu[L(t)](1 - \nu^*) \leq 0, \quad t \geq 0,$$

implies (9.7.16).

Moreover, if

$$\int_0^\infty \mu[L(x)]\mathrm{d}x = -\infty \quad \text{and} \quad \omega^*(t) + \rho\mu[L(t)](1 - \nu^*) \leq 0$$

for some $\rho \in (0, 1)$ and all $t \geq 0$, asymptotic stability is guaranteed.

For the constant coefficients instance

$$\begin{cases} y'(t) = Ly(t) + My(qt) + Ny'(qt), & t \geq 0, \\ y(0) = y_0, \end{cases}$$

the two approaches provide exactly the same conditions shown in (9.5.12) and (9.5.13) for contractivity and asymptotic stability, respectively.

However, condition (9.5.13) for asymptotic stability is sufficient, but not necessary at all. In fact, it was proved by Iserles [145] that, confined to the class of matrices N such that the problem is well-posed (i.e. no eigenvalue of N has the form q^{-s}, $s \in Z^+$), asymptotic stability for all $q \in (0, 1)$ is assured by the condition

$$\begin{cases} \lambda \in \sigma[L] \Rightarrow \Re(\lambda) < 0, \\ \rho[L^{-1}M] < 1. \end{cases} \tag{9.7.17}$$

Surprisingly, in the class of well-posed problems, the neutral term N does not play any role in the asymptotic stability of the solution and condition (9.7.17) coincides with (9.7.11), which characterizes the non-neutral case.

9.7.3 *Pure DDEs*

Now we particularize the stability analysis to the class of *pure* DDEs

$$\begin{cases} y'(t) = f(t, y(t-\tau)), & t \geq t_0, \\ y(t) = \phi(t), & t \leq t_0, \end{cases}$$

and, in particular, of its linear scalar versions

$$\begin{cases} y'(t) = \mu(t)y(t-\tau), & t \geq t_0, \\ y(t) = \phi(t), & t \leq t_0, \end{cases} \tag{9.7.18}$$

and

$$\begin{cases} y'(t) = \mu y(t-\tau), & t \geq t_0, \\ y(t) = \phi(t), & t \leq t_0. \end{cases} \tag{9.7.19}$$

It is clear that it is not possible to get any information by setting λ or $\lambda(t)$ equal to zero in Theorems 9.2.1 and 9.2.2, where delay independent contractivity and asymptotic stability are investigated. In fact, the sufficient conditions for μ and $\mu(t)$ are never verified in Theorem 9.2.2 and would imply $\mu = 0$ and $\mu(t) \equiv 0$ in Theorem 9.2.1. Therefore, for pure DDEs, we can get only delay dependent conditions.

By analyzing the characteristic equation

$$\zeta - \mu e^{\zeta\tau} = 0,$$

Barwell [19] proved that the linear scalar equation (9.7.19) with constant delay τ is asymptotically stable for any initial function $\phi(t)$ if

$$\Re(\mu) < 0 \quad (\text{that is } \pi/2 < \arg(\mu) < 3\pi/2),$$

and

$$0 < \tau|\mu| < \min\{3\pi/2 - \arg(\mu),\ \arg(\mu) - \pi/2\}.$$

For $\mu \in \mathbb{R}$, the condition reduces to

$$0 < -\tau\mu < \pi/2.$$

The complex region of asymptotic stability of (9.7.19) for a fixed delay τ is shown in Figure 9.9.

For the non-autonomous equation (9.7.18) with real coefficient $\mu(t)$, Torelli and Vermiglio [249] proved that the solutions satisfy

$$|y(t)| \leq 2 \max_{x \leq t_0} |\phi(x)|, \quad t \geq t_0,$$

whenever

$$-1/\tau \leq \mu(t) < 0, \quad t \geq t_0.$$

The previous results for complex scalar coefficients μ immediately yield the conditions for the asymptotic stability of systems such as

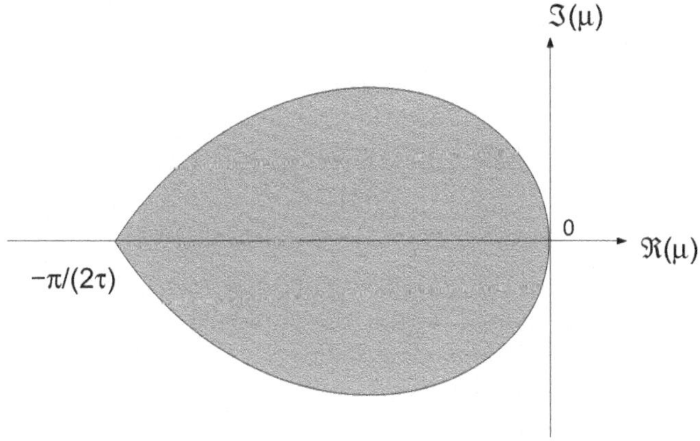

FIG. 9.9. Asymptotic stability region for equation (9.7.19) for a fixed delay τ in the complex μ-plane.

$$y'(t) = My(t - \tau).$$

In fact, for the characteristic polynomial $\det(\zeta I - e^{\zeta\tau} M)$, the stability condition

$$\det(\zeta - e^{\zeta\tau} M) = 0 \quad \Longrightarrow \quad \Re(\zeta) < 0$$

is equivalent to

$$\{\mu \in \sigma[M] \quad \text{and} \quad \zeta - e^{\zeta\tau}\mu = 0\} \quad \Longrightarrow \quad \Re(\zeta) < 0.$$

In other words, the whole spectrum of M must lie in the stability region of the scalar equation (9.7.19).

10

STABILITY ANALYSIS OF RUNGE–KUTTA METHODS FOR DDES

In this chapter we analyze the stability of the RK methods for DDEs proposed in Chapter 6 and implemented within the standard approach. The analysis is confined to the case of time dependent delays $\tau(t)$, even if some results could be extended to state dependent delays.

We begin with the simplest test equation

$$y'(t) = \lambda y(t) + \mu y(t - \tau),$$

which leads to generalizations of the concept of A-stability for ODEs. This concept spreads out in many directions as we are interested, for example, in *contractivity* or *asymptotic stability*, which for DDEs do not necessarily coincide. Moreover, the concepts are different according to whether the constant stepsize is required to be a submultiple of the delay or not. For the numerical methods, the two occurrences are not equivalent. Finally, we shall be interested in characterizing methods that are stable *for all delays* (*delay independent* stability) or *for fixed delay* (*delay dependent* stability).

Then we consider the test equation

$$y'(t) = \lambda(t)y(t) + \mu(t)y(t - \tau),$$

which leads to generalizations of the concept of AN-stability for ODEs and the general non-linear system

$$y'(t) = f\left(t, y(t), y(t - \tau)\right),$$

which leads to generalizations of the concept of BN-stability for ODEs.

To date, the stability analysis of numerical methods for DDEs is almost completely restricted to equations with constant delay and to methods advancing with constant stepsize. Nevertheless, some results for variable delays have been obtained by using the theory of Section 8.5. Moreover, a thorough investigation has been carried out for the pantograph equation, where the delay is variable and vanishes at the initial point $t_0 = 0$, so that overlapping is necessarily present at least in the first step $[t_0, t_1]$.

We shall also consider other kinds of test equations for further generalizations, such as linear systems with constant coefficients, equations with variable delays, DDEs of neutral type, etc.

In doing this we will be forced to introduce many hyphenated definitions of stability according to the large variety of significantly different cases, such as real and complex equations, constant and variable delay, free and constrained meshes, as well as to different goals such as contractivity and asymptotic stability for fixed delay and for all delays, etc.

10.1 Linear error growth

By repeating the introductory arguments of Section 8.1 on the concepts of stability and error propagation in the numerical integration of ODEs, we can say that, also for DDEs such as

$$\begin{cases} y'(t) = f\left(t, y(t), y(t-\tau(t))\right), & t_0 \le t \le t_f, \\ y(t) = \phi(t), \quad t \le t_0, \end{cases} \tag{10.1.1}$$

one of the goals is to control the propagation of the error in terms of the length of the integration interval $[t_0, t_f]$ along a mesh $\Delta = \{t_0, t_1, \ldots, t_n, \ldots, t_N = t_f\}$ with as large as possible stepsizes.

According to the notation we adopted in Chapter 4 for the standard approach, assume that, in the step $[t_n, t_{n+1}]$, the one-step method provides the discrete approximation y_{n+1} and the continuous approximation $\eta(t_n + \theta h_{n+1})$, $0 \le \theta \le 1$, for the local problem

$$\begin{cases} w'_{n+1}(t) = f\big(t, w_{n+1}(t), x(t-\tau(t))\big), & t_n \le t \le t_{n+1}, \\ w_{n+1}(t_n) = \eta(t_n), \end{cases} \tag{10.1.2}$$

where

$$x(s) = \begin{cases} \phi(s) & \text{for } s \le t_0, \\ \eta(s) & \text{for } t_0 \le s \le t_n, \\ w_{n+1}(s) & \text{for } t_n \le s \le t_{n+1}, \end{cases}$$

and $\eta(t)$ is the continuous approximate solution previously computed by the method itself.

Moreover, consider also the fictitious local problem

$$\begin{cases} z'_{n+1}(t) = f\big(t, z_{n+1}(t), u(t-\tau(t))\big), & t_n \le t \le t_{n+1}, \\ z_{n+1}(t_n) = y(t_n), \end{cases} \tag{10.1.3}$$

where

$$u(s) = \begin{cases} \phi(s) & \text{for } s \le t_0, \\ y(s) & \text{for } t_0 \le s \le t_n, \\ z_{n+1}(s) & \text{for } t_n \le s \le t_{n+1}, \end{cases}$$

the solution of which is just $z_{n+1}(t) \equiv y(t)$, the solution of (10.1.1). Denote by z_{n+1} and $\zeta_{n+1}(t_n + \theta h_{n+1})$, $0 \le \theta \le 1$, the discrete and continuous approximations of (10.1.3), respectively, and observe that the local problem (10.1.2) differs from (10.1.3) by the error $y(t_n) - y_n$ in the initial value and by the error $y(t-\tau(t)) - \eta(t-\tau(t))$ in the forcing term relevant to some arguments $t - \tau(t) < t_n$.

Therefore, unlike the ordinary case, it is convenient to carry out the stability analysis by also controlling the maximum uniform global error

$$E_n = \max_{0 \le t \le t_n} \|y(t) - \eta(t)\|,$$

rather than the sole maximum discrete global error

$$e_n = \max_{0 \le i \le n} \|y(t_i) - y_i\|.$$

There are two approaches to achieve the desired result on the linear error growth of the global error with respect to the length of the integration interval $[t_0, t_f]$.

The simpler one consists in a straightforward generalization of the analysis made at the beginning of Chapter 8 for ODE methods. In this approach, the stability condition for the uniform propagated error is

$$\max_{t_n \le t \le t_{n+1}} \|\zeta_{n+1}(t) - \eta(t)\| \le E_n \tag{10.1.4}$$

for all $n = 0, 1, \ldots, N-1$.

The second approach is more sophisticated and assumes a stability condition apparently stronger than (10.1.4), that is

$$\max_{t_n \le t \le t_{n+1}} \|\zeta_{n+1}(t) - \eta(t)\| \le (1 - h_{n+1}\gamma_{n+1})\|y(t_n) - y_n\| + h_{n+1}\gamma_{n+1}R \cdot E_n \tag{10.1.5}$$

for all $n = 0, 1, \ldots, N-1$, where

$$0 \le \gamma_{n+1} \le \gamma \quad \text{and} \quad h_{n+1}\gamma_{n+1} \le 1 \tag{10.1.6}$$

for some constant $\gamma > 0$ independent of the mesh Δ (see (8.5.48) in Section 8.5.3), and

$$0 < R < 1 \tag{10.1.7}$$

is a constant depending only on the underlying DDE (10.1.1) (see Section 9.1). Indeed, for DDEs (10.1.1) such that $0 < R < 1$ exists, the theory developed in Section 8.5.4 assures that (10.1.4) and (10.1.5) are equivalent.

In analogy to the notation introduced in Chapters 7 and 8, let

$$\sigma^*_{n+1} = \|y(t_{n+1}) - z_{n+1}\|/h_{n+1}$$

be the discrete local error per unit step and let

$$\Sigma^*_{n+1} = \max_{t_n \le t \le t_{n+1}} \|y(t) - \zeta_{n+1}(t)\|$$

be the uniform local error (with respect to the fictitious local problem (10.1.3)). Then let

$$\sigma^*_\Delta = \max_{1 \le n \le N} \sigma^*_n \quad \text{and} \quad \Sigma^*_\Delta = \max_{1 \le n \le N} \Sigma^*_n.$$

With this notation, since $E_0 = 0$, the triangular inequality

$$\max_{t_n \le t \le t_{n+1}} \|y(t) - \eta(t)\| \le \max_{t_n \le t \le t_{n+1}} \|y(t) - \zeta_{n+1}(t)\| + \max_{t_n \le t \le t_{n+1}} \|\zeta_{n+1}(t) - \eta(t)\|, \tag{10.1.8}$$

along with (10.1.4) and simple recursion arguments, implies

$$E_N \le (t_f - t_0) \max_{1 \le n \le N} \Sigma_n^* / h_n.$$

This result shows the linear growth of the uniform global error with the coefficient given by the *maximum uniform local error per unit step*. Note that it is not completely consistent with the theory developed in Sections 6.1 and 7.2, in view of which one would expect the coefficient given by a suitable linear combination of σ_Δ^* and Σ_Δ^*. Indeed, condition (10.1.5) serves to this end.

The inequalities (10.1.8) and (10.1.5) imply

$$\max_{t_n \le t \le t_{n+1}} \|y(t) - \eta(t)\| \le \Sigma_{n+1}^* + (1 - h_{n+1}\gamma_{n+1})e_n + h_{n+1}\gamma_{n+1}R \cdot E_n \tag{10.1.9}$$

for all $n = 0, 1, \ldots, N-1$.

Now let the integer k, $0 \le k \le n$, be such that $E_k < E_{k+1} = \cdots = E_{n+1}$. It certainly exists in that $E_0 = 0$. Moreover, $E_k < E_{k+1}$ implies

$$E_{k+1} = \max_{t_k \le t \le t_{k+1}} \|y(t) - \eta(t)\|.$$

Thus, by (10.1.6) for $n = k$ and by (10.1.7), the inequality (10.1.9) applied to the step $[t_k, t_{k+1}]$ yields

$$E_k < E_{k+1} = \cdots = E_{n+1} \le \frac{1}{1 - h_{k+1}\gamma_{k+1}R} \Sigma_{k+1}^* + \frac{1 - h_{k+1}\gamma_{k+1}}{1 - h_{k+1}\gamma_{k+1}R} e_k$$

and consequently, since $e_k \le e_n$,

$$E_n \le \frac{1}{1-R} \Sigma_\Delta^* + e_n \tag{10.1.10}$$

for all $n = 0, 1, \ldots, N$.

On the other hand, the triangular inequality

$$\|y(t_{n+1}) - y_{n+1}\| \le \|y(t_{n+1}) - z_{n+1}\| + \|z_{n+1} - y_{n+1}\|$$

and condition (10.1.5) imply

$$\|y(t_{n+1}) - y_{n+1}\| \le \sigma_{n+1}^* + (1 - h_{n+1}\gamma_{n+1})e_n + h_{n+1}\gamma_{n+1}R \cdot E_n.$$

Therefore, by (10.1.6), substituting (10.1.10) for E_n leads us to

$$e_{n+1} \le e_n + h_{n+1}\Big(\sigma_\Delta^* + \frac{\gamma R}{1-R} \Sigma_\Delta^*\Big).$$

Since $e_0 = 0$, a simple recursion allows us to conclude that

$$e_N \leq (t_f - t_0)\Big(\sigma_\Delta^* + \frac{\gamma R}{1-R}\Sigma_\Delta^*\Big).$$

Moreover, substituting into (10.1.10) yields

$$E_N \leq (t_f - t_0)\sigma_\Delta^* + \frac{1 + (t_f - t_0)\gamma R}{1-R}\Sigma_\Delta^*.$$

The last two inequalities represent the desired result in terms of σ_Δ^* and Σ_Δ^*.

In conclusion, the linear growth of the uniform global error is guaranteed by condition (10.1.4) or (10.1.5) on the uniform propagated error $\max_{t_n \leq t \leq t_{n+1}} \|\eta(t) - \zeta_{n+1}(t)\|$. It depends on two different error sources: the initial error $\|y(t_n) - y_n\|$ and the error $\|y(s) - \eta(s)\|$ at some points in the past. This means that the numerical method, as applied to the local problem (10.1.2), must keep under control the effects due to both perturbations in the initial value and in the forcing term. In other words, instability similar to the one related to the stiffness of ODEs may arise also from perturbations in the past values of the approximate solution, and the use of a method which is contractive for ODEs does not guarantee the required stability condition (10.1.4) or (10.1.5) for DDEs. Finding methods which, for stable (contractive) problems, satisfy (10.1.4) or (10.1.5) *unconditionally* is much harder than for the ordinary case. Here "unconditionally" means that the conditions are expected to hold for arbitrarily large stepsizes with possible overlapping.

In order to illustrate this point, consider the equation

$$\begin{cases} y'(t) = -50y(t) + 45y(t-\tau), & t \geq 0, \\ y(t) = 1, & t \leq 0. \end{cases} \tag{10.1.11}$$

On the basis of Theorem 9.2.1, the equation is contractive for any $\tau \geq 0$.

Now consider the midpoint rule which is A-stable, i.e. contractive for any stepsize and for any stable constant coefficient linear scalar equation such as (10.1.11) with $\tau = 0$. It turns out that, in general, integration of equation (10.1.11) by the midpoint rule extended by linear interpolation does not satisfy (10.1.4) and the error increases more than linearly with respect to the length of the integration interval. This is shown in Figure 10.1, where (10.1.11) for $\tau = 1$ is integrated with stepsize $h = 0.15$. On the other hand, contractivity is recovered for sufficiently small stepsizes.

10.2 Generalizations of A-stability to DDEs

The straightforward generalization of the scalar linear autonomous test equation $y'(t) = \lambda y(t)$, $y(t_0) = y_0$, to DDEs is given by

$$\begin{cases} y'(t) = \lambda y(t) + \mu y(t-\tau), & t \geq t_0, \\ y(t) = \phi(t), & t \leq t_0, \end{cases} \tag{10.2.1}$$

where $\lambda, \mu \in \mathbb{C}$ and τ is a constant delay. Equation (10.2.1) is the simplest test equation suitable for generalizing the concept of A-stability to DDEs. As widely

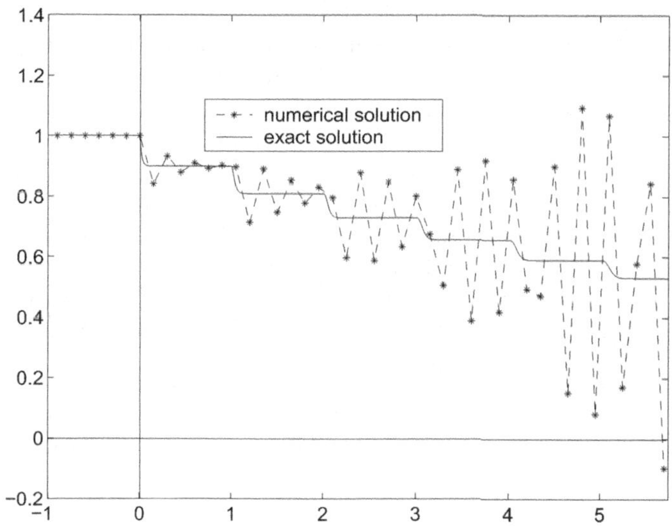

FIG. 10.1. Exact and numerical solution of (10.1.11), for $\tau = 1$, by the midpoint rule with stepsize $h = 0.15$.

explained in Chapter 9, the stability concepts, essentially asymptotic stability and contractivity, may be stated in the delay dependent or delay independent sense, i.e. for fixed τ or for all τ. Whereas for contractivity the two concepts do not differ for equation (10.2.1), the classes of equations that are asymptotically stable in the two senses are actually different and the same is expected for the numerical methods.

In fact, in Section 9.2 we have seen that, for equation (10.2.1), the condition

$$\Re(\lambda) + |\mu| \leq 0 \tag{10.2.2}$$

characterizes the class of equations fulfilling the contractivity property

$$|y(t)| \leq \max_{x \leq t_0} |\phi(x)|, \quad t \geq t_0, \tag{10.2.3}$$

for any fixed value of the delay τ. Because condition (10.2.2) is independent of τ, it also characterizes the contractivity for all delays.

Moreover, we have seen that the stronger condition

$$\Re(\lambda) + |\mu| < 0 \tag{10.2.4}$$

essentially characterizes the asymptotic stability of (10.2.1) for all delays τ, i.e.

$$\lim_{t \to +\infty} y(t) = 0$$

for all initial functions $\phi(t)$.

On the contrary, the asymptotic stability of (10.2.1) for a fixed value of τ is ensured if and only if the coefficients λ and μ belong to the larger region $\mathcal{S}_\tau$ described in Section 9.2 for both the real and complex cases.

Therefore, it is natural to ask the numerical method for the preservation of the asymptotic stability property $\lim_{n\to\infty} y_n = 0$ and/or the contractivity property $|y_n| \leq \max_{x\leq t_0} |\phi(x)|$, $n \geq 0$, under the same conditions that guarantee such properties for the exact solution. We present the definitions and related results given in the literature and complete the frame with new definitions and results. Moreover, we investigate the relationships between them.

Dealing with the asymptotic stability properties of numerical methods for DDEs, Barwell [19] gave the following definitions of P-stability and GP-stability.

Definition 10.2.1 *The* P-stability region *of a numerical method for DDEs is the set S_P of pairs of complex numbers (α, β), $\alpha = h\lambda$, $\beta = h\mu$, such that the discrete numerical solution $\{y_n\}_{n\geq 0}$ of (10.2.1), obtained with constant stepsize h under the constraint*

$$h = \tau/m, \quad m \geq 1, \ m \text{ integer}, \tag{10.2.5}$$

satisfies

$$\lim_{n\to\infty} y_n = 0 \tag{10.2.6}$$

for all constant delays τ and all initial functions $\phi(t)$.

Definition 10.2.2 *A numerical method for DDEs is* P-stable *if*

$$S_P \supseteq \{(\alpha, \beta) \in \mathbb{C}^2 \mid \Re(\alpha) + |\beta| < 0\}.$$

In other words, a numerical method is P-stable if, for any stepsizes h satisfying (10.2.5), it preserves the asymptotic stability of (10.2.1) whenever $\Re(\lambda) < -|\mu|$, i.e. whenever (10.2.1) is asymptotically stable for all delays. Note that this definition disregards the part of the stability region of (10.2.1) lying on the boundary (see (9.2.14)).

Removing the constraint (10.2.5) leads to the following stronger concept of stability.

Definition 10.2.3 *The* GP-stability region *of a numerical method for DDEs is the set S_{GP} of pairs of complex numbers (α, β), $\alpha = h\lambda$, $\beta = h\mu$, such that the discrete numerical solution $\{y_n\}_{n\geq 0}$ of (10.2.1), obtained with constant stepsize h, satisfies (10.2.6) for all constant delays τ and all initial functions $\phi(t)$.*

Definition 10.2.4 *A numerical method for DDEs is* GP-stable *if*

$$S_{GP} \supseteq \{(\alpha, \beta) \in \mathbb{C}^2 \mid \Re(\alpha) + |\beta| < 0\}.$$

Observe that the constraint (10.2.5) defines, indeed, a constrained mesh (see Chapter 6 and, in particular, Definition 6.3.2). Therefore, the study of the P-stability properties is confined to methods implemented under the constrained mesh strategy. On the other hand, when we remove constraint (10.2.5) the study of the GP-stability properties might be more appropriate. However, in both cases, the stepsize is assumed to be constant, and hence the stability properties of the numerical method implemented with varying stepsize are not completely captured by such investigations.

In order to fill in this gap, we give a definition of stability in which the stepsize is allowed to be variable. Moreover, in order to generalize our stability analysis further, in this new definition we also allow the delay to be variable. Therefore, we consider the more general test equation with variable delay

$$\begin{cases} y'(t) = \lambda y(t) + \mu y(t - \tau(t)), & t \geq t_0, \\ y(t) = \phi(t), & t \leq t_0. \end{cases} \tag{10.2.7}$$

In Chapter 9 we saw that, under hypotheses (H_1), (H_2) and (H_4), condition (10.2.4) also implies asymptotic stability for the variable delay equation (10.2.7).

Definition 10.2.5 *A numerical method for DDEs is* fully P-stable *(in short,* FP-stable*) if the discrete numerical solution of (10.2.7), obtained with any mesh* Δ*, satisfies (10.2.6) for all initial functions* $\phi(t)$ *and all delays* $\tau(t)$ *satisfying* (H_1)*,* (H_2) *and* (H_4)*, whenever (10.2.4) holds.*

It is clear that $S_{GP} \subseteq S_P$, that an FP-stable method is also GP-stable and that a GP-stable method is P-stable, too. Moreover, a P-stable method for DDEs is A-stable for ODEs.

FP-stability $\Rightarrow$ GP-stability $\Rightarrow$ P-stability $\Rightarrow$ A-stability

Dealing with the contractivity properties of the numerical method, Bellen and Zennaro [38] gave the definitions of P-contractivity and GP-contractivity. They considered only implicit methods and confined themselves to the demand for full preservation of the property $|y(t)| \leq \max_{x \leq t_0} |\phi(x)|$ for all delays. Here we introduce also the regions of contractivity and extend the concepts to the case of variable delays and variable stepsizes.

Definition 10.2.6 *The* P-contractivity region *of a numerical method for DDEs is the set* C_P *of pairs of complex numbers* (α, β)*,* $\alpha = h\lambda$*,* $\beta = h\mu$*, such that the discrete numerical solution* $\{y_n\}_{n \geq 0}$ *of (10.2.1), obtained with constant stepsize* h *under the constraint (10.2.5), satisfies*

$$|y_n| \leq \max_{x \leq t_0} |\phi(x)|, \quad n \geq 0, \tag{10.2.8}$$

for all constant delays τ *and all initial functions* $\phi(t)$*.*

Definition 10.2.7 *A numerical method for DDEs is* P-contractive *if*

$$C_P \supseteq \{(\alpha, \beta) \in \mathbb{C}^2 \mid \Re(\alpha) + |\beta| \leq 0\}.$$

Removing constraint (10.2.5) leads to the following stronger concept of contractivity.

Definition 10.2.8 *The* GP-contractivity region *of a numerical method for DDEs is the set C_{GP} of pairs of complex numbers (α, β), $\alpha = h\lambda$, $\beta = h\mu$, such that the discrete numerical solution $\{y_n\}_{n\geq 0}$ of (10.2.1) obtained with constant stepsize h satisfies (10.2.8) for all constant delays τ and all initial functions $\phi(t)$.*

Definition 10.2.9 *A numerical method for DDEs is* GP-contractive *if*

$$C_{GP} \supseteq \{(\alpha, \beta) \in \mathbb{C}^2 \mid \Re(\alpha) + |\beta| \leq 0\}.$$

Recall that in Chapter 9 we saw that, under hypotheses (H_1) and (H_4), the solution $y(t)$ of (10.2.7) satisfies (10.2.3) if (10.2.2) holds.

Definition 10.2.10 *A numerical method for DDEs is* fully P-contractive *(in short,* FP-contractive*) if the discrete numerical solution $\{y_n\}_{n\geq 0}$ of (10.2.7), obtained with any mesh Δ, satisfies (10.2.8) for all initial functions $\phi(t)$ and all variable delays $\tau(t)$ under hypotheses (H_1) and (H_4), whenever (10.2.2) holds.*

It is clear that $C_{GP} \subseteq C_P$, that an FP-contractive method is also GP-contractive and that a GP-contractive method is P-contractive, too.

FP-contractivity $\Rightarrow$ GP-contractivity $\Rightarrow$ P-contractivity

It is worth remarking that, in the foregoing definitions of a stability region, the conditions on the parameters h, λ and μ were given in terms of the pair $(h\lambda, h\mu)$. The correctness of such a choice is not obvious a priori despite being the straight generalization of the definition of the A-stability region for ODEs. Indeed, we shall see that, since the conditions on the numerical solution $\{y_n\}_{n\geq 0}$ are required for all delays τ, the choice is consistent.

This is not the case in the definition of the stronger property of asymptotic stability for fixed delay. This was the first concept of stability considered for numerical methods for DDEs and was first presented by Brayton and Willoughby [49] and subsequently by Wiederholdt [267] and Cryer [72] in the early 1970s. Contrary to the concept of P-stability, it is relevant to the larger class of equations (10.2.1) which are asymptotically stable for a fixed value of the delay (see Definition 9.2.4). In the early literature it was referred to as DA-stability and later on as τ-stability in Guglielmi [106] and subsequent papers. Here we adopt a notation closer to the original one that seems more suitable to make the concepts clear and to avoid ambiguity.

Definition 10.2.11 *For any fixed τ, the* D$_\tau$-stability region *of a numerical method for DDEs is the set S_{D_τ} of triplets (h, λ, μ), with*

$$h = \tau/m, \quad m \geq 1, \ m \ integer, \tag{10.2.9}$$

such that the discrete numerical solution $\{y_n\}_{n\geq 0}$ of (10.2.1), obtained with constant stepsize h, satisfies

$$\lim_{n\to\infty} y_n = 0$$

for all initial functions $\phi(t)$.

Definition 10.2.12 *A numerical method for DDEs is* D-stable *if, for any fixed τ,*

$$(\lambda, \mu) \in \mathcal{S}_\tau \quad \Longrightarrow \quad (h, \lambda, \mu) \in S_{D_\tau}$$

for any constant stepsize h satisfying (10.2.9).

Observe that the concept of D-stability is stronger than P-stability. In fact, $S_{D_\tau} \supset \{(h,\lambda,\mu) \in \mathbb{R}^+ \times \mathbb{C}^2 \mid \Re(\lambda) + |\mu| < 0\}$ for all $\tau > 0$ and, hence, D-stability requires preservation of the asymptotic stability on a larger subclass of equations (10.2.1).

D-stability $\Rightarrow$ P-stability

All the definitions throughout this section may be restricted to equations with real coefficients λ and μ and, in this case, the notations modify according to the notations for real ODEs, e.g. the *P(0)-stability region*, the *D(0)-stable* method, etc.

Not so obvious are the relationships among the various concepts of asymptotic stability and contractivity. However, in what follows we shall see that the sufficient conditions for contractivity are essentially sufficient to also guarantee asymptotic stability.

10.2.1 *P-stability*

Investigating the P-stability properties of numerical methods for DDEs is not easy since the final product is always a difference equation of arbitrary order m depending on the ratio τ/h. However, for RK methods, Zennaro [280] was able to reduce significantly the complexity of the problem and found a general procedure to compute their P-stability regions. Most of the results in this subsection are taken from his paper.

The RK method for DDEs (6.2.1), (6.2.2), applied to the test equation (10.2.1) with constant stepsize h satisfying the constraint (10.2.5), takes the form

$$Y_{n+1}^i = y_n + h \sum_{j=1}^{s} a_{ij}(\lambda Y_{n+1}^j + \mu \eta(t_{n-m+1}^j)), \quad i = 1, \ldots, s, \tag{10.2.10}$$

$$\eta(t_n + \theta h) = y_n + h \sum_{i=1}^{s} b_i(\theta)(\lambda Y_{n+1}^i + \mu\eta(t_{n-m+1}^i)). \tag{10.2.11}$$

With $b = [b_1, \ldots, b_s]^T$, $b(\theta) = [b_1(\theta), \ldots, b_s(\theta)]^T$, $e = [1, \ldots, 1]^T$ the unit s vector, I the s-dimensional identity matrix, $A = [a_{ij}]_{i,j=1}^s$, $\alpha = h\lambda$ and $\beta = h\mu$, after elimination of the stage values Y_{n+1}^i from (10.2.10) and computation of (10.2.11) for $\theta = c_1, \ldots, c_s, 1$, we get

$$\eta(t_{n+1}^i) = R_i(\alpha)y_n + \beta \sum_{j=1}^{s} (b(c_i)^T(I - \alpha A)^{-1})_j \eta(t_{n-m+1}^j), \quad i = 1, \ldots, s, \tag{10.2.12}$$

where

$$R_i(\alpha) = R^{(c_i)}(\alpha) = 1 + \alpha b(c_i)^T(I - \alpha A)^{-1}e \tag{10.2.13}$$

(see (8.4.2)) and $(x)_j$ is to be understood as the jth component of the row vector x. Moreover, by (10.2.11) for $\theta = 1$, we get

$$y_{n+1} = R(\alpha)y_n + \beta \sum_{j=1}^{s} (b^T(I - \alpha A)^{-1})_j \eta(t_{n-m+1}^j), \tag{10.2.14}$$

where $R(\alpha)$ is the A-stability function (8.2.7).

This pair of equations reduces to the recurrence relation with constant coefficients

$$\mathbf{H}_{n+1} = P(\alpha)\mathbf{H}_n + \beta Q(\alpha)\mathbf{H}_{n-m+1} \tag{10.2.15}$$

for the sequence of $(s+1)$-dimensional vectors $\mathbf{H}_n = [\eta(t_n^1), \ldots, \eta(t_n^s), y_n]^T$, where

$$P(\alpha) = \begin{bmatrix} 0 & e + \alpha B(I - \alpha A)^{-1}e \\ 0^T & 1 + \alpha b^T(I - \alpha A)^{-1}e \end{bmatrix}, \quad Q(\alpha) = \begin{bmatrix} B(I - \alpha A)^{-1} & 0 \\ b^T(I - \alpha A)^{-1} & 0 \end{bmatrix},$$

and $B = [b_j(c_i)]_{i,j=1}^s$.

The asymptotic behavior of the solutions of (10.2.15) is determined by the roots ζ of its characteristic equation

$$\det[\zeta^m I - \zeta^{m-1}P(\alpha) - \beta Q(\alpha)] = 0, \tag{10.2.16}$$

where, this time, I is the $(s+1)$-dimensional identity matrix (see Gohberg, Lancaster and Rodman [103]). In [280] the K notation for the RK formula is used, and hence slightly different formulae are obtained. By a small direct calculation, it is easy to check that, for all $\zeta \neq 0$ such that $\det[I - \alpha A - (\beta/\zeta^m)B] \neq 0$, the left-hand side of (10.2.16) can be factorized as follows:

$$\det[\zeta^m I - \zeta^{m-1}P(\alpha) - \beta Q(\alpha)]$$

$$= \zeta^{ms+m-1}\det[I - \alpha A - (\beta/\zeta^m)B] \cdot (\zeta - R^*(\alpha, \beta/\zeta^m)). \quad (10.2.17)$$

Therefore, instead of the roots ζ of the characteristic equation (10.2.16), we can equivalently consider the solutions of the algebraic equation

$$\zeta = R^*(\alpha, \beta/\zeta^m), \quad (10.2.18)$$

where the rational function

$$R^*(\alpha, z) = 1 + (\alpha + z)b^T(I - \alpha A - zB)^{-1}e$$

is called the *P-stability function*, provided that the following very mild assumption on the RK method holds.

Assumption 10.2.13 *The matrix $I - \alpha A - z^*B$ is singular if and only if z^* is a pole of $R^*(\alpha, z)$.*

Observe that, by (8.2.9) and since $R^*(\alpha, 0) = R(\alpha)$, the A-stability region of the underlying discrete RK method (5.0.2), (5.0.3) is

$$S_A = \{\alpha \in \mathbb{C} \mid |R^*(\alpha, 0)| < 1\}.$$

As with the A-stability function $R(\alpha)$ (see (8.2.8)), it can easily be checked that the P-stability function can be rewritten as

$$R^*(\alpha, z) = \frac{\det[I - \alpha A - zB + (\alpha + z)eb^T]}{\det[I - \alpha A - zB]}.$$

Now, for the P-stability function $R^*(\alpha, z)$, consider the curve

$$\Gamma_\alpha = \{z \in \mathbb{C} \mid |R^*(\alpha, z)| = 1\}$$

in the complex plane and let

$$\sigma_\alpha = \inf_{z \in \Gamma_\alpha} |z|$$

be the distance of the curve Γ_α from the origin of the complex plane.

The proof of the following lemma is very long and technical and, therefore, we refer the interested reader to [280].

Lemma 10.2.14 *Consider the following three statements:*

(a) $\alpha \in S_A$ *and* $|\beta| < \sigma_\alpha$*;*

(b) *all the roots ζ of (10.2.18) are inside the unit circle for all $m \geq 1$;*

(c) $\alpha \in S_A$ *and* $|\beta| \leq \sigma_\alpha$.

Then (a) $\Longrightarrow$ (b) $\Longrightarrow$ (c).

If we define the set

$$\Sigma_P = \{(\alpha, \beta) \in \mathbb{C}^2 \mid \alpha \in S_A \quad \text{and} \quad |\beta| < \sigma_\alpha\},$$

we get the following characterization of the P-stability region.

Theorem 10.2.15 *The P-stability region S_P of the RK method for DDEs (6.2.1), (6.2.2) is such that $\Sigma_P \subseteq S_P$. Moreover, under the mild assumption on the method that $\lim_{n\to\infty} \mathbf{H}_n = 0$ if and only if $\lim_{n\to\infty} y_n = 0$, it also holds that $S_P \subseteq \overline{\Sigma_P}$.*

Proof. Let $(\alpha, \beta) \in \Sigma_P$ and assume, by contradiction, that $(\alpha, \beta) \notin S_P$. Then the characteristic equation (10.2.16) has a root $\overline{\zeta}$ with $|\overline{\zeta}| \geq 1$ for some $m \geq 1$. On the other hand, by Lemma 10.2.14 all the roots of (10.2.18) are inside the unit circle for all $m \geq 1$. Moreover, by the definition of σ_α and since $|R^*(\alpha, 0)| < 1$, $\beta/\overline{\zeta}^m$ cannot be a pole of $R^*(\alpha, z)$. Therefore, by Assumption 10.2.13, equality (10.2.17) gives a contradiction.

Conversely, let $(\alpha, \beta) \in S_P$. Since $\lim_{n\to\infty} \mathbf{H}_n = 0$ for all possible solutions $\{\mathbf{H}_n\}_{n\geq 0}$ of (10.2.15), all the roots of the characteristic equation (10.2.16) are inside the unit circle for all $m \geq 1$. If, by contradiction, we assume that $(\alpha, \beta) \notin \overline{\Sigma_P}$, by Lemma 10.2.14 there exists a root $\overline{\zeta}$ of (10.2.18) with $|\overline{\zeta}| \geq 1$ for some $m \geq 1$. It is clear that $\beta/\overline{\zeta}^m$ cannot be a pole of $R^*(\alpha, z)$ and thus, again, by Assumption 10.2.13 equality (10.2.17) gives a contradiction. ■

We can conclude that, in general, the study of the P-stability properties of the RK methods for DDEs is reduced to a constrained minimum problem in the complex plane, namely to compute σ_α for $\alpha \in S_A$.

Observe that, if the method is natural, i.e. $A = B$ (see Definition 5.1.1), we have $R^*(\alpha, z) = R^*(\alpha + z, 0)$ and, hence,

$$R^*(\alpha, z) = R(\alpha + z).$$

Then, the set Γ_α is given by the complex numbers z such that $\alpha + z$ lies on ∂S_A, the boundary of the stability region S_A. Now, it is easy to see that

$$\sigma_\alpha = \text{dist}(\alpha, \partial S_A),$$

where $\text{dist}(\cdot, \cdot)$ denotes the Euclidean distance in the complex plane. Moreover, if the underlying discrete RK method is A-stable, that is $S_A \supseteq \{\alpha \in \mathbb{C} \mid \Re(\alpha) < 0\}$, then clearly it holds that

$$\sigma_\alpha \geq -\Re(\alpha),$$

and we have the following result.

Corollary 10.2.16 *A natural RK method for DDEs (6.2.3), (6.2.4) is P-stable if the underlying discrete RK method (5.0.2), (5.0.3) is A-stable.*

By virtue of Corollary 10.2.16, we can claim that the one-step collocation method at Gaussian points for DDEs is P-stable, since it is obviously natural and, as is well known, A-stable for ODEs. However, for a direct proof of this fact, see Zennaro [278].

Example 10.2.17 As an application of the previous analysis, let us determine the P(0)-stability regions of the Θ-method

$$\begin{array}{c|cc} 0 & 0 & 0 \\ 1 & 1-\Theta & \Theta \\ \hline & 1-\Theta & \Theta \end{array}$$

extended by linear interpolation. For constant stepsize h, it provides the nodal approximation

$$y_{n+1} = y_n + h\lambda\Big((1-\Theta)y_n + \Theta y_{n+1}\Big) + h\mu\Big((1-\Theta)\eta(t_n-\tau) + \Theta\eta(t_{n+1}-\tau)\Big). \tag{10.2.19}$$

In particular, with $\alpha = h\lambda$ and $\beta = h\mu$, for $h = \tau/m$ it takes the form

$$y_{n+1} = y_n + \alpha\Big((1-\Theta)y_n + \Theta y_{n+1}\Big) + \beta\Big((1-\Theta)y_{n-m} + \Theta y_{n+1-m}\Big).$$

As with linear ODEs, the Θ-method coincides with the collocation method at the point Θ

$$y_{n+1} = y_n + \alpha\Big((1-\Theta)y_n + \Theta y_{n+1}\Big) + \beta\eta(t_n + \Theta h - \tau), \tag{10.2.20}$$

where, by using linear interpolation, the retarded value $\eta(t_n + \Theta h - \tau)$ is still given by $\Big((1-\Theta)\eta(t_{n-m}) + \Theta\eta(t_{n+1-m})\Big)$.

Because the one-stage collocation method is a natural RK method with $A = B = \Theta$, the resulting P(0)-stability function turns out to be

$$R^*(\alpha, z) = 1 + \frac{\alpha + z}{1 - \alpha\Theta - z\Theta} = \frac{1 + (\alpha+z)(1-\Theta)}{1 - (\alpha+z)\Theta}$$

and satisfies Assumption 10.2.13.

The A(0)-stability region of the Θ-method is

$$S_{A(0)} = \left(\frac{2}{2\Theta - 1}, 0\right) \quad \text{for } 0 \le \Theta < \tfrac{1}{2}$$

and

$$S_{A(0)} = (-\infty, 0) \cup \left(\frac{2}{2\Theta - 1}, \infty\right) \quad \text{for } \tfrac{1}{2} \le \Theta \le 1.$$

Moreover, for any $\Theta \in [0,1]$, the boundary ∂S_A of the A-stability region is given, in the complex plane, by the circle centered in the real point $1/(2\Theta - 1)$ with radius equal to $|1/(2\Theta-1)|$.

Therefore, for real α,

$$\sigma_\alpha = \min\left\{|\alpha|, |\alpha - \frac{2}{2\Theta - 1}|\right\}$$

and

$$\Sigma_{P(0)} = \left\{(\alpha, \beta) \in \mathbb{R}^2 \mid \alpha \in S_{A(0)} \text{ and } |\beta| < \sigma_\alpha\right\}.$$

The regions of P(0)-stability are depicted in Figure 10.2 for $0 \le \Theta < \frac{1}{2}$, and in Figure 10.3 for $\frac{1}{2} \le \Theta \le 1$. We can conclude that both the Θ-method and collocation at one point Θ are P(0)-stable for all $\frac{1}{2} \le \Theta \le 1$. ◇

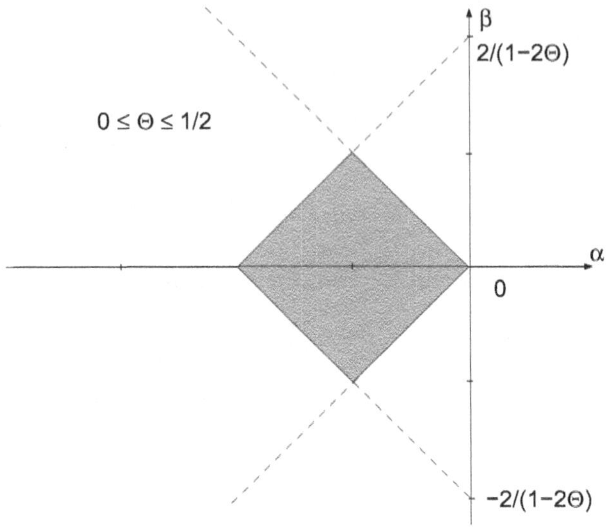

FIG. 10.2. P(0)-stability region of the Θ-methods and the one-stage collocation method at the point Θ for $0 \leq \Theta < \frac{1}{2}$.

A-stable CRK methods that are also P-stable are not necessarily confined to the subclass of natural CRK methods. If the CRK method $(A, b(\theta))$ is not natural, it is $A \neq B$ and the computation of σ_α for $\alpha \in S_A$ becomes more complicated. Koto [175] faced the problem of finding non-natural A-stable CRK methods that are also P-stable, and proved that Radau IA and Lobatto IIIC methods are so. Koto [177] used a different approach involving the characterization of the A-stable RK methods given by Theorem 8.2.3. On this basis, he provided a criterion for P-stability of RK methods for DDEs and, in particular, proved P-stability of some *singly implicit RK* (SIRK) and *singly diagonally implicit RK* (SDIRK) methods.

10.2.2 *P-contractivity*

In this section we investigate the regions of P-contractivity of the RK methods for DDEs and the relationships with the regions of P-stability.

In order to treat properly the general case of possibly confluent methods, we consider the confluency partition $\{I_1, \ldots, I_\rho\}$ of the index set $\{1, \ldots, s\}$ (see Definition 8.5.17) and observe that $i, j \in I_k \Longrightarrow R_i(\alpha) = R_j(\alpha)$ (see (10.2.13)).

First, we find the following necessary condition, which depends only on the underlying discrete RK method and is independent of the particular underlying interpolant.

Lemma 10.2.18 *Assume that $(\alpha, \beta) \in C_P$ with $\beta \neq 0$. Then $\alpha \in S_A$ and*

$$|\beta| \leq \frac{1 - |R(\alpha)|}{\sum_{k=1}^{\rho} \left| \sum_{i_k \in I_k} (b^T (I - \alpha A)^{-1})_{i_k} \right|}.$$

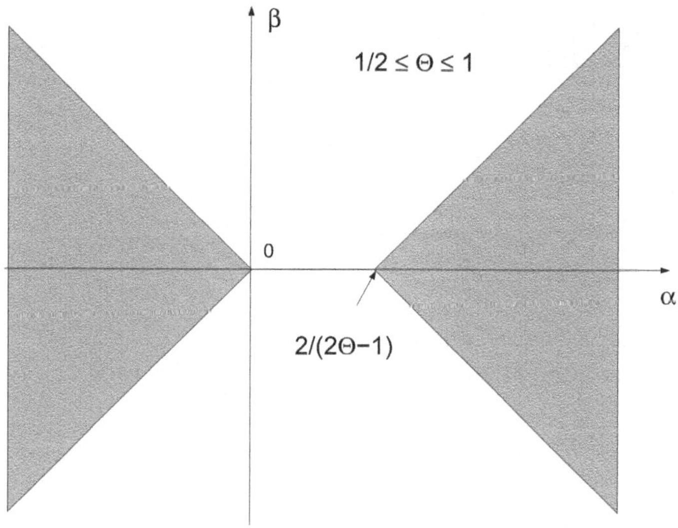

FIG. 10.3. P(0)-stability region of the Θ-methods and the one-stage collocation method at the point Θ for $\frac{1}{2} \leq \Theta \leq 1$.

Proof. Taking the confluency partition into account, formula (10.2.14) becomes

$$y_{n+1} = R(\alpha) y_n + \beta \sum_{k-1}^{\rho} \sum_{i_k \in I_k} (b^T (I - \alpha A)^{-1})_{i_k} \eta(t_{n-m+1}^{i_k}). \tag{10.2.21}$$

By assuming $m = 2$, in the first step $[t_0, t_1]$ we obtain

$$y_1 = R(\alpha)\phi(t_0) + \beta \sum_{k=1}^{\rho} \sum_{i_k \in I_k} (b^T (I - \alpha A)^{-1})_{i_k} \phi(t_{-1}^{i_k}),$$

where the initial values $\phi(t_0)$ and $\phi(t_{-1}^{i_k})$, $k = 1, \ldots, r$, may be chosen arbitrarily. Therefore, our hypothesis implies

$$|R(\alpha)| + |\beta| \sum_{k=1}^{\rho} \left| \sum_{i_k \in I_k} (b^T (I - \alpha A)^{-1})_{i_k} \right| \leq 1$$

and hence, since $\beta \neq 0$, the proof is complete. ■

We now give a sufficient condition for a pair (α, β) to belong to the P-contractivity region C_P.

Lemma 10.2.19 *Assume that $\alpha \in S_A$ and that there exist non-negative real numbers $\pi_k, k = 1, \ldots, \rho$, and $\sigma > 0$ such that*

$$|R(\alpha)| + \sigma \sum_{k=1}^{\rho} \left| \sum_{i_k \in I_k} (b^T (I - \alpha A)^{-1})_{i_k} \right| \leq 1, \tag{10.2.22}$$

$$|R_{i_k}(\alpha)| + \sigma \sum_{j=1}^{\rho} \left| \sum_{i_j \in I_j} (b(c_{i_k})^T (I - \alpha A)^{-1})_{i_j} \right| \leq \pi_k, \quad k = 1, \ldots, \rho, \tag{10.2.23}$$

$$|R(\alpha)| + \sigma \sum_{k=1}^{\rho} \left| \sum_{i_k \in I_k} (b^T (I - \alpha A)^{-1})_{i_k} \right| \pi_k \leq 1, \tag{10.2.24}$$

$$|R_{i_k}(\alpha)| + \sigma \sum_{j=1}^{\rho} \left| \sum_{i_j \in I_j} (b(c_{i_k})^T (I - \alpha A)^{-1})_{i_j} \right| \pi_j \leq \pi_k, \quad k = 1, \ldots, \rho. \tag{10.2.25}$$

Then $(\alpha, \beta) \in C_P$ *whenever* $|\beta| \leq \sigma$.

Proof. In the first interval $[t_0, t_0 + \tau]$, formula (10.2.21) reads

$$y_{n+1} = R(\alpha) y_n + \beta \sum_{k=1}^{\rho} \sum_{i_k \in I_k} (b^T (I - \alpha A)^{-1})_{i_k} \phi(t_{n-m+1}^{i_k}).$$

Thus, since $|y_0| \leq \max_{x \leq t_0} |\phi(x)|$, by using induction on n, condition (10.2.22) and $|\beta| \leq \sigma$ immediately yield

$$|y_n| \leq \max_{x \leq t_0} |\phi(x)| \tag{10.2.26}$$

for $1 \leq n \leq m$. On the other hand, in $[t_0, t_0 + \tau]$ we also have

$$\eta(t_{n+1}^{i_k}) = R_{i_k}(\alpha) y_n + \beta \sum_{j=1}^{\rho} \sum_{i_j \in I_j} (b(c_i)^T (I - \alpha A)^{-1})_{i_j} \phi(t_{n-m+1}^{i_j}), \quad k = 1, \ldots, \rho, \tag{10.2.27}$$

(see (10.2.12)) so that (10.2.23) and (10.2.27) imply

$$|\eta(t_n^{i_k})| \leq \pi_k \max_{x \leq t_0} |\phi(x)|, \quad k = 1, \ldots, \rho, \tag{10.2.28}$$

for $1 \leq n \leq m$.

In the next intervals $[t_0 + q\tau, t_0 + (q+1)\tau]$, $q \geq 1$, by using induction on both n and q, formulae (10.2.21) and (10.2.12) and conditions (10.2.24) and (10.2.25), along with inequalities (10.2.26) and (10.2.28) for $1 \leq n \leq m$, easily yield (10.2.26) and (10.2.28) also for $qm + 1 \leq n \leq (q+1)m$, $q \geq 1$, i.e. for all n.

We can conclude that $(\alpha, \beta) \in C_P$. ■

Observe that, for any choice of the variables π_k, $k = 1, \ldots, \rho$, such that $|R_{i_k}(\alpha)| < \pi_k$, $k = 1, \ldots, \rho$, there exists $\sigma > 0$ that satisfies the system (10.2.22), (10.2.23), (10.2.24), (10.2.25). The greatest of such numbers σ will be denoted by $\sigma_\alpha(\pi_1, \ldots, \pi_\rho)$.

It is clear that there exist choices of the vector of variables $\pi = (\pi_1, \ldots, \pi_\rho)$ that maximize $\sigma_\alpha(\pi)$. If we define

$$\Pi_\alpha = \left\{ \pi \in \mathbb{R}^\rho \;\middle|\; |R_{i_k}(\alpha)| < \pi_k, \; k = 1, \ldots, \rho \right\},$$

$$\hat{\sigma}_\alpha = \max_{\pi \in \Pi_\alpha} \sigma_\alpha(\pi),$$

$$\hat{\Sigma}_P = \{(\alpha, \beta) \in \mathbb{C}^{\,2} \mid \alpha \in S_A \text{ and } |\beta| < \hat{\sigma}_\alpha\}$$

and

$$\Sigma^{\max} = \left\{ (\alpha, \beta) \in \mathbb{C}^{\,2} \mid \alpha \in S_A \text{ and } |\beta| < \frac{1 - |R(\alpha)|}{\sum_{k=1}^{\rho} \left| \sum_{i_k \in I_k} (b^T (I - \alpha A)^{-1})_{i_k} \right|} \right\},$$

we can summarize the previous Lemmas 10.2.18 and 10.2.19 in the following theorem.

Theorem 10.2.20 *The P-contractivity region C_P of the RK method for DDEs (6.2.1), (6.2.2) is such that*

$$\overline{\hat{\Sigma}_P} \subseteq C_P \subseteq \overline{\Sigma^{\max}}.$$

Corollary 10.2.21 *If the underlying interpolant is linear, then*

$$C_P = \overline{\hat{\Sigma}_P} = \overline{\Sigma^{\max}}.$$

Proof. When the interpolant is linear, it turns out that

$$b(c_i) = c_i b, \quad i = 1, \ldots, s.$$

As a consequence, we have

$$\sum_{i_j \in I_j} (b(c_{i_k})^T (I - \alpha A)^{-1})_{i_j} = c_{i_k} \sum_{i_j \in I_j} (b^T (I - \alpha A)^{-1})_{i_j},$$

whereas (10.2.13) and (8.2.7) yield

$$R_{i_k}(\alpha) = 1 + c_{i_k} \alpha b^T (I - \alpha A)^{-1} e = 1 - c_{i_k} + c_{i_k} R(\alpha).$$

Since $c_{i_k} \in [0, 1]$ (see Assumption 6.1.1), we can conclude that

$$\begin{aligned} &|R_{i_k}(\alpha)| + \sigma \sum_{j=1}^{\rho} \left| \sum_{i_j \in I_j} (b(c_{i_k})^T (I - \alpha A)^{-1})_{i_j} \right| \\ &\quad \le 1 + c_{i_k} \left(|R(\alpha)| + \sigma \sum_{j=1}^{\rho} \left| \sum_{i_j \in I_j} (b^T (I - \alpha A)^{-1})_{i_j} \right| - 1 \right). \end{aligned}$$

This means that, for

$$\sigma = \frac{1 - |R(\alpha)|}{\sum_{k=1}^{\rho} \left| \sum_{i_k \in I_k} (b^T (I - \alpha A)^{-1})_{i_k} \right|},$$

the obvious validity of (10.2.22) implies also the validity of (10.2.23) with $\pi_k = 1$, $k = 1, \ldots, \rho$.

Finally, we just observe that, for $\pi_k = 1$, $k = 1, \ldots, \rho$, conditions (10.2.24) and (10.2.25) reduce to (10.2.22) and (10.2.23). Therefore, Lemma 10.2.19 applies and the result is proved. ■

Now observe that if $(\alpha, \beta) \in \overline{\hat{\Sigma}_P}$, then the characteristic equation (10.2.16) cannot have any root ζ such that $|\zeta| > 1$. Otherwise, there would exist a particular initial function $\phi(t)$ for which the numerical solution y_n would blow up as $n \to \infty$, against Theorem 10.2.20. Therefore, we can easily conclude that the P-stability region S_P verifies the inclusion

$$\hat{\Sigma}_P \subseteq S_P.$$

From the above results we can extract the following theorem.

Theorem 10.2.22 *If the RK method for DDEs (6.2.1), (6.2.2) is P-contractive, then the underlying discrete RK method (5.0.2), (5.0.3) is* A_f*-stable.*

Conversely, if the underlying CRK method (5.0.2), (5.1.6), (5.1.1) is semi-A_f*-stable, then the RK method for DDEs (6.2.1), (6.2.2) is P-contractive.*

Proof. If the RK method for DDEs is P-contractive, then it holds that $(\alpha, |\Re(\alpha)|) \in C_P$ whenever $\Re(\alpha) < 0$. Therefore, Lemma 10.2.18 and Proposition 8.5.18 imply the A_f-stability of the underlying discrete RK method.

Conversely, if the underlying CRK method is semi-A_f-stable, then Proposition 8.5.19 implies that conditions (10.2.22), (10.2.23), (10.2.24) and (10.2.25) are satisfied with $\pi_k = 1$, $k = 1, \ldots, \rho$, for $\sigma = |\Re(\alpha)|$ whenever $\Re(\alpha) < 0$. Therefore, Lemma 10.2.19 yields the P-contractivity of the RK method for DDEs. ■

10.2.3 *GP-stability*

For LM methods, Bickart [42] and Watanabe and Roth [265] established a sort of equivalence between A-stability and GP-stability in the sense that each A-stable LM formula for ODEs endowed with suitable Lagrange interpolation is GP-stable for DDEs. For RK methods, In 't Hout and Spijker [138] generalized the tools developed by Zennaro [278] and found a method that allows the computation of the GP-stability regions. Most of the results in this subsection are taken from their paper.

For simplicity, we assume that the abscissae have non-decreasing indices, i.e. $0 \le c_1 \le c_2 \le \cdots \le c_s \le 1$ (see also Assumption 6.1.1). Since the constant stepsize h does not necessarily satisfy the constraint (10.2.5), there exists $\delta \in [0, 1)$ such that

$$h = \tau/(m - \delta),$$

where m is a positive integer.

Observe that, for $m = 1$ and $\delta > 0$, the stepsize h is larger that τ and, hence, overlapping occurs.

If r is the integer such that $c_r + \delta \le 1 < c_{r+1} + \delta$ with the convention $c_0 = 0$ and $c_{s+1} = 1$, the RK method for DDEs (6.2.1), (6.2.2), applied to the test equation (10.2.1), takes the form

$$Y_{n+1}^i = y_n + h\sum_{j=1}^{r} a_{ij}(\lambda Y_{n+1}^j + \mu\eta(t_{n-m+1}^j + \delta h))$$
$$+h\sum_{j=r+1}^{s} a_{ij}(\lambda Y_{n+1}^j + \mu\eta(t_{n-m+2}^j + (\delta-1)h)), \quad i = 1,\dots,s, \tag{10.2.29}$$

$$\eta(t_n+\theta h) = y_n + h\sum_{i=1}^{r} b_i(\theta)(\lambda Y_{n+1}^i + \mu\eta(t_{n-m+1}^i + \delta h))$$
$$+h\sum_{i=r+1}^{s} b_i(\theta)(\lambda Y_{n+1}^i + \mu\eta(t_{n-m+2}^i + (\delta-1)h)). \tag{10.2.30}$$

As with equations (10.2.10), (10.2.11), after formal elimination of the stage values Y_{n+1}^i from (10.2.29) and computation of (10.2.30) for $\theta = c_1+\delta,\dots,c_r+\delta, c_{r+1}+\delta-1,\dots,c_s+\delta-1, 1$, we get the following equivalent recurrence relation with constant coefficients for the sequence of $(s+1)$-dimensional vectors $\mathbf{H}_n^{(\delta)} = [\eta(t_n^1+\delta h)),\dots,\eta(t_n^r+\delta h)), \eta(t_n^{r+1}+(\delta-1)h),\dots,\eta(t_n^s+(\delta-1)h), y_n]^T$:

$$\mathbf{H}_{n+1}^{(\delta)} = P^{(\delta)}(\alpha)\mathbf{H}_n^{(\delta)} + \beta T^{(\delta)}(\alpha)\mathbf{H}_{n-m+2}^{(\delta)} + \beta Q^{(\delta)}(\alpha)\mathbf{H}_{n-m+1}^{(\delta)}, \tag{10.2.31}$$

where

$$P^{(\delta)}(\alpha) = \begin{bmatrix} 0 & e+\alpha B^{(\delta)}(I-\alpha A)^{-1}e \\ 0\dots 0 & 1+\alpha b^T(I-\alpha A)^{-1}e \end{bmatrix},$$

$$Q^{(\delta)}(\alpha) = \begin{bmatrix} & 0 \\ B_1^{(\delta)} + \alpha B^{(\delta)}(I-\alpha A)^{-1}A_1^{(\delta)} & \vdots \\ & 0 \\ b_1^{(\delta)T} + \alpha b^T(I-\alpha A)^{-1}A_1^{(\delta)} & 0 \end{bmatrix},$$

$$T^{(\delta)}(\alpha) = \begin{bmatrix} & 0 \\ B_2^{(\delta)} + \alpha B^{(\delta)}(I-\alpha A)^{-1}A_2^{(\delta)} & \vdots \\ & 0 \\ b_2^{(\delta)T} + \alpha b^T(I-\alpha A)^{-1}A_2^{(\delta)} & 0 \end{bmatrix},$$

$$A_1^{(\delta)} = \begin{bmatrix} a_{11} & \dots & a_{1r} & 0 & \dots & 0 \\ \vdots & & \vdots & \vdots & & \vdots \\ a_{s1} & \dots & a_{sr} & 0 & \dots & 0 \end{bmatrix}, \quad A_2^{(\delta)} = \begin{bmatrix} 0 & \dots & 0 & a_{1,r+1} & \dots & a_{1s} \\ \vdots & & \vdots & \vdots & & \vdots \\ 0 & \dots & 0 & a_{s,r+1} & \dots & a_{ss} \end{bmatrix},$$

$$B^{(\delta)} = \begin{bmatrix} b_1(c_1+\delta) & \dots & b_s(c_1+\delta) \\ \vdots & & \vdots \\ b_1(c_r+\delta) & \dots & b_s(c_r+\delta) \\ b_1(c_{r+1}+\delta-1) & \dots & b_s(c_{r+1}+\delta-1) \\ \vdots & & \vdots \\ b_1(c_s+\delta-1) & \dots & b_s(c_s+\delta-1) \end{bmatrix},$$

$$B_1^{(\delta)} = \begin{bmatrix} b_1(c_1+\delta) & \dots & b_r(c_1+\delta) & 0\dots 0 \\ \vdots & & \vdots & \vdots \quad \vdots \\ b_1(c_r+\delta) & \dots & b_r(c_r+\delta) & 0\dots 0 \\ b_1(c_{r+1}+\delta-1) & \dots & b_r(c_{r+1}+\delta-1) & 0\dots 0 \\ \vdots & & \vdots & \vdots \quad \vdots \\ b_1(c_s+\delta-1) & \dots & b_r(c_s+\delta-1) & 0\dots 0 \end{bmatrix},$$

$$B_2^{(\delta)} = \begin{bmatrix} 0\dots 0 & b_{r+1}(c_1+\delta) & \dots & b_s(c_1+\delta) \\ \vdots \quad \vdots & \vdots & & \vdots \\ 0\dots 0 & b_{r+1}(c_r+\delta) & \dots & b_s(c_r+\delta) \\ 0\dots 0 & b_{r+1}(c_{r+1}+\delta-1) & \dots & b_s(c_{r+1}+\delta-1) \\ \vdots \quad \vdots & \vdots & & \vdots \\ 0\dots 0 & b_{r+1}(c_s+\delta-1) & \dots & b_s(c_s+\delta-1) \end{bmatrix},$$

$b_1^{(\delta)} = [b_1(1),\dots,b_r(1),0,\dots,0]^T$ and $b_2^{(\delta)} = [0,\dots,0,b_{r+1}(1),\dots,b_s(1)]^T$.

Remark 10.2.23 *For $\delta = 0$, we have $\mathbf{H}_n^{(0)} = \mathbf{H}_n$, $P^{(0)}(\alpha) = P(\alpha)$, $Q^{(0)}(\alpha) = Q(\alpha)$, $T^{(0)}(\alpha) = 0$, the zero matrix, and (10.2.31) reduces to (10.2.15). In fact, for $\delta = 0$ the constraint (10.2.5) is satisfied.*

As with the recurrence (10.2.15), the asymptotic behavior of the solutions of (10.2.31) is determined by the roots ζ of its characteristic equation

$$\det[\zeta^m I - \zeta^{m-1}P^{(\delta)}(\alpha) - \beta\zeta T^{(\delta)}(\alpha) - \beta Q^{(\delta)}(\alpha)] = 0. \tag{10.2.32}$$

In [138] the K notation for the RK formula is used, and hence slightly different formulae are obtained. However, the essence of the problem is that, for $\delta \neq 0$, the characteristic roots ζ are no longer the solutions of an algebraic equation similar to (10.2.18). Therefore, a more general result than Lemma 10.2.14 is needed. To this end, observe that the characteristic equation (10.2.32) is a particular case of

$$\det[\zeta^m U(\zeta) + V(\zeta)] = 0, \tag{10.2.33}$$

where $U(\zeta)$ and $V(\zeta)$ are matrix polynomials.

Lemma 10.2.24 *Let $U(\zeta)$ and $V(\zeta)$ be matrix polynomials with* $\det[U(\zeta)]$ *non-identically zero, and let $m_0 \geq 0$ be the smallest integer such that* $\deg(\det[\zeta^m U(\zeta) + V(\zeta))] = \deg(\det[\zeta^m U(\zeta))]$ *for all integers $m \geq m_0$. Consider the following three statements:*

(a) *$U(\zeta)$ is invertible whenever $|\zeta| \geq 1$ and $\sup_{|\zeta|=1} \rho[V(\zeta)U(\zeta)^{-1}] < 1$;*

(b) *all the roots ζ of (10.2.33) are inside the unit circle for all $m \geq m_0$;*

(c) *$U(\zeta)$ is invertible whenever $|\zeta| \geq 1$ and $\sup_{|\zeta|=1} \rho[V(\zeta)U(\zeta)^{-1}] \leq 1$.*

Then (a) $\Longrightarrow$ (b) $\Longrightarrow$ (c).

If we define the sets

$$\Sigma(\delta) = \Big\{(\alpha,\beta) \in \mathbb{C}^2 \mid \rho[P^{(\delta)}(\alpha)] < 1 \text{ and}$$

$$|\beta| < \Big(\sup_{|\zeta|=1} \rho\big[(\zeta T^{(\delta)}(\alpha) + Q^{(\delta)}(\alpha))(\zeta I - P^{(\delta)}(\alpha))^{-1}\big]\Big)^{-1}\Big\}$$

and

$$\Sigma_{GP} = \bigcap_{0<\delta<1} \Sigma(\delta),$$

then by Lemma 10.2.24 we get the following characterization of the GP-stability region.

Theorem 10.2.25 *The GP-stability region S_{GP} of the RK method for DDEs (6.2.1), (6.2.2) is such that $\Sigma_{GP} \subseteq S_{GP}$. Moreover, under the mild assumption on the method that $\lim_{t\to+\infty} \eta(t) = 0$ if and only if $\lim_{n\to\infty} y_n = 0$, it also holds that $S_{GP} \subseteq \overline{\Sigma_{GP}}$.*

Let us consider again the Θ-method (10.2.19) introduced in Example 10.2.17 and observe that, for the stepsize $h = \tau/(m-\delta)$, $m \geq 1$ integer and $0 \leq \delta < 1$, the deviated arguments $\eta(t_n - \tau)$ and $\eta(t_{n+1} - \tau)$ take the form

$$\eta(t_n - \tau) = \eta(t_{n-m} + \delta h)$$

and

$$\eta(t_{n+1} - \tau) = \eta(t_{n-m+1} + \delta h).$$

Therefore, by linear interpolation, we get

$$\eta(t_n - \tau) = (1-\delta) y_{n-m} + \delta y_{n-m+1}$$

and

$$\eta(t_{n+1} - \tau) = (1-\delta) y_{n-m+1} + \delta y_{n-m+2},$$

and the Θ-method reads

$$\begin{aligned} y_{n+1} &= y_n + \alpha\big((1-\Theta) y_n + \Theta y_{n+1}\big) \\ &\quad + \beta\Big((1-\Theta)(1-\delta) y_{n-m} + \big((1-\Theta)\delta + \Theta(1-\delta)\big) y_{n-m+1} + \Theta\delta y_{n-m+2}\Big). \end{aligned}$$

On the contrary, the deviated term $\eta(t_n + \Theta h - \tau)$ in the collocation method (10.2.20) takes the form

$$\eta(t_n + \Theta h - \tau) = \eta(t_n + \Theta h - mh + \delta h) = \eta\big(t_{n-m} + (\Theta + \delta) h\big)$$

and, by linear interpolation,

$$\eta(t_n + \Theta h - \tau) = \begin{cases} (1 - (\Theta+\delta)) y_{n-m} + (\Theta+\delta) y_{n-m+1} & \text{for } \Theta + \delta \leq 1, \\ (2 - (\Theta+\delta)) y_{n-m+1} + (\Theta+\delta-1) y_{n-m+2} & \text{for } \Theta + \delta > 1. \end{cases}$$

In this case, the Θ-method and the collocation method at the point Θ no longer coincide apart for $\Theta = 0$ and $\Theta = 1$, corresponding to forward and backward Euler, respectively. In particular, for $\Theta = 1/2$ the midpoint rule differs from the trapezoidal rule and, indeed, the stability properties are different (see

Figures 1.12 and 1.13 in Chapter 1). Liu and Spijker [188] were able to compute the GP-stability regions of Θ-methods by using a much less general form of Lemma 10.2.24 and showed that they are GP-stable if and only if $\frac{1}{2} \leq \Theta \leq 1$.

As a negative result, by using the potential of Lemma 10.2.24, In 't Hout [132] proved the following theorem.

Theorem 10.2.26 *No one-step collocation method with abscissae in* $[0, 1)$ *can be GP-stable. In particular, no Gaussian collocation method is GP-stable.*

On the other hand, as a positive result, In 't Hout [133] defined a new type of interpolation procedure for the term $\eta(t^j_{n-m+1} + \delta h)$ in (10.2.29). The interpolation, called *equistage interpolation*, is given, for suitable values of r and q, by

$$\eta(t^j_{n-m+1} + \delta h) = \sum_{k=-r}^{q} \ell_k(\delta) Y^j_{n-m+1+k} \tag{10.2.34}$$

where $q \leq m - 1$ and

$$\ell_k(\epsilon) = \prod_{i=-r, i \neq k}^{q} \left(\frac{\epsilon - i}{k - i} \right).$$

Observe that, for $\delta = 0$, we get

$$\eta(t^j_{n-m+1}) = Y^j_{n-m+1}.$$

With this interpolation, any A-stable RK method gives rise to a DDE method that is GP-stable, provided that $r \leq q \leq r+2$. The stability analysis of equistage interpolation has been successfully extended to systems of DDEs and NDDEs. However, this interpolation procedure makes the overall method of multistep type and, therefore, it undergoes all the thoeretical and practical difficulties outlined in Section 4.6.

10.2.4 *GP-contractivity*

In this section we adapt the results of Section 10.2.2 to the analysis of the GP-contractivity regions of RK methods for DDEs.

To this end, in addition to the notation introduced in Section 10.2.3, we define

$$R^{(\delta)}_i(\alpha) = \begin{cases} 1 + \alpha b(c_i + \delta)^T (I - \alpha A)^{-1} e & \text{for } i = 1, \ldots, r, \\ 1 + \alpha b(c_i + \delta - 1)^T (I - \alpha A)^{-1} e & \text{for } i = r + 1, \ldots, s, \end{cases}$$

$$V^{(\delta)}_{ij}(\alpha) = \begin{cases} b_j(c_i + \delta) + \alpha \Big(b(c_i + \delta)^T (I - \alpha A)^{-1} A^{(\delta)}_1 \Big)_j & \text{for } j = 1, \ldots, r, \\ b_j(c_i + \delta) + \alpha \Big(b(c_i + \delta)^T (I - \alpha A)^{-1} A^{(\delta)}_2 \Big)_j & \text{for } j = r + 1, \ldots, s, \end{cases}$$

for $i = 1, \ldots, r$, and

$$V^{(\delta)}_{ij}(\alpha) = \begin{cases} b_j(c_i + \delta - 1) + \alpha \Big(b(c_i + \delta - 1)^T (I - \alpha A)^{-1} A^{(\delta)}_1 \Big)_j \\ \qquad\qquad\qquad\qquad\qquad\qquad \text{for } j = 1, \ldots, r, \\ b_j(c_i + \delta - 1) + \alpha \Big(b(c_i + \delta - 1)^T (I - \alpha A)^{-1} A^{(\delta)}_2 \Big)_j \\ \qquad\qquad\qquad\qquad\qquad\qquad \text{for } j = r + 1, \ldots, s, \end{cases}$$

for $i = r+1, \dots, s$,

$$v_j^{(\delta)}(\alpha) = \begin{cases} b_j + \alpha\Big(b^T(I-\alpha A)^{-1}A_1^{(\delta)}\Big)_j & \text{for } j = 1, \dots, r, \\ b_j + \alpha\Big(b^T(I-\alpha A)^{-1}A_2^{(\delta)}\Big)_j & \text{for } j = r+1, \dots, s. \end{cases}$$

Moreover, with reference to the confluency partition, let $\hat{\rho}$ be the index such that $c_r \in I_{\hat{\rho}}$.

Now we are in a position to state the analog of Lemma 10.2.19, the proof of which is similar to the previous one, but just a little bit more tricky from the technical point of view.

Lemma 10.2.27 *Assume that $\alpha \in S_A$ and that there exist non-negative real numbers $\pi_k, k = 1, \dots, \rho$, and $\sigma > 0$ such that*

$$|R(\alpha)| + \sigma \sum_{k=1}^{\rho} \left| \sum_{i_k \in I_k} v_{i_k}^{(\delta)}(\alpha) \right| \le 1,$$

$$|R_{i_k}(\alpha)| + \sigma \sum_{j=1}^{\rho} \left| \sum_{i_j \in I_j} V_{i_k,i_j}^{(\delta)}(\alpha) \right| \le \pi_k, \quad k = 1, \dots, \rho,$$

$$|R(\alpha)| + \sigma \sum_{k=1}^{\rho} \left| \sum_{i_k \in I_k} v_{i_k}^{(\delta)}(\alpha) \right| \pi_k \le 1,$$

$$|R_{i_k}(\alpha)| + \sigma \sum_{j=1}^{\rho} \left| \sum_{i_j \in I_j} V_{i_k,i_j}^{(\delta)}(\alpha) \right| \pi_j \le \pi_k, \quad k = 1, \dots, \rho,$$

$$|R(\alpha)| + \sigma \left(\sum_{k=1}^{\hat{\rho}} \left| \sum_{i_k \in I_k} v_{i_k}^{(\delta)}(\alpha) \right| + \sum_{k=\hat{\rho}+1}^{\rho} \left| \sum_{i_k \in I_k} v_{i_k}^{(\delta)}(\alpha) \right| \pi_k \right) \le 1,$$

$$|R_{i_k}(\alpha)| + \sigma \left(\sum_{j=1}^{\hat{\rho}} \left| \sum_{i_j \in I_j} V_{i_k,i_j}^{(\delta)}(\alpha) \right| + \sum_{j=\hat{\rho}+1}^{\rho} \left| \sum_{i_j \in I_j} V_{i_k,i_j}^{(\delta)}(\alpha) \right| \pi_j \right) \le \pi_k, \quad k = 1, \dots, \rho.$$

Then $(\alpha, \beta) \in C_{GP}$ whenever $|\beta| \le \sigma$.

As with the hypotheses of Lemma 10.2.19, for any choice of the variables π_k, $k = 1, \dots, \rho$, such that $|R_{i_k}(\alpha)| < \pi_k$, $k = 1, \dots, \rho$, there exists $\sigma > 0$ that satisfies them. The greatest of such numbers σ will be denoted by $\sigma_\alpha^{(\delta)}(\pi_1, \dots, \pi_\rho)$.

It is clear that there exist choices of the vector of variables $\pi = (\pi_1, \dots, \pi_\rho)$ that maximize $\sigma_\alpha^{(\delta)}(\pi)$. If we define

$$\Pi_\alpha^{(\delta)} = \Big\{ \pi \in \mathbb{R}^\rho \mid |R_{i_k}(\alpha)| < \pi_k, \ k = 1, \dots, \rho \Big\},$$

$$\hat{\sigma}_\alpha^{(\delta)} = \max_{\pi \in \Pi_\alpha^{(\delta)}} \sigma_\alpha^{(\delta)}(\pi),$$

$$\Sigma_{GP}^{(\delta)} = \{(\alpha, \beta) \in \mathbb{C}^2 \mid \alpha \in S_A \text{ and } |\beta| < \hat{\sigma}_\alpha^{(\delta)}\}$$

and

$$\hat{\Sigma}_{GP} = \bigcap_{0 \le \delta \le 1} \Sigma_{GP}^{(\delta)},$$

then by Lemmas 10.2.18 and 10.2.27 we obtain the following theorem.

Theorem 10.2.28 *The GP-contractivity region C_{GP} of the RK method for DDEs (6.2.1), (6.2.2) is such that*

$$\overline{\hat{\Sigma}_{GP}} \subseteq C_{GP} \subseteq \overline{\Sigma^{\max}}.$$

Similarly as in Section 10.2.2, we can obtain the following results.

Corollary 10.2.29 *If the underlying interpolant is linear, then*

$$C_{GP} = \overline{\hat{\Sigma}_{GP}} = \overline{\Sigma^{\max}}.$$

Theorem 10.2.30 *If the RK method for DDEs (6.2.1), (6.2.2) is GP-contractive, then the underlying discrete RK method (5.0.2), (5.0.3) is A_f-stable.*

Conversely, if the underlying CRK method (5.0.2), (5.1.6), (5.1.1) is A_f-stable, then the RK method for DDEs (6.2.1), (6.2.2) is GP-contractive.

Now observe that, if $(\alpha, \beta) \in \overline{\hat{\Sigma}_{GP}}$, then the characteristic equation (10.2.32) cannot have any root ζ such that $|\zeta| > 1$. Otherwise, there would exist a particular initial function $\phi(t)$ for which the numerical solution y_n would blow up as $n \to \infty$, against Theorem 10.2.28. Therefore, we can easily conclude that the GP-stability region S_{GP} verifies the inclusion

$$\hat{\Sigma}_{GP} \subseteq S_{GP}.$$

10.2.5 *FP-stability and FP-contractivity*

In this section we analyze the asymptotic stability and contractivity properties of RK methods for DDEs applied to the test equation

$$y'(t) = \lambda y(t) + \mu y(t - \tau(t)), \tag{10.2.35}$$

where the delay $\tau(t)$ satisfies hypotheses (H_1) and (H_4) and, possibly, also (H_2) and (H_3). However, as in Chapter 9, some of the forthcoming results could be proved just with some increase in complexity without assuming the monotonicity hypothesis (H_4).

We recall that, if besides hypotheses (H_1) and (H_4) the delay $\tau(t)$ satisfies (H_2), then it determines a set of discontinuity points $\{\xi_k\}_{k \ge 0}$, $\xi_0 = t_0$, which is monotonically increasing and diverging as $k \to \infty$. Moreover, if also the hypothesis (H_3) is verified, then $\xi_{k+1} - \xi_k \le \tau_1$ for all $k \ge 0$.

In order to better understand the statement of the forthcoming Lemma 10.2.32 that generalizes Theorem 8.3.23 to the DDE case (see Zennaro [284] for a less general version which excludes the possibility of overlapping), we analyze the possible images in the deviated argument $\alpha(t) = t - \tau(t)$ of a mesh interval $[t_n, t_{n+1}]$ for an arbitrary mesh Δ, possibly allowing overlapping.

To this end, define $n_0 = k_0 = 0$ and, for each $l \geq 1$, consider the integer $n_l \geq 1$ for which there exists $k_l \geq 1$ such that

$$t_{n_l} \leq \xi_{k_l} \quad \text{and} \quad t_{n_l+1} > \xi_{k_l}. \tag{10.2.36}$$

It is evident that, under the assumption of hypotheses (H_1) and (H_4), the integers n_l and k_l are uniquely determined for any $l \geq 1$. Therefore, a subset of discontinuity points ξ_{k_l} is defined, which are *significant* for the mesh Δ.

Moreover, it is also easy to see that, if the mesh Δ does not allow overlapping, then all the discontinuity points ξ_k are significant for Δ (i.e. $k_l = l$ for all $l \geq 1$).

Remark 10.2.31 *If the mesh Δ includes all the discontinuity points, in which case overlapping never occurs, then it is clear that $[t_n, t_{n+1}] \subseteq [\xi_{k-1}, \xi_k]$ for some $k \geq 1$ and that, if $[t_{m_1}, t_{m_2}]$ is the minimum union of mesh intervals that contain all the internal points of the image $\alpha([t_n, t_{n+1}])$, then we have $[t_{m_1}, t_{m_2}] \subseteq [\xi_{k-2}, \xi_{k-1}]$.*

On the contrary, if the mesh Δ does not necessarily include the discontinuity points, then the mesh interval $[t_n, t_{n+1}]$ may contain at most one significant discontinuity point, say $\xi_{k_{l-1}}$, as an internal point. Moreover, $[t_{m_1}, t_{m_2}]$ may also contain internal points of $[\xi_{k_{l-4}}, \xi_{k_{l-3}}]$, but the right-hand extremum t_{m_1+1} of the first image-step $[, t_{m_1+1}]$ is always $> \xi_{k_{l-3}}$.

The following lemma, stated in abstract form, provides the basic tool for carrying out in a unified way the asymptotic stability analysis relevant for all the other test equations we shall consider in the following sections. Solely for technical reasons, define the additional (significant) discontinuity points $\xi_{-1} = \xi_0 - \tau(\xi_0)$ and $\xi_{-2} = \xi_{-1}$ and extend backwards any mesh Δ defined in $[t_0, +\infty)$ to the points $t_{-2} = t_{-1} = \xi_{-2} = \xi_{-1}$.

Lemma 10.2.32 *Given the interval $[t_0, +\infty)$ and the set of discontinuity points $\{\xi_k\}_{k \geq -2}$ determined by a delay $\tau(t)$ satisfying hypotheses (H_1), (H_2) and (H_4), let F and G be two functionals defined on the set of all the meshes Δ (possibly allowing overlapping) that associate real numbers $F_n(\Delta) \geq 0$ and $G_n(\Delta) \geq 0$ such that $F_n(\Delta) \leq G_n(\Delta)$ to each point $t_n \in \Delta$, with $G_{-1}(\Delta) = G_0(\Delta)$. Moreover, let the functionals F and G be such that, for some non-negative real number $R < 1$,*

$$F_{n+1}(\Delta) \leq \Psi_{n+1} F_n(\Delta) + (1 - \Psi_{n+1}) R \max_{\xi_{k_{l-3}} < t_\nu \leq t_{n+1}} G_\nu(\Delta), \tag{10.2.37}$$

where

$$0 \leq \Psi_{n+1} \leq \Phi(h_{n+1}), \tag{10.2.38}$$

$\Phi(h)$ being an asymptotically negative superexponential function (see Section 8.3.3), and

$$G_{n+1}(\Delta) \leq \max\left\{F_n(\Delta), R \max_{\xi_{k_{l-3}} < t_\nu \leq t_{n+1}} G_\nu(\Delta)\right\} \tag{10.2.39}$$

for all $t_{n+1} \in (\xi_{k_{l-1}}, \xi_{k_l}]$ *and* $l \geq 1$. *Then*

$$\lim_{n\to\infty} F_n(\Delta) = \lim_{n\to\infty} G_n(\Delta) = 0. \tag{10.2.40}$$

Moreover, if the additional hypothesis (H_3) *is satisfied and if there exists* $h_0 > 0$ *such that* $h_{n+1} \leq h_0$ *for all* $n \geq 0$, *then the rate of convergence is at least of exponential type, i.e. at least as* $\mathrm{e}^{-\alpha(t_n - t_0)}$ *for some* $\alpha > 0$.

Proof. We shall make a kind of mixture of the proofs of Theorems 9.1.4 and 8.3.23.

Define $n_{-2} = -2$, $n_{-1} = -1$ and the numbers ϵ_{-2}, ϵ_{-1}, ϵ_0, E_{-2}, E_{-1} and E_0 to be all equal to $G_0(\Delta)$. Observe that the interesting case is $G_0(\Delta) > 0$, otherwise (10.2.40) is obvious, since (10.2.37) and (10.2.39) would imply $F_n(\Delta) = G_n(\Delta) = 0$ for all n.

In view of the definition of the significant discontinuity points in (10.2.36), for $l \geq 1$ define inductively (with respect to l) the numbers

$$E_l = \epsilon_{n_l}, \tag{10.2.41}$$

where, for $n_{l-1} + 1 \leq n \leq n_l$,

$$\epsilon_n = RE_{l-3} + \left(\prod_{\nu=n_{l-1}+1}^{n} \Psi_\nu\right)(E_{l-1} - RE_{l-3}). \tag{10.2.42}$$

Now, as an inductive hypothesis, assume that $E_{l-1} - RE_{l-3} \geq 0$, $E_{l-3} - E_{l-2} \geq 0$ and $E_{l-2} - E_{l-1} \geq 0$. These inequalities clearly hold for $l = 1$. Then, with

$$\Pi_l = \prod_{\nu=n_{l-1}+1}^{n_l} \Psi_\nu, \tag{10.2.43}$$

by (10.2.42) and (10.2.41) we have

$$\begin{aligned} E_l - RE_{l-2} &= RE_{l-3} + \Pi_l(E_{l-1} - RE_{l-3}) - RE_{l-2} \\ &= R(E_{l-3} - E_{l-2}) + \Pi_l(E_{l-1} - RE_{l-3}) \geq 0 \end{aligned}$$

and, since $\Psi_\nu < 1$ for all ν, for $n_{l-1} + 1 \leq n \leq n_l$ we have also

$$E_{l-1} - \epsilon_n = E_{l-1} - RE_{l-3} - \left(\prod_{\nu=n_{l-1}+1}^{n} \Psi_\nu\right)(E_{l-1} - RE_{l-3})$$

$$= (E_{l-1} - RE_{l-3}) \left(1 - \prod_{\nu=n_{l-1}+1}^{n} \Psi_\nu\right) \geq 0.$$

In particular, for $n = n_l$ we obtain

$$E_{l-1} - E_l \geq 0, \quad l \geq 0. \tag{10.2.44}$$

Therefore, we have proved that the sequence $\{\epsilon_n\}$ is non-increasing and that

$$E_l - RE_{l-2} \geq 0, \quad l \geq 0. \tag{10.2.45}$$

Now we want to prove that

$$G_\nu(\Delta) \leq \epsilon_{\nu-1}, \quad \nu \geq -1,$$

and

$$F_\nu(\Delta) \leq \epsilon_\nu, \quad \nu \geq -1.$$

The above inequalities clearly hold for $\nu = -1, 0$. Now assume by induction that they hold for $\nu \leq n$ with $n_{l-1} \leq n \leq n_l - 1$ for some $l \geq 1$. Therefore, by the inductive hypothesis and by (10.2.39), we get

$$G_{n+1}(\Delta) \leq \max\left\{\epsilon_n, RG_{n+1}(\Delta), R \max_{\xi_{k_{l-3}} < t_\nu \leq t_n} \epsilon_{\nu-1}\right\}$$

and thus, since $\xi_{k_{l-3}} < t_\nu \leq t_n$ implies $n_{l-3} + 1 \leq \nu \leq n$, since the sequence $\{\epsilon_n\}$ is non-increasing and since $R < 1$, by (10.2.41) and (10.2.42) we obtain

$$G_{n+1}(\Delta) \leq \max\{\epsilon_n, RG_{n+1}(\Delta), RE_{l-3}\} = \epsilon_n. \tag{10.2.46}$$

Analogously, by (10.2.37), by the inductive hypothesis and by (10.2.46), we get

$$F_{n+1}(\Delta) \leq \Psi_{n+1}\epsilon_n + (1 - \Psi_{n+1})R \max_{\xi_{k_{l-3}} < t_\nu \leq t_{n+1}} \epsilon_{\nu-1}$$

which, by (10.2.41) and (10.2.42), yields

$$F_{n+1}(\Delta) \leq \Psi_{n+1}\epsilon_n + (1 - \Psi_{n+1})RE_{l-3} = \epsilon_{n+1}.$$

To conclude the proof, we have to show that, under hypothesis (H_2), the sequence $\{\epsilon_n\}$ vanishes as $n \to \infty$. Indeed, hypothesis (H_2) implies that, as $n \to \infty$, $t_n \in (\xi_{k_{l-1}}, \xi_{k_l}]$ with $l \to \infty$. Therefore, it is sufficient to prove that $E_l \to 0$.

To this end, observe that, by (10.2.42), (10.2.41) and (10.2.43), the sequence $\{E_l\}$ is such that

$$\begin{cases} E_l = \Pi_l E_{l-1} + R(1 - \Pi_l)E_{l-3}, & l \geq 1, \\ E_{-2} = E_{-1} = E_0 > 0. \end{cases} \tag{10.2.47}$$

We need to give a uniform upper bound $\Pi < 1$ to the sequence $\{\Pi_l\}$.

Since (10.2.38) holds and $\Phi(h)$ is asymptotically negative superexponential, no problems are caused if the stepsizes h_{n+1} are not uniformly bounded from above.

On the contrary, one can easily construct a mesh Δ containing a subsequence of steps that vanishes at infinity, with steps of the type $[t_{n_{l-1}}, t_{n_l}]$, where $n_l = n_{l-1} + 1$. In this case the corresponding subsequence in $\{\Pi_l\}$ also goes to 1 at infinity, and hence the uniform upper bound $\Pi < 1$ does not exist.

In order to overcome this inconvenience, observe that (10.2.36) implies that the sum of the lengths of two consecutive intervals $[t_{n_{l-1}}, t_{n_l}]$ and $[t_{n_l}, t_{n_{l+1}}]$ is greater than the length of the interval $[\xi_{k_{l-1}}, \xi_{k_l}]$, which is $\geq \tau_0$ by hypothesis (H_1). Therefore, since $\Phi(h)$ satisfies (nse) and (anse) in Section 8.3.3, we obtain

$$\begin{aligned}\Pi_{l+1}\Pi_l &= \prod_{\nu=n_{l-1}+1}^{n_{l+1}} \Psi_\nu \leq \prod_{\nu=n_{l-1}+1}^{n_{l+1}} \Phi(h_\nu) \\ &\leq \Phi(t_{n_{l+1}} - t_{n_{l-1}}) \leq \Pi = \max_{h\geq\tau_0} \Phi(h) < 1. \end{aligned} \tag{10.2.48}$$

Now, by (10.2.44), (10.2.45) and (10.2.48), iterating (10.2.47) twice easily yields

$$\begin{cases} E_{l+1} \leq \Pi E_{l-1} + R(1-\Pi)E_{l-3}, & k \geq 1, \\ E_{-2} = E_{-1} = E_0 > 0, \quad E_1 > 0. \end{cases}$$

This is a fourth-order difference inequality with constant coefficients. Since the coefficients are positive and also all the numbers E_l are positive, it follows that $E_l \leq \Gamma_l$, $l \geq -2$, where the sequence $\{\Gamma_l\}$ satisfies the fourth-order difference equation with constant coefficients

$$\begin{cases} \Gamma_{l+1} = \Pi\Gamma_{l-1} + R(1-\Pi)\Gamma_{l-3}, & l \geq 1, \\ \Gamma_i = E_i, \quad i = -2, \ldots, 1. \end{cases}$$

Since the sum of the (positive) coefficients in the right-hand side is < 1, all four characteristic roots are < 1 in modulus. Thus, if β is the biggest of them, we can conclude that $E_l \to 0$ at least as $|\beta|^l$.

Finally, if hypothesis (H_3) is satisfied, for a given $t_n \geq t_0$, we have that $t_n \in (\xi_{k-1}, \xi_k]$ for some

$$k \geq \frac{t_n - t_0}{\tau_1}.$$

On the other hand, since hypothesis (H_1) holds, if $h_{n+1} \leq h_0$, $h_0 > 0$, for all $n \geq 0$, then the macro interval $[\xi_{k-1}, \xi_k]$ is necessarily included between two consecutive significant discontinuity points $\xi_{k_{l-1}}$ and ξ_{k_l}, where

$$l \geq \frac{k}{r_0},$$

with

$$r_0 = \left\lceil \frac{h_0}{\tau_0} \right\rceil.$$

Therefore, since $\{\epsilon_n\}$ is non-increasing, it follows that $\epsilon_n \to 0$ at least like $e^{-\alpha(t_n-t_0)}$, where $\alpha = -\log(|\beta|)/(r_0\tau_1)$. ■

Theorem 10.2.33 *If the underlying CRK method (5.0.2), (5.1.6), (5.1.1) is A_f-stable, then the RK method for DDEs (6.2.1), (6.2.2) is FP-stable. Moreover, if the additional hypothesis (H_3) is verified and if there exists $h_0 > 0$ such that $h_{n+1} \leq h_0$ for all $n \geq 0$, then the convergence rate in (10.2.6) is at least of exponential type, i.e. at least like $e^{-\alpha(t-t_0)}$ for some $\alpha > 0$.*

Proof. We apply the RK method (5.0.2), (5.1.6), (5.1.1) to the test equation (10.2.35) that satisfies (10.2.4). Observe that the numerical solution in the mesh interval $[t_n, t_{n+1}]$ is the numerical solution of the local problem

$$\begin{cases} w'_{n+1}(t) = \lambda w_{n+1}(t) + \mu x(t - \tau(t)), & t_n \leq t \leq t_{n+1}, \\ w_{n+1}(t_n) = y_n, \end{cases} \tag{10.2.49}$$

where

$$x(s) = \begin{cases} \phi(s) & \text{for } s \leq t_0, \\ \eta(s) & \text{for } t_0 \leq s \leq t_n, \\ w_{n+1}(s) & \text{for } t_n \leq s \leq t_{n+1}. \end{cases}$$

Since (10.2.4) is satisfied, we have that $|\mu| = -R \cdot \Re(\lambda)$, with $0 \leq R < 1$, and that $\Re(\lambda) < 0$. Therefore, since the underlying CRK method is A_f-stable, we have

$$|\eta(t)| \leq \max\left\{|y_n|, R \max_{1\leq i\leq s} |\eta(t^i_{n+1} - \tau(t^i_{n+1}))|\right\}, \quad t_n \leq t \leq t_{n+1},$$

and, moreover, Theorem 8.5.23 yields

$$|y_{n+1}| \leq \Psi^A_{n+1}|y_n| + (1 - \Psi^A_{n+1})R \max_{1\leq i\leq s} |\eta(t^i_{n+1} - \tau(t^i_{n+1}))|.$$

Assuming that $t_{n+1} \in (\xi_{k_{l-1}}, \xi_{k_l}]$, as observed in Remark 10.2.31, the delayed argument $t^i_{n+1} - \tau(t^i_{n+1}) \in [t_{m_1}, t_{m_2}]$ with $t_{m_1+1} > \xi_{k_{l-3}}$ and, of course, $t_{m_2} \leq t_{n+1}$. Therefore, with $F_n(\Delta) = |y_n|$ and $G_n(\Delta) = \max_{t_{n-1}\leq t\leq t_n} |\eta(t)|$, by (8.5.37), Proposition 8.5.22 and Theorem 8.3.22, the hypotheses of Lemma 10.2.32 are satisfied and, hence, the result is proved. ■

In Section 10.2.2 (Theorem 10.2.22), we have already seen that P-contractivity can be (almost) characterized in terms of the semi-A_f-stability of the underlying CRK method. Here we adapt the analysis to the case of FP-contractivity.

Theorem 10.2.34 *If the underlying CRK method (5.0.2), (5.1.6), (5.1.1) is A_f-stable, then the RK method for DDEs (6.2.1), (6.2.2) is FP-contractive.*

Proof. We apply the RK method (5.0.2), (5.1.6), (5.1.1) to the test equation (10.2.35) that satisfies (10.2.2). Thus, the numerical solution in the step $[t_n, t_{n+1}]$ is the numerical solution of the local problem (10.2.49).

We assume by induction that

$$|\eta(t)| \leq \max_{x \leq t_0} |\phi(x)|, \quad t \leq t_n, \tag{10.2.50}$$

which obviously holds for $n = 0$.

Since (10.2.2) is satisfied, we have that $|\mu| = -R \cdot \Re(\lambda)$, with $0 \leq R \leq 1$, and that $\Re(\lambda) \leq 0$. Therefore, since the underlying CRK method is A_f-stable, we have

$$|\eta(t)| \leq \max\left\{|y_n|, R \max_{1 \leq i \leq s} |\eta(t_{n+1}^i - \tau(t_{n+1}^i))|\right\}, \quad t_n \leq t \leq t_{n+1}.$$

If overlapping does not occur, i.e. if $t_{n+1}^i - \tau(t_{n+1}^i)) \leq t_n$, $i = 1, \ldots, s$, then (10.2.50) clearly holds also for $t \leq t_{n+1}$ and the proof is complete.

On the contrary, if overlapping occurs, then the inductive hypothesis (10.2.50) just implies

$$|\eta(t)| \leq \max\left\{\max_{x \leq t_0} |\phi(x)|, R \max_{t_n \leq x \leq t_{n+1}} |\eta(x)|\right\}, \quad t_n \leq t \leq t_{n+1}.$$

Now, if $R < 1$, again (10.2.50) clearly holds also for $t \leq t_{n+1}$. Otherwise, if $R = 1$, we can consider the one-parameter family of local problems

$$\begin{cases} w_{n+1,\rho}'(t) = \lambda w_{n+1,\rho}(t) + \rho\mu x(t - \tau(t)), & t_n \leq t \leq t_{n+1}, \\ w_{n+1,\rho}(t_n) = y_n, \end{cases} \tag{10.2.51}$$

where

$$x(s) = \begin{cases} \phi(s) & \text{for } s \leq t_0, \\ \eta(s) & \text{for } t_0 \leq s \leq t_n, \\ w_{n+1,\rho}(s) & \text{for } t_n \leq s \leq t_{n+1}, \end{cases}$$

for $1 - \epsilon \leq \rho < 1$, $\epsilon > 0$.

For the local problem (10.2.51) it holds that $R = \rho < 1$. Thus, in view of the previous case, the numerical solution $\eta_\rho(t)$ satisfies

$$|\eta_\rho(t)| \leq \max_{x \leq t_0} |\phi(x)|, \quad t_n \leq t \leq t_{n+1}.$$

Since continuity arguments easily show that $\eta_\rho(t)$ converges uniformly in $[t_n, t_{n+1}]$ to the numerical solution $\eta(t)$ of (10.2.49) as $\rho \to 1$, again we can conclude that (10.2.50) holds also for $t \leq t_{n+1}$. ■

Remark 10.2.35 *From the proof of Theorem 10.2.34, we can see that A_f-stability of the underlying CRK method implies even more than FP-contractivity, namely*

$$|\eta(t)| \leq \max_{x \leq t_0} |\phi(x)|, \quad t \geq t_0.$$

In other words, the contractivity property (10.2.8) extends to the continuous numerical solution.

10.2.6 *D-stability*

Compared with the delay independent analysis, the asymptotic stability analysis for a fixed value of the delay is much more difficult and, apart some early approaches by Cryer [72] and Barwell [19] and later on by van der Houwen [139] for the case $\lambda = 0$, the analysis up to the late 1990s reduced to some numerical experiments carried out by Al-Mutib [2] for some popular RK methods. For some integer m, the author plotted, in the real plane (λ, μ), the border of the region where $y_n \to 0$ with stepsize $h = \tau/m$. A systematic analysis was started by Guglielmi [106, 107, 108] and continued by Guglielmi and Hairer [110, 111] for real λ and μ, and by Maset [194, 195, 197] for the complex case and systems.

The analysis is carried out for the class of natural RK methods for DDEs for which equation (10.2.18) takes the form $\zeta = R(\alpha + \beta/\zeta^m)$ and hence, for $h = \tau/m$,

$$\zeta = R\left(\frac{1}{m}\left(\tau\lambda + \frac{\tau\mu}{\zeta^m}\right)\right).$$

So, the D_τ-stability region S_{D_τ} is described by the triplets (h, λ, μ), with $h = \tau/m$, such that

$$\zeta = R\left(\frac{1}{m}\left(\tau\lambda + \frac{\tau\mu}{\zeta^m}\right)\right) \quad \Longrightarrow \quad |\zeta| < 1.$$

By a change of variable $\xi = \tau\lambda + \tau\mu/\zeta^m$, the last condition is equivalent to

$$\xi - \tau\lambda - \frac{\tau\mu}{\left(R\left(\frac{\xi}{m}\right)\right)^m} = 0 \quad \Longrightarrow \quad \left|R\left(\frac{\xi}{m}\right)\right| < 1. \tag{10.2.52}$$

In the forthcoming analysis it is convenient to consider the following property for the A-stability function $R(\alpha)$ of an A-stable RK method:

(P) The stability condition (10.2.52) holds for all $(\lambda, \mu) \in \mathcal{S}_\tau$.

We then observe that the RK method is D-stable if and only if (P) holds for all $m \geq 1$.

Note that P-stability requires the validity of (10.2.52) for all $m \geq 1$ on the restricted set of pairs (λ, μ) such that $|\mu| < -\Re(\lambda)$.

The increase in difficulty in passing from the P-stability to the D-stability analysis essentially originates from the fact that the set $\mathcal{S}_\tau$ is larger, depends on τ and is more complicated to describe. Actually, the success of the analysis depends on the parameterization adopted for describing the set $\mathcal{S}_\tau$ and the D_τ-stability regions S_{D_τ}.

Two alternative approaches have been pursued in the literature. The first is based on the set

$$Q_*[\tau\lambda] = \{\tau\mu \mid (\lambda, \mu) \in \mathcal{S}_\tau\}$$

and on the sets

$$Q_m[\tau\lambda] = \{\tau\mu \mid (10.2.52) \text{ holds}\},$$

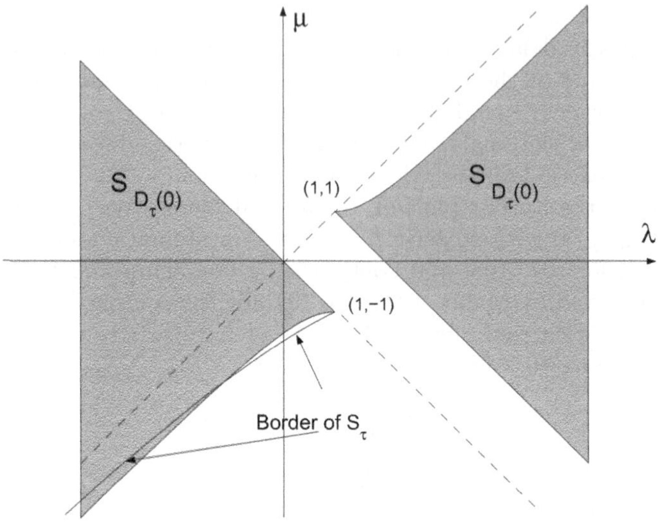

FIG. 10.4. The stability region $\mathcal{S}_{D_\tau(0)}$ of the Lobatto IIIC method for $h = \tau/m$, $\tau = 1$, $m = 1$, versus the asymptotic stability region $\mathcal{S}_\tau$ in the real (λ, μ)-plane.

and seems more suitable for analyzing the real case $(\lambda, \mu) \in \mathbb{R}^2$. Then D-stability is equivalent to

$$Q_*[\tau\lambda] \subseteq \bigcap_{m=1}^{\infty} Q_m[\tau\lambda] \quad \text{for all } \lambda \in \mathbb{C}.$$

Conversely, the second approach describes the set $\mathcal{S}_\tau$ and the regions S_{D_τ} in terms of $\tau\lambda$.

By following the first approach, Guglielmi [108] proved that the class of Θ-methods is D(0)-stable for all $\frac{1}{2} \leq \Theta \leq 1$ and that the trapezoidal rule is not D-stable. Moreover, Guglielmi [107] provided a necessary condition for D(0)-stability based on the analysis of the behavior of the $\mathrm{D}_\tau(0)$-stability region in a neighborhood of the cusp point $(1/\tau, -1/\tau)$ on the border of the stability set $\mathcal{S}_\tau$. On this basis he proved that the Lobatto IIIC method is not D(0)-stable (see Figure 10.4).

Guglielmi and Hairer [110] refined and extended the analysis to the class of A-stable symmetric methods by using the *order stars* (see Wanner, Hairer and Nørsett [264]). They proved the following result for A-stable methods with the symmetric A-stability function $R(\alpha)$ (i.e. $R(-\alpha)R(\alpha) = 1$).

Theorem 10.2.36 *Let us consider an A-stable RK method with the symmetric A-stability function $R(\alpha) = P(\alpha)/Q(\alpha)$ such that $P(\alpha)$ and $Q(\alpha)$ are polynomials of degree $\leq s$. If $R(\alpha) = \mathrm{e}^\alpha - C\alpha^{p+1} + O(\alpha^{p+2})$, with $p \geq 2s - 2$, and if the error constant satisfies $(-1)^{p/2}C > 0$, then the corresponding natural RK* *method for DDEs is D(0)-stable. In particular, all Gaussian collocation methods are D(0)-stable.*

In order to extend the analysis to the class of Radau methods, the authors proved the following sufficient condition.

Theorem 10.2.37 *Suppose that the RK method is A-stable and the A-stability region S_A is connected. Furthermore, let $z(u) = x(u) + iy(u)$, $u \geq 0$, be a parameterization of ∂S_A and let the argument $\psi(u)$ of $R(z(u))$ be such that $\psi(0) = 0$, i.e.*

$$R(z(u)) = e^{i\psi(u)} \qquad \text{and} \qquad R(z(0)) = 1.$$

If for all $u \geq 0$ with $y(u) > 0$ at least one of the three conditions

$$\psi(u) \leq y(u), \qquad mx(u) \geq 2, \qquad \sin(m\psi(u)) \leq 0,$$

is satisfied, then (P) holds.

By using Theorem 10.2.37 the authors obtained the following corollary.

Corollary 10.2.38 *The s-stage Radau IIA methods are D(0)-stable for $s = 2, 3$.*

On the basis of the same theorem, they obtained the numerical evidence that Radau IIA methods are D(0)-stable also for $4 \leq s \leq 7$.

As a negative result, Guglielmi and Hairer [111] proved the following result on D-stability.

Theorem 10.2.39 *A natural RK method for DDEs with symmetric A-stability function cannot be D-stable. In particular, no Gaussian collocation method is D-stable.*

In order to obtain some positive results on D-stability, the second approach seems to be more effective. With this approach Maset [195] proved the following results.

Theorem 10.2.40 *If (P) holds with $m = 1$, then it holds for any integer $m \geq 1$.*

In other words, for any asymptotically stable equation, the method is D-stable if the numerical solution y_n tends to zero for the largest admissible stepsize $h = \tau$.

The next theorem gives a characterization of the stability functions $R(\alpha)$ satisfying (P) with $m = 1$.

Theorem 10.2.41 *For every $c \geq 0$, let us define the set*

$$H_c = \{h \in \mathbb{C} \mid \forall v \in \mathbb{C}, \exists z \in \mathbb{C} \text{ such that } \Re(z) \geq -c \text{ and } z + vh - ve^z = 0\}.$$

The A-stability function $R(\alpha)$ satisfies property (P) with $m = 1$ if and only if

$$\left\{ \frac{e^{\xi}}{R(\xi)} \;\middle|\; \xi \in S'_A,\ \Re(\xi) = c \right\} \subseteq H_c \quad \text{for all } c \geq 0,$$

where S'_A is the complementary set of S_A.

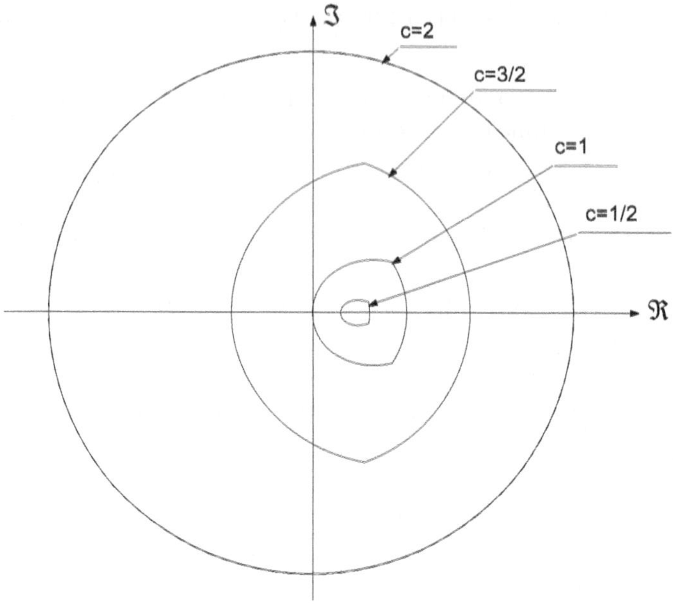

FIG. 10.5. The sets H_c for $c = \frac{1}{2}, 1, \frac{3}{2}$ and 2.

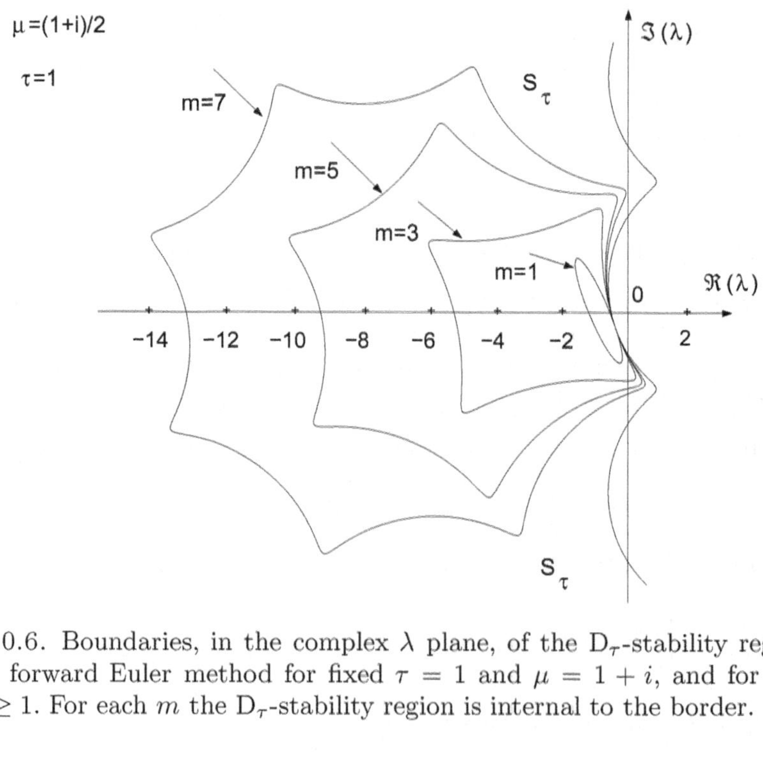

FIG. 10.6. Boundaries, in the complex λ plane, of the D_τ-stability regions of the forward Euler method for fixed $\tau = 1$ and $\mu = 1 + i$, and for various $m \geq 1$. For each m the D_τ-stability region is internal to the border.

Proof. For $\mu \in \mathbb{C}$ define the sets

$$\tilde{S}_\mu = \left\{ \xi - \frac{\tau\mu}{R(\xi)} \;\middle|\; \xi \in S'_A \right\}$$

and

$$S_\mu = \{\xi - \tau\mu e^{-\xi} \mid \xi \in \mathbb{C},\ \Re(\xi) \geq 0\}.$$

It is easy to see that the equation is asymptotically stable if and only if $\tau\lambda \notin S_\mu$. Analogously, condition (10.2.52) holds for $m = 1$ if and only if $\tau\lambda \notin \tilde{S}_\mu$. Therefore, property (P) for $m = 1$ holds if and only if $\tilde{S}_\mu \subseteq S_\mu$ for all $\mu \in \mathbb{C}$ or, equivalently, if

$$\forall \xi \in S'_A,\ \forall \tau\mu \in \mathbb{C},\ \exists x \in \mathbb{C} \text{ such that}$$
$$\Re(x) \geq 0 \text{ and } \xi - \frac{\tau\mu}{R(\xi)} = x - \tau\mu e^{-x}.$$

By setting $z = x - \xi$ and $v = \tau\mu e^{-\xi}$, we have

$$\xi - \frac{\tau\mu}{R(\xi)} = x - \tau\mu e^{-x} \Longleftrightarrow z + v\frac{e^{\xi}}{R(\xi)} - v e^{-z} = 0$$

and, hence, the theorem is proved. ■

The next theorem gives an explicit description of the sets H_c in polar representation.

Theorem 10.2.42 *If $c \geq 2$, then*

$$H_c = \{re^{i\phi} \mid 0 \leq r \leq e^c\}.$$

If $1 \leq c < 2$, then

$$H_c = \Big\{ re^{i\phi} \mid 0 \leq r \leq e^c \text{ and}$$
$$r > e^c(c-1) \Longrightarrow |\phi| \leq \arccos\left(\frac{e^c(1-c)}{r}\right) - \sqrt{r^2 e^{-2c} - (1-c)^2} \Big\}.$$

If $0 \leq c < 1$, then

$$H_c = \Big\{ re^{i\phi} \mid 0 \leq r \leq e^c,\ \ r \geq e^c(c-1) \text{ and}$$
$$|\phi| \leq \arccos\left(\frac{e^c(1-c)}{r}\right) - \sqrt{r^2 e^{-2c} - (1-c)^2} \Big\}.$$

Observe that $H_0 = 1$ and that $H_{c_1} \subset H_{c_2}$ if $c_1 \leq c_2$. The sets H_c are depicted in Figure 10.5 for some values of c.

On the basis of Theorems 10.2.40 and 10.2.41, Maset [194] proved the following result for the backward Euler method.

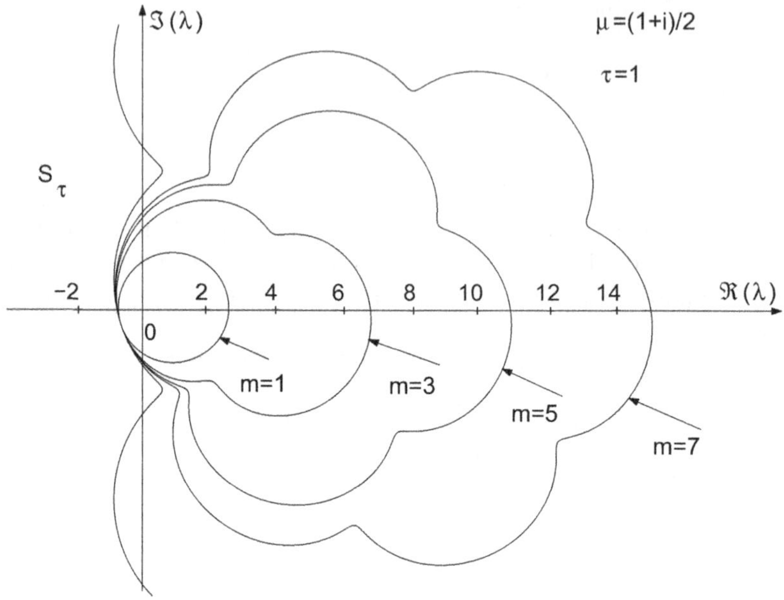

FIG. 10.7. Boundaries, in the complex λ-plane, of the D_τ-stability regions of the backward Euler method for fixed $\tau = 1$ and $\mu = 1 + i$, and for various $m \geq 1$. For each m the D_τ-stability region is external to the border.

Theorem 10.2.43 *The backward Euler method satisfies the property (P) for $m = 1$ and, hence, it is D-stable.*

Figures 10.6 and 10.7 show the D_τ-stability regions of the forward and backward Euler methods for various $m \geq 1$.

As for s-stage Radau IIA methods with $s > 1$, Figure 10.8 displays, for some values of c, the sets

$$\left\{ \frac{e^{\xi}}{R(\xi)} \mid \xi \in S'_A,\ \Re(\xi) = c \right\} \tag{10.2.53}$$

relevant to $s = 1, 2, 3$. For $s = 1$ we have the backward Euler method, which has been proved to be D-stable. In fact, the set (10.2.53) is always included in H_c. On the contrary, for $s > 1$ a number c exists such that the inclusion does not hold. So we have the numerical evidence that the property (P) is not satisfied and that, by Theorem 10.2.41, the methods are not D-stable.

Theorem 10.2.44 *In the class of Radau IIA methods the only one that is D-stable is the backward Euler method.*

10.3 Generalizations of A-stability to NDDEs

A-stability has also been generalized to DDEs of neutral type

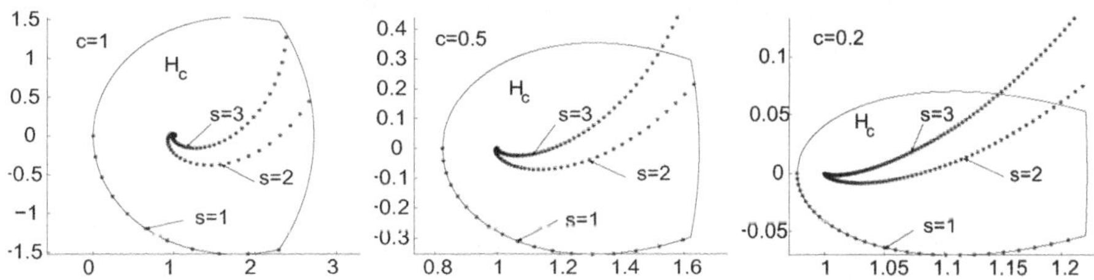

FIG. 10.8. The sets (10.2.53) and H_c for the s-stage Radau IIA methods with $s = 1, 2, 3$.

$$\begin{cases} y'(t) = \lambda y(t) + \mu y(t-\tau) + \nu y'(t-\tau), & t \geq t_0, \\ y(t) = \phi(t), & t \leq t_0, \end{cases} \tag{10.3.1}$$

where $\lambda, \mu, \nu \in \mathbb{C}$ and τ is a constant delay.

Also in this case we have two concepts of asymptotic stability, one for all delays and another for fixed delay (see Definition 9.6.1).

Definition 10.3.1 *The* NP-stability region *of a numerical method for NDDEs is the set S_{NP} of triplets of complex numbers (α, β, ν), $\alpha = h\lambda$, $\beta = h\mu$, such that the discrete numerical solution $\{y_n\}_{n\geq 0}$ of (10.3.1), obtained with constant stepsize h under the constraint*

$$h = \tau/m, \quad m \geq 1, \ m \ \textit{integer}, \tag{10.3.2}$$

satisfies

$$\lim_{n\to\infty} y_n = 0 \tag{10.3.3}$$

for all constant delays τ and all initial functions $\phi(t)$.

Definition 10.3.2 *A numerical method for NDDEs is* NP-stable *if*

$$S_{NP} \supseteq \left\{ (\alpha, \beta, \nu) \in \mathbb{C}^{\,3} \;\middle|\; \Re(\alpha) < 0 \ \textit{and} \ |\alpha\nu + \beta| < \big|2\Re(\alpha) + |\alpha\bar{\nu} - \bar{\beta}|\big| \right\}.$$

In other words, a numerical method is NP-stable if, for any stepsize h satisfying (10.3.2), it preserves the asymptotic stability on the whole class of asymptotically stable equations (10.3.1) characterized by condition (9.6.2).

Removing constraint (10.3.2) leads to the following stronger concept of stability.

Definition 10.3.3 *The* GNP-stability region *of a numerical method for NDDEs is the set S_{GNP} of triplets of complex numbers (α, β, ν), $\alpha = h\lambda$, $\beta = h\mu$, such that the discrete numerical solution $\{y_n\}_{n\geq 0}$ of (10.3.1), obtained with constant stepsize h, satisfies (10.3.3) for all constant delays τ and all initial functions $\phi(t)$.*

Definition 10.3.4 *A numerical method for NDDEs is* GNP-stable *if*

$$S_{GNP} \supseteq \left\{ (\alpha, \beta, \nu) \in \mathbb{C}^{\,3} \;\middle|\; \Re(\alpha) < 0 \ \textit{and} \ |\alpha\nu + \beta| < \big|2\Re(\alpha) + |\alpha\bar{\nu} - \bar{\beta}|\big| \right\}.$$

GNP-stability ⇒ NP-stability ⇒ P-stability

GNP-stability ⇒ GP-stability

Definition 10.3.5 *For any fixed* τ*, the* ND$_\tau$-stability region *of a numerical method for NDDEs is the set* S_{ND_τ} *of quaternaries* (h, λ, μ, ν)*, with* h *under the constraint (10.3.2), such that the discrete numerical solution* $\{y_n\}_{n\geq 0}$ *of (10.3.1), obtained with constant stepsize* h*, satisfies (10.3.3) for all initial functions* $\phi(t)$.

Definition 10.3.6 *A numerical method for NDDEs is* ND-stable *if, for any fixed* τ,

$$(\lambda, \mu, \nu) \in \mathcal{NS}_\tau \quad \Longrightarrow \quad (h, \lambda, \mu, \nu) \in S_{ND_\tau}$$

for any constant stepsize h *satisfying (10.3.2).*

Observe that, for $\nu = 0$, NP-stability reduces to P-stability and ND-stability reduces to D-stability. Moreover, the concept of ND-stability is stronger than NP-stability.

ND-stability ⇒ D-stability

ND-stability ⇒ NP-stability

10.3.1 *NP-stability*

By extending in a natural way the arguments developed in Section 10.2.1 for the P-stability analysis, we consider the NDDE (10.3.1) and, according to the notation of Section 6.5, consider the class of RK methods for NDDEs based on the option (6.5.2), i.e. $y'(t-\tau) \approx \eta'(t-\tau)$, which reads

$$Y_{n+1}^i = y_n + h \sum_{j=1}^{s} a_{ij} (\lambda Y_{n+1}^j + \mu \eta(t_{n-m+1}^j) + \nu \eta'(t_{n-m+1}^j)), \quad i = 1, \dots, s, \tag{10.3.4}$$

$$\eta(t_n + \theta h) = y_n + h \sum_{i=1}^{s} b_i(\theta) (\lambda Y_{n+1}^i + \mu \eta(t_{n-m+1}^i) + \nu \eta'(t_{n-m+1}^i)), \tag{10.3.5}$$

$$\eta'(t_n + \theta h) = \sum_{i=1}^{s} b_i'(\theta) (\lambda Y_{n+1}^i + \mu \eta(t_{n-m+1}^i) + \nu \eta'(t_{n-m+1}^i)). \tag{10.3.6}$$

With $b = [b_1, \dots, b_s]^T$, $b(\theta) = [b_1(\theta), \dots, b_s(\theta)]^T$, $e = [1, \dots, 1]^T$ the unit s-vector, I the s-dimensional identity matrix, $A = [a_{ij}]_{i,j=1}^s$, $\alpha = h\lambda$ and $\beta = h\mu$, after elimination of the stage values Y_{n+1}^i from (10.3.4) and computation of (10.3.5) and (10.3.6) for $\theta = c_1, \dots, c_s, 1$, we get

$$\eta(t_{n+1}^i) = R_i(\alpha)y_n + \beta\sum_{j=1}^{s}(b(c_i)^T(I-\alpha A)^{-1})_j\eta(t_{n-m+1}^j)$$
$$+\nu\sum_{j=1}^{s}(b(c_i)^T(I-\alpha A)^{-1})_j h\eta'(t_{n-m+1}^j), \quad i=1,\dots,s,$$

and

$$h\eta'(t_{n+1}^i) = \bar{R}_i(\alpha)y_n + \beta\sum_{j=1}^{s}(b'(c_i)^T(I-\alpha A)^{-1})_j\eta(t_{n-m+1}^j)$$
$$+\nu\sum_{j=1}^{s}(b'(c_i)^T(I-\alpha A)^{-1})_j h\eta'(t_{n-m+1}^j), \quad i=1,\dots,s,$$

where

$$R_i(\alpha) = 1 + \alpha b(c_i)^T(I-\alpha A)^{-1}e,$$
$$\bar{R}_i(\alpha) = \alpha b'(c_i)^T(I-\alpha A)^{-1}e,$$

and $(x)_j$ is understood to be the jth component of the row vector x. Moreover, by (10.3.5) for $\theta = 1$, we get

$$y_{n+1} = R(\alpha)y_n + \beta\sum_{j=1}^{s}(b^T(I-\alpha A)^{-1})_j\eta(t_{n-m+1}^j)$$
$$+\nu\sum_{j=1}^{s}(b(c_i)^T(I-\alpha A)^{-1})_j h\eta'(t_{n-m+1}^j),$$

where $R(\alpha)$ is the A-stability function (8.2.7).

Then consider the sequence of $(2s+1)$-dimensional vectors

$$\mathbf{H}_n = [h\eta'(t_n^1), \dots, h\eta'(t_n^s), \eta(t_n^1), \dots, \eta(t_n^s), y_n]^T$$

and the relevant recurrence relation with constant coefficients

$$\mathbf{H}_{n+1} = P(\alpha)\mathbf{H}_n + (\beta Q(\alpha) + \nu S(\alpha))\mathbf{H}_{n-m+1}, \tag{10.3.7}$$

where

$$P(\alpha) = \begin{bmatrix} 0 & 0 & \alpha C(I-\alpha A)^{-1}e \\ 0 & 0 & e+\alpha B(I-\alpha A)^{-1}e \\ 0^T & 0^T & 1+\alpha b^T(I-\alpha A)^{-1}e \end{bmatrix},$$

$$Q(\alpha) = \begin{bmatrix} 0 & C(I-\alpha A)^{-1} & 0 \\ 0 & B(I-\alpha A)^{-1} & 0 \\ 0^T & b^T(I-\alpha A)^{-1} & 0 \end{bmatrix},$$

$$S(\alpha) = \begin{bmatrix} C(I-\alpha A)^{-1} & 0 & 0 \\ B(I-\alpha A)^{-1} & 0 & 0 \\ b^T(I-\alpha A)^{-1} & 0^T & 0 \end{bmatrix},$$

$B = [b_j(c_i)]_{i,j=1}^s$ and $C = [b'_j(c_i)]_{i,j=1}^s$.

Observe that the inclusion of the elements $h\eta'(t_n^i)$ in the vector $\mathbf{H}_n$ is simply due to the formal request to reproduce a three-term recurrence relation (10.3.7), as was done in the non-neutral case. In the original paper by Bellen, Jackiewicz and Zennaro [31], where the K notation is used, a four-term recurrence relation was obtained for the shorter vector $[y_n, hK_n^1, \ldots, hK_n^s]^T$.

The characteristic equation of (10.3.7) is

$$\det\left[\zeta^m I - \zeta^{m-1}P(\alpha) - \beta Q(\alpha) - \nu S(\alpha)\right] = 0 \tag{10.3.8}$$

and, hence, direct computations, something longer than for the non-neutral case, lead to the factorization

$$\begin{aligned} &\det\left[\zeta^m I - \zeta^{m-1}P(\alpha) - \beta Q(\alpha) - \nu S(\alpha)\right] \\ &\quad = \zeta^{(2s+1)m-1}\det\left[I - \alpha A - (\beta/\zeta^m)B - (\nu/\zeta^m)C\right] \\ &\quad \times\left(\zeta - R^{**}(\alpha, \beta/\zeta^m, \nu/\zeta^m)\right), \end{aligned} \tag{10.3.9}$$

where

$$R^{**}(\alpha, w, z) = 1 + (\alpha + w)b^T\left(I - \alpha A - wB - zC\right)^{-1}e$$

is called the *NP-stability function.*

Now assume $\nu \neq 0$ and, as in [31], define

$$q = \frac{\beta}{\nu}$$

so that the factor $\zeta - R^{**}(\alpha, \beta/\zeta^m, \nu/\zeta^m)$ in (10.3.9) can be rewritten as $\zeta - \hat{R}^{**}(\alpha, q, \nu/\zeta^m)$, where

$$\hat{R}^{**}(\alpha, q, z) = 1 + (\alpha + qz)b^T\left(I - \alpha A - z(qB + C)\right)^{-1}e.$$

Therefore, instead of the roots ζ of the characteristic equation (10.3.8), we can equivalently consider the solutions of the algebraic equation

$$\zeta = \hat{R}^{**}(\alpha, q, \nu/\zeta^m), \tag{10.3.10}$$

provided the following assumption on the RK method holds.

Assumption 10.3.7 *The matrix $I - \alpha A - z^*(qB + C)$ is singular if and only if z^* is a pole of $\hat{R}^{**}(\alpha, q, z)$.*

Alternatively, assume $\beta \neq 0$ and define

$$\tilde{q} = \frac{\nu}{\beta}.$$

Observe that $\tilde{q} = q^{-1}$ for $\nu \neq 0$. Then the factor $\zeta - R^{**}(\alpha, \beta/\zeta^m, \nu/\zeta^m)$ in (10.3.9) can be rewritten as $\zeta - \tilde{R}^{**}(\alpha, \beta/\zeta^m, \tilde{q})$, where

$$\tilde{R}^{**}(\alpha, w, \tilde{q}) = 1 + (\alpha + w)b^T\big(I - \alpha A - w(B + \tilde{q}C)\big)^{-1}e,$$

so that (10.3.10) changes to

$$\zeta = \tilde{R}^{**}(\alpha, \beta/\zeta^m, \tilde{q}) \tag{10.3.11}$$

and Assumption 10.3.7 modifies as follows.

Assumption 10.3.8 *The matrix $I - \alpha A - w^*(B + \tilde{q}C)$ is singular if and only if w^* is a pole of $\tilde{R}^{**}(\alpha, w, \tilde{q})$.*

Observe that, for $\beta \neq 0$ and $\nu = 0$, equation (10.3.11) equals (10.2.18) and, hence, the approach is consistent with the non-neutral case.

Bellen, Jackiewicz and Zennaro [31] found the conditions such that the roots of (10.3.10) satisfy $|\zeta| < 1$ for all $m \geq 1$. Then, by using arguments similar to those of Section 10.2.1, they proved that any A-stable natural RK method for NDDEs with $C = I$ is asymptotically stable on the class of equations satisfying the condition

$$|\lambda\bar{\nu} - \bar{\mu}| + |\lambda\nu + \mu| < -2\Re(\lambda),$$

whenever the constant stepsize h is subject to constraint (10.3.2). The result is not sharp in that, as remarked in Section 9.6, such a condition does not characterize the whole class of asymptotically stable equations with complex coefficients, but it does whenever $\lambda \in \mathbb{R}$. On the other hand, for $\lambda, \mu, \nu \in \mathbb{R}$, the result is optimal and the method is NP(0)-stable. However, the NP-stability of A-stable natural RK methods for NDDEs with $C = I$ is proved in the forthcoming Theorem 10.6.1 for the larger class of neutral systems and, therefore, we omit any proof here.

10.3.2 *GNP-stability*

There are no specific results based on the scalar equation (10.3.1). On the contrary, there are more general results based on linear systems of NDDEs that are reported in the forthcoming Section 10.6.

10.3.3 *ND-stability*

To date, the analysis of ND-stability and ND(0)-stability is at a very early stage and we propose this section just for the sake of completeness. To the authors' knowledge, the only paper on this topic is due to Guglielmi [109], who proved that:

- no natural RK method for NDDEs of order $p \geq 1$ with p odd can be ND(0)-stable;

- all natural RK methods for NDDEs with symmetric A-stability function, whose order star $S^* = \{\alpha \in \mathbb{C} \mid |R(\alpha)| > |e^{\alpha}|\}$ has the whole imaginary axis as a boundary with S^* lying to the left (such as, for example, Gaussian collocation), are ND(0)-stable.

10.4 Generalizations of AN-stability and BN-stability to DDEs

The straightforward generalization of the scalar linear non-autonomous test equation $y'(t) = \lambda(t)y(t)$, $y(t_0) = y_0$, to DDEs is given by

$$\begin{cases} y'(t) = \lambda(t)y(t) + \mu(t)y(t-\tau), & t \geq t_0, \\ y(t) = \phi(t), & t \leq t_0, \end{cases} \tag{10.4.1}$$

where $\lambda(t)$ and $\mu(t)$ are continuous complex valued functions and τ is a constant delay.

In Chapter 9, we saw that the condition

$$\Re(\lambda(t)) + |\mu(t)| \leq 0, \quad t \geq t_0, \tag{10.4.2}$$

guarantees that the solution $y(t)$ satisfies

$$|y(t)| \leq \max_{x \leq t_0} |\phi(x)|, \quad t \geq t_0. \tag{10.4.3}$$

Therefore, a numerical method could be sought for the preservation of the contractivity property (10.4.3) for all delays under condition (10.4.2). Again, we present the definitions and related results given in the literature and complete the frame with new definitions and results.

Dealing with the contractivity properties of numerical methods for DDEs, Torelli gave the following definitions of PN-stability and GPN-stability (see [247] and Bellen and Zennaro [38]).

Definition 10.4.1 *A numerical method for DDEs is* PN-stable *if the discrete numerical solution* $\{y_n\}_{n\geq 0}$ *of any test equation (10.4.1) satisfying (10.4.2) is such that*

$$|y_n| \leq \max_{x \leq t_0} |\phi(x)|, \quad n \geq 0, \tag{10.4.4}$$

for all constant delays τ *and all initial functions* $\phi(t)$ *and for any constant stepsize* h *under the constraint*

$$h = \tau/m, \quad m \geq 1, \ m \text{ integer}. \tag{10.4.5}$$

Definition 10.4.2 *A numerical method for DDEs is* GPN-stable *if the discrete numerical solution* $\{y_n\}_{n\geq 0}$ *of any test equation (10.4.1) satisfying (10.4.2) is such that (10.4.4) holds for all constant delays* τ *and all initial functions* $\phi(t)$ *and for any constant stepsize* h.

Also, for the test equation (10.4.1) we distinguish between a constrained and a free mesh by imposing condition (10.4.5) or not. Moreover, this time we do not consider the regions of PN-stability and GPN-stability (although they have been defined in [247]), but consider directly those methods that completely preserve the contractivity properties of the exact solutions.

It is clear that a GPN-stable method is also PN-stable and that a PN-stable method for DDEs is AN-stable for ODEs. Moreover, PN-stability and GPN-stability are stronger concepts than P-contractivity and GP-contractivity because they are based on a more general test equation.

As with the generalization of A-stability, we consider the test equation

$$\begin{cases} y'(t) = \lambda(t)y(t) + \mu(t)y(t-\tau(t)), & t \geq t_0, \\ y(t) = \phi(t), & t \leq t_0, \end{cases} \tag{10.4.6}$$

more general than (10.4.1), with variable delay $\tau(t)$ and a related stability concept that allows a variable stepsize.

In Chapter 9 we saw that, under hypotheses (H_1) and (H_4), the solution $y(t)$ satisfies (10.4.3) if (10.4.2) is satisfied.

Definition 10.4.3 *A numerical method for DDEs is* fully PN-stable *(in short,* FPN-stable*) if the discrete numerical solution* $\{y_n\}_{n\geq 0}$ *of (10.4.6), obtained with any mesh* Δ*, satisfies (10.4.4) for all initial functions* $\phi(t)$ *and all variable delays* $\tau(t)$ *under hypotheses* (H_1) *and* (H_4)*, whenever (10.4.2) holds.*

FPN-stability ⇒ GPN-stability ⇒ PN-stability ⇒ AN-stability

The straightforward generalization of the ODE system $y'(t) = g(t, y(t))$, $y(t_0) = y_0$, to DDEs is given by

$$\begin{cases} y'(t) = f\left(t, y(t), y(t-\tau)\right), & t \geq t_0, \\ y(t) = \phi(t), & t \leq t_0, \end{cases} \tag{10.4.7}$$

where τ is a constant delay.

In order to introduce the relevant stability concepts, consider another system

$$\begin{cases} \tilde{y}'(t) = f\left(t, \tilde{y}(t), \tilde{y}(t-\tau)\right), & t \geq t_0, \\ \tilde{y}(t) = \tilde{\phi}(t), & t \leq t_0. \end{cases} \tag{10.4.8}$$

In Chapter 9, Theorem 9.1.3, we saw that the condition

$$Y(t) + X(t) \leq 0, \quad t \geq t_0, \tag{10.4.9}$$

where $Y(t)$ and $X(t)$ are the continuous functions (8.5.3) and (8.5.4), guarantees that the DDE system (10.4.7) is contractive, i.e.

$$\|\delta y(t)\| \leq \max_{x \leq t_0} \|\delta\phi(x)\|, \quad t \geq t_0, \tag{10.4.10}$$

where

$$\delta y(t) = \tilde{y}(t) - y(t)$$

and

$$\delta\phi(t) = \tilde{\phi}(t) - \phi(t).$$

Let $\{y_n\}_{n\geq 0}$ and $\{\tilde{y}_n\}_{n\geq 0}$ be the numerical solutions of (10.4.7) and (10.4.8), respectively, and define

$$\delta y_n = \tilde{y}_n - y_n.$$

Torelli gave the following definitions of RN-stability and GRN-stability (see again [247] and [38]).

Definition 10.4.4 *A numerical method for DDEs is* RN-stable *if the difference $\{\delta y_n\}_{n\geq 0}$ in the discrete numerical solutions of the systems (10.4.8) and (10.4.7) satisfying (10.4.9) is such that*

$$\|\delta y_n\| \leq \max_{x\leq t_0} \|\delta\phi(x)\|, \quad n \geq 0, \tag{10.4.11}$$

for all initial functions $\phi(t)$ and $\tilde{\phi}(t)$, for all constant delays τ and for any constant stepsize h under constraint (10.4.5).

Definition 10.4.5 *A numerical method for DDEs is* GRN-stable *if the difference $\{\delta y_n\}_{n\geq 0}$ in the discrete numerical solutions of the systems (10.4.8) and (10.4.7) satisfying condition (10.4.9) is such that (10.4.11) holds for all initial functions $\phi(t)$ and $\tilde{\phi}(t)$, for all constant delays τ and for any constant stepsize h.*

It is clear that RN-stability and GRN-stability are stronger concepts than PN-stability and GPN-stability, respectively, since thay are based on more general test equations.

RN-stability ⇒ PN-stability ⇒ P-contractivity

GRN-stability ⇒ GPN-stability ⇒ GP-contractivity

Again, we consider the more general test systems

$$\begin{cases} y'(t) = f\left(t, y(t), y(t-\tau(t))\right), & t \geq t_0, \\ y(t) = \phi(t), & t \leq t_0, \end{cases} \tag{10.4.12}$$

and

$$\begin{cases} \tilde{y}'(t) = f\left(t, \tilde{y}(t), \tilde{y}(t-\tau(t))\right), & t \geq t_0, \\ \tilde{y}(t) = \tilde{\phi}(t), & t \leq t_0, \end{cases} \tag{10.4.13}$$

with variable delay $\tau(t)$ satisfying hypotheses (H_1) and (H_4). Under condition (10.4.9), they still satisfy the contractivity property (10.4.10).

Definition 10.4.6 *A numerical method for DDEs is* fully RN-stable *(in short,* FRN-stable*) if the difference $\{\delta y_n\}_{n\geq 0}$ in the discrete numerical solutions of the systems (10.4.13) and (10.4.12) satisfying condition (10.4.9) is such that (10.4.11) holds for all initial functions $\phi(t)$ and $\tilde{\phi}(t)$, for all variable delays $\tau(t)$ under hypotheses (H_1) and (H_4) and for any mesh Δ.*

FRN-stability ⇒ GRN-stability ⇒ RN-stability ⇒ BN-stability

FRN-stability ⇒ FPN-stability ⇒ FP-contractivity

10.4.1 *PN-stability, GPN-stability and FPN-stability*

The next theorem represents the link, almost an equivalence result, between the theory of PN-stability, GPN-stability and FPN-stability concerning DDE methods and the theory of AN$_f$-stability concerning ODE methods.

Theorem 10.4.7 *If the RK method for DDEs (6.2.1), (6.2.2) is PN-stable, then the underlying discrete RK method (5.0.2), (5.0.3) is AN_f-stable.*

Conversely, if the underlying CRK method (5.0.2), (5.1.6), (5.1.1) is semi-AN_f-stable, then the RK method for DDEs (6.2.1), (6.2.2) is PN-stable. Moreover, if the underlying CRK method (5.0.2), (5.1.6), (5.1.1) is AN_f-stable, then the RK method for DDEs (6.2.1), (6.2.2) is FPN-stable and, hence, GPN-stable too.

Proof. Let the RK method for DDEs (6.2.1), (6.2.2) be PN-stable. Therefore, for the particular DDE with constant delay

$$\begin{cases} y'(t) = \lambda(t)y(t) + \Re(\lambda(t))y(t-\tau), & t \geq t_0, \\ y(t) = \phi(t), & t \geq t_0, \end{cases} \tag{10.4.14}$$

with $\Re(\lambda(t)) \leq 0$, we have that $|y_1| \leq \max_{x\leq t_0} |\phi(x)|$ for all $h = \tau/m$, m a positive integer, and for all initial functions $\phi(t)$.

In particular, set $h = \tau/2$, $y_0 = \phi(t_0)$ and, for $t_0 \leq t \leq t_0 + h$, $g(t) = \phi(t-\tau)$. Therefore, for $t_0 \leq t \leq t_0 + h$ the DDE (10.4.14) reduces to the ODE with forcing term (8.5.13) and, moreover, $|y_1| \leq \max\{|y_0|, \max_{t_0\leq x\leq t_0+h} |g(x)|\}$. Since $\phi(t)$ is arbitrary, y_0 and $g(t)$ are also arbitrary and thus, in view of Remark 8.5.13, the underlying discrete RK method (5.0.2), (5.0.3) is AN$_f$-stable.

The converse implication is proved by induction. Let the underlying CRK method (5.0.2), (5.1.6), (5.1.1) be semi-AN$_f$-stable and apply the RK method for DDEs (6.2.1), (6.2.2) with constant stepsize $h = \tau/m$, m a positive integer, to the DDE (10.4.1) satisfying (10.4.2). Since $\eta(t) = \phi(t)$ for $t \leq t_0$, by extending the mesh Δ also to the interval $[t_0 - \tau, t_0]$, we have that

$$|y_k| \leq \max_{x\leq t_0} |\phi(x)| \quad \text{and} \quad |\eta(t_k^i)| \leq \max_{x\leq t_0} |\phi(x)| \tag{10.4.15}$$

for all $k = -m+1, \ldots, 0$ and $i = 1, \ldots, s$, where $y_k = \eta(t_k)$.

Now assume that (10.4.15) holds for all $k \leq n$ and consider the numerical solution of (10.4.1) in the step $[t_n, t_{n+1}]$. This is the numerical solution of the local problem

$$\begin{cases} w'_{n+1}(t) = \lambda(t)w_{n+1}(t) + \mu(t)\eta(t-\tau), & t_n \leq t \leq t_{n+1}, \\ w_{n+1}(t_n) = y_n. \end{cases} \tag{10.4.16}$$

Since (10.4.2) is satisfied, we have that

$$\mu(t) = r(t)\Re(\lambda(t)) \quad \text{with} |r(t)| \leq R \leq 1 \tag{10.4.17}$$

and that $\Re(\lambda(t)) \leq 0$. Therefore, since the underlying CRK method is semi-AN$_\mathrm{f}$-stable, we can conclude that

$$|y_{n+1}| \leq \max\left\{|y_n|, \max_{1\leq i\leq s} |\eta(t^i_{n+1} - \tau)|\right\}$$

and

$$|\eta(t^i_{n+1})| \leq \max\left\{|y_n|, \max_{1\leq i\leq s} |\eta(t^i_{n+1} - \tau)|\right\}, \quad i = 1, \ldots, s.$$

On the other hand, since $\eta(t^i_{n+1} - \tau) = \eta(t^i_{n-m+1})$ for some $m \geq 1$, by the inductive hypothesis we can conclude that (10.4.15) holds also for $k = n+1$. Therefore the RK method for DDEs (6.2.1), (6.2.2) is PN-stable.

We are left to prove the FPN-stability of the RK method for DDEs (6.2.1), (6.2.2) if the underlying CRK method (5.0.2), (5.1.6), (5.1.1) is AN$_\mathrm{f}$-stable. In this case we apply the RK method to the DDE (10.4.6) and observe that the numerical solution in the step $[t_n, t_{n+1}]$ is the numerical solution of the local problem

$$\begin{cases} w'_{n+1}(t) = \lambda(t)w_{n+1}(t) + \mu(t)x(t-\tau(t)), & t_n \leq t \leq t_{n+1}, \\ w_{n+1}(t_n) = y_n, \end{cases} \tag{10.4.18}$$

where

$$x(s) = \begin{cases} \phi(s) & \text{for } s \leq t_0, \\ \eta(s) & \text{for } t_0 \leq s \leq t_n, \\ w_{n+1}(s) & \text{for } t_n \leq s \leq t_{n+1}. \end{cases}$$

We assume by induction that

$$|\eta(t)| \leq \max_{x \leq t_0} |\phi(x)|, \quad t \leq t_n, \tag{10.4.19}$$

which obviously holds for $n = 0$. By (10.4.17), since the underlying CRK method is AN$_\mathrm{f}$-stable, we have

$$|\eta(t)| \leq \max\left\{|y_n|, R \max_{1\leq i\leq s} |\eta(t^i_{n+1} - \tau(t^i_{n+1}))|\right\}, \quad t_n \leq t \leq t_{n+1}.$$

If overlapping does not occur, i.e. if $t^i_{n+1} - \tau(t^i_{n+1})) \leq t_n$, $i = 1, \ldots, s$, then (10.4.19) clearly holds also for $t \leq t_{n+1}$ and the proof is complete.

On the contrary, if overlapping occurs, then the inductive hypothesis (10.4.19) just implies

$$|\eta(t)| \leq \max\left\{\max_{x\leq t_0}|\phi(x)|, R \max_{t_n\leq x\leq t_{n+1}}|\eta(x)|\right\}, \quad t_n \leq t \leq t_{n+1},$$

but the proof can be completed in the same way as for Theorem 10.2.34 by considering a one-parameter family of local problems perturbed from (10.4.18) if $R = 1$. ■

Remark 10.4.8 *From the proof of Theorem 10.4.7, we can see that AN_f-stability of the underlying CRK method implies even more than FPN-stability, namely*

$$|\eta(t)| \leq \max_{x\leq t_0}|\phi(x)|, \quad t \geq t_0.$$

In other words, the contractivity property (10.4.4) extends to the continuous numerical solution.

In the light of the results reviewed in Section 8.4, we can say that the only one-stage PN-stable RK method for DDEs of order 1 is the backward Euler method with linear interpolation, that is

$$Y_{n+1}^1 = y_n + h_{n+1}f(t_{n+1}, Y_{n+1}^1, \eta(t_{n+1}-\tau)),$$

$$\eta(t_n+\theta h_{n+1}) = y_n + \theta h_{n+1}f(t_{n+1}, Y_{n+1}^1, \eta(t_{n+1}-\tau)).$$

In the class of two-stage RK methods for DDEs of order 2, the only one that is PN-stable is the Lobatto IIIC method with linear interpolation, that is

$$\begin{aligned} Y_{n+1}^1 &= y_n + \tfrac{1}{2}h_{n+1}\left(f(t_n, Y_{n+1}^1, \eta(t_n-\tau)) - f(t_{n+1}, Y_{n+1}^2, \eta(t_{n+1}-\tau))\right), \\ Y_{n+1}^2 &= y_n + \tfrac{1}{2}h_{n+1}\left(f(t_n, Y_{n+1}^1, \eta(t_n-\tau)) + f(t_{n+1}, Y_{n+1}^2, \eta(t_{n+1}-\tau))\right), \end{aligned}$$

$$\begin{aligned} \eta(t_n+\theta h_{n+1}) = y_n + \tfrac{1}{2}\theta h_{n+1}\big(&f(t_n, Y_{n+1}^1, \eta(t_n-\tau)) \\ &+f(t_{n+1}, Y_{n+1}^2, \eta(t_{n+1}-\tau))\big). \end{aligned}$$

The first part of Theorem 10.4.7 already says that AN-stability of the underlying discrete RK method is not sufficient to guarantee PN-stability. But there is something more. In fact, Torelli [248] showed how one-step collocation at one Gaussian point (the midpoint rule), which is AN-stable, gives rise to a numerical solution $\{y_n\}_{n\geq 0}$ that blows up as $n \to \infty$ for a particular test equation (10.4.1) with $\lambda(t) = -\mu(t) \leq 0$. Other examples of this phenomenon have been illustrated numerically in Chapter 1. In line with these counter-examples, we give the following slight generalization of a result by In 't Hout [131], saying that AN_f-stability of the underlying discrete RK method is necessary even for just boundedness.

Theorem 10.4.9 *Let the RK method for DDEs (6.2.1), (6.2.2) be such that* $c_i \in [0,1], i = 1,\dots,s$, *with* 0 *and* 1 *both not abscissae of the method. If we assume that, when it is applied to any test equation (10.4.1) satisfying (10.4.2) with any constant delay* τ *and any initial function* $\phi(t)$, *it produces a discrete numerical solution* $\{y_n\}_{n\geq 0}$ *which is bounded for any constant stepsize* $h > 0$ *under the constraint (10.4.5), then the underlying discrete RK method (5.0.2), (5.0.3) is* AN_f*-stable.*

Proof. It is not restrictive to assume that the abscissae are counted in increasing order. Therefore there are two possible cases: either $0 \leq c_1 \leq \cdots \leq c_s < 1$ or $0 < c_1 \leq \cdots \leq c_s \leq 1$.

Consider a particular test equation (10.4.1) where the coefficients $\lambda(t)$ and $\mu(t)$ are 2τ-periodic continuous functions such that $\lambda(t) \equiv \mu(t) \equiv 0$ for $t \in [t_0, t_0 + c_s\tau]$ in the first case and for $t \in [t_0 + c_1, t_0 + \tau]$ in the second case. Moreover, they are assumed to satisfy condition (10.4.2).

Then apply the RK method with constant stepsize $h = \tau$ so as to get

$$Y^i_{n+1} = \eta(t^i_{n+1}) = y_{n+1} = y_n, \quad i = 1,\dots,s,$$

for all $n = 0, 2, 4, \dots$, and

$$Y^i_{n+1} = y_n + h\sum_{j=1}^{s} a_{ij}(\lambda^j Y^j_{n+1} + \mu^j \eta(t^j_n)), \quad i = 1,\dots,s,$$

$$y_{n+1} = y_n + h\sum_{i=1}^{s} b_i(\lambda^i Y^i_{n+1} + \mu^i \eta(t^i_n)),$$

$$\eta(t^i_{n+1}) = y_n + h\sum_{j=1}^{s} b_j(c_i)(\lambda^j Y^j_{n+1} + \mu^j \eta(t^j_n)), \quad i = 1,\dots,s,$$

for all $n = 1, 3, 5, \dots$, where, due to the periodicity of $\lambda(t)$ and $\mu(t)$, we have set $\lambda^i = \lambda(t^i_2) = \lambda(t^i_{n+1})$ and $\mu^i = \mu(t^i_2) = \mu(t^i_{n+1})$.

Therefore, for all $n = 1, 3, 5, \dots$, we get the following two-step recurrence relation:

$$Y^i_{n+1} = y_{n-1} + h\sum_{j=1}^{s} a_{ij}(\lambda^j Y^j_{n+1} + \mu^j y_{n-1}), \quad i = 1,\dots,s,$$

$$y_{n+1} = y_{n-1} + h\sum_{i=1}^{s} b_i(\lambda^i Y^i_{n+1} + \mu^i y_{n-1}).$$

After elimination of the stage values Y^i_{n+1}, it can be written as

$$y_{n+1} = \left(1 + b^T(I - ZA)^{-1}(Z + W)e\right) y_{n-1},$$

where $Z = \mathrm{diag}(h\lambda^1, \dots, h\lambda^s)$ and $W = \mathrm{diag}(h\mu^1, \dots, h\mu^s)$ satisfy the restriction that $i, j \in I_k \Longrightarrow \{h\lambda^i = h\lambda^j \text{ and } h\mu^i = h\mu^j\}$, where $\{I_1, \dots, I_r\}$ is the confluency partition of the index set $\{1, \dots, s\}$.

The boundedness of the sequence $\{y_n\}$ implies $|1+b^T(I-ZA)^{-1}(Z+W)e| \leq 1$. Finally, the arbitrariness of $\lambda^i = \lambda(t_2^i)$ and $\mu^i = \mu(t_2^i)$ under condition (10.4.2) implies (8.5.33), which is the AN$_\mathrm{f}$-stability of the underlying discrete RK method. ■

This result is important, since it states that, for a wide class of RK methods for DDEs, AN$_\mathrm{f}$-stability of the underlying discrete RK method is a necessary condition even just to ensure the boundedness of the numerical solutions of the test equation (10.4.1) satisfying (10.4.2), no matter what interpolation procedure is used. A less general version of Theorem 10.4.9, valid for natural RK methods for DDEs (in particular, for collocation methods), was given by In 't Hout [134].

However, the above result does not apparently refer to those methods, such as the trapezoidal rule, that have both 0 and 1 in the set of abscissae. Nevertheless, we recall that the numerical examples given in Section 1.2.2 revealed the same instability phenomenon also for the trapezoidal rule, which is not AN$_\mathrm{f}$-stable.

10.4.2 *RN-stability, GRN-stability and FRN-stability*

An analogous result to Theorem 10.4.7 holds, which shows the link between the theory of BN$_\mathrm{f}$-stability and the stability concepts for DDE methods with respect to non-linear test equations.

Theorem 10.4.10 *If the RK method for DDEs (6.2.1), (6.2.2) is RN-stable, then the underlying discrete RK method (5.0.2), (5.0.3) is BN$_f$-stable.*

Conversely, if the underlying CRK method (5.0.2), (5.1.6), (5.1.1) is semi-BN$_f$-stable, then the RK method for DDEs (6.2.1), (6.2.2) is RN-stable. Moreover, if the underlying CRK method (5.0.2), (5.1.6), (5.1.1) is BN$_f$-stable, then the RK method for DDEs (6.2.1), (6.2.2) is FRN-stable and, hence, GRN-stable too.

Proof. Consider the systems of ODEs with forcing terms (8.5.1) and (8.5.2) satisfying (8.5.5) and (8.5.6) in the interval $[t_0, t_0+h]$ for some stepsize $h > 0$. Observe that if $r(t^*) = 0$ for some t^*, then $X(t^*) = 0$ and, thus, $f(t^*, y, x)$ is constant with respect to x. Therefore, if we define the function

$$f^*(t,y,x) = \begin{cases} f(t,y,x/r(t)) & \text{if } r(t) \neq 0, \\ f(t,y,0) & \text{if } r(t) = 0, \end{cases}$$

then we have that

$$f(t,y,x) = f^*(t,y,r(t)x)$$

for all $t \geq t_0$ and $x, y \in \mathbb{C}^d$.

In the interval $[t_0 - 2h, t_0]$, define the continuous functions

$$\phi(t) = \begin{cases} r(t+2h)u(t+2h) & \text{if } t_0 - 2h \leq t \leq t_0 - h, \\ \frac{t_0-t}{h} r(t_0+h)u(t_0+h) + (1 - \frac{t_0-t}{h})y_0 & \text{if } t_0 - h \leq t \leq t_0, \end{cases}$$

and

$$\tilde{\phi}(t) = \begin{cases} r(t+2h)\tilde{u}(t+2h) & \text{if } t_0 - 2h \le t \le t_0 - h, \\ \frac{t_0-t}{h} r(t_0+h)\tilde{u}(t_0+h) + (1 - \frac{t_0-t}{h})\tilde{y}_0 & \text{if } t_0 - h \le t \le t_0, \end{cases}$$

so that

$$\max_{t_0-2h\le x\le t_0} \|\delta\phi(x)\| = \max\left\{\|\delta y_0\|, \max_{t_0\le x\le t_0+h} (|r(x)| \cdot \|\delta u(x)\|)\right\}. \tag{10.4.20}$$

Thus, the systems of ODEs (8.5.1) and (8.5.2) can be equivalently rewritten in $[t_0, t_0+h]$ as the system of DDEs with constant delay $\tau = 2h$:

$$\begin{cases} y'(t) = f^*\left(t, y(t), y(t-2h)\right), & t_0 \le t \le t_0 + h, \\ y(t) = \phi(t), & t \le t_0, \end{cases}$$

and

$$\begin{cases} \tilde{y}'(t) = f^*\left(t, \tilde{y}(t), \tilde{y}(t-2h)\right), & t_0 \le t \le t_0 + h, \\ \tilde{y}(t) = \tilde{\phi}(t), & t \le t_0. \end{cases}$$

It is immediately seen that their numerical solutions y_1 and $\tilde{y}_1$ at the point $t_0 + h$, obtained with the RK method for DDEs (6.2.1), (6.2.2), equal the numerical solutions of the original systems of ODEs obtained with the underlying discrete RK method.

By (8.5.5) and (8.5.6), the quantities (8.5.3) and (8.5.4) relevant to f^* can be chosen to be

$$Y^*(t) = Y(t) \qquad \text{and} \qquad X^*(t) = -Y(t)$$

and, hence, they satisfy condition (10.4.9). Therefore, since the RK method for DDEs (6.2.1), (6.2.2) is RN-stable, (10.4.20) yields

$$\|\delta y_1\| \le \max\left\{\|\delta y_0\|, \sup_{t_0\le x\le t_0+h} (|r(x)| \cdot \|\delta u(x)\|)\right\},$$

i.e. in view of Remark 8.5.16, the underlying discrete RK method is $\mathrm{BN_f}$-stable.

The converse implication is proved by induction. Let the underlying CRK method (5.0.2), (5.1.6), (5.1.1) be semi-$\mathrm{BN_f}$-stable and let us apply the RK method for DDEs (6.2.1), (6.2.2) with constant stepsize $h = \tau/m$, m a positive integer, to the systems (10.4.8) and (10.4.7) satisfying (10.4.9). Since $\eta(t) = \phi(t)$ and $\tilde{\eta}(t) = \tilde{\phi}(t)$ for $t \le t_0$, by also extending the mesh Δ to the interval $[t_0-\tau, t_0]$, we have that

$$\|\delta y_k\| \le \max_{x\le t_0} \|\delta\phi(x)\| \quad \text{and} \quad \|\delta\eta(t_k^i)\| \le \max_{x\le t_0} \|\delta\phi(x)\| \tag{10.4.21}$$

for all $k = -m+1, \ldots, 0$ and $i = 1, \ldots, s$, where $\delta y_k = \delta\eta(t_k)$ and $\delta\eta(t) = \tilde{\eta}(t) - \eta(t)$.

Now assume that (10.4.21) holds for all $k \leq n$ and consider the numerical solutions of (10.4.8) and (10.4.7) in the step $[t_n, t_{n+1}]$. They are the numerical solutions of the local problems

$$\begin{cases} w'_{n+1}(t) = f\left(t, w_{n+1}(t), \eta(t-\tau)\right), & t_n \leq t \leq t_{n+1}, \\ w_{n+1}(t_n) = y_n, \end{cases}$$

and

$$\begin{cases} \tilde{w}'_{n+1}(t) = f\left(t, \tilde{w}_{n+1}(t), \tilde{\eta}(t-\tau)\right), & t_n \leq t \leq t_{n+1}, \\ \tilde{w}_{n+1}(t_n) = \tilde{y}_n. \end{cases}$$

Since (10.4.9) is satisfied, we have that

$$X(t) = r(t)Y(t) \quad \text{with } |r(t)| \leq R \leq 1 \tag{10.4.22}$$

and $Y(t) \leq 0$. Moreover, since the underlying CRK method is semi-BN$_{\mathrm{f}}$-stable, we also have that

$$\|\delta y_{n+1}\| \leq \max\left\{\|\delta y_n\|, \max_{1\leq i\leq s} \|\delta\eta(t^i_{n+1} - \tau)\|\right\}$$

and

$$\|\delta\eta(t^i_{n+1})\| \leq \max\left\{\|\delta y_n\|, \max_{1\leq i\leq s} \|\delta\eta(t^i_{n+1} - \tau)\|\right\}, \quad i = 1, \ldots, s.$$

On the other hand, since $\delta\eta(t^i_{n+1} - \tau) = \delta\eta(t^i_{n-m+1})$ for some $m \geq 1$, by the inductive hypothesis we can conclude that (10.4.21) also holds for $k = n+1$. Therefore, the RK method for DDEs (6.2.1), (6.2.2) is RN-stable.

We are left to prove the FRN-stability of the RK method for DDEs (6.2.1), (6.2.2) if the underlying CRK method (5.0.2), (5.1.6), (5.1.1) is BN$_{\mathrm{f}}$-stable. In this case we apply the RK method to the systems of DDEs (10.4.12) and (10.4.13) and observe that the numerical solutions in the step $[t_n, t_{n+1}]$ are the numerical solutions of the local problems

$$\begin{cases} w'_{n+1}(t) = f\left(t, w_{n+1}(t), x(t-\tau(t))\right), & t_n \leq t \leq t_{n+1}, \\ w_{n+1}(t_n) = y_n, \end{cases} \tag{10.4.23}$$

where

$$x(s) = \begin{cases} \phi(s) & \text{for } s \leq t_0, \\ \eta(s) & \text{for } t_0 \leq s \leq t_n, \\ w_{n+1}(s) & \text{for } t_n \leq s \leq t_{n+1}, \end{cases}$$

and

$$\begin{cases} \tilde{w}'_{n+1}(t) = f\left(t, \tilde{w}_{n+1}(t), \tilde{x}(t-\tau(t))\right), & t_n \leq t \leq t_{n+1}, \\ \tilde{w}_{n+1}(t_n) = \tilde{y}_n, \end{cases} \tag{10.4.24}$$

where

$$\tilde{x}(s) = \begin{cases} \tilde{\phi}(s) & \text{for } s \leq t_0, \\ \tilde{\eta}(s) & \text{for } t_0 \leq s \leq t_n, \\ \tilde{w}_{n+1}(s) & \text{for } t_n \leq s \leq t_{n+1}, \end{cases}$$

respectively.

We assume by induction that

$$\|\delta\eta(t)\| \le \max_{x \le t_0} \|\delta\phi(x)\|, \quad t \le t_n, \tag{10.4.25}$$

which obviously holds for $n = 0$. By (10.4.22), since the underlying CRK method is $\mathrm{BN_f}$-stable, we have

$$\|\delta\eta(t)\| \le \max\left\{\|\delta y_n\|, R \max_{1 \le i \le s} \|\delta\eta(t_{n+1}^i - \tau(t_{n+1}^i))\|\right\}, \quad t_n \le t \le t_{n+1}.$$

If overlapping does not occur, i.e. if $t_{n+1}^i - \tau(t_{n+1}^i)) \le t_n$, $i = 1, \dots, s$, then (10.4.25) clearly holds also for $t \le t_{n+1}$ and the proof is complete.

On the contrary, if overlapping occurs, then the inductive hypothesis (10.4.25) just implies

$$\|\delta\eta(t)\| \le \max\left\{\max_{x \le t_0} \|\delta\phi(x)\|, R \max_{t_n \le x \le t_{n+1}} \|\delta\eta(x)\|\right\}, \quad t_n \le t \le t_{n+1},$$

but the proof can be completed in the same way as for Theorem 10.2.34 by considering two one-parameter families of local problems perturbed from (10.4.23) and (10.4.24) if $R = 1$. ■

Remark 10.4.11 *From the proof of Theorem 10.4.10 we can see that BN_f-stability of the underlying CRK method implies even more than FRN-stability, namely*

$$\|\delta\eta(t)\| \le \max_{x \le t_0} \|\delta\phi(x)\|, \quad t \ge t_0.$$

In other words, the contractivity property (10.4.11) extends to the continuous numerical solution.

It is known (see Section 8.5.3) that the backward Euler method with linear interpolation and the Lobatto IIIC method with linear interpolation are $\mathrm{BN_f}$-stable. Therefore, they are also FRN-stable.

10.4.3 *Asymptotic stability via contractivity*

In Section 9.2 we saw that, for any variable delay $\tau(t)$ that satisfies hypotheses (H_1), (H_2) and (H_4), the conditions

$$\Re(\lambda(t)) \le \Lambda_0 < 0, \quad t \ge t_0, \tag{10.4.26}$$

and

$$R \cdot \Re(\lambda(t)) + |\mu(t)| \le 0, \quad t \ge t_0, \tag{10.4.27}$$

for some non-negative real number $R < 1$, guarantee that equation (10.4.6) is asymptotically stable, i.e.

$$\lim_{t \to +\infty} y(t) = 0$$

for any initial function $\phi(t)$.

More generally, under the same hypotheses on the delay, we have seen that the conditions

$$Y(t) \leq Y_0 < 0, \quad t \geq t_0, \tag{10.4.28}$$

and

$$R \cdot Y(t) + X(t) \leq 0, \quad t \geq t_0, \tag{10.4.29}$$

for some non-negative real number $R < 1$, guarantee that the the non-linear system of DDEs (10.4.12) is asymptotically stable, i.e.

$$\lim_{t \to +\infty} \delta y(t) = 0$$

for all initial functions $\phi(t)$ and $\tilde{\phi}(t)$ (see Definition 9.1.2).

Moreover, in both cases the additional hypothesis (H_3) on the delay ensures that the convergence rate to zero is at least of exponential type.

Now we give sufficient conditions in order that the numerical solutions furnished by the RK method preserve such asymptotic stability properties.

Theorem 10.4.12 *If the underlying CRK method (5.0.2), (5.1.6) - (5.1.1) is semi-AN$_f$-stable, then the discrete numerical solution $\{y_n\}_{n\geq 0}$ of any test equation (10.4.1) satisfying (10.4.26) and (10.4.27), furnished by the RK method for DDEs (6.2.1), (6.2.2) with constant stepsize h under the constraint (10.4.5), is asymptotically stable, that is*

$$\lim_{n \to \infty} y_n = 0 \tag{10.4.30}$$

for all initial functions $\phi(t)$, for all constant delays τ. Moreover, the convergence rate in (10.4.30) is at least of exponential type, i.e. at least like $e^{-\alpha(t_n - t_0)}$ for some $\alpha > 0$.

If the underlying CRK method (5.0.2), (5.1.6), (5.1.1) is AN$_f$-stable, then the continuous numerical solution $\eta(t)$ of any test equation (10.4.6) satisfying (10.4.26) and (10.4.27), furnished by the RK method for DDEs (6.2.1), (6.2.2) with any mesh Δ, is asymptotically stable, that is

$$\lim_{t \to +\infty} \eta(t) = 0 \tag{10.4.31}$$

for all initial functions $\phi(t)$, for all variable delays $\tau(t)$ satisfying hypotheses (H_1), (H_2) and (H_4). Moreover, if the additional hypothesis (H_3) is verified and if there exists $h_0 > 0$ such that $h_{n+1} \leq h_0$ for all $n \geq 0$, then the convergence rate in (10.4.31) is at least of exponential type, i.e. at least like $e^{-\alpha(t-t_0)}$ for some $\alpha > 0$.

Proof. In order to prove the first part, observe that the numerical solution y_{n+1} of (10.4.1) in the step $[t_n, t_{n+1}] \subseteq [\xi_{k-1}, \xi_k]$ is the numerical solution of the local problem (10.4.16). Since (10.4.27) is satisfied, we have that

$$\mu(t) = r(t)\Re(\lambda(t)) \quad \text{with } |r(t)| \leq R < 1 \tag{10.4.32}$$

and that $\Re(\lambda(t)) \leq 0$. Therefore, since the underlying CRK method is semi-AN$_\text{f}$-stable, we have

$$|\eta(t_{n+1}^i)| \leq \max\left\{|y_n|, R \max_{1\leq i\leq s} |\eta(t_{n+1}^i - \tau)|\right\}, \quad i = 1, \ldots, s,$$

and, moreover, Theorem 8.5.24 yields

$$|y_{n+1}| \leq \Psi_{n+1}^{AN}|y_n| + (1 - \Psi_{n+1}^{AN})R \max_{1\leq i\leq s} |\eta(t_{n+1}^i - \tau)|. \tag{10.4.33}$$

On the other hand, because of (10.4.5) we have $t_{n+1}^i - \tau = t_{n-m+1}^i \in [t_{n-m}, t_{n-m+1}] \subseteq [\xi_{k-2}, \xi_{k-1}]$. Therefore, if we define $F_n(\Delta) = |y_n|$ and $G_n(\Delta) = \max_{1\leq i\leq s} |\eta(t_n^i)|$, by (10.4.26), (8.5.38), Proposition 8.5.22 and Theorem 8.3.22, the hypotheses of Lemma 10.2.32 are satisfied. Finally, observe that constant delays clearly verify hypothesis (H_3) and that constant stepsizes satisfying the constraint (10.4.5) are definitely uniformly bounded. Therefore, the first part is proved.

In order to prove the second part, observe that the numerical solution of (10.4.6) in the step $[t_n, t_{n+1}]$ is the numerical solution of the local problem (10.4.18).

Because of (10.4.32), the inequality (10.4.33) still holds with $\tau = \tau(t_{n+1}^i)$ along with

$$|\eta(t)| \leq \max\left\{|y_n|, R \max_{1\leq i\leq s} |\eta(t_{n+1}^i - \tau(t_{n+1}^i))|\right\}, \quad t_n \leq t \leq t_{n+1}.$$

Assuming that $t_{n+1} \in (\xi_{k_{l-1}}, \xi_{k_l}]$, as observed in Remark 10.2.31, the delayed argument $t_{n+1}^i - \tau(t_{n+1}^i) \in [t_{m_1}, t_{m_2}]$ with $t_{m_1+1} > \xi_{k_{l-3}}$ and, of course, $t_{m_2} \leq t_{n+1}$. Therefore, with $F_n(\Delta) = |y_n|$ and $G_n(\Delta) = \max_{t_{n-1}\leq t\leq t_n} |\eta(t)|$, again the hypotheses of Lemma 10.2.32 are satisfied, and the second part is proved, too. ■

The proof of the next theorem can be found in [284] for meshes without overlapping and, since it is quite similar to the one of the previous Theorem 10.4.12, we do not give it here.

Theorem 10.4.13 *If the underlying CRK method (5.0.2), (5.1.6), (5.1.1) is semi-BN_f-stable, then the difference $\{\delta y_n\}_{n\geq 0}$ in the discrete numerical solutions of the systems (10.4.8) and (10.4.7) satisfying (10.4.28) and (10.4.29), furnished by the RK method for DDEs (6.2.1), (6.2.2) with constant stepsize h under the constraint (10.4.5), is asymptotically stable, that is*

$$\lim_{n\to\infty} \delta y_n = 0 \tag{10.4.34}$$

for all initial functions $\phi(t)$ and $\tilde{\phi}(t)$, for all constant delays τ. Moreover, the convergence rate in (10.4.34) is at least of exponential type, i.e. at least like $e^{-\alpha(t_n - t_0)}$ for some $\alpha > 0$.

If the underlying CRK method (5.0.2), (5.1.6), (5.1.1) is BN_f-stable, then the difference $\delta\eta(t)$ in the continuous numerical solutions of the systems (10.4.13) *and (10.4.12) satisfying (10.4.28) and (10.4.29), furnished by the RK method for DDEs (6.2.1), (6.2.2) with any mesh* Δ*, is asymptotically stable, that is*

$$\lim_{t\to+\infty} \delta\eta(t) = 0 \tag{10.4.35}$$

for all initial functions $\phi(t)$ *and* $\tilde{\phi}(t)$*, for all variable delays* $\tau(t)$ *satisfying hypotheses* (H_1)*,* (H_2) *and* (H_4)*. Moreover, if the additional hypothesis* (H_3) *is verified and if there exists* $h_0 > 0$ *such that* $h_{n+1} \leq h_0$ *for all* $n \geq 0$*, then the convergence rate in (10.4.35) is at least of exponential type, i.e. at least like* $e^{-\alpha(t-t_0)}$ *for some* $\alpha > 0$*.*

In view of the results of the present section, we can conclude that possible differences in the various stability properties (namely, contractivity, asymptotic stability and uniform boundedness) of a large class of RK methods for DDEs with respect to non-autonomous linear scalar test equations and general non-linear systems may be determined only by the stability properties of the underlying interpolant, the underlying discrete RK method being in any case necessarily AN_f-stable.

Indeed, an interesting open problem is whether or not the interpolant is also subject to a necessary condition for boundedness of the numerical solution similar to the one expressed by Theorem 10.4.9 for the discrete underlying method.

10.5 Boundedness and asymptotic stability for NDDEs

In this section we illustrate some boundedness and asymptotic stability results for the CRK methods (6.5.17), (6.5.18), (6.5.20) or (6.5.21) and (6.5.28), (6.5.29), (6.5.26) or (6.5.27), implemented within the method of steps (see Algorithms 4.5.1 and 4.5.3 in Section 4.5) for the numerical solution of non-linear systems of NDDEs rewritten as the ODE

$$\begin{cases} \Phi'(t) = K\left(t, \Phi(t), y(t-\tau(t))\right), & t \geq t_0, \\ \Phi(t_0) = \phi(t_0) - G(t_0, \phi(t_0 - \tau(t_0))), \end{cases} \tag{10.5.1}$$

coupled with the algebraic recursion

$$y(t) = \begin{cases} \phi(t), & t \leq t_0, \\ \Phi(t) + G(t, y(t-\tau(t))), & t \geq t_0, \end{cases} \tag{10.5.2}$$

and as the ODE

$$\begin{cases} y'(t) = F\left(t, y(t), \Psi(t)\right), & t \geq t_0, \\ y(t_0) = \phi(t_0), \end{cases} \tag{10.5.3}$$

coupled with the algebraic recursion

$$\Psi(t) = \begin{cases} G\Big(t, \phi(t-\tau(t)), \phi'(t-\tau(t))\Big), & t-\tau(t) \leq t_0, \\ H\Big(t, y(t-\tau(t)), \Psi(t-\tau(t))\Big), & t-\tau(t) > t_0, \end{cases} \tag{10.5.4}$$

respectively, where the delay $\tau(t)$ satisfies hypotheses (H_1) and (H_4) (see Section 6.5.1). The boundedness and stability analysis of the true solutions to such equations was done in Section 9.4.

We start with the DDAE system (10.5.1), (10.5.2). To this end, consider another system of DDAEs

$$\begin{cases} \tilde{\Phi}'(t) = K\left(t, \tilde{\Phi}(t), \tilde{y}(t-\tau(t))\right), & t \geq t_0, \\ \tilde{\Phi}(t_0) = \tilde{\phi}(t_0) - G(t_0, \tilde{\phi}(t_0 - \tau(t_0))), \end{cases} \tag{10.5.5}$$

$$\tilde{y}(t) = \begin{cases} \tilde{\phi}(t), & t \leq t_0, \\ \tilde{\Phi}(t) + G(t, \tilde{y}(t-\tau(t))), & t \geq t_0, \end{cases} \tag{10.5.6}$$

with the same right-hand sides $K(t, y, x)$ and $G(t, x)$, but a different initial function $\tilde{\phi}(t)$, and define

$$\delta\Phi(t) = \tilde{\Phi}(t) - \Phi(t),$$
$$\delta\phi(t) = \tilde{\phi}(t) - \phi(t).$$

Moreover, define the differences

$$\delta\eta(t) = \tilde{\eta}(t) - \eta(t),$$
$$\delta y_n = \tilde{y}_n - y_n,$$
$$\delta\lambda(t) = \tilde{\lambda}(t) - \lambda(t),$$
$$\delta\Phi_n = \tilde{\Phi}_n - \Phi_n,$$

between the relevant continuous and discrete numerical solutions obtained by formulae (6.5.17), (6.5.18), (6.5.20) or (6.5.21).

We shall assume that, given a norm $\|\cdot\|$ in $\mathbb{C}^d$, the interpolation operator $\mathcal{P}$ in (6.5.21) satisfies the bound

$$\max_{t_n \leq t \leq t_{n+1}} \|\mathcal{P}(u(\cdot))(t)\| \leq \max_{t_n \leq t \leq t_{n+1}} \|u(t)\| \tag{10.5.7}$$

for all continuous functions $u(t)$. Observe that this bound holds true for constant and linear interpolation of nodal values.

The following theorem was proved by Torelli and Vermiglio [250] (who assume a slightly stronger hypothesis than (10.5.9)).

Theorem 10.5.1 *Assume that the underlying CRK method of formulae (6.5.17), (6.5.18) is BN_f-stable and that, for a given inner product $\langle\cdot,\cdot\rangle$ and induced norm $\|\cdot\|$ in $\mathbb{C}^d$, the hypotheses*

$$X(t) = r(t)Y(t), \quad t \geq t_0, \tag{10.5.8}$$

with $r(t) \leq 0$ and $r(t_0) < 0$, and

$$\sup_{t_0 \leq s \leq t} |r(s)| + \sup_{t_0 \leq s \leq t} Z(s) \leq \rho \leq 1, \quad t \geq t_0, \tag{10.5.9}$$

slightly stronger than (9.4.8) of Theorem 9.4.1, hold. Moreover, assume that the algebraic recursions are approximately solved by (6.5.20) or by (6.5.21), where *the interpolation operator $\mathcal{P}$ is subject to (10.5.7). Then, for all delays $\tau(t)$ satisfying hypotheses (H_1) and (H_4) and for all initial functions $\phi(t)$ and $\tilde{\phi}(t)$, the difference $\delta\eta(t)$ in the continuous numerical solutions of the DDAE systems (10.5.5), (10.5.6) and (10.5.1), (10.5.2), obtained within the method of steps, satisfies the bound (9.4.9), i.e.*

$$\|\delta\eta(t)\| \leq \max\left\{\sup_{x\leq t_0}\|\delta\phi(x)\|, \frac{\|\delta\Phi(t_0)\|}{-r(t_0)}\right\}, \quad t \geq t_0. \tag{10.5.10}$$

Moreover, if $\rho < 1$ and

$$Y(t) \leq Y_0 < 0, \quad t \geq t_0, \tag{10.5.11}$$

then the method is asymptotically stable, that is

$$\lim_{t\to+\infty} \delta\eta(t) = 0$$

for any initial functions $\phi(t)$ and $\tilde{\phi}(t)$, for all delays $\tau(t)$ satisfying also hypothesis (H_2). Finally, if also hypothesis (H_3) holds, then the convergence rate is at least of exponential type, i.e. at least as $\mathrm{e}^{-\alpha(t-t_0)}$ for some $\alpha > 0$.

Proof. The proof is analogous to the one of Theorem 9.4.1 for the true solutions and also shows some similarities to the proof of Lemma 10.2.32.

Since the mesh Δ includes all the discontinuity points, we can relabel the mesh points in each macro interval $I_k = [\xi_{k-1}, \xi_k]$ and define the submesh $\Delta_k = \{t_{k,0} = \xi_{k-1}, t_{k,1}, \ldots, t_{k,N_k} = \xi_k\}$.

Now, for any vector function $v(x)$, set

$$|||v|||_0 = \max_{x\leq t_0}\|v(x)\|$$

and, for any integer $l \geq 1$,

$$|||v|||_l = \max_{x\in I_l}\|v(x)\|.$$

Moreover, for any scalar function $u(x)$, set

$$u^*(t) = \sup_{t_0\leq x\leq t}|u(x)|.$$

By Theorem 8.5.25 and by hypothesis (H_4), for the BN$_\mathrm{f}$-stable RK method (6.5.17), (6.5.19) we have

$$\begin{aligned}\|\delta\Phi_{k,n+1}\| \leq\ & \Psi^{BN}_{k,n+1}\|\delta\Phi_{k,n}\| \\ & +(1-\Psi^{BN}_{k,n+1})\max_{1\leq i\leq\nu}\left(|r(t^i_{k,n+1})|\cdot\|\delta\eta(t^i_{k,n+1}-\tau(t^i_{k,n+1}))\|\right)\end{aligned}$$

$$\leq \Psi^{BN}_{k,n+1}\|\delta\Phi_{k,n}\| + (1-\Psi^{BN}_{k,n+1})r^*(t_{k,n+1})|||\delta\eta|||_{k-1}$$

for all $n = 0, 1, \ldots, N_k - 1$. Then, in view of Definition 8.3.16, by (8.5.40), (8.3.2), Theorem 8.3.22 and by the monotonicity of $r^*(t)$, easy standard manipulations yield

$$\|\delta\Phi_{k,n+1}\| \leq \bar{\Psi}_{k,n+1}\|\delta\lambda(\xi_{k-1})\| + (1-\bar{\Psi}_{k,n+1})r^*(t_{k,n+1})|||\delta\eta|||_{k-1} \tag{10.5.12}$$

for all $n = 0, 1, \ldots, N_k - 1$, where

$$\bar{\Psi}_{k,n+1} = \prod_{l=1}^{n+1} \Psi^{BN}_{k,l} \leq \Phi_{BN}(Y_0, t_{k,n+1} - \xi_{k-1}) \leq 1 \tag{10.5.13}$$

and, with reference to (10.5.8), we have $Y(t) \leq Y_0 \leq 0$. In particular, for $n = N_k - 1$,

$$\|\delta\lambda(\xi_k)\| \leq \bar{\Psi}_{k,N_k}\|\delta\lambda(\xi_{k-1})\| + (1-\bar{\Psi}_{k,N_k})r^*(\xi_k)|||\delta\eta|||_{k-1}$$

and thus, again by the monotonicity of $r^*(t)$,

$$\frac{\|\delta\lambda(\xi_k)\|}{r^*(\xi_k)} \leq \bar{\Psi}_{k,N_k}\frac{\|\delta\lambda(\xi_{k-1})\|}{r^*(\xi_{k-1})} + (1-\bar{\Psi}_{k,N_k})|||\delta\eta|||_{k-1}. \tag{10.5.14}$$

Moreover, the CRK method (6.5.17), (6.5.18) being BN$_\mathrm{f}$-stable, by (10.5.12) we have

$$\begin{aligned}\max_{t_{k,n}\leq t\leq t_{k,n+1}} \|\delta\lambda(t)\| &\leq \max\Big\{\|\delta\Phi_{k,n}\|,\\ &\qquad \max_{1\leq i\leq \nu}(|r(t^i_{k,n+1})|\cdot\|\delta\eta(t^i_{k,n+1}-\tau(t^i_{k,n+1}))\|)\Big\}\\ &\leq \max\Big\{\|\delta\Phi_{k,n}\|, r^*(t_{k,n+1})|||\delta\eta|||_{k-1}\Big\}\\ &\leq \max\Big\{\bar{\Psi}_{k,n}\|\delta\lambda(\xi_{k-1})\| + (1-\bar{\Psi}_{k,n})r^*(t_{k,n})|||\delta\eta|||_{k-1},\\ &\qquad r^*(t_{k,n+1})|||\delta\eta|||_{k-1}\Big\}\end{aligned} \tag{10.5.15}$$

for all $n = 0, 1, \ldots, N_k - 1$, where $\bar{\Psi}_{k,0} = 1$.

Now (6.5.20), or (6.5.21) under condition (10.5.7), yields

$$\max_{t_{k,n}\leq t\leq t_{k,n+1}} \|\delta\eta(t)\| \leq \max_{t_{k,n}\leq t\leq t_{k,n+1}} \|\delta\lambda(t)\| + Z^*(t_{k,n+1})|||\delta\eta|||_{k-1}$$

and hence, by (10.5.15), (10.5.9) and, again, by ther monotonicity of $r^*(t)$,

$$\max_{t_{k,n}\leq t\leq t_{k,n+1}} \|\delta\eta(t)\| \leq \max\Big\{\bar{\Psi}_{k,n}r^*(t_{k,n+1})\frac{\|\delta\lambda(\xi_{k-1})\|}{r^*(\xi_{k-1})}$$

$$+(\rho - \bar{\Psi}_{k,n} r^*(t_{k,n+1}))|||\delta\eta|||_{k-1}, \rho|||\delta\eta|||_{k-1}\Big\}.$$

Let n_k be the integer that maximizes the right-hand side of the above inequality for $n = 0, 1, \ldots, N_k - 1$. Then we obtain

$$|||\delta\eta|||_k \le \max\Big\{\bar{\Psi}_{k,n_k} r^*(t_{k,n_k+1})\frac{\|\delta\lambda(\xi_{k-1})\|}{r^*(\xi_{k-1})} \\ +(\rho - \bar{\Psi}_{k,n_k} r^*(t_{k,n_k+1}))|||\delta\eta|||_{k-1}, \rho|||\delta\eta|||_{k-1}\Big\}. \quad (10.5.16)$$

Now assume $\rho = 1$. Then, by (10.5.14), (10.5.16) and (10.5.13), we get

$$\frac{\|\delta\lambda(\xi_k)\|}{r^*(\xi_k)} \le \max\left\{\frac{\|\delta\lambda(\xi_{k-1})\|}{r^*(\xi_{k-1})}, |||\delta\eta|||_{k-1}\right\}$$

and

$$|||\delta\eta|||_k \le \max\left\{\frac{\|\delta\lambda(\xi_{k-1})\|}{r^*(\xi_{k-1})}, |||\delta\eta|||_{k-1}\right\},$$

so that the sequence of numbers

$$\max\left\{\frac{\|\delta\lambda(\xi_k)\|}{r^*(\xi_k)}, |||\delta\eta|||_k\right\}, \quad k = 0, 1, \ldots,$$

is non-increasing, proving (10.5.10).

Finally, assume $\rho < 1$ and that (10.5.11) holds. Similarly to the proof of Theorem 9.4.1, the inequalities (10.5.14) and (10.5.16) can be rewritten as

$$Q_k \le W_k Q_{k-1},$$

where $Q_k = \left(\frac{\|\delta\lambda(\xi_k)\|}{r^*(\xi_k)}, |||\delta\eta|||_k\right)^T$ and

$$W_k = \begin{bmatrix} \bar{\Psi}_{k,N_k} & 1 - \bar{\Psi}_{k,N_k} \\ \bar{\Psi}_{k,n_k} r^*(t_{k,n_k+1}) & \rho - \bar{\Psi}_{k,n_k} r^*(t_{k,n_k+1}) \end{bmatrix}$$

if

$$\bar{\Psi}_{k,n_k} r^*(t_{k,n_k+1})\frac{\|\delta\lambda(\xi_{k-1})\|}{r^*(\xi_{k-1})} + (\rho - \bar{\Psi}_{k,n_k} r^*(t_{k,n_k+1}))|||\delta\eta|||_{k-1} \ge \rho|||\delta\eta|||_{k-1}$$

in (10.5.16), or

$$W_k = \begin{bmatrix} \bar{\Psi}_{k,N_k} & 1 - \bar{\Psi}_{k,N_k} \\ 0 & \rho \end{bmatrix}$$

otherwise.

We define the 2×2 matrix norm

$$\|W_k\| = \|P^{-1}W_kP\|_\infty \quad \text{with } P = \begin{bmatrix} 1 & 0 \\ 0 & \epsilon \end{bmatrix} \text{ and } \epsilon < 1.$$

For both the possible values of W_k, direct computations lead to

$$\|W_k\| \le \max\left\{\epsilon + (1-\epsilon)\bar{\Psi}_{k,N_k}\ ,\ (\epsilon^{-1}-1)\bar{\Psi}_{k,n_k}r^*(t_{k,n_k+1}) + \rho\right\}.$$

By hypothesis (H_1), Corollary 8.3.12, (10.5.13), (10.5.11), Proposition 8.5.22 and Theorem 8.3.22, we have

$$\bar{\Psi}_{k,N_k} \le \Phi_{BN}(Y_0, \xi_k - \xi_{k-1}) \le \Phi_{BN}(Y_0, \tau_0) < 1$$

for all $k \ge 1$. Therefore, by using (10.5.9), with the choice $\epsilon^* = (1+\rho)/2$ we obtain

$$\|W_k\| \le \nu_* = \max\left\{\frac{1+\rho}{2} + \frac{1-\rho}{2}\Phi_{BN}(Y_0,\tau_0), \frac{2\rho}{1+\rho}\right\} < 1 \quad \forall k \ge 1,$$

and then $\|Q_k\| \to 0$ as $k \to +\infty$ at least like ν_*^k. Since the sequence of ξ_k's diverges because of hypotheses (H_1) and (H_2), the asymptotic stability is proved.

Finally, assume also that hypothesis (H_3) holds. In this case, for a given $t \ge t_0$, we have that $t \in I_k$ for some $k \ge (t-t_0)/\tau_1$. Therefore, it follows that $\|\delta\eta(t)\| \to 0$ at least as $\mathrm{e}^{-\alpha(t-t_0)}$, where $\alpha = -\log(\nu_*)/\tau_1$. ■

Along with the NDDEs rewritten as the DDAE system (10.5.3), (10.5.4), we consider another system of DDAEs

$$\begin{cases} \tilde{y}'(t) = F\left(t, \tilde{y}(t), \tilde{\Psi}(t)\right), & t \ge t_0, \\ \tilde{y}(t_0) = \tilde{\phi}(t_0), \end{cases} \tag{10.5.17}$$

$$\tilde{\Psi}(t) = \begin{cases} G\Big(t, \tilde{\phi}(t-\tau(t)), \tilde{\phi}'(t-\tau(t))\Big), & t-\tau(t) \le t_0, \\ H\Big(t, \tilde{y}(t-\tau(t)), \tilde{\Psi}(t-\tau(t))\Big), & t-\tau(t) > t_0, \end{cases} \tag{10.5.18}$$

with the same right-hand sides $F(t,y,x)$, $H(t,y,x)$ and $G(t,x)$, but a different initial function $\tilde{\phi}(t)$, and define

$$\begin{aligned} \delta\Psi(t) &= \tilde{\Psi}(t) - \Psi(t), \\ \delta\phi(t) &= \tilde{\phi}(t) - \phi(t). \end{aligned}$$

Also in this case we shall assume that, given a norm $\|\cdot\|$ in $\mathbb{C}^d$, the interpolation operator $\mathcal{P}$ in (6.5.27) satisfies the bound

$$\max_{t_{k_n} \le t \le t_{k_n+1}} \|\mathcal{P}(u(\cdot))(t)\| \le \max_{t_{k_n} \le t \le t_{k_n+1}} \|u(t)\| \tag{10.5.19}$$

for all continuous functions $u(t)$.

The following theorem can be found in Bellen, Guglielmi and Zennaro [30] and its proof is similar to the the proof of Theorem 10.5.1.

Theorem 10.5.2 *Assume that the underlying CRK method of formulae (6.5.28), (6.5.29) is BN_f-stable and that, for a given inner product $\langle\cdot,\cdot\rangle$ and induced norm $\|\cdot\|$ in $\mathbb{C}^d$, the hypotheses*

$$X(t) = r(t)Y(t), \quad t \geq t_0,$$

with $r(t) < 0$, and

$$-r(t)\Big(W(t) + \frac{Z(t)}{-r(t-\tau(t))}\Big) \leq 1, \quad t \geq t_0,$$

of Theorem 9.4.2 hold. Moreover, assume that the algebraic recursions are approximately solved by (6.5.25) and by (6.5.26) or (6.5.27), where the interpolation operator $\mathcal{P}$ is subject to (10.5.19). Then, for all delays $\tau(t)$ satisfying hypotheses (H_1) and (H_4) and for all initial functions $\phi(t)$ and $\tilde{\phi}(t)$, the difference $\delta\eta(t) = \tilde{\eta}(t) - \eta(t)$ in the continuous numerical solutions of the DDAE systems (10.5.17), (10.5.18) and (10.5.3), (10.5.4), obtained within the method of steps, satisfies the bound (9.4.25), i.e.

$$\|\delta\eta(t)\| \leq \max\left\{\|\delta\phi(t_0)\|, \sup_{t_0\leq x\leq \xi_1}\big(|r(x)|\cdot\|\delta\Psi(x)\|\big)\right\}, \quad t \geq t_0,$$

where ξ_1 is the first primary discontinuity point (i.e. $\xi_1 - \tau(\xi_1) = t_0$).

Moreover, if

$$Y(t) \leq Y_0 < 0, \quad t \geq t_0,$$

and if there exist two non-negative constants $\nu < 1$ and $k < 1$ such that

$$\frac{r(t)}{r(t-\tau(t))}Z(t) \leq \nu, \quad t \geq t_0,$$

and

$$-r(t)W(t) \leq k\Big(1 - \frac{r(t)}{r(t-\tau(t))}Z(t)\Big), \quad t \geq t_0,$$

then the method is asymptotically stable, that is

$$\lim_{t\to+\infty} \delta\eta(t) = 0$$

for any initial functions $\phi(t)$ and $\tilde{\phi}(t)$, for all delays $\tau(t)$ satisfying also hypothesis (H_2). Finally, if also hypothesis (H_3) holds, then the convergence rate is at least of exponential type, i.e. at least as $\mathrm{e}^{-\alpha(t-t_0)}$ for some $\alpha > 0$.

It is easy to specialize the results of Theorems 10.5.1 and 10.5.2 for linear systems (see Section 9.5) and for scalar equations, and this is left to the interested reader. However, we just observe that the only change in the hypotheses to make to the CRK method lies in the possible relaxation of the property of BN_f-stability in AN_f-stability for scalar non-autonomus equations and in A_f-stability for scalar autonomous equations.

10.6 Delay independent asymptotic stability for linear systems

Asymptotic stability analysis of numerical methods for the linear NDDE

$$\begin{cases} y'(t) = Ly(t) + My(t-\tau) + Ny'(t-\tau), & t \geq t_0, \\ y(t) = \phi(t), \quad t \leq t_0, \end{cases} \tag{10.6.1}$$

where L, M and N are complex $d \times d$ matrices and τ is a constant delay, cannot be derived by the scalar case as happens for ODEs. This can be done only in some exceptional cases such as, for instance, when L, M and N are simultaneously diagonalizable.

So, it is natural to extend the definitions of P- and GP-stability as well as NP- and GNP-stability to linear systems fulfilling the conditions described in Chapter 9, which characterize the asymptotic stability of (10.6.1) for all constant delays τ, namely condition (9.3.4) for DDEs (i.e. $N = 0$) and condition (9.5.15) or, equivalently, (9.5.17) and (9.5.18) for NDDEs.

In particular for DDEs, Lu [192] obtained results on Θ-methods when the matrices L and M are diagonal and reverse diagonal, respectively, whereas In 't Hout [135] considered the general case where L and M are arbitrary matrices. He found that, if a constant stepsize $h > 0$ is used, the Θ-methods with $\frac{1}{2} \leq \Theta \leq 1$ are P-stable, i.e. the numerical solution preserves the asymptotic stability whenever the sufficient condition (9.3.4) holds. To do this, he again used Lemma 10.2.24. Later, Koto [174] extended the P-stability results for systems to all A-stable natural RK methods for DDEs and, in [175, 176], he proved P- and NP-stability for some non-natural A-stable RK methods such as Radau IA and Lobatto IIIC methods equipped with particular continuous extensions. Moreover, by using the equistage interpolation (10.2.34), In 't Hout [137] proved GP-stability results for systems of DDEs.

According to Section 6.5 and by using the K notation, the RK method for NDDEs based on the option (6.5.2), i.e. $y'(t-\tau) \approx \eta'(t-\tau)$ in (10.6.1), reads

$$\begin{aligned} K_{n+1}^i = {} & L\left(y_n + h\sum_{j=1}^s a_{ij}K_{n+1}^j\right) + M\left(y_{n-m} + h\sum_{j=1}^s b_j(c_i)K_{n-m+1}^j\right) \\ & + N\left(\sum_{j=1}^s b_j'(c_i)K_{n-m+1}^j\right), \quad i = 1, \dots, s, \end{aligned} \tag{10.6.2}$$

$$y_{n+1} = y_n + h\sum_{i=1}^s b_i K_{n+1}^i, \tag{10.6.3}$$

where the constant stepsize h is assumed to satisfy the constraint

$$h = \tau/m, \quad m \geq 1, \ m \text{ integer}. \tag{10.6.4}$$

With $b = [b_1, \dots, b_s]^T$, $b(\theta) = [b_1(\theta), \dots, b_s(\theta)]^T$, $e = [1, \dots, 1]^T$ the unit s vector, I_r the r-dimensional identity matrix, $A = [a_{ij}]_{i,j=1}^s$, $B = [b_j(c_i)]_{i,j=1}^s$, $C = [b_j'(c_i)]_{i,j=1}^s$ and

$$\mathbf{H}_{n+1} = [K^1_{n+1}, \ldots, K^s_{n+1}, y_{n+1}]^T,$$

the method (10.6.2), (10.6.3) can be rewritten as

$$D\mathbf{H}_{n+1} = E\mathbf{H}_n + F\mathbf{H}_{n-m+1} + G\mathbf{H}_{n-m}, \tag{10.6.5}$$

where

$$D = \begin{bmatrix} I_{sd} - h(A \otimes L) & 0 \\ hb^T \otimes I_d & I_d \end{bmatrix},$$

$$E = \begin{bmatrix} 0 & (e \otimes L) \\ 0 & I_d \end{bmatrix},$$

$$F = \begin{bmatrix} h(B \otimes M) + C \otimes N & 0 \\ 0 & 0 \end{bmatrix},$$

$$G = \begin{bmatrix} 0 & (e \otimes M) \\ 0 & 0 \end{bmatrix}.$$

The characteristic equation is

$$\det\left[\zeta^{m+1} D - \zeta^m E - \zeta F - G\right] = 0,$$

and, in short,

$$\det[\mathcal{P}] = 0,$$

where

$$\mathcal{P} = \begin{bmatrix} T_1 & T_2 \\ T_3 & T_4 \end{bmatrix},$$

with

$$\begin{aligned} T_1 &= \zeta^{m+1}\big((I_{sd} - h(A \otimes L)) - \zeta h(B \otimes M) + C \otimes N\big), \\ T_2 &= -\zeta^m e \otimes (L + \zeta^{-m} M), \\ T_3 &= -\zeta^{m+1} h(b^T \otimes I_d), \\ T_4 &= (\zeta^{m+1} - \zeta^m) I_d. \end{aligned}$$

For the simplified case of natural RK methods for NDDEs ($A = B$) with interpolation satisfying the additional condition $C = I$, Hu and Mitsui [140] proved the following theorem.

Theorem 10.6.1 *Let the linear NDDE (10.6.1) satisfy the conditions*

$$\rho[N] < 1 \tag{10.6.6}$$

and

$$\gamma \in \sigma\left[((I - \xi N)^{-1}(L + \xi M))\right] \Longrightarrow \Re(\gamma) < 0, \tag{10.6.7}$$

for all $\xi \in \mathbb{C}$ such that $|\xi| \le 1$. Then, for any A-stable natural RK method for NDDEs such that $C = I$, the discrete numerical solution of (10.6.1) is asymptotically stable for any delay τ and for any constant stepsize h satisfying the constraint (10.6.4). In short, the method is NP-stable for systems.

In particular, for $N = 0$, any A-stable natural RK method for DDEs is P-stable for systems.

Proof. It is sufficient to prove that, under hypothesis (10.6.7),

$$|\zeta| \geq 1 \Rightarrow \det[\mathcal{P}] \neq 0.$$

Hence, assume $|\zeta| \geq 1$. Direct computation of T_1 yields

$$\begin{aligned} T_1 &= \zeta^{m+1}\big((I_{sd} - h(A \otimes L)) - \zeta\,(h(A \otimes M) + I_s \otimes N)\big) \\ &= \zeta^{m+1}\big((I_s \otimes (I_d - \zeta^{-m}N)) - A \otimes h(L + \zeta^{-m}M)\big). \end{aligned}$$

Hypothesis (10.6.6) implies $\det[I_d - \zeta^{-m}N] \neq 0$ and, thus,

$$\begin{aligned} T_1 &= \zeta^{m+1}\left(I_s \otimes (I_d - \zeta^{-m}N)\right)\left(I_{sd} - (I_s \otimes \zeta^{-m}N)^{-1}(A \otimes h(L + \zeta^{-m}M))\right) \\ &= \zeta^{m+1}\big(I_s \otimes (I_d - \zeta^{-m}N)\big)\left(I_{sd} - hA \otimes (Q(\zeta^{-m}))\right), \end{aligned}$$

where

$$Q(\zeta) = \left((I - \zeta N)^{-1}(L + \zeta M)\right).$$

Therefore

$$\det[T_1] = \zeta^{(m+1)sd}\det\left[I_d - \zeta^{-m}N\right]^s \cdot \det\left[I_{sd} - hA \otimes (Q(\zeta^{-m}))\right].$$

Under the mild assumption that $\det\left[I_{sd} - hA \otimes (Q(\zeta^{-m}))\right] \neq 0$, we have $\det[T_1] \neq 0$ and, hence,

$$\det[\mathcal{P}] = \det[T_1] \cdot \det[T_4 - T_3 T_1^{-1} T_2].$$

Now it remains to prove that $\det[T_4 - T_3T_1^{-1}T_2] \neq 0$ for $|\zeta| \geq 1$. Direct computation of $T_4 - T_3T_1^{-1}T_2$ yields

$$T_4 - T_3T_1^{-1}T_2 = \zeta^m\left[\zeta I_d - I_d - \left(b^T \otimes Q(\zeta^{-m})\right)\left(I_{sd} - hA \otimes Q(\zeta^{-m})\right)^{-1}(he \otimes I_d)\right],$$

which implies

$$\det[T_4 - T_3T_1^{-1}T_2] = \zeta^{md}\det\left[\zeta I_d - R(hQ(\zeta^{-m}))\right],$$

where $R(\alpha)$ is the A-stability function of the RK method.

By hypothesis (10.6.7) and by the A-stability of the RK method, we have that

$$|R(\gamma)| < 1$$

for any $\gamma \in \sigma[hQ(\zeta^{-m})]$ and any $h > 0$. Therefore, by the spectral mapping theorem, we also have

$$\rho[R(hQ(\zeta^{-m}))] < 1.$$

The last inequality prevents $\det\left[\zeta I_d - R(hQ(\zeta^{-m}))\right]$ from vanishing and, hence, the proof is complete. ■

In order to remove the constraint (10.6.4) on the stepsize h, Qiu, Yang and Kuang [221] considered the method

$$K_{n+1}^i = L\left(y_n + h\sum_{j=1}^{s} a_{ij}K_{n+1}^j\right) + M\left(y_{n-m+1+\delta} + h\sum_{j=1}^{s} a_{ij}K_{n-m+1+\delta}^j\right)$$
$$+NK_{n-m+1+\delta}^i, \quad i = 1,\ldots,s, \tag{10.6.8}$$

$$y_{n+1} = y_n + h\sum_{i=1}^{s} b_i K_{n+1}^i, \tag{10.6.9}$$

where $h = \tau/(m-\delta)$, $m \geq 1$ integer, and $0 \leq \delta < 1$, and $y_{n-m+1+\delta}$ and $K_{n-m+1+\delta}$ are given by the following equistage interpolation procedure:

$$y_{n-m+1+\delta} = \sum_{k=-r}^{q} \ell_k(\delta) y_{n-m+1+k},$$

$$K_{n-m+1+\delta}^j = \sum_{k=-r}^{q} \ell_k(\delta) K_{n-m+1+k}^j,$$

with $q \leq m-1$ and

$$\ell_k(\epsilon) = \prod_{i=-r, i\neq k}^{q} \left(\frac{\epsilon - i}{k-i}\right).$$

Theorem 10.6.2 *Let the system (10.6.1) satisfy the same hypotheses of Theorem 10.6.1. Moreover, assume that the underlying discrete RK method is A-stable and the equistage interpolation procedure satisfies $r \leq q \leq r+2$. Then the discrete numerical solution obtained by the method (10.6.8), (10.6.9) is asymptotically stable for any delay τ and for any constant stepsize $h = \tau/(m-\delta)$, $m \geq 1$ integer, and $0 \leq \delta < 1$. In short, the method is GNP-stable for systems.*

In particular, for $N = 0$, the method is GP-stable for systems.

10.7 Delay dependent asymptotic stability for linear systems

Delay dependent asymptotic stability of numerical methods for systems is a real challenge. We saw in Section 10.2.6 that, for scalar equations with real coefficients, only some low order Radau IIA methods have been proved to be D(0)-stable, even though it is reasonable to conjecture it for any order. The case of complex coefficients is worse, where the only numerical method that is known to be D-stable is backward Euler.

For systems of linear DDEs,

$$\begin{cases} y'(t) = Ly(t) + My(t-\tau), & t \geq t_0, \\ y(t) = \phi(t), & t \leq t_0, \end{cases} \tag{10.7.1}$$

the asymptotic stability analysis is even more disappointing. In fact, Maset [197] proved that, for any A-stable natural RK method for DDEs and for any stepsize

$h = \tau/m$, there exists an asymptotically stable linear system of dimension $d = 4$ for which $\lim_{n\to\infty} y_n = 0$ does not hold. Later, Guglielmi [109] proved the same negative result for $d = 2$. So we can conclude as follows.

Theorem 10.7.1 *No A-stable natural RK method for DDEs is asymptotically stable on the whole class of asymptotically stable linear systems. In short, no natural RK method for DDEs is D-stable for systems.*

Theorem 10.7.1, along with the previous negative results presented in this chapter, makes the search for numerical methods, other than those based on the standard approach, more and more pressing. In this regard, the methods of lines, MOL and TMOL, presented in Section 3.5 for the discretization of the equivalent partial differential equation

$$\frac{\partial u(t,\theta)}{\partial t} = \frac{\partial u(t,\theta)}{\partial \theta}, \quad t \geq 0, \quad -\tau \leq \theta \leq 0,$$

with initial boundary conditions related to the given DDE, have been successfully investigated from the stability point of view.

In particular, for the real scalar equation

$$y'(t) = \lambda y(t) + \mu y(t-\tau)$$

with $(\lambda, \mu) \in \mathcal{S}_\tau$, Bellen and Maset [32] proved that, for any fixed delay, the backward difference approximation in the MOL leads to the ODE system (see Section 3.5)

$$\begin{cases} \frac{\mathrm{d}}{\mathrm{d}t} U_N(t) = \mathcal{A}_N U_N(t), & t \geq 0, \\ U_N(0) = \phi_N, \end{cases} \tag{10.7.2}$$

which is asymptotically stable for any choice of the space discretization parameter $h = \tau/N$. Despite some D(0)-stable Radau IIA methods existing in the standard approach, the MOL is of some interest in that the resulting linear ODE system (10.7.2) can be integrated by any A-stable RK method with free stepsize even larger than τ. Some experiments with a scheme of order 2 that seem to preserve the property have been carried out in [32]. Later, Koto [179] extended the property to the class of Lobatto IIIA methods.

Subsequently, Maset [198] proved the negative result that, for scalar DDEs with complex coefficients $(\lambda, \mu) \in \mathcal{S}_\tau$, discretization of the MOL cannot lead to asymptotically stable ODE systems. On the other hand, a full discretization (in time and in space) of the equivalent PDE may lead to asymptotically stable systems, provided that the methods and the discretization parameters adopted, $h = \tau/N$ and Δt, satisfy the condition of the following theorem.

Theorem 10.7.2 *Let the space discretization be derived by an s-stage A-stable RK method (A, b) of order p with A invertible, $[a_{s,1}, \ldots, a_{s,s}] = b^T$ and $0 < c_1 < \cdots < c_s = 1$. Moreover, let the RK method (A^*, b^*) for integrating the resulting*

ODE system be A-stable, of odd order p^* and A-stability region S_A such that its complementary set is bounded. If $\frac{p+1}{p^*+1} \geq 2$ and

$$h \leq \kappa \Delta t^{\left(\frac{p+1}{p} \frac{p^*}{p^*+1}\right)},$$

where the constant κ depends only on the RK methods, then, for every τ, the resulting fully discretized method is asymptotically stable for any DDE with complex coefficients $(\lambda, \mu) \in \mathcal{S}_\tau$.

Observe that the constraint on h and Δt in Theorem 10.7.2 does not reduce the stability region of the overall method because it is independent of the equation. More severe is the condition on the order of the methods which can probably be weakened. In fact, the use of the backward Euler method in both time and space discretization provides a stable procedure despite the two orders $p = p^* = 1$ not fulfilling the hypothesis of Theorem 10.7.2. However, the MOL does not give any help in the numerical solution of systems where, on the contrary, the TMOL (see Section 3.5) looks like more promising.

In particular, the TMOL

$$U^{k+1} = U^k + \Delta t \mathcal{A} U^{k+1}, \quad k = 0, 1, 2, \ldots, \quad U^0 = \phi,$$

corresponding to the abstract implicit Euler method for (3.5.2), requires solving the sequence of two-point boundary value problems (3.5.8), (3.5.10), that is

$$\begin{cases} U^{k+1}(\theta) = U^k(\theta) + \Delta t \left(U^{k+1}\right)'(\theta), & -\tau \leq \theta \leq 0, \\ U^{k+1}(0) = U^k(0) + \Delta t \left(LU^{k+1}(0) + MU^{k+1}(-\tau)\right). \end{cases} \tag{10.7.3}$$

To this end, we consider, for example, the discretized version

$$\begin{cases} U_n^{k+1} = U_n^k + \Delta t \frac{U_{n+1}^{k+1} - U_n^{k+1}}{h}, & n = -N, \ldots, -1, \\ U_0^{k+1} = U_0^k + \Delta t \left(LU_0^{k+1} + MU_N^{k+1}\right), \end{cases} \tag{10.7.4}$$

where U_n^i, $i = k, k+1$, $n = -N, \ldots, -1, 0$, is an approximation of $U^i(nh)$, $h = \tau/N$.

Observe that the scheme (10.7.4) is nothing but the backward Euler method applied to the ODE (3.5.5) with stepsize Δt. Since it is impossible to have unconditional preservation of asymptotic stability for all scalar complex equations when we consider the ODE (3.5.5), it is reasonable to conjecture that the full discretization scheme (10.7.4) does not preserve the asymptotic stability for all Δt and h when applied to complex scalar equations and, *a fortiori*, to systems.

Since the nice property of unconditional preservation of asymptotic stability seems lost when a discretization scheme for the boundary value problem (10.7.3) is introduced, we solve it exactly. This is feasible for significant classes of initial functions $\phi(t)$.

For the method (10.7.3) Maset [196] proved the following theorem.

Theorem 10.7.3 *The abstract implicit Euler method (10.7.3) is asymptotically stable, that is*

$$\lim_{k\to+\infty} U^k(\theta) = 0, \qquad -\tau \le \theta \le 0,$$

for any linear system (10.7.1) that is asymptotically stable.

It is worth remarking that, so far, the abstract implicit Euler method is the only known numerical scheme that preserves the delay dependent stability for systems.

10.8 The pantograph equation

The pantograph ($N = 0$) and generalized pantograph ($N \neq 0$) equations

$$\begin{cases} y'(t) = Ly(t) + My(qt) + Ny'(qt), & t \ge 0, \\ y(0) = y_0, \end{cases} \tag{10.8.1}$$

are characterized by a variable delay $\tau(t) = (1-q)t$ which is proportional to t, vanishing at $t = 0$ and unbounded as $t \to +\infty$.

The numerics of the pantograph equation develops along various lines, each one presenting merits and demerits with respect to the various issues such as convergence and superconvengence, preservation of boundedness, contractivity or asymptotic stability and, last but not least, storage capacity in the long-term integration.

A thorough investigation of the dynamics of the analytical and numerical solution of the pantograph equation began in a series of papers by Buhmann and Iserles [58, 59, 60], Iserles [146, 147] and Buhmann, Iserles and Nørsett [61], where some low order numerical methods are analyzed.

In order to avoid the need for unbounded computer memory in the long-term integration, Bellen, Guglielmi and Torelli [28] proposed proceeding on a mesh with an essentially geometric growth in the stepsize. They proved that such a choice also preserves the asymptotic stability of the method.

Alternatively, by the changes of variable

$$t = e^x \qquad \text{and} \qquad u(x) = y(e^x),$$

(10.8.1) can be transformed into the DDE

$$u'(x) = e^x Lu(x) + e^x Mu(x-\tau) + Nu'(x-\tau), \tag{10.8.2}$$

with constant delay $\tau = -\log(q)$, to be integrated by constant stepsize. Integrating (10.8.1) in $[0, h]$ is equivalent to integrating (10.8.2) in $(-\infty, \log(h)]$ with solution $u(x) = y(e^x)$. So, once this is done, it remains to integrate

$$\begin{cases} u'(x) = e^x Lu(x) + e^x Mu(x-\tau) + Nu'(x-\tau), & x \ge \log(h), \\ u(x) = y(e^x), \quad x \le \log(h). \end{cases}$$

For more general equations with proportional delay, such a transformation is not advisable because the resulting equation with constant delay has variable coefficients and this makes the stability analysis of the exact and numerical approximations more difficult. Nevertheless, the procedure, already considered by Jackiewicz [156] and Liu [190], is effective for the generalized pantograph equation, and the P-stability analysis of the numerical solution has been successfully carried out by Koto [178].

We shall give up using such a transformation and provide the stability results acting on the original form (10.8.1). All our approaches have the common feature of being based on some splitting of the integration interval $[0, +\infty)$ into $[0, h] \cup [h, +\infty)$, where the analysis is carried out separatly on the two subintervals.

10.8.1 *Contractivity and boundedness*

A suitable approach for analyzing the contractivity and boundedness of RK methods for (10.8.1) is based on the results of Sections 10.4.2, 10.4.3 and 10.5 for DDEs and NDDEs with non-vanishing time dependent delays and their adaptation in the neighborhood of $t = 0$.

For the non-neutral case ($N = 0$), Theorem 10.4.10 tells us that, if the underlying CRK method is $\mathrm{BN_f}$-stable, then the DDE method preserves contractivity with respect to the initial function $\phi(t) \equiv \eta(t)$ (the continuous numerical solution computed in $[0, h]$ by means of formulae (6.4.1), (6.4.2), (6.4.3), where overlapping occurs), that is

$$\|\eta(t)\| \le \max_{0 \le x \le h} \|\eta(x)\|, \quad t \ge h,$$

for all equations satisfying the condition

$$\|M\| \le -\mu[L].$$

On the other hand, it is not difficult to prove (see Bellen, Maset and Torelli [33] for the general neutral case) that, on the first step $[0, h]$, a $\mathrm{BN_f}$-stable RK method preserves the contractivity with respect to the initial value y_0 as well. Therefore, the overall procedure satisfies

$$\|\eta(t)\| \le \|y_0\|, \quad t \ge t_0, \tag{10.8.3}$$

in agreement with the behavior of the true solution (see (9.7.10) in Section 9.7).

For the neutral equation (10.8.1) ($N \ne 0$), under the conditions

$$\mu[L] < 0 \qquad \text{and} \qquad \frac{\|NL + M\|}{-\mu[L]} + \|N\| \le 1,$$

Theorem 10.5.2 implies that, if the underlying CRK method is $\mathrm{BN_f}$-stable, then the CRK method (6.5.28), (6.5.29), (6.5.26) or (6.5.27) satisfies the bound

$$\|\eta(t)\| \le \max\left\{ \|y_0\|, \max_{0 \le x \le h} \frac{\|M\eta(qx) + N\lambda(qx)\|}{-\mu[L]} \right\}, \quad t \ge h. \tag{10.8.4}$$

Here $\eta(t)$ and $\lambda(t)$ are the continuous numerical approximations of $y(t)$ and $y'(t)$, respectively, computed in $[0, h]$ by means of the DDE method (6.5.4),

(6.5.5), (6.5.6) or (6.5.8) with overlapping, or by means of a suitable adaptation of the CRK method (6.5.28), (6.5.29), (6.5.26) or (6.5.27) itself to the case of overlapping (see [33]).

Observe that the quantity $\max_{0 \le x \le h} \frac{\|M\eta(qx)+N\lambda(qx)\|}{-\mu[L]}$ in the bound (10.8.4) is not necessarily $\le \|y_0\|$. Thus, we cannot conclude with (10.8.3), which is satisfied by the true solution $y(t)$ (see (9.7.16)). Nevertheless, since

$$\lim_{t \to 0^+} \eta(t) = y_0 \qquad \text{and} \qquad \lim_{t \to 0^+} \lambda(t) = y'(0),$$

the same arguments used in Section 9.7.2 allow us to conclude that

$$\lim_{h \to 0^+} \max_{0 \le x \le h} \frac{\|M\eta(qx) + N\lambda(qx)\|}{-\mu[L]} \le \|y_0\|.$$

Therefore, the shorter the first step $[0, h]$, the closer the bound (10.8.4) to $\|y_0\|$.

10.8.2 *Asymptotic stability*

Another approach to the integration on $[h, +\infty)$ is based on the use of the quasi-geometric mesh defined in Section 6.4.1, which we briefly recall here. Fix an initial macro interval $H_1 = [h, h/q]$ of length $d_1 = h(1-q)/q$ and, for a given $m \ge 1$, define the equidistant nodal points

$$t_0 = h, \quad t_n = t_0 + n\frac{d_1}{m}, \qquad n = 1, \dots, m.$$

Then, define the subsequent nodal points t_n, $m < n \le 2m$, by simply repeating the procedure on the next macro interval $H_2 = [h/q, h/(q^2)]$ (see Figure 6.2). The resulting mesh is evidently a constrained mesh satisfying

$$qt_n = t_{n-m} \qquad \text{and} \qquad qh_{n+1} = h_{n-m+1},$$

and is such that

$$h_n = O(1/q^{\lfloor n/m \rfloor}).$$

Moreover, $q(t_n + c_i h_{n+1}) = t_{n-m} + c_i h_{n-m+1}$ and, therefore, the CRK method (10.6.2), (10.6.3) reads

$$\begin{aligned} K_{n+1}^i = L &\left(y_n + h_{n+1} \sum_{j=1}^s a_{ij} K_{n+1}^j \right) \\ &+ M \left(y_{n-m} + h_{n-m+1} \sum_{j=1}^s b_j(c_i) K_{n-m+1}^j \right) \\ &+ N \left(\sum_{j=1}^s b_j'(c_i) K_{n-m+1}^j \right) \qquad i = 1, \dots, s, \end{aligned}$$

$$y_{n+1} = y_n + h_{n+1} \sum_{i=1}^{s} b_i K_{n+1}^i.$$

As in Section 10.6, with

$$\mathbf{H}_{n+1} = [K_{n+1}^1, \ldots, K_{n+1}^s, y_{n+1}]^T,$$

we obtain the difference equation

$$D\mathbf{H}_{n+1} = E\mathbf{H}_n + F\mathbf{H}_{n-m+1} + G\mathbf{H}_{n-m},$$

where the matrices D, E, F and G depend on h_{n+1} and h_{n-m+1} as follows:

$$D = \begin{bmatrix} I_{sd} - h_{n+1}(A \otimes L) & 0 \\ h_{n+1} b^T \otimes I_d & I_d \end{bmatrix},$$

$$E = \begin{bmatrix} 0 & (e \otimes L) \\ 0 & I_d \end{bmatrix},$$

$$F = \begin{bmatrix} h_{n-m+1}(B \otimes M) + C \otimes N & 0 \\ 0 & 0 \end{bmatrix},$$

$$G = \begin{bmatrix} 0 & (e \otimes M) \\ 0 & 0 \end{bmatrix}.$$

After some computations, multiplication by the matrix

$$\begin{bmatrix} \frac{(I_s \otimes L^{-1})}{h_{n+1}} & 0 \\ 0 & I_d \end{bmatrix}$$

yields the non-autonomous system

$$\begin{cases} \frac{(I_s \otimes L^{-1})}{h_{n+1}} \big(I_{sd} \otimes \mathbf{U}_{n+1} - q^{-1}(C \otimes N)\mathbf{U}_{n-m+1} \big) = e \otimes y_n + (A \otimes I_d)\mathbf{U}_{n+1} \\ \qquad + (I_s \otimes L^{-1} M)\big(e \otimes y_{n-m} + (B \otimes I_d)\mathbf{U}_{n-m+1} \big), \\ y_{n+1} - y_n - (b^T \otimes I_d)\mathbf{U}_{n+1} = 0, \end{cases} \tag{10.8.5}$$

where

$$\mathbf{U}_{n+1} = [h_{n+1} K_{n+1}^1, \ldots, h_{n+1} K_{n+1}^s]^T$$

and the right-hand side does not depend on h_{n+1} and h_{n-m+1}.

Observe that the relevant stationary difference equation

$$\begin{cases} e \otimes u_n + (A \otimes I_d)\mathbf{U}_{n+1} + (I_s \otimes L^{-1}M)\big(e \otimes u_{n-m} + (B \otimes I_d)\mathbf{U}_{n-m+1}\big) = 0, \\ u_{n+1} - u_n - (b^T \otimes I_d)\mathbf{U}_{n+1} = 0, \end{cases} \tag{10.8.6}$$

is just the RK method for the difference equation

$$u(t) + L^{-1}Mu(qt) = 0, \tag{10.8.7}$$

which, in turn, is the stationary version of the neutral DDAE

$$L^{-1}\left(y'(t) - Ny'(qt)\right) = y(t) + L^{-1}My(qt),$$

equivalent to (10.8.1). Observe also that condition $\rho[L^{-1}M] < 1$, sufficient for the asymptotic stability of (10.8.1), is necessary and sufficient for the asymptotic stability of (10.8.7).

The forthcoming Theorem 10.8.2 provides the relationship between the numerical stability of the stationary and the non-stationary equations above. It is based on the following lemma, which can be obtained by standard arguments on perturbed difference equations (see Elaydi [79, Exercise 4.5.9]).

Lemma 10.8.1 *Let us consider the first-order system of difference equations*

$$x_{n+1} = Ax_n$$

along with the perturbed system

$$y_{n+1} = (A + B_n)y_n, \tag{10.8.8}$$

where the matrix A is non-singular. If $\lim_{n\to\infty} x_n = 0$, then also

$$\lim_{n\to\infty} y_n = 0,$$

provided that $\sum_{n=0}^{\infty} \|B_n\| < +\infty$.

Theorem 10.8.2 *If the solution $\{u_n\}_{n\geq 0}$ of (10.8.6) is asymptotically stable, then also the solution $\{y_n\}_{n\geq 0}$ of (10.8.5) is asymptotically stable.*

Proof. The difference equation (10.8.5) is perturbed from (10.8.6) by means of the non-stationary term

$$\frac{(I_s \otimes L^{-1})}{h_{n+1}}\left(I_{sd} \otimes \mathbf{U}_{n+1} - q^{-1}(C \otimes N)\mathbf{U}_{n-m+1}\right)$$

in the left-hand side.

Since $h_n = O(1/q^{\lfloor n/m \rfloor})$ for any fixed $q \in (0,1)$ and $m \geq 1$, the perturbing term is bounded by

$$\kappa q^{\lfloor n/m \rfloor} \max\{\|\mathbf{U}_{n+1}\|, \|\mathbf{U}_{n-m+1}\|\}$$

for some costant κ depending only on q and N. Therefore, the difference equation (10.8.5) can easily be set in the form (10.8.8) with matrices B_n satisfying the hypothesis of Lemma 10.8.1. ■

So the problem is reduced to the search for CRK methods that preserve the asymptotic stability of the difference equation (10.8.7) independently of $q \in (0,1)$. It is convenient to consider the following definition.

Definition 10.8.3 *A CRK method $(A, b(\theta))$ is called P_∞-stable if the solution $\{u_n\}_{n\geq 0}$ of (10.8.7) asymptotically vanishes for any $q \in (0,1)$, any L and M such that $\rho[L^{-1}M] < 1$, and any $m \geq 1$.*

In order to characterize the class of P_∞-stable CRK methods, we define the matrix $R = L^{-1}M$ and consider the characteristic equation of (10.8.6)

$$\det\left[\zeta^{m+1}P + \zeta^m Q + \zeta S(R) + T(R)\right] = 0,$$

where

$$P = \begin{bmatrix} (A \otimes I_d) & 0 \\ -b^T \otimes I_d & I_d \end{bmatrix},$$

$$Q = \begin{bmatrix} 0 & (e \otimes I_d) \\ 0 & -I_d \end{bmatrix},$$

$$S(R) = \begin{bmatrix} (I_s \otimes R)(B \otimes I_d) & 0 \\ 0 & 0 \end{bmatrix},$$

$$T(R) = \begin{bmatrix} 0 & (I_s \otimes R)(e \otimes I_d) \\ 0 & 0 \end{bmatrix}.$$

Long and tedious calculations lead to the equivalent form

$$\zeta^{(m+s)d} \det\left[\zeta I_d - r_\infty(1/\zeta^m R)\right] \cdot \det\left[\zeta^m A \otimes I_d - B \otimes R\right] = 0,$$

where

$$r_\infty(z) = 1 - (1+z)b^T(A + zB)^{-1}e.$$

Koto [178] proved the following result.

Theorem 10.8.4 *Assume that the CRK method $(A, b(\theta))$ satisfies $\rho[A^{-1}B] \leq 1$. Then the method is P_∞-stable if and only if*

$$|r_\infty(z)| < 1 \quad \textit{whenever } |z| < 1.$$

If the CRK is natural ($A = B$) and A^{-1} exists, then

$$r_\infty(z) \equiv R(\infty) = 1 - b^T A^{-1} e,$$

which is the value of the A-stability function $R(\alpha)$ at infinity (see (8.3.10)). So, we immediately get the following result.

Theorem 10.8.5 *The natural CRK methods based on the Radau IIA methods are* P_∞*-stable and, hence, asymptotically stable for the generalized pantograph equation (10.8.1).*

The result does not apply to the class of Θ-methods, where A is singular. However, for the linearity of (10.8.1), the Θ-method is equivalent to collocation at the point Θ for which $A = \Theta$ and $r_\infty = (\Theta - 1)/\Theta$. So, P_∞-stability of Θ-methods is proved for $\frac{1}{2} < \Theta \leq 1$. For the scalar equation, this was first proved by Bellen, Guglielmi and Torelli [28]. The technique, still based on separating the stationary and non-stationary parts of the difference equations, extends to the non-neutral pantograph equation with variable scalar coefficients.

A more general result for (10.8.1) was proved by Koto [178] for RK methods with A singular. Assume

$$A = \begin{bmatrix} 0 & 0 \\ \hat{a} & \hat{A} \end{bmatrix},$$

where

$$\hat{a} = [a_{21}, a_{31}, \ldots, a_{s1}]^T \qquad \text{and} \qquad \hat{A} = [a_{ij}]_{i,j=2}^{s},$$

with $\hat{A}$ non-singular.

Theorem 10.8.6 *If the RK method* (A, b) *is stiffly accurate (i.e.,* $b_j = a_{sj}$*,* $j = 1, \ldots, s$*) and* $\hat{A}$ *is non-singular, then*

$$R(\infty) = (\hat{A}^{-1}\hat{a})_{s-1}.$$

Moreover, if $|R(\infty)| \leq 1$*, then the natural CRK method based on* (A, b) *is* P_∞*-stable and, hence, asymptotically stable for the generalized pantograph equation (10.8.1).*

The Lobatto IIIA methods and the Θ-methods are typical examples of RK methods that satisfy the hypotheses of Theorem 10.8.6.

Observe that the quasi-geometric mesh is not the only constrained mesh fulfilling the exponential growth of the stepsizes. To see this, for $\gamma = q^{-1/m}$ define the mesh points in the first macro interval $H_1 = [h, h/q]$ as

$$t_0 = h, \; t_1 = \gamma h, \; \ldots, \; t_m = \gamma^m h = h/q,$$

and build up the constrained mesh on subsequent macro intervals H_k, $k \geq 2$, by $t_n = t_{n-m}/q$. The resulting mesh Δ_m is a *fully geometric mesh* (see Figure 10.9)

$$\Delta_m = \{t_0 = h, t_1 = \gamma h, \ldots, t_n = \gamma^n h, \ldots\}$$

with stepsizes satisfying the exponential growth $h_{n+1} = \gamma^n h_1$, where $h_1 = t_1 - t_0 = (\gamma - 1)h$.

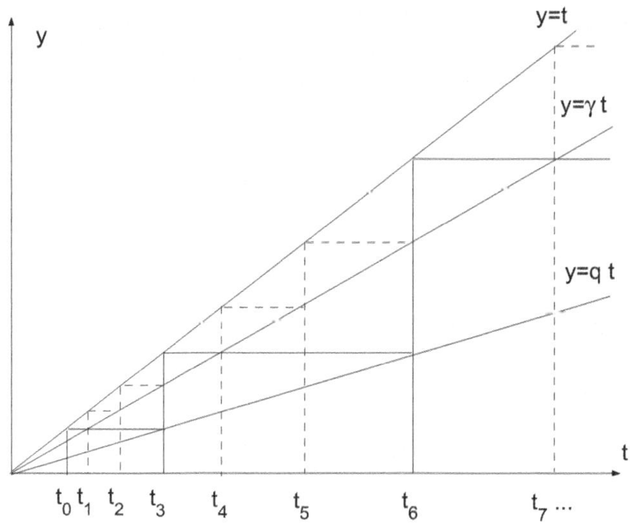

FIG. 10.9. The fully geometric mesh Δ_m for $m = 3$.

10.9 Further stability investigations

In this section we review a few topics that expand the contents of the present chapter, either as extensions of the results to other classes of test equations or as more specific stability analyses related to further aspects of the numerical methods.

Our aim is simply to give a rough idea of such investigations and to provide some of the most significant references.

10.9.1 *Using the Kreiss resolvent*

The stability analysis based on the Kreiss resolvent condition for bounding the error growth in RK schemes for ODEs has been extended to DDEs in a set of papers by Spijker [237] and Van den Heuvel [254, 255, 256]. Roughly speaking, the aim is to bound, in a given norm, the powers $\|B^n\|$ of a matrix B, with $\rho[B] < 1$, representing the map which transforms two successive states of a discretized evolutionary system

$$U_{n+1} = BU_n, \quad n = 0, 1, \ldots,$$

or, equivalently,

$$U_n = B^n U_0, \quad n = 1, 2, \ldots.$$

Here $\|\cdot\|$ is the matrix norm induced by the norm on the state elements U_n.

The analysis, which goes beyond the classical spectral analysis, is based on the use of suitable modifications of the well-known *Kreiss matrix theorem* (see Kreiss [180], or Hairer and Wanner [118]) where bounds for the powers $\|B^n\|$ are determined by assuming some conditions involving the *resolvent* $(\zeta I - B)^{-1}$.

The approach has been used extensively for bounding the effect of a perturbation E_0 in the initial data of processes modeled by high order difference equations, formulated as one-step methods in a higher dimension state space. For a comprehensive review of variations to the original Kreiss resolvent condition, see Van Dorsselaer, Kraajievanger and Spjiker [253].

In particular, Spjiker [237] proved the following theorem.

Theorem 10.9.1 *Suppose that the $\mathcal{N} \times \mathcal{N}$ matrix B is such that $\zeta I - B$ is invertible for all $\zeta \in \mathbb{C}$ with $|\zeta| > 1$ and, in a given norm $\|\cdot\|$, the inverse satisfies the condition*

(S) $\quad \|(\zeta I - B)^{-1}\| \leq \frac{R|\zeta|^{k\mathcal{N}}}{|\zeta|-1} \quad$ *for all* $\quad \zeta \in \mathbb{C}$ *with* $\quad |\zeta| \in (1, \rho)$,

where $R > 0$, $\rho > 1$ and $k \geq 0$ are constants. Then $\gamma > 0$, depending only on ρ and k, exists such that

$$\|B^n\| \leq \gamma R \min\{n + k\mathcal{N}, (1+k)\mathcal{N}\}.$$

In the analysis of the RK method (10.6.2), (10.6.3) with constant stepsize $h = \tau/m$ for linear systems (10.6.1), the vector of states

$$\mathbf{H}_{n+1} = [K^1_{n+1}, \ldots, K^s_{n+1}, y_{n+1}]^T$$

is enlarged to

$$U_{n+1} = [\mathbf{H}_{n+1}, \ldots, \mathbf{H}_{n-m+1}]^T$$

and the companion matrix generated by the recurrence relation (10.6.5) is the $\mathcal{N} \times \mathcal{N}$ matrix

$$B = \begin{bmatrix} D^{-1}E & 0 & \cdots & 0 & D^{-1}F & D^{-1}G \\ I & 0 & \cdots & 0 & 0 & 0 \\ \vdots & \vdots & & \vdots & \vdots & \vdots \\ 0 & 0 & \cdots & 0 & I & 0 \end{bmatrix},$$

where $\mathcal{N} = d(s+1)(m+1)$ and the blocks have dimension $d(s+1)$.

By using condition (S), Spjiker [237] analyzed the growth of $\|B^n\|_\infty$ relevant to the trapezoidal rule for the scalar equation

$$y'(t) = \lambda y(t) + \mu y(t - \tau),$$

with λ and μ satisfying $\Re(\lambda) \leq -|\mu|$, and proved that

$$\|B^n\|_\infty \leq c \min\{n, m+1\}.$$

Later, Van den Heuvel [254, 255], extended the result to Θ-methods and more general RK methods, including Radau IA, Radau IIA and Lobatto IIIC.

Furthermore, for non-neutral systems ($N = 0$) and integration stepsize $h = \tau/(m - \delta)$, $m \geq 1$ integer, and $0 \leq \delta < 1$, Van den Heuvel [256] extended the analysis to the method (10.6.8), (10.6.9) which, for $\delta \neq 0$, leads to a slightly different matrix B. His result is still based on the following Kreiss resolvent conditions involving the matrices $X = hL$ and $Y = hM$:

(a) $(\alpha I - X)^{-1}$ and $(\zeta I - (\alpha I - X)^{-1}Y)^{-1}$ exist and

$$\|(\zeta I - (\alpha I - X)^{-1}Y)^{-1}\| \leq D_1/(|\zeta| - 1)$$

for all $\zeta \in \mathbb{C}$ with $|\zeta| > 1$ and all $\alpha \in \mathbb{C} \setminus S_A$;

(aa) $(\alpha I - X - \zeta Y)^{-1}$ exists and

$$\|(\alpha I - X - \zeta Y)^{-1}\| \leq D_2/(\mathrm{dist}(\alpha, S_A))$$

for all $\zeta \in \mathbb{C}$ with $|\zeta| < 1$ and all $\alpha \in \mathbb{C} \setminus S_A$;

for suitable constants $D_1, D_2 > 0$.

Define the vector norms

$$\|\mathbf{H}_{n+1}\|_\infty^{d(s+1)} = \max\{\|K_{n+1}^1\|, \ldots, \|K_{n+1}^s\|, \|y_{n+1}\|\},$$

$$\|U_{n+1}\|_\infty^{\mathcal{N}} = \max\{\|\mathbf{H}_{n+1}\|_\infty^{d(s+1)}, \ldots, \|\mathbf{H}_{n-m+1}\|_\infty^{d(s+1)}\},$$

and the induced matrix norm $\|B\|_\infty^{\mathcal{N}\times\mathcal{N}}$.

In this setting, Van den Heuvel proved the following theorem.

Theorem 10.9.2 *If the CRK method (10.6.8), (10.6.9) with equistage interpolation such that $r \leq q \leq r+2$ satisfies the conditions*

(i) $|R(\infty)| = \lim_{|\alpha|\to+\infty} |R(\alpha)| \neq 1$;

(ii) $R'(\alpha) \neq 0$ *for all* $\alpha \in \partial S_A$, *the border of the A-stability region;*

(iii) $(I - \alpha A)^{-1}$ *exists and* $\|\alpha b^T(I - \alpha A)^{-1}\|_\infty$ *is uniformly bounded for all* $\alpha \in S_A$;

(iv) $(I - \alpha A)^{-1}$ *exists and* $\|(I - \alpha A)^{-1}\|_\infty$ *is uniformly bounded for all* $\alpha \in S_A$;

and if the stepsize h is such that conditions (a) and (aa) hold, then

$$\|B^n\|_\infty^{\mathcal{N}\times\mathcal{N}} \leq C_1 + C_2 \min\{n+1, \mathcal{N}\}$$

for suitable constants $C_1, C_2 > 0$.

10.9.2 *Pure DDEs*

Here we briefly mention some stability investigations based on the pure delay test equations

$$\begin{cases} y'(t) = \mu(t)y(t-\tau), & t \geq t_0, \\ y(t) = \phi(t), & t \leq t_0, \end{cases} \tag{10.9.1}$$

and

$$\begin{cases} y'(t) = \mu y(t-\tau), & t \geq t_0, \\ y(t) = \phi(t), & t \leq t_0, \end{cases} \tag{10.9.2}$$

which were introduced in Section 9.7.3.

Barwell [19] gave the definition of *Q-stability* and *GQ-stability* with respect to (10.9.2) with $\mu \in \mathbb{C}$. This test equation was considered also by van der Houwen and Sommeijer [139], who analyzed the stability properties of LM methods.

Subsequently, Torelli and Vermiglio [249] gave the definition of *QN(0)-stability* with respect to (10.9.1) with real coefficient $\mu(t)$ and found some sufficient conditions to guarantee this property for the *continuous quadrature rule*

$$\eta(t_n + \theta h) = y_n + h_{n+1} \sum_{i=1}^{s} b_i(\theta)\mu(t_{n+1}^i)\eta(t_{n+1}^i - \tau), \quad 0 \le \theta \le 1.$$

10.9.3 *Multiple delays*

Most of the results presented in Chapters 9 and 10 on the stability of the exact and numerical solutions of DDEs and NDDEs readily extend to equations with multiple delays. In particular, for the equation

$$\begin{cases} y'(t) = f\Big(t, y(t), y\big(\alpha_1(t)\big), \ldots, y\big(\alpha_r(t)\big)\Big), & t \ge t_0, \\ y(t) = \phi(t), \quad t \le t_0, \end{cases}$$

Theorem 9.1.3 holds under the condition, slightly more general than (9.1.5),

$$Y(t) + \sum_{j=1}^{r} X_j(t) \ge 0, \quad t \ge t_0,$$

where, for $j = 1, \ldots, r$,

$$X_j(t) \ge \sup_{y, x_1, \ldots, x_r, \tilde{x}_j \ne x_j} \frac{\|(f(t, y, x_1, \ldots, x_j, \ldots, x_r) - f(t, y, x_1, \ldots, \tilde{x}_j, \ldots, x_r)\|}{\|x_j - \tilde{x}_j\|}.$$

Similarly, Theorem 9.1.4 holds under the condition

$$R \cdot Y(t) + \sum_{j=1}^{r} X_j(t) \ge 0, \quad t \ge t_0.$$

With these definitions, the contractivity and asymptotic stability results of Section 10.4, relevant to the numerical methods, can also be adapted in a straightforward way.

For the general non-linear NDDE

$$\begin{cases} y'(t) = f\Big(t, y(t), y\big(\alpha_1(t)\big), \ldots, y\big(\alpha_r(t)\big), y'\big(\beta_1(t)\big), \ldots, y'\big(\beta_r(t)\big)\Big), & t \ge t_0, \\ y(t) = \phi(t), \quad t \le t_0, \end{cases}$$

and for the alternative formulations (9.4.2) and (9.4.18), the results on contractivity, boundedness and asymptotic stability described in Section 9.4, as well as the numerical counterparts in Section 10.5, can be suitably adapted to the presence of multiple delays. As with DDEs, the extensions are straightforward and we leave them to the reader.

Also the spectral analysis of Theorems 9.5.3 and 10.6.1 can be extended to linear systems with multiple delays

$$y'(t) = Ly(t) + \sum_{j=1}^{r} M_j y(t - j\tau) + \sum_{j=1}^{r} N_j y'(t - j\tau), \tag{10.9.3}$$

for which Hu and Mitsui [140] proved the asymptotic stability of the exact solutions under the condition

$$\begin{cases} \lambda \in \sigma[L] \Longrightarrow \Re(\lambda) < 0, \\ \\ \zeta \in \sigma\Big[(I - \sum\limits_{j=1}^{r} \xi^j N_j)^{-1}(L + \sum\limits_{j=1}^{r} \xi^j M_j)\Big] \Longrightarrow \Re(\zeta) < 0 \quad \text{for all } \xi \in \mathbb{C},\ |\xi| = 1. \end{cases}$$

Correspondingly, for any natural A-stable CRK method $(A, b(\theta))$ the numerical solution satisfies $\lim_{n\to\infty} y_n = 0$ for any stepsize $h = \tau/m$, $m \geq 1$ integer. Note that the delays in (10.9.3) have the particular form $j\tau$, $j = 1, 2, \ldots, r$, and that h is a common submultiple of all of them.

Yang, Qiu and Mitsui [275] considered DDEs with more general delays τ_j in the simplified form (2.1.7), i.e.

$$y'(t) = Ly(t) + My(t - \tau),$$

where $y(t-\tau) = \big[y_1(t-\tau_1), \ldots, y_d(t-\tau_d)\big]^T$, and proved the asymptotic stability of any A-stable RK method equipped with equistage interpolation (10.2.34) for any constant stepsize h.

10.9.4 *Regularity*

An autonomous DDE

$$\begin{cases} y'(t) = f\left(y(t), y(t-\tau)\right), & t \geq t_0, \\ y(t) = \phi(t), & t \leq t_0, \end{cases} \tag{10.9.4}$$

with constant delay τ is said to have an *asymptotic value* y^* if

$$f(y^*, y^*) = 0.$$

Let us denote by $\mathcal{A}$ the set of all asymptotic values of (10.9.4).

Moreover, an asymptotic value y^* of (10.9.4) is *attractive* if there exists a non-empty open set $\mathcal{U}$ in the space $C^0([-\tau, 0])$ of continuous initial functions $\phi(t)$ such that

$$\lim_{t\to+\infty} y(t) = y^* \quad \text{for all } \phi \in \mathcal{U}.$$

The theory of *regularity* of the RK methods for DDEs concerns the set of asymptotic values of the relevant discretized equations and their possible attractiveness. This issue was considered by Jackiewicz, Vermiglio and Zennaro [163], who extended the analogous theory developed for ODEs by Iserles [144], Hairer, Iserles and Sanz-Serna [116] and Jackiewicz, Vermiglio and Zennaro [162].

The essence of regularity of an RK method for DDEs (6.2.1), (6.2.2) is that if the sequence of nodal values y_n, obtained by constant stepsize $h = \tau/m$, $m \geq 1$ integer, for the autonomous system (10.9.4), converges to some y^* as $n \to \infty$, then such an asymptotic value of the method belongs to the set $\mathcal{A}$ of the asymptotic values of the DDE (10.9.4) itself. In other words, no spurious asymptotic values exist for a regular method. On the other hand, it is easy to see that, for any method, every asymptotic value y^* of the DDE is also an asymptotic value of the method.

In [163] the authors found necessary and sufficient conditions for the regularity of RK methods for DDEs. The general results need a lot of definitions and technicalities in order to be stated. Therefore, in this brief outline, we just refer the interested reader to their paper. However, we remark that, once again, natural RK methods for DDEs (i.e. $A = B$) play a privileged role in that they preserve the regularity properties of the underlying discrete RK method for ODEs.

Jackiewicz, Vermiglio and Zennaro [163] also gave sufficient conditions for the attractiveness of the asymptotic values of both the DDE (10.9.4) and the numerical scheme. More precisely, they proved that an asymptotic value y^* of the DDE (10.9.4) is attractive if the linearized equation

$$\begin{cases} z'(t) = Lz(t) + Mz(t-\tau), & t \geq t_0, \\ z(t) = \phi(t), & t \leq t_0, \end{cases} \tag{10.9.5}$$

where

$$L = \frac{\partial f}{\partial y}(y^*, y^*) \quad \text{and} \quad M = \frac{\partial f}{\partial x}(y^*, y^*),$$

is asymptotically stable, and that y^* is attractive also for the RK method implemented with a given constant stepsize $h = \tau/m$, $m \geq 1$ integer, if the discrete numerical solution $\{z_n\}$ of the linearized equation (10.9.5), obtained with the stepsize h, is asymptotically stable as well.

10.9.5 *Stability of the stepsize control mechanism*

In Chapter 7 we discussed local error estimations and stepsize control strategies for the production of adaptive numerical algorithms for DDEs based on CRK methods. In particular, we went deep into the analysis of the required order of accuracy q of the continuous extension $\eta(t)$ in order that the resulting procedure perform to the expected order p (the order of the discrete RK formula) and show a good proportional response to the user-supplied tolerance TOL. We also stressed that, in some cases, it is necessary to control also the continuous local error, and not just the discrete local error.

Here, last but not least, we briefly mention another aspect of the stepsize control mechanisms for DDE solvers, which contributes to making a code actually effective: the existence of *equilibrium states* and their *stability*. This issue was considered by Higham and Famelis [129], who adapted Hall's theory [124] of equilibrium states for RK methods for ODEs.

Recall that an equilibrium state exists for an explicit RK method with stepsize control applied to a stable ODE if the map $\mathcal{G}$ that associates the new pair $[y_{n+1}, h_{n+2}]^T$ with the previous pair $[y_n, h_{n+1}]^T$ has a fixed point $[y_D, h_D]^T$. Such a fixed point is important since it provides a reasonable solution to the ODE, it uses the largest "stable" stepsize and produces a global error that is proportional to the tolerance TOL. The fixed point $[y_D, h_D]^T$ may arise in practice if it is *stable* with respect to small perturbations. This stability is determined by the Jacobian $\mathcal{G}'\left([y_D, h_D]^T\right)$.

As for DDEs, in [129] the authors considered the test equation

$$\begin{cases} y'(t) = \lambda y(t) + \mu y(t-\tau), & t \geq t_0, \\ y(t) = \phi(t), & t \leq t_0, \end{cases} \tag{10.9.6}$$

with constant delay, where $\lambda, \mu \in \mathbb{R}$ and satisfy the asymptotic stability condition $\lambda + |\mu| < 0$.

An explicit RK method for DDEs (6.2.1), (6.2.2) of order p with stepsize control making use of a formula similar to (7.3.5), namely

$$h_{n+2} = h_{\text{new}} = \rho\left(\frac{TOL}{\tilde{\sigma}_{n+1}}\right)^{1/p} \cdot h_{n+1}, \tag{10.9.7}$$

applied to the test equation (10.9.6), has an equilibrium state $[y_D, h_D]^T$ if the relevant formulae

$$\eta(t_n + \theta h_{n+1}) = y_n + h_{n+1} \sum_{i-1}^{s} b_i(\theta)\left(\lambda Y_{n+1}^i + \mu\eta(t_{n+1}^i - \tau)\right), \quad 0 \leq \theta \leq 1,$$

$$y_{n+1} = y_n + h_{n+1} \sum_{i=1}^{s} b_i\left(\lambda Y_{n+1}^i + \mu\eta(t_{n+1}^i - \tau)\right),$$

$$Y_{n+1}^i = y_n + h_{n+1} \sum_{j=1}^{s} a_{ij}\left(\lambda Y_{n+1}^j + \mu\eta(t_{n+1}^j - \tau)\right), \quad i = 1, \ldots, s,$$

coupled to (10.9.7) have the constant solution $[y_{n+1}, h_{n+2}]^T = [y_n, h_{n+1}]^T \equiv [y_D, h_D]^T$.

By defining the integer $m \geq 1$ and the number $\delta \in [0, 1)$ such that

$$h_D = \tau/(m - \delta),$$

we can write the $2(m+1)$ vector $[y_{n+1}, h_{n+2}, y_n, h_{n+1}, \ldots, y_{n-m+1}, h_{n-m+2}]^T$ as the image of the previous vector $[y_n, h_{n+1}, y_{n-1}, h_n, \ldots, y_{n-m}, h_{n-m+1}]^T$ in a suitable map $\mathcal{G}$, provided that the interpolant $\eta(t)$ is obtained by using only the nodal values (for example, Lagrange or Hermite interpolation). The stability of the equilibrium state is governed by the Jacobian $\mathcal{G}'$ computed in the $2(m+1)$ vector fixed point $[y_D, h_D, \ldots, y_D, h_D]^T$.

In [129] some explicit low order RK methods ($p = 1, 2$) with linear Lagrange and cubic Hermite interpolation are investigated. It turns out that, as in the ODE case, every stable equilibrium state is reproducible in practice and its stability is independent of the tolerance TOL. However, making the difference from the ODE case, the stability of the equilibrium state depends on the parameters λ and μ of (10.9.6) and on the interpolant $\eta(t)$. In perfect agreement with the conclusions we already arrived at in Chapter 7, the message from the analysis in [129] is that, in order to obtain a smooth behavior of the overall method, it is important that the stepsize control mechanism also takes the interpolation procedure into account.

10.9.6 *Numerical approximation of the characteristic roots*

Computation of the characteristic roots of linear (or linearized) DDEs and NDDEs is important for the asymptotic stability analysis of solutions and *steady-state solutions* (previously called asymptotic values in Section 10.9.4) and to detect bifurcation for general autonomous functional differential equations

$$y'(t) = f(y_t)$$

(see Engelborghs, Luzyanina and Roose [82] and the bibliography therein). The root with a maximum real part is of particular interest. For the linear DDE

$$y'(t) = Ly(t) + \sum_{j=1}^{r} M_j y(t - \tau_j),$$

where L and M_i are $d \times d$ matrices, the characteristic equation is the quasi-polynomial equation

$$\lambda I - L - \sum_{j=1}^{r} M_j \mathrm{e}^{-\lambda \tau_j} = 0,$$

and computation of the roots requires the use of a root-finder for non-linear systems. In general, this technique is not advisable since the roots are very sensitive to changes in the coefficients.

Efficient approaches have been proposed by Engelborghs and Roose [83, 84] for the computation of the rightmost characteristic root, based on discretization of the delay differential operator by means of linear k-step formulae with constant stepsize $h = \tau_{\max}/m$, $m \geq 1$ integer. Roughly speaking, as m increases, more and more roots λ_i of the quasi-polynomial, starting from the rightmost one, are approximated by $\lambda^*_{m,i} = \log(\mu_{m,i})/h$, where $\mu_{m,i}$, $i = 1, \ldots, md$, are the characteristic roots of the resulting d vector difference equation of order m. The accuracy of $\lambda^*_{m,i}$ increases with m and decreases with i. The characteristic roots of the difference equations are computed as eigenvalues of the relevant companion matrices.

Breda, Maset and Vermiglio [50] modified the procedure by discretizing the infinitesimal generator $\mathcal{A}$ in the abstract Cauchy problem (3.5.2) derived from the more general functional differential equation

$$y'(t) = Ly(t) + \sum_{j=1}^{r} M_j y(t - \tau_j) + \int_{-\tau_a}^{-\tau_b} M(s) y(t+s) \mathrm{d}s.$$

They discretized $\mathcal{A}$ by the matrix $\mathcal{A}_{N,\nu}$ in (3.5.7) using a ν-stage RK method. Unlike the previous approach, the eigenvalues of this matrix directly provide approximations of the roots of the quasi-polynomial.

REFERENCES

[1] A.N. Al-Mutib, One-step implicit methods for solving delay differential equations, *Internat. J. Computer Math.* **16** (1984), 157–168.

[2] A.N. Al-Mutib, Stability properties of numerical methods for solving delay differential equations, *J. Comput. Appl. Math.* **10** (1984), 71–79.

[3] H. Arndt, The influence of interpolation on the global error in retarded differential equations, in *Differential-Difference Equations, ISNM* **62** (1983), 9–17.

[4] H. Arndt, Numerical solution of retarded initial value problems: local and global error and stepsize control, *Numer. Math.* **43** (1984), 343–360.

[5] U. Ascher and L. Petzold, The numerical solution of delay-differential-algebraic equations of retarded and neutral type, *SIAM J. Numer. Anal.* **32** (1995), 1635–1657.

[6] C.T.H. Baker, Numerical analysis of Volterra functional and integral equations, in *The State of the Art in Numerical Analysis*, I.S. Duff and G.A. Watson Eds., Clarendon Press, Oxford, 1996.

[7] C.T.H. Baker, Retarded differential equations, *J. Comput. Appl. Math.* **125** (2000), 309–335.

[8] C.T.H. Baker, J.C. Butcher and C.A.H. Paul, Experience of STRIDE applied to delay differential equations, *NA Report* **208**, Dept of Mathematics, University of Manchester, 1992.

[9] C.T.H. Baker and N. J. Ford, Convergence of linear multistep methods for a class of delay-integro-differential equations, *ISNM* **86** (1988), 47–59.

[10] C.T.H. Baker and N. J. Ford, Some applications of the boundary locus method and the method of D-Partitions, *IMA J. Numer. Anal.* **11** (1991), 143–158.

[11] C.T.H. Baker and C.A.H. Paul, Parallel continuous Runge–Kutta methods and vanishing lag delay differential equations, *Adv. Comput. Math.* **1** (1993), 367–394.

[12] C.T.H. Baker and C.A.H. Paul, A global convergence theorem for a class of PCERK methods and vanishing lag DDEs, *SIAM J. Numer. Anal.* **33** (1996), 1559–1576.

[13] C.T.H. Baker and C.A.H. Paul, Computing stability regions – Runge–Kutta methods for delay differential equations, *IMA J. Numer. Anal.* **14** (1994), 347–362.

[14] C.T.H. Baker, C.A.H. Paul and D.R. Willé, Issues in the numerical solution of evolutionary delay differential equations, *Adv. Comput. Math.* **3** (1995), 171–196.

[15] C.T.H. Baker, C.A.H. Paul and D.R. Willé, A bibliography on the numerical solution of delay differential equations, *NA Report* **269**, Dept of Mathematics, University of Manchester, 1995.

[16] V.L. Bakke and Z. Jackiewicz, Stability analysis of linear multistep methods for delay differential equations, *Internat. J. Math. Math. Sci.* **9** (1986), 447–458.

[17] H. T. Banks and F. Kappel, Spline approximations for Functional Differential Equations, *J. Differential Equations* **34** (1979), 496–522.

[18] Z. Bartoszewski and M. Kwapisz, On error estimates for waveform methods for delay differential equations, *SIAM J. Numer. Anal.* **34** (2001), 639–659.

[19] V.K. Barwell, Special stability problems for functional differential equations, *BIT* **15** (1975), 130–135.

[20] A. Bellen, Cohen's iteration process for boundary value problems for delay differential equations, *Rend. Istit. Matem. Univ. Trieste* **11** (1979), 55–66.

[21] A. Bellen, One-step collocation for delay differential equations, *J. Comput. Appl. Math.* **10** (1984), 275–283.

[22] A. Bellen, Constrained mesh methods for functional differential equations, in *Delay Equations, Approximation and Application, ISNM* **74** (1985), 52–70.

[23] A. Bellen, Contractivity of continuous Runge–Kutta methods for delay differential equations, *Appl. Numer. Math.* **24** (1997), 219–232.

[24] A. Bellen, Preservation of superconvergence in the numerical integration of delay differential equations with proportional delay, *IMA J. Numer. Anal.* **22** (2002), 529–536.

[25] A. Bellen, N. Guglielmi, S. Maset and M. Zennaro, Recent trends in the numerical solution of retarded functional differential equations, *Acta Numerica* **18** (2009), 1–110, Cambridge University Press.

[26] A. Bellen, N. Guglielmi and A. Ruehli, On a class of stable methods for linear systems of delay differential equations of neutral type, *IBM Research Report* **RC21253(94633)**, 1998.

[27] A. Bellen, N. Guglielmi and A. Ruehli, Methods for linear systems of circuit delay differential equations of neutral type, *IEEE Trans. Circuits Systems–I: Fund. Theory Appl.* **46** (1999), 212–216.

[28] A. Bellen, N. Guglielmi and L. Torelli, Asymptotic stability properties of θ-methods for the pantograph equations, *Appl. Numer. Math.* **24** (1997), 275–293.

[29] A. Bellen, N. Guglielmi and M. Zennaro, On the contractivity and asymptotic stability of systems of delay differential equations of neutral type, *BIT* **39** (1999), 1–24.

[30] A. Bellen, N. Guglielmi and M. Zennaro, Numerical stability of nonlinear delay differential equations of neutral type, *J. Comput. Appl. Math.* **125** (2000), 251–263.

[31] A. Bellen, Z. Jackiewicz and M. Zennaro, Stability analysis of one-step methods for neutral delay-differential equations, *Numer. Math.* **52** (1988), 605–619.

[32] A. Bellen and S. Maset, Numerical solution of constant coefficient linear delay differential equations as abstract Cauchy problems, *Numer. Math.* **84** (2000), 351–374.

[33] A. Bellen, S. Maset and L. Torelli, Contractive initialising methods for the pantograph equations of neutral type, in *Recent Trends in Numerical Analysis*, D. Trigiante Ed., *Advances in the Theory of Computational Mathematics* **3** (2000), 35–41.

[34] A. Bellen and M. Zennaro, Sulla ricerca di soluzioni periodiche di equazioni e disequazioni differenziali ordinarie e con ritardo, *Boll. Un. Mat. Ital. (B)* **2** (1983), 803–817.

[35] A. Bellen and M. Zennaro, Maximum principles for periodic solutions of linear delay differential equations, in *Differential-Difference Equations, ISNM* **62** (1983), 19–24.

[36] A. Bellen and M. Zennaro, Numerical solution of delay differential equations by uniform corrections to an implicit Runge–Kutta method, *Numer. Math.* **47** (1985), 301–316.

[37] A. Bellen and M. Zennaro, Stability properties of interpolants for Runge–Kutta methods, *SIAM J. Numer. Anal.* **25** (1988), 411–432.

[38] A. Bellen and M. Zennaro, Strong contractivity properties of numerical methods for ordinary and delay differential equations, *Appl. Numer. Math.* **9** (1992), 321–346.

[39] R. Bellman, On the computational solution of differential-difference equations, *J. Math. Anal. Appl.* **2** (1961), 108–110.

[40] R. Bellman and K.L. Cooke, Differential-Difference Equations, Academic Press, New York–London, 1963.

[41] R. Bellman and K.L. Cooke, On the computational solution of a class of functional differential equations, *J. Math. Anal. Appl.* **12** (1965), 495–500.

[42] T.A. Bickart, P-stable and P$[\alpha, \beta]$-stable integration/interpolation methods in the solution of retarded differential-difference equations, *BIT* **22** (1982), 464–476.

[43] G. Birkhoff, Numerical solution of reactor kinetics equations, in *Numerical Solution of Nonlinear Differential Equations*, D. Greespan Ed., Wiley, New York, 1966.

[44] M. Bjørhus, On dynamic iterations for delay differential equations, *BIT* **34** (1994), 325–336.

[45] G.A. Bocharov, G.I. Marchuk and A.A. Romanyukha, Numerical solution by LMMs of stiff delay-differential system modelling an immune-reponse, *Numer. Math.* **73** (1996), 131–148.

[46] H.G. Bock and J. Schlöder, Numerical solution of retarded differential equations with state dependent time lags, *ZAMM* **61** (1981), 269–271.

[47] P. Bogacki and L.F. Shampine, Interpolating high-order Runge–Kutta formulas, *Comput. Math. Appl.* **20** (1990), 15–24.

[48] C. de Boor and B. Swartz, Collocation at Gaussian points, *SIAM J. Numer. Anal.* **10** (1973), 582–606.

[49] R.K. Brayton and R.A. Willoughby, On the numerical integration of a symmetric system of difference-differential equations of neutral type, *J. Math. Anal. Appl.* **18** (1967), 182–189.

[50] D. Breda, S. Maset and R. Vermiglio, Computing the characteristic roots for delay differential equations, *IMA J. Numer. Anal.* **24** (2004), 1–19.

[51] H. Brunner, The numerical solution of initial value problems for integro-differential equations, in *Proceedings of Dundee Conference on Numerical Analysis*, D.F. Griffiths and G.A. Watson Eds., *Pitman Res. Notes Math. Ser.* **170**, Longman, Harlow, 1988.

[52] H. Brunner, The approximate solution of initial value problems for general Volterra integro-differential equation, *Computing* **40** (1988), 125–137.

[53] H. Brunner, Open problems in the discretization of Volterra integral equation, *Numer. Funct. Anal. Optim.* **17** (1996), 717–736.

[54] H. Brunner, On the discretization of differential and Volterra integral equation with variable delay, *BIT* **37** (1997), 1–12.

[55] H. Brunner, The use of splines in the numerical solution of differential and Volterra integral equation, in *Spline Functions and the Theory of Wavelets*, S. Dubuc and G. Deslauries Eds., *CRM Proceedings and Lecture Notes* **18**, AMS, Providence RI, 1999.

[56] H. Brunner, Collocation Methods for Volterra Integral and Related Functional Differential Equations, Cambridge University Press, Cambridge, 2004.

[57] H. Brunner, Q. Hu and Q. Lin, Geometric meshes in collocation methods for Volterra integral equations with proportional delays, *IMA J. Numer. Anal.* **21** (2001), 783–798.

[58] M.D. Buhmann and A. Iserles, Numerical analysis of functional equations with a variable delay, in *Numerical Analysis*, D.F. Griffiths and G.A. Watson Eds., Longman, Harlow, 1991.

[59] M.D. Buhmann and A. Iserles, On the dynamics of a discretized neutral equation, *IMA J. Numer. Anal.* **12** (1992), 339–363.

[60] M.D. Buhmann and A. Iserles, Stability of the discretized pantograph differential equation, *Math. Comput.* **60** (1993), 575–589.

[61] M.D. Buhmann, A. Iserles and S.P. Nørsett, Runge–Kutta methods for neutral differential equation, in *Contribution in Numerical Mathematics*, R.P. Agarwal Ed., *World Sci. Ser. Appl. Anal.* **2** (1993), 85–98.

[62] J.C. Butcher, The Numerical Analysis of Ordinary Differential Equations, Wiley, London, 1987.

[63] J.C. Butcher, The adaptation of STRIDE to delay differential equations, *Appl. Numer. Math.* **9** (1992), 415–425.

[64] M. Calvo, J.I. Montijano and L. Randez, A fifth-order interpolant for the Dormand and Prince Runge–Kutta method, *J. Comput. Appl. Math.* **29** (1990), 91–100.

[65] M. Calvo, J.I. Montijano and L. Randez, New continuous extensions for fifth-order RK formulas, *Rev. Acad. Cienc. Zaragoza* **45** (1990), 69–81.

[66] M. Calvo, J.I. Montijano and L. Randez, A new embedded pair of Runge–Kutta formulae of order 5 and 6, *Comput. Math. Appl.* **20** (1990), 15–24.

[67] R.N. Castleton and L.J. Grimm, A first order method for differential equations of neutral type, *Math. Comput.* **27** (1973), 571–577.

[68] B.A. Chartes and R.S. Stepleman, Order of convergence of Linear Multistep methods for functional differential equations, *SIAM J. Numer. Anal.* **6** (1975), 876–886.

[69] S.P. Corwin, D. Sarafyan and R. Thompson, DKLAG6: a code based on continuously imbedded sixth-order Runge–Kutta methods for the solution of state-dependent functional differential equations, *Appl. Numer. Math.* **24** (1997), 319–330.

[70] S.P. Corwin and R. Thompson, Error estimation and step-size control for delay differential equation solvers based on continuously imbedded Runge–Kutta–Sarafyan methods, *Comput. Math. Appl.* **31** (1996), 1–11.

[71] C.W. Cryer, Numerical methods for functional differential equations, in *Delay and Functional Differential Equations and Their Applications*, K. Schmitt Ed., Academic Press, New York, 1972.

[72] C. W. Cryer, Highly stable multistep methods for retarded differential equations, *SIAM J. Numer. Anal.* **11** (1974), 788–797.

[73] C.W. Cryer and L. Tavernini, The numerical solution of Volterra functional differential equations by Euler's method, *SIAM J. Numer. Anal.* **9** (1972), 105–129.

[74] K. Dekker and J.G. Verwer, Stability of Runge–Kutta Methods for Stiff Nonlinear Differential Equations, North-Holland, Amsterdam, 1984.

[75] O. Diekmann, S.A. van Gils, S.M. Verduyn Lunel and H.O. Walther, Delay Equations: Functional-, Complex-, and Nonlinear Analysis, *AMS series* **110**, Springer-Verlag, Berlin, 1995.

[76] J.R. Dormand and P.J. Prince, A family of embedded Runge–Kutta formulae, *J. Comput. Appl. Math.* **6** (1980), 19–26.

[77] R.D. Driver, Existence theory for a delay-differential system, *Contributions to Differential Equations* **1** (1963), 317–336.

[78] R.D. Driver, Ordinary and Delay Differential Equations, Springer-Verlag, Berlin, 1977.

[79] S.N. Elaydi, An Introduction to Difference Equations, Springer-Verlag, Berlin–Heidelberg–New York, 1996.

[80] L.E. El'sgol'ts, Qualitative Methods in Mathematical Analysis, Amer. Math. Soc., Providence R.I., 1964.

[81] L.E. El'sgol'ts and S.B. Norkin, Introduction to the Theory and Application of Differential Equations with Deviating Arguments, Academic Press, New York, 1973.

[82] K. Engelborghs, T. Luzyanina and D. Roose, Numerical bifurcation analysis of delay differential equations, *J. Comput. Appl. Math.* **125** (2000), 265–275.

[83] K. Engelborghs and D. Roose, Numerical computation of stability and detection of Hopf bifurcation of steady state solutions of delay differential equations, *Adv. Comput. Math.* **10** (1999), 271–289.

[84] K. Engelborghs and D. Roose, On stability of LMS-methods and characteristic roots of delay differential equations, *SIAM J. Numer. Anal.* **40** (2002), 629–650.

[85] W.H. Enright, Analysis of error control strategies for continuous Runge–Kutta methods, *SIAM J. Numer. Anal.* **26** (1989), 588–599.

[86] W.H. Enright, A new error-control for initial value solvers, *Appl. Math. Comput.* **31** (1989), 288–301.

[87] W.H. Enright, Continuous numerical methods for ODEs and the implication for delay differential equations, in *Proceedings of HERMIS94 Conference*, Athens, 1994.

[88] W.H. Enright and H. Hayashi, The evaluation of numerical software for delay differential equations, in *Proceedings of The IFIP TC2/WG2.5 Working Conference on the Quality of Numerical Software, Assessment and Enhancement*, R.F. Boisvert Ed., Chapman & Hall, London, 1997.

[89] W.H. Enright and H. Hayashi, A delay differential equation solver based on a continuous Runge–Kutta method with defect control, *Numer. Algorithms* **16** (1997), 349–364.

[90] W.H. Enright and H. Hayashi, Convergence analysis of the solution of retarded and neutral differential equations by continuous numerical methods, *SIAM J. Numer. Anal.* **35** (1998), 572–585.

[91] W.H. Enright, D.J. Higham, B. Owren and P. Sharp, A survey of the explicit Runge–Kutta method, *Technical Report* **291/94**, Dept of Computer Science, University of Toronto, 1994.

[92] W.H. Enright and M. Hu, Interpolating Runge–Kutta methods for vanishing lag delay differential equations, *Computing* **55** (1995), 223–236.

[93] W.H. Enright, K.R. Jackson, S.P. Nørsett and P.G. Thomsen, Interpolants for Runge–Kutta formulas, *ACM Trans. Math. Software* **12** (1986), 193–218.

[94] E. Fehlberg, Low order classical Runge–Kutta formulae with stepsize control and their application to some heat transfer problems, *Technical Report* **TR R-135**, NASA, 1969.

[95] A. Feldstein, Discretization methods for retarded ordinary differential equation, *Ph.D. Thesis*, Dept of Mathematics, UCLA, Los Angeles, 1964.

[96] A. Feldstein, A. Iserles and D. Levin, Embedding of delay equations into an infinite-dimensional ODE system, *J. Differential Equations* **117** (1995), 127–150.

[97] A. Feldstein and K.W. Neves, High order methods for state-dependent delay differential equations with nonsmooth solutions, *SIAM J. Numer. Anal.* **21** (1984), 844–863.

[98] S. Filippi and U. Buchacker, Stepsize control for delay differential equations using a pair of Runge–Kutta formulae, *J. Comput. Appl. Math.* **26** (1989), 339–343.

[99] L. Fox, D.F. Mayers, J.R. Ockendon and A.B. Tayler, On a functional differential equation, *J. Inst. Math. Appl.* **8** (1971), 271–307.

[100] G.W. Gear, Numerical Initial Value Problems in Ordinary Differential Equations, Prentice Hall, Englewood Cliffs, New York, 1971

[101] M. de Gee, Linear multistep methods for functional differential equations, *Math. Comput.* **48** (1987), 633–649.

[102] I. Gladwell, L.F. Shampine, L.S. Baca and R.W. Brankin, Practical aspects of interpolation in Runge–Kutta codes, *SIAM J. Sci. Statist. Comput.* **8** (1987), 322–341.

[103] I. Gohberg, P. Lancaster and L. Rodman, Matrix Polynomials, Academic Press, New York, 1982.

[104] A.D. Gorbunov and V.N. Popov, On methods of Adams type for the approximate solution of the Cauchy problem for ordinary differential equations with time lag, *Z. Vycisl. Mat. Fiz.* **4** (1964), suppl., 135–148.

[105] R.B. Grafton, Lienard equation with delay: existence and stability of periodic solution, in *Delay and Functional Differential Equations and Their Applications*, K. Schmitt Ed., Academic Press, New York, 1972.

[106] N. Guglielmi, Sulla stabilità dei metodi a un passo per la soluzione numerica di equazioni differenziali con ritardo, *Ph.D. Thesis*, University of Padova, 1996.

[107] N. Guglielmi, On asymptotic stability properties for Runge–Kutta methods for delay differential equations, *Numer. Math.* **77** (1997), 467–485.

[108] N. Guglielmi, Delay dependent stability regions of θ-methods for delay differential equations, *IMA J. Numer. Anal.* **18** (1998), 339–418.

[109] N. Guglielmi, Asymptotic stability barriers for natural Runge–Kutta processes for delay equations, *SIAM J. Numer. Anal.* **39** (2001), 763–783.

[110] N. Guglielmi and E. Hairer, Order stars and stability for delay differential equations, *Numer. Math.* **83** (1999), 371–383.

[111] N. Guglielmi and E. Hairer, Geometric proofs of numerical stability for delay equations, *IMA J. Numer. Anal.* **21** (2001), 439–450.

[112] N. Guglielmi and E. Hairer, Implementing Radau IIA methods for stiff delay differential equations, *Computing* **67** (2001), 1–12.

[113] A. Guillou and F.L. Soulé, La risolution numerique de problèmes differentielles aux conditions initials par des méthodes de collocation, *RAIRO* **3** (1969), 17–44.

[114] K. Gustafsson, Control theoretic techniques for stepsize selection in explicit Runge–Kutta methods, *ACM Trans. Math. Software* **17** (1991), 533–554.

[115] K. Gustafsson, M. Lundh and G. Söderlind, A PI stepsize control for the numerical solution of ordinary differential equations, *BIT* **28** (1988), 270–287.

[116] E. Hairer, A. Iserles and J.M. Sanz-Serna, Equilibria of Runge–Kutta methods, *Numer. Math.* **58** (1990), 243–254.

[117] E. Hairer, S.P. Nørsett and G. Wanner, Solving Ordinary Differential Equations I, Nonstiff Problems, Springer-Verlag, Berlin, 1993.

[118] E. Hairer and G. Wanner, Solving Ordinary Differential equations II, Stiff and Differential Algebraic Problems, Springer-Verlag, New York, 1991.

[119] E. Hairer and G. Wanner, On a generalization of a theorem of Von Neumann, *ZAMM* **76** (1996), suppl. 1, 95–98.

[120] E. Hairer and M. Zennaro, On error growth functions of Runge–Kutta methods, *Appl. Numer. Math.* **22** (1996), 205–216.

[121] J.K. Hale, Theory of Functional Differential Equations, Springer-Verlag, New York, 1977.

[122] J.K. Hale, Homoclinic orbits and chaos in delay equations, in *Proceedings of The Ninth Dundee Conference on Ordinary and Partial Differential Equations*, B.D. Sleeman and R.J. Jarvis Eds., Wiley, New York, 1986.

[123] J.K. Hale and S.M. Verduyn Lunel, Introduction to Functional Differential Equations, *Applied Mathematical Sciences* **99**, Springer-Verlag, New York, 1977.

[124] G. Hall, Equilibrium states of Runge–Kutta methods, *ACM Trans. Math. Software* **11** (1985), 289–301.

[125] H. Hayashi, Numerical solution of retarded and neutral delay differential equations using continuous Runge–Kutta methods, *Ph.D. Thesis*, University of Toronto, 1996.

[126] H. Hayashi and W.H. Enright, A new algorithm for vanishing delay problems, in *Procedings of Canadian Applied Mathematical Society Annual Meeting, University of York*, 1993.

[127] N.D. Hayes, Roots of the transcendental equations associated with certain difference-differential equations, *J. London Math. Soc.* **25** (1950), 226–232.

[128] D. Higham, Error control for initial value problems with discontinuities and delays, *Appl. Numer. Math.* **12** (1993), 315–330.

[129] D. Higham and I.Th. Famelis, Equilibrium states of adaptive algorithms for delay differential equations, *J. Comput. Appl. Math.* **58** (1995), 151–169.

[130] M.K. Horn, Fourth- and fifth-order scaled Runge–Kutta algorithms for treating dense output, *SIAM J. Numer. Anal.* **20** (1983), 558–568.

[131] K.J. in 't Hout, *unpublished handwritten notes*, 1989.

[132] K.J. in 't Hout, The stability of a class of Runge–Kutta methods for delay differential equations, *Appl. Numer. Math.* **9** (1992), 347–355.

[133] K.J. in 't Hout, A new interpolation procedure for adapting Runge–Kutta methods to delay differential equations, *BIT* **32** (1992), 634–649.

[134] K.J. in 't Hout, Runge–Kutta methods in the numerical solution of delay differential equations, *Ph.D. Thesis*, University of Leiden, 1992.

[135] K.J. in 't Hout, The stability of θ-methods for systems of delay differential equations, *Ann. Numer. Math.* **1** (1994), 323–334.

[136] K.J. in 't Hout, A note on unconditional maximum norm contractivity of waveform relaxation Runge–Kutta methods, *SIAM J. Numer. Anal.* **33** (1996), 1125–1134.

[137] K.J. in 't Hout, Stability of Runge–Kutta methods for systems of delay differential equations, *IMA J. Numer. Anal.* **17** (1997), 17–27.

[138] K.J. in 't Hout and M.N. Spijker, Stability analysis of numerical methods for delay differential equations, *Numer. Math.* **59** (1991), 807–814.

[139] P.J. van der Houwen and B.P. Sommeijer, Stability in linear multistep methods for pure delay equations, *J. Comput. Appl. Math.* **10** (1984), 55–63.

[140] G.D. Hu and T. Mitsui, Stability analysis of numerical methods for systems of neutral delay-differential equations, *BIT* **35** (1995), 504–515.

[141] G.D. Hu and T. Mitsui, Stability of linear delay differential systems with matrices having common eigenvectors, *Japan J. Indust. Appl. Math.* **13** (1996), 487–494.

[142] B. L. Hulme, One-step piecewise polynomial Galerkin methods for initial value problems, *Math. Comput.* **26** (1972), 415–426.

[143] J. Hutchinson, Finite difference solution to delay differential equation, *Ph.D. Thesis*, Rensselaer Polytechnic Institute, Troy, New York, 1971.

[144] A. Iserles, Stability and dynamics of numerical methods for nonlinear ordinary differential equations, *IMA J. Numer. Anal.* **10** (1990), 1–30.

[145] A. Iserles, On the generalized pantograph functional-differential equation, *Europ. J. Appl. Math.* **4** (1993), 1–38.

[146] A. Iserles, On nonlinear delay differential equations, *Trans. Amer. Math. Soc.* **344** (1994), 441–477.

[147] A. Iserles, Numerical analysis of delay differential equations with variable delays, *Ann. Numer. Math.* **1** (1994), 133–152.

[148] A. Iserles, Exact and discretized stability of the pantograph equation, *Appl. Numer. Math.* **24** (1997), 295–308.

[149] A. Iserles, Ordinary differential equations, in *The State of the Art in Numerical Analysis*, I.S. Duff and G.A. Watson Eds., Clarendon Press, Oxford, 1997.

[150] A. Iserles and J. Terjéki, Stability and asymptotic stability of functional-differential equations, *J. London Math. Soc.* **51** (1995), 559–572.

[151] E. Ishiwata, On the attainable order of collocation methods for the neutral functional-differential equation with proportional delay, *Computing* **64** (2000), 207–222.

[152] Z. Jackiewicz, Convergence of multistep methods for Volterra functional differential equations, *Numer. Math.* **32** (1979), 307–322.

[153] Z. Jackiewicz, The numerical solution of Volterra functional differential equations of neutral type, *SIAM J. Numer. Anal.* **18** (1981), 615–626.

[154] Z.Jackiewicz, Adams methods for neutral functional differential equations, *Numer. Math.* **39** (1982), 221–230.

[155] Z. Jackiewicz, One-step methods of any order for neutral functional differential equations, *SIAM J. Numer. Anal.* **21** (1984), 486–511.

[156] Z.Jackiewicz, Asymptotic stability analysis of θ-methods for functional differential equations, *Numer. Math.* **43** (1984), 389–396.

[157] Z.Jackiewicz, Quasilinear multistep methods and variable step predictor–corrector methods for neutral functional differential equations, *SIAM J. Numer. Anal.* **23** (1986), 423–452.

[158] Z.Jackiewicz, Variable-step variable-order algorithms for the numerical solution of neutral functional differential equations, *Appl. Numer. Math.* **3** (1987), 317–329.

[159] Z. Jackiewicz, M. Kwapisz and E. Lo, Waveform relaxation methods for functional differential systems of neutral type, *J. Math. Anal. Appl.* **207** (1997), 255–285.

[160] Z. Jackiewicz and E. Lo, The numerical solution of neutral functional differential equations by Adams predictor–corrector methods, *Appl. Numer. Math.* **8** (1991), 477–491.

[161] Z. Jackiewicz and E. Lo, The numerical integration of neutral functional differential equations by fully implicit one-step methods, *ZAMM* **75** (1995), 207–221.

[162] Z. Jackiewicz, R. Vermiglio and M. Zennaro, Regularity properties of Runge–Kutta methods for ordinary differential equations, *Appl. Numer. Math.* **22** (1996), 251–262.

[163] Z. Jackiewicz, R. Vermiglio and M. Zennaro, Regularity properties of Runge–Kutta methods for delay differential equations, *Appl. Numer. Math.* **24** (1997), 265–278.

[164] G.A. Kamenskii, Existence, uniqueness and continuous dependence on initial conditions of systems of differential equations with a deviating argument of neutral type, *Math. Sb. (N.S.)* **55** (1961), 363–378.

[165] F. Kappel and K. Kunisch, Spline approximation for neutral functional differential equations, *SIAM J. Numer. Anal.* **18** (1981), 1058–1080.

[166] A. Karoui and R. Vaillancourt, Computer solution of state-dependent delay differential equations, *Comput. Math. Appl.* **27** (1994), 37–51.

[167] A. Karoui and R. Vaillancourt, A numerical method for vanishing-lag delay differential equations, *Appl. Numer. Math.* **17** (1995), 383–395.

[168] T. Kato and J.B. McLeod, The functional-differential equation $y'(x) = ay(\lambda x) + by(x)$, *Bull. Amer. Math. Soc.* **77** (1971), 891–937.

[169] G.A. Kemper, Linear multistep methods for a class of functional differential equations, *Numer. Math.* **19** (1972), 361–372.

[170] A.V. Kim et al., Time-Delay System Toolbox, User's Guide, Inst. of Mathematics and Mechanics, Russian Academy of Sciences, and Engineering Research Center for Advanced Control and Instrumentation, Seoul National University, 1999.

[171] A.V. Kim and V.G. Pimenov, Numerical methods for delay differential equations, *Lecture Notes Series* **44**, Seoul National University, 1999.

[172] V. Kolmanovskii and A. Myshkis, Applied Theory of Functional Differential Equations, Kluwer, Dordrecht, 1992.

[173] V.Kolmanovskii and V. Nosov, Stability of Functional Differential Equations, Academic Press, London, 1986.

[174] T. Koto, A stability property of A-stable natural Runge–Kutta methods for systems of delay differential equations, *BIT* **34** (1994), 262–267.

[175] T. Koto, NP-stability of Runge Kutta methods based on classical quadrature, *BIT* **37** (1997), 870–884.

[176] T. Koto, Stability of Radau IA and Lobatto IIIC methods for delay differential equations, *Numer. Math.* **79** (1998), 569–580.

[177] T. Koto, A criterion for P-stability of Runge Kutta methods, *BIT* **38** (1998), 737–750.

[178] T. Koto, Stability of Runge–Kutta methods for the generalized pantograph equation, *Numer. Math.* **84** (1999), 233–247.

[179] T. Koto, Method of lines approximations of delay differential equations, *Comput. Math. Appl.* **48** (2004), 45–59.

[180] H.O. Kreiss, Über die Stabilitätsdefinition für Differenzengleichungen die partielle Differentialgleichungen approximiereng, *BIT* **2** (1962), 153–181.

[181] J. Kuang, J. Xiang and H. Tian, On the stability of one-parameter methods for systems of neutral differential equations, *BIT* **34** (1994), 400–408.

[182] Y. Kuang, On neutral delay logistic Gause-type predator–prey systems, *Dynam. Stability Systems* **6** (1991), 173–189.

[183] Y. Kuang, Delay Differential Equations with Applications in Population Dynamics, Academic Press, Boston, 1993.

[184] J.D. Lambert, Numerical Methods for Ordinary Differential Equations, Wiley, Chichester, 1991.

[185] E. Lelaresmee, A. Ruehli and A. Sangiovanni-Vincentelli, The waveform relaxation method for time domain analysis of large scale integrated circuits, *IEEE Trans. CAD IC Syst.* **1** (1982), 131–145.

[186] L.M. Li, Stability of linear neutral delay-differential systems, *Bull. Austral. Math. Soc.* **38** (1988), 339–344.

[187] P. Linz, Linear multistep methods for Volterra integro-differential equations, *J. Assoc. Comput. Mach.* **16** (1969), 295–301.

[188] M.Z. Liu and M.N. Spijker, The stability of the θ-methods in the numerical solution of delay differential equations, *IMA J. Numer. Anal.* **10** (1990), 31–48.

[189] Y. Liu, Functional differential equations with proportional delays, *Ph.D. Thesis*, University of Cambridge, 1996.

[190] Y. Liu, On the θ-method for delay differential equations with infinite lag, *J. Comput. Appl. Math.* **71** (1996), 177–190.

[191] Y. Liu, Runge–Kutta-collocation methods for systems of functional-differential and functional equations, *Adv. Comput. Math.* **11** (1999), 315–329.

[192] L. Lu, Numerical stability of the θ-methods for systems of differential equations with several delay terms, *J. Comput. Appl. Math.* **34** (1991), 291–304.

[193] M.C. Mackey and L. Glass, Oscillations and chaos in physiological control systems, *Science* **197** (1977), 287–289.

[194] S. Maset, Asymptotic stability in the numerical solution of linear pure delay differential equations as abstract Cauchy problems, *J. Comput. Appl. Math.* **111** (1999), 163–172.

[195] S. Maset, Stability of Runge Kutta methods for linear delay differential equations, *Numer. Math.* **87** (2000), 355–371.

[196] S. Maset, An abstract implicit Euler method for retarded functional differential equations, *Quaderno Matematico, Serie II* **484**, Dipartimento di Scienze Matematiche, University of Trieste, 2000.

[197] S. Maset, Instability of Runge Kutta methods when applied to linear systems of delay differential equations, *Numer. Math.* **90** (2002), 555–562.

[198] S. Maset, Numerical solution of retarded functional differential equations as abstract Cauchy problems, *J. Comput. Appl. Math.* **161** (2003), 259–282.

[199] R.M. May, The dynamics of natural and managed populations, in *The Mathematical Theory of the Dynamics of Biological Populations II*, R.W. Hiorns and D. Cooke Eds., Academic Press, London–New York, 1981.

[200] G. Meinardus and G. Nürnberger, Approximation theory and numerical methods for delay differential equations, in *Delay Equations, Approximation and Application, ISNM* **74** (1985), 13–40.

[201] V. Miekkala and O. Nevanlinna, Convergence of dynamic iteration methods for initial value problems, *SIAM J. Sci. Statist. Comput.* **8** (1987), 459–482.

[202] W.L. Miranker, Existence, uniqueness and stability of solutions of systems of nonlinear difference differential equations, *J. Math. Mech.* **11** (1962), 101–108.

[203] J.L. Naimark, D-subdivisions and spaces of quasi-polynomials, *Prikl. Mat. Mech.* **13** (1949), 349–380.

[204] K.W. Neves, Automatic integration of functional differential equations: an approach, *ACM Trans. Math. Software* **1** (1975), 357–368.

[205] K.W. Neves, Algorithm 497 – Automatic integration of functional differential equations, *ACM Trans. Math. Software* **1** (1975), 369–371.

[206] K.W. Neves and A. Feldstein, Characterization of jump discontinuities for state dependent delay differential equations, *J. Math. Anal. Appl.* **56** (1976), 689–707.

[207] K.W. Neves and S. Thompson, Software for the numerical solution of systems of functional differential equations with state dependent delays, *Appl. Numer. Math.* **9** (1992), 385–401.

[208] S. Niculescu, Delay Effects on Stability: A Robust Control Approach, Springer-Verlag, Heidelberg, 2001.

[209] S.P. Nørsett and G. Wanner, The real-pole sandwich for rational approximation and oscillation equations, *BIT* **19** (1979), 79–94.

[210] H.J. Oberle and H.J. Pesh, Numerical treatment of delay differential equations by Hermite interpolation, *Numer. Math.* **37** (1981), 235–255.

[211] J.R. Ockendon and A.B. Tayler, The dynamics of a current collection system for an electric locomotive, *Proc. Royal Soc. A* **322** (1971), 447–468.

[212] J. Oppelstrup, The RKFHB4 method for delay-differential equations, in *Numerical Treatment of Differential Equations*, R. Bulirsch, R. Grigorieff and J. Schröder Eds., *Lecture Notes in Mathematics* **631** (1978), 133–146.

[213] B. Owren and M. Zennaro, Order barriers for continuous explicit Runge–Kutta methods, *Math. Comput.* **56** (1991), 645–661.

[214] B. Owren and M. Zennaro, Derivation of efficient, continuous, explicit Runge–Kutta methods, *SIAM J. Sci. Statist. Comput.* **13** (1992), 1488–1501.

[215] G. Papageorgiou, T. Simos and C. Tsitouras, Some new Runge–Kutta methods with interpolation properties and their application to the magnetic binary problem, *Celestial Mechanics* **44** (1988), 167–177.

[216] C.A.H. Paul, Developing a delay differential equation solver, *Appl. Numer. Math.* **9** (1992), 403–414.

[217] C.A.H. Paul, A test set of functional differential equation, *NA Report* **243**, Dept of Mathematics, University of Manchester, 1994.

[218] C.A.H. Paul, A user guide to ARCHI – An explicit (Runge–Kutta) code for solving delay and neutral differential equations, *NA Report* **283**, Dept of Mathematics, University of Manchester, 1995.

[219] C.A.H. Paul, Designing efficient software for solving delay differential equation, *J. Comput. Appl. Math.* **125** (2000), 287–295.

[220] A. Pazy, Semigroups of Linear Operators and Applications to Partial Differential Equations, Springer-Verlag, New York, 1983.

[221] L. Qiu, B. Yang and J. Kuang, The NGP-stability of Runge Kutta methods for systems of neutral delay differential equations, *Numer. Math.* **81** (1999), 451–459.

[222] M. Santo, Metodi continui ad un passo per la risoluzione numerica di equazioni differenziali ordinarie, *Tesi di Laurea*, University of Udine, 1991 (in Italian).

[223] D. Sarafyan, Improved sixth-order Runge–Kutta formulas and approximate continuous solutions of ordinary differential equations, *J. Math. Anal. Appl.* **40** (1972), 436–445.

[224] D. Sarafyan, Continuous approximate solution of ordinary differential equations and their systems, *Comput. Math. Appl.* **10** (1984), 139–159.

[225] D. Sarafyan, New algorithms for the continuous solutions of ordinary differential equations and the upgrading of the order of the process, *Comput. Math. Appl.* **20** (1990), 77–100.

[226] R. Scherer and M. Müller, Algebraic conditions for A-stable Runge–Kutta methods, in *Computational Ordinary Differential Equations*, J.R. Cash and I. Gladwell Eds., Clarendon Press, Oxford, 1992.

[227] L.F. Shampine, Interpolation for Runge–Kutta methods, *SIAM J. Numer. Anal.* **22** (1985), 1014–1027.

[228] L.F. Shampine, Some practical Runge–Kutta formulas, *Math. Comput.* **46** (1986), 135–150.

[229] L.F. Shampine and S. Thompson, Event location for ordinary differential equations, *Comput. Math. Appl.* **39** (2000), 43–54.

[230] L.F. Shampine and S. Thompson, Solving DDEs in MATLAB, *Appl. Numer. Math.* **37** (2001), 441–458.

[231] P.W. Sharp and J.H. Verner, Explicit Runge–Kutta 4-5 pairs with interpolants, *Mathematical Preprint* **1995-03**, Dept of Mathematics and Statistics, Queen's University, Kingston, 1995.

[232] R. Skeel, Waveform iteration and the shifted Picard splitting, *SIAM J. Sci. Statist. Comput.* **10** (1989), 756–776.

[233] M. Slemrod and E.F. Infante, Asymptotic stability criteria for linear systems of difference-differential equations of neutral type and their discrete analogues, *J. Math. Anal. Appl.* **38** (1972), 399–415.

[234] H.L. Smith and Y. Kuang, Slowly oscillating periodic solutions of autonomous state-dependent delay equations, *Nonlinear Anal.* **19** (1992), 855–872.

[235] W. Snow, Existence, uniqueness and stability for nonlinear difference-differential equations in the neutral case, *Math. Sci. Rep. IMM-NYU* **328**, N.Y.U. Courant Inst., 1965.

[236] M.N. Spijker, Contractivity in the numerical solution of initial value problems, *Numer. Math.* **42** (1983), 271–290.

[237] M.N. Spijker, Numerical stability, resolvent conditions and delay differential equations, *Appl. Numer. Math.* **24** (1997), 233–246.

[238] H.J. Stetter, Numerische Lösung von Differentialgleichungen mit nacheilendem Argument, *ZAMM* **45** (1965), 79–80.

[239] H.J. Stetter, Considerations concerning a theory for ODE-solvers, in *Numerical Treatment of Differential Equations*, R. Bulirsch, R. Grigorieff and J. Schröder Eds., *Lecture Notes in Mathematics* **631** (1978), 188–200.

[240] H.J. Stetter, Tolerance proportionality in ODE-codes, in *Proceedings of The Second Conference on Numerical Treatment of Ordinary Differential Equations*, R. März Ed., Humboldt University, Berlin, 1980.

[241] N. Takama, Y. Muroya and E. Ishiwata, On the attainable order of collocation methods for the delay differential equation with proportional delay, *BIT* **40** (2000), 374–394.

[242] L. Tavernini, One-step methods for the numerical solution of Volterra functional differential equations, *SIAM J. Numer. Anal.* **4** (1971), 786–795.

[243] L. Tavernini, Linear multistep methods for the numerical solution of Volterra functional differential equations, *Applicable Anal.* **3** (1973), 169–185.

[244] L. Tavernini, The approximate solution of Volterra differential systems with state dependent time lags, *SIAM J. Numer. Anal.* **15** (1978), 1039–1052.

[245] L. Tavernini, A user-friendly interactive Turbo Pascal simulation toolkit, *Simulation* **53** (1989), 45–55.

[246] L. Tavernini, Continuous-time Modelling and Simulation: Using Turbo Pascal and CTMS/TP, Gordon & Breach, Amsterdam, 1996.

[247] L. Torelli, Stability of numerical methods for delay differential equations, *J. Comput. Appl. Math.* **25** (1989), 15–26.

[248] L. Torelli, A sufficient condition for GPN-stability for delay differential equations, *Numer. Math.* **59** (1991), 311–320.

[249] L. Torelli and R. Vermiglio, On the stability of continuous quadrature rules for differential equations with several constant delays, *IMA J. Numer. Anal.* **13** (1993), 291–302.

[250] L. Torelli and R. Vermiglio, A stable numerical approach for implicit non-linear neutral delay differential equations, *BIT* **43** (2003), 195–215.

[251] C. Tsitouras and G. Papageorgiou, New interpolants for Runge–Kutta algorithms using second derivatives, *Internat. J. Computer Math.* **31** (1989), 105–113.

[252] C. Tsitouras and G. Papageorgiou, Runge–Kutta interpolants based on values from two successive integration steps, *Computing* **43** (1990), 255–266.

[253] J.L.M. Van Dorsselaer, J.F.B.M. Kraaijevanger and M.N. Spijker, Linear stability analysis in the numerical solution of initial value problems, *Acta Numerica* (1993), 199–237.

[254] E.G. Van den Heuvel, New stability results for Runge–Kutta methods adapted to delay differential equations, *Appl. Numer. Math.* **34** (2000), 293–308.

[255] E.G. Van den Heuvel, Using resolvent conditions to obtain new stability results for θ-methods for delay differential equations, *IMA J. Numer. Anal.* **21** (2001), 421–438.

[256] E.G. Van den Heuvel, A new approach to the stability of Runge–Kutta methods for solving systems of ordinary and delay differential equations, *Ph.D. Thesis*, University of Leiden, 2002.

[257] R. Vermiglio, A one-step subregion method for delay differential equations, *Calcolo* **22** (1985), 429–455.

[258] R. Vermiglio, Multistep high order interpolants of Runge–Kutta methods, *J. Comput. Appl. Math.* **45** (1993), 75–88.

[259] R. Vermiglio and M. Zennaro, Multistep natural continuous extensions of Runge–Kutta methods: the potential for stable interpolation, *Appl. Numer. Math.* **12** (1993), 521–546.

[260] J.H. Verner, Differentiable interpolants for high-order Runge–Kutta methods, *SIAM J. Numer. Anal.* **30** (1993), 1446–1466.

[261] J.H. Verner and M. Zennaro, Continuous explicit Runge–Kutta methods of order 5, *Math. Comput.* **64** (1995), 1123–1146.

[262] J.H. Verner and M.Zennaro, The order of embedded continuous explicit Runge–Kutta methods, *BIT* **35** (1995), 406–416.

[263] P. Waltman, A threshold model of antigen-stimulated antibody production, *Theoretical Immunology, Immunology Series* **8** (1978), 437–453, Dekker, New York.

[264] G. Wanner, E. Hairer and S.P. Nørsett, Order stars and stability theorems, *BIT* **18** (1978), 475–489.

[265] D.S. Watanabe and M.G. Roth, The stability of difference formulas for delay differential equations, *SIAM J. Numer. Anal.* **22** (1985), 132–145.

[266] R. Weiner and K. Strehmel, A type insensitive code for delay differential equations based on adaptive and explicit Runge–Kutta interpolation methods, *Computing* **40** (1988), 255–265.

[267] L. F. Wiederholt, Stability of multistep methods for delay differential equations, *Math. Comput.* **30** (1976), 283–290.

[268] D.R. Willé, The numerical solution of delay-differential equations, *Ph.D. Thesis*, University of Manchester, 1989.

[269] D.R. Willé and C.T.H. Baker, A short note on the propagation of derivative discontinuities in Volterra-delay integro-differential equations, *NA Report* **187**, Dept of Mathematics, University of Manchester, 1990.

[270] D.R. Willé and C.T.H. Baker, The tracking of derivative discontinuities in systems of delay differential equations, *Appl. Numer. Math.* **9** (1992), 209–222.

[271] D.R. Willé and C.T.H. Baker, DELSOL – a numerical code for the solution of systems of delay differential equations, *Appl. Numer. Math.* **9** (1992), 223–234.

[272] D.R. Willé and C.T.H. Baker, Stepsize control and continuity consistency for state-dependent delay-differential equations, *J. Comput. Appl. Math.* **53** (1994), 163–170.

[273] D.R. Willé and C.T.H. Baker, Some issues in the detection and location of derivative discontinuities in delay differential equations, *NA Report* **238**, Dept of Mathematics, University of Manchester, 1994.

[274] K. Wright, Some relationships between implicit Runge–Kutta collocation and Lanczos τ-methods, and their stability properties, *BIT* **10** (1970), 217–227.

[275] B. Yang, L. Qiu and T. Mitsui, GP_G-stability of Runge–Kutta methods for generalized delay differential systems, *Comput. Math. Appl.* **37** (1999), 89–97.

[276] M. Zennaro, Maximum principles for difference-differential operators in periodic function spaces, *Numer. Math.* **43** (1984), 121–139.

[277] M. Zennaro, One-step collocation: uniform superconvergence, predictor–corrector method, local error estimate, *SIAM J. Numer. Anal.* **22** (1985), 1135–1152.

[278] M. Zennaro, On the P-stability of one-step collocation for delay differential equations, in *Delay Equations, Approximation and Application, ISNM* **74** (1985), 334–343.

[279] M. Zennaro, Natural continuous extensions of Runge–Kutta methods, *Math. Comput.* **46** (1986), 119–133.

[280] M. Zennaro, P-stability properties of Runge–Kutta methods for delay differential equations, *Numer. Math.* **49** (1986), 305–318.

[281] M. Zennaro, Natural Runge–Kutta and projection methods, *Numer. Math.* **53** (1988), 423–438.

[282] M. Zennaro, Contractivity of Runge–Kutta methods with respect to forcing terms, *Appl. Numer. Math.* **10** (1993), 321–345.

[283] M. Zennaro, Delay differential equations: theory and numerics, in *Theory and Numerics of Ordinary and Partial Differential Equations*, M. Ainsworth, J. Levesley, W.A. Light and M. Marletta Eds., Clarendon Press, Oxford, 1995.

[284] M. Zennaro, Asymptotic stability analysis of Runge–Kutta methods for nonlinear systems of delay differential equations, *Numer. Math.* **77** (1997), 549–563.

[285] B. Zubik-Kowal and S. Vandewalle, Waveform relaxation for functional differential equations, *SIAM J. Sci. Comput.* **21** (1999), 207–226.

[286] T.S. Zwerkina, Approximate solution of differential equation with retarded argument and differential equations with discontinuous right-hand side, *Trudy Sem. Teor. Differential Uravnenii s Otklon. Argumemtom Univ. Druzby Narodov Patrisa Lumumby* **1** (1962), 76–93.

[287] T.S. Zwerkina, A modification of finite difference methods for integrating ordinary differential equations with non-smooth solutions, *Z. Vycisl. Mat. Fiz.* **4** (1964), suppl., 149–160.

[288] T.S. Zwerkina, A modified Adams' formula for the integration for equations with deviating argument, *Trudy Sem. Teor. Differential Uravnenii s Otklon. Argumemtom Univ. Druzby Narodov Patrisa Lumumby* **3** (1965), 221–232.

[289] T.S. Zwerkina, On the question of the choice of the method of integration for equations with deviating argument, *Trudy Sem. Teor. Differential Uravnenii s Otklon. Argumemtom Univ. Druzby Narodov Patrisa Lumumby* **7** (1969), 75–81.

INDEX

Printed and bound by CPI Group (UK) Ltd, Croydon, CR0 4YY

The manufacturer's authorised representative in the EU for product safety is Oxford University Press España S.A. of El Parque Empresarial San Fernando de Henares, Avenida de Castilla, 2 - 28830 Madrid (www.oup.es/en or product.safety@oup.com). OUP España S.A. also acts as importer into Spain of products made by the manufacturer.
Printed and bound by CPI Group (UK) Ltd, Croydon, CR0 4YY
11/08/2026
02193547-0004